Topics in Transport Phenomena

ADVANCES IN THERMAL ENGINEERING

EDITORS:

JAMES P. HARTNETT
THOMAS F. IRVINE, JR.

I	Blackshear	• **Heat Transfer in Fires: Thermophysics, Social Aspects, Economic Impact***
II	Afgan and Beer	• **Heat Transfer in Flames***
III	de Vries and Afgan	• **Heat and Mass Transfer in the Biosphere: Part I Transfer Processes in Plant Environment***
IV	Vargaftik	• **Tables on the Thermophysical Properties of Liquids and Gases**
V	Veziroglu	• **Remote Sensing: Energy-Related Studies**
VI	Gutfinger	• **Topics in Transport Phenomena: Bioprocesses, Mathematical Treatment, Mechanisms**

IN PREPARATION

Begell	• **Glossary of Terms in Heat Transfer and Related Topics: English, Russian, German, French, and Japanese**
Chang	• **Control of Flow Separation: Energy Conservation, Operational Efficiency and Safety**
Chi	• **Heat Pipe Theory and Practice**
Denton	• **Future Energy Production***
Eckert and Goldstein	• **Measurements in Heat Transfer 2nd revised and augmented edition**
Ginoux	• **Two-Phase Flows with Application to Nuclear Reactor Design Problems†**
Hartnett	• **Heat and Mass Transfer Problems in Future Energy Production***
Hsu and Graham	• **Transport Processes in Boiling and Two-Phase Systems, Including Near-Critical Fluids**
Pfender	• **High-Temperature Phenomena in Electric Arcs**
Richards	• **Measurement of Unsteady Fluid Dynamic Phenomena†**
Sieverding	• **Two-Phase Fluid Dynamics in Turbines: Engineering, Instrumentation, Theory†**
Yovanovich	• **Advanced Heat Conduction**

*A publication of the International Centre for Heat and Mass Transfer, Belgrade.
†A von Karman Institute Book, Brussels.

Topics in Transport Phenomena

bioprocesses, mathematical treatment, mechanisms

Edited by
Chaim Gutfinger

Department of Mechanical Engineering
Technion—Israel Institute of Technology
Haifa, Israel

HEMISPHERE PUBLISHING CORPORATION

Washington London

A HALSTED PRESS BOOK

JOHN WILEY & SONS

New York London Sydney Toronto

Hemisphere Publishing Corporation
1025 Vermont Ave., N.W., Washington, D.C. 20005

Distributed solely by Halsted Press, a Division of John Wiley & Sons, Inc., New York.

Library of Congress Cataloging in Publication Data

Main entry under title:

Topics in transport phenomena.

Includes index.
1. Heat–Transmission–Mathematical models–Congresses. 2. Mass transfer–Mathematical models–Congresses. 3. Turbulence–Mathematical models–Congresses. 4. Thermal biology–Mathematical models–Congresses. I. Gutfinger, Chaim.
QC319.8.T66 1975 536′.2 75-15867
ISBN 0-470-33712-5

Printed in the United States of America

This volume is dedicated to the memory of
Professor Yeshaiahu Winograd
Head of the Energy Laboratory
Department of Mechanical Engineering
Technion–Israel Institute of Technology
who was killed in action in the Yom Kippur War

תנצב״ה

Contents

Preface

The topics covered in this volume were presented in the Third Israel Symposium on Thermal Sciences held in Haifa on July 18, 1974. These topics comprise comprehensive review articles that reflect the interests of the respective authors and the personal choice of the editor. They are grouped into three general categories:

1. Transport mechanisms, devoted to models of turbulence and their application to heat and mass transfer at interfaces.
2. Biotransport, covering heat transfer and blood circulation in mammals.
3. Mathematical methods, covering finite differences on the one hand and asymptotic methods on the other.

Contrary to compilations of original articles which are usually intended for the specialist, the review chapters in the present volume are aimed toward the research worker who has specialized in the field discussed but still wants to maintain an up-to-date knowledge. It will also be of interest to graduate students and practicing engineers, who will find valuable the experts' general view on the subject matter covered.

I would like first to thank the contributors to this volume, who have compiled, sorted, discussed, criticized, emphasized, summarized, and condensed the material of hundreds of articles and presented them to the reader as clear, up-to-date reviews. Thanks are also due to colleagues who have reviewed the chapters and given unselfishly from their experience and knowledge toward lightening the burdens of editorship. These include Professor James P. Hartnett, University of Illinois, Chicago Circle; Professors Eli Ruckenstein and Thomas F. Irvine, Jr., The State University of New York at Buffalo and Stony Brook, respectively; Dr. Leonard Y. Cooper, Bell Telephone Laboratories, Whippany, N.J.; Professor Gad Hetsroni, Dr. David Degani, and Dr. Yoram Zvirin of Technion–Israel Institute of Technology, Haifa; and Mr. Meir Toren of the Israel Defense Industry.

Chaim Gutfinger

PART ONE
Transport Mechanisms

1
Models of Turbulence

M. Wolfshtein
Department of Aeronautical Engineering
Technion—Israel Institute of Technology
Haifa, Israel

D. Naot
Department of Mechanical Engineering
Ben-Gurion University of Negev
Beer Sheva, Israel

A. Lin
Department of Aeronautical Engineering
Technion—Israel Institute of Technology
Haifa, Israel

CONTENTS

1 INTRODUCTION

1.1 Aim of the paper

The rapid growth of computer speed and storage capacity in the last decade has made it possible to solve many problems in fluid mechanics that were not amenable to theoretical treatment only a few years ago. Today it is possible to investigate fairly complex turbulent flow fields and to obtain results that compare favorably with experimental data. Much of this progress is connected with the development of the so-called "models of turbulence." It is the aim of the present paper to describe some of these developments and to indicate some future directions of research.

Owing to the rapid development of the field as well as the limited space available for the present contribution, it has not been attempted to describe all the existing methods in detail. Instead, it was preferred to compare a number of representative approaches in a unified way and to draw some general conclusions.

1.2 Survey of the literature

The beginning of turbulence modeling may be traced back to Kolmogorov (1942), who suggested a two-variable model of turbulence but never used it owing to the inadequacy of computing capabilities. At about the same time Prandtl (1945) and Chou (1945) published papers on this subject. Two papers by Rotta (1951) may be considered as marking the termination of the initial period in turbulence modeling. This period is characterized by an imparity between the very complex models and the crude computers and computing methods available at that time.

The second period did not commence until the late Sixties, when computers were already available in most research institutions. Two-variable models were formulated and tested by Glushko (1972), Ng and Spalding (1972), Harlow and Nakayama (1967, 1968), and others. Successes were recorded, but new questions were raised as well. It was realized that sooner or later a direct calculation of the Reynolds stresses would be required if the applicability of the methods was to be extended. At this point, two comprehensive studies of the subject by Reynolds (1970) and Launder and Spalding (1972) appeared, which may be considered the starting point of the third period.

In the third period the Reynolds stress equations were investigated, and important research was conducted, leading to a better understanding of the subject. Against this background, a comparative study of existing models was made from a unified point of view, which has been drawn upon by the authors in recent years.

1.3 Brief description of turbulence

A turbulent flow field is characterized by random fluctuations of velocity and pressure. These fluctuations vary in scale, but even the smallest scales encountered in experiments appear to be much larger than the mean free path. Therefore, the fluid in a turbulent flow field may still be regarded as a continuum, and the pressure and velocity are governed by the Navier-Stokes and continuity equations.

In practice, it is impossible to solve the nonsteady Navier-Stokes equations for a turbulent flow field analytically. Even numerical solutions may be obtained only for low velocity and simple flow fields. On the other hand, it usually suffices, for most applications, to specify the mean values of the turbulent flow field only. Osborn Reynolds suggested that the velocity and pressure be divided into mean and fluctuating components. In this way, he was able to derive the equations of motion for the mean values, which are known as the Reynolds equations. Unfortunately, the nonlinear terms of the Navier-Stokes equations introduce second-order correlations of the fluctuating velocities, known as the Reynolds stresses, into the equations of motion. As these correlations are unknown, the equations of motion cannot be solved. A considerable part of the research on turbulent flows has been directed towards the measurement and calculation of the Reynolds stresses. Ultimately, the aim is to get a closed system of well-posed equations, and to "close" the problem. This problem of "closure" will be dealt with in the following sections.

1.4 What is a turbulence model?

As a stochastic phenomenon, a turbulent flow field may be characterized by its mean values. Yet an infinite number of correlation coefficients for the turbulent velocities and pressure is required to obtain a complete description of the field. On the other hand, experience shows that it usually suffices to know only a limited number of correlations to enable prediction of the practical characteristics of the field.

A turbulence model is therefore based on the assumption that a turbulent flow field may be characterized by a finite number of statistical correlations. These correlations, referred to as the *model variables,* are identified by investigating the particular flow field in question.

The governing equations for these correlations may usually be derived from the Navier-Stokes equations by suitable averaging procedures. Owing to the nonlinearity of the Navier-Stokes equations the resulting model equations always contain new unknown correlations which should be replaced by functions of the model variables. This procedure is referred to as *closure,* and results in a well-posed set of equations. In the present paper we shall deal mainly with two families of models: two-variable models and Reynolds stress models.

"Two-variable models" include all closure approximations that contain transport equations for the turbulence energy and for a typical length scale. In these models the Reynolds stresses are related to the mean rate of strain through an eddy viscosity, which depends on the length scale and the turbulence energy and thus varies from point to point in the field. The turbulence energy and the length scale are governed by differential equations and are therefore subjected to convection and diffusion processes in addition to local processes. Thus these models are useful for all flows where turbulence is influenced by previous effects. Typical examples include developing boundary layers, turbulent stagnation flows, and turbulent recirculating flow fields. In such flow fields the convection or turbulent diffusion, or both, have a significant influence on the turbulent stresses.

"Reynolds stress models" include closure approximations in which the model variables are the Reynolds stresses. In addition, these models also contain a typical length scale. These models are useful in cases where a scalar eddy viscosity cannot be found. For example, in wall jets, flows through annuli, and flows between two plates of different roughnesses, the surfaces of maximum velocity and zero shear stress do not coincide. In these cases, the eddy viscosity as used in the two-variable models will fail to describe the stress. Another case of interest is that of a flow along a corner or a discontinuity in roughness which is characterized by a secondary mean flow perpendicular to the main flow. This secondary flow is related to the components of the Reynolds stress tensor. The calculation of such flow fields depends on our ability to calculate the Reynolds stresses.

1.5 Outline of the paper

At the present time the two-variable models and the Reynolds stress models appear to be the most useful tools for the characterization of complex turbulent flow fields. These two groups of models will be discussed in Secs. 2 and 3, respectively. In Sec. 2 a number of two-variable models will be compared.

The Reynolds stress equations have still not been given their final form. Therefore we chose to describe some of the derivation of the Reynolds stress equations in Sec. 3. Here the emphasis is placed on a critical comparison of some recent proposals, with an effort to find the features common to all models. In both Secs. 2 and 3 numerical values for the empirical constants are given, and the cases where they were shown to yield reasonable results are enumerated.

In Sec. 4 some future prospects are discussed. These are based on the current work of the authors and represent possible improvements in the Reynolds stress models.

The present paper is restricted to steady incompressible flows. Thermal and atmospheric turbulence are not dealt with low turbulence flows such as in the viscous sublayer or in transitional flows. In addition, certain models are not discussed. This applies in particular to mixing length models and models in which only one differential equation is used. We realize that such models are very successful for the solution of certain problems. However, the principal aim of this work is the description of the two-variable and Reynolds stress models, as we feel that such models have greater potential for future development.

2 TWO-VARIABLE MODELS

The simplest models of turbulence are the algebraic models, which are based on the assumption that the turbulence stresses are uniquely dependent on the velocity field. As simplified and nonuniversal as this assumption may be, it has still produced some remarkable calculation methods, such as the Prandtl mixing length hypothesis. Yet such models may be expected to be realistic only when the turbulence is influenced by the local properties of the velocity field. In many cases turbulence is not in local equilibrium with its immediate neighborhood, but depends on processes and events that occur at a considerable distance from the point in question. In such cases we must consider the transport of the turbulence by means of differential equations. Thus we turn our attention to differential models.

As a fluctuating phenomenon, turbulence is characterized by its amplitude and frequency. Usually the amplitude varies with the frequency, and the details of this variation are of considerable significance. Yet if we wish to describe the turbulence in terms of two parameters, the mean amplitude and frequency are the most appropriate ones. This line of reasoning, together with the realization that the turbulent variables must, in general, be governed by transport equations, leads to the so-called two-variable models of turbulence. The universality of such a model depends on the variety of turbulent flow phenomena which may be described by the same set of equations and constants. Experience has shown that two-variable models are more universal than algebraic models.

The number of turbulence variables in a flow field exceeds two. Usually we are interested in the six Reynolds stresses which must be related to the amplitude and frequency of the turbulence. Other variables that may be of interest are the turbulence length scale, the turbulence dissipation, etc. The selection of the two representative variables and their relation to the other turbulence variables will be discussed in the following section.

2.1 Selection of the variables

In a two-variable model of turbulence the selection of the main variables is a matter of convenience, provided that the mean amplitude and frequency of the fluctuations can be given in terms of the two-model variables. Many pairs of variables are suitable, and the best criterion for their selection is the simplicity of the governing equations and their boundary conditions.

2.1.1 Basic definitions Generally, turbulence is nonisotropic, and thus the mean amplitude and frequency of the fluctuations may have some directional properties. Definition of turbulence by scalar amplitude and frequency implies isotropy of these two properties. Therefore, if a two-variable model is used, the nonisotropy of the turbulence must be uniquely dependent on the nonisotropy of the mean flow field, usually through an eddy viscosity, ν_t. (Clearly this is not a true representation of turbulence, and may be regarded as the main deficiency of the two-variable model.)

On the basis of the above, the Reynolds stresses are given by

$$\overline{u_i u_j} = \frac{1}{3}\,\overline{u_l u_l}\,\delta_{ij} - \nu_t\left(\frac{\partial U_i}{\partial x_j} + \frac{\partial U_j}{\partial x_i}\right) \tag{2.1}$$

where the first term in the right hand side has been added in order to make the equation applicable to homogeneous and isotropic turbulence. With this formulation the equations of motion will retain their form for incompressible flow with a nonuniform viscosity. Dimensional considerations suggest that the turbulence viscosity is a product of the turbulence velocity, $k^{1/2}$, and the turbulence length scale, l representing the amplitude and frequency respectively:

$$\nu_t = C_\mu k^{1/2} l \tag{2.2}$$

where C_μ is a universal function of the Reynolds number of turbulence defined by

$$\mathrm{Re}_t = \frac{k^{1/2} l}{\nu} \tag{2.3}$$

Dimensional reasoning also suggests that C_μ tends to a constant value at a high Reynolds number of turbulence.

The quantities k and l must be properly defined. Fortunately, it is easy to define typical velocity and scale, which are measurable. Thus, for the velocity we may choose the amplitude of the velocity fluctuations. Usually, the following definition is used:

$$k = \frac{1}{2}\overline{u_i^2} \tag{2.4}$$

This definition calls to mind kinetic energy, and therefore k is termed the turbulence energy.

The typical length scale represents the distance over which the turbulence velocities are correlated with one another. This distance may be calculated by using the two point correlation $Q_{ij} = \overline{u_i(x)u_j(x+r)}$. When the two points are far from one another, $r \to \infty$, the correlation Q_{ij} vanishes, and when the distance tends to zero, $r \to 0$, Q_{ij} tends to the Reynolds stress $\overline{u_i u_j}$. The correlation Q_{ij} may be used to define the length scale l in various ways. A convenient definition is the "macro" or "integral" scale. The isotropic macro-scale is given by

$$2kl = \int_v Q_{ll} \frac{dv}{r^2} \tag{2.5}$$

By this definition the averaging process, yielding the length scale, is performed equally in all directions.

2.1.2 Exact equations for the turbulence properties The quantities k and l may serve as the two model variables. It should be emphasized that these are not the only possible variables, and in fact others will be described below. Yet it is convenient to formulate the governing equations for k and l at this point. These equations are obtained by suitable manipulations of the Navier-Stokes equations. While the method is fairly straightforward the amount of algebraic labor involved is significant. Therefore the equations will simply be presented here without going into all the mathematical details. In addition, it will be convenient to present here all the transport equations used in the paper.

The starting point is the equation for the two-point correlation (Wolfshtein, 1970) given as

$$\begin{aligned}\frac{D}{Dt} Q_{ij} =& [\bar{U}_k(\mathbf{x}) - \bar{U}_k(\mathbf{x}+\mathbf{r})] \frac{\partial}{\partial r_k} Q_{ij} \\ & - \left[Q_{ik} \frac{\partial}{\partial x_k} \bar{U}_j(\mathbf{x}+\mathbf{r}) + Q_{kj} \frac{\partial}{\partial x_k} \bar{U}_i(\mathbf{x})\right] \\ & - \left\{\frac{\partial}{\partial x_k} \overline{u_k(\mathbf{x}) u_i(\mathbf{x}) u_j(\mathbf{x}+\mathbf{r})} + \frac{\partial}{\partial r_k} \left[\overline{u_i(\mathbf{x}) u_k(\mathbf{x}+\mathbf{r}) u_j(\mathbf{x}+\mathbf{r})}\right.\right. \\ & \left.\left. - \overline{u_i(\mathbf{x}) u_k(\mathbf{x}) u_j(\mathbf{x}+\mathbf{r})}\right]\right\} - \frac{1}{\rho}\left[\frac{\partial}{\partial x_i} \overline{p(\mathbf{x}) u_j(\mathbf{x}+\mathbf{r})}\right. \\ & \left. + \frac{\partial}{\partial r_j} \overline{u_i(\mathbf{x}) p(\mathbf{x}+\mathbf{r})} - \frac{\partial}{\partial r_i} \overline{p(\mathbf{x}) u_j(\mathbf{x}+\mathbf{r})}\right] \\ & - \nu\left(\frac{\partial^2}{\partial x_k{}^2} Q_{ij} - 2 \frac{\partial^2}{\partial x_k \partial r_k} Q_{ij} + 2 \frac{\partial^2}{\partial r_k{}^2} Q_{ij}\right)\end{aligned} \tag{2.6}$$

where $Q_{ij} = \overline{u_i(\mathbf{x}) u_j(\mathbf{x}+\mathbf{r})}$ (2.6a)

Q_{ij} is a function of the radius vectors **x** and **r**. It should be noted that Eq. (2.6a) is not the only definition of Q_{ij}. Hinze (1959) defines the two points as lying at equal distances on both sides of the central point **x**, but this formulation is inconvenient for present purpose.

Contraction of Eq. (2.6) yields

$$\begin{aligned}\frac{D}{Dt} Q_{ii} =& [\bar{U}_k(\mathbf{x}) - \bar{U}_k(\mathbf{x}+\mathbf{r})] \frac{\partial}{\partial r_k} Q_{ii} \\ & - \left[Q_{ik} \frac{\partial}{\partial x_k} \bar{U}_i(\mathbf{x}+\mathbf{r}) + Q_{ki} \frac{\partial}{\partial x_k} \bar{U}_i(\mathbf{x})\right] \\ & - \left\{\frac{\partial}{\partial x_k} \overline{u_k u_i(\mathbf{x}) u_i(\mathbf{x}+\mathbf{r})} + \frac{\partial}{\partial r_k} \left[\overline{u_i(\mathbf{x}) u_k(\mathbf{x}+\mathbf{r}) u_j(\mathbf{x}+\mathbf{r})}\right.\right. \\ & \left.\left. - \overline{u_i(\mathbf{x}) u_k(\mathbf{x}) u_j(\mathbf{x}+\mathbf{r})}\right]\right\} - \frac{1}{\rho}\left\{\frac{\partial}{\partial x_i} \overline{p(\mathbf{x}) u_i(\mathbf{x}+\mathbf{r})}\right. \\ & \left. + \frac{\partial}{\partial r_i} \left[\overline{u_i(\mathbf{x}) p(\mathbf{x}+\mathbf{r})} - \overline{p(\mathbf{x}) u_i(\mathbf{x}+\mathbf{r})}\right]\right\} \\ & + \nu\left(\frac{\partial^2}{\partial x_k{}^2} Q_{ii} - 2 \frac{\partial^2}{\partial x_k \partial r_k} Q_{ii} + 2 \frac{\partial^2}{\partial r_k{}^2} Q_{ii}\right)\end{aligned} \tag{2.7}$$

By virtue of Eq. (2.5) the volume integral of (2.7) yields a transport equation for the product kl. We shall refer to this equation later.

Then the separation radius r tends to zero, we obtain from (2.6) the Reynolds stress equations, as follows:

$$\frac{D}{Dt}\overline{u_i u_j} = -\left(\overline{u_i u_k}\frac{\partial \bar{U}_j}{\partial x_k} + \overline{u_j u_k}\frac{\partial \bar{U}_j}{\partial x_k}\right) - \left[\frac{\partial}{\partial x_k}\overline{u_k u_i u_j} + \frac{1}{\rho}\left(\frac{\partial}{\partial x_i}\overline{p u_j} + \frac{\partial}{\partial x_j}\overline{p u_i}\right)\right] + \frac{1}{\rho}\left(\overline{p\frac{\partial u_i}{\partial x_j}} + \overline{p\frac{\partial u_j}{\partial x_i}}\right) + \nu\left(\frac{\partial^2}{\partial {x_k}^2}\overline{u_i u_j} - 2\overline{\frac{\partial u_j}{\partial x_k}}\,\overline{\frac{\partial u_j}{\partial x_k}}\right) \tag{2.8}$$

Finally, contraction of Eq. (2.8) yields the turbulence energy equation:

$$\frac{D}{Dt}k = -\overline{u_i u_n}\frac{\partial \bar{U}_i}{\partial x_n} - \frac{\partial}{\partial x_n}\left(\frac{1}{2}\overline{u_n u_i u_i} + \frac{1}{\rho}\overline{p u_n}\right) + \nu\left(\frac{\partial^2}{\partial x^2}k - \overline{\frac{\partial u_j}{\partial x_n}}\,\overline{\frac{\partial u_i}{\partial x_n}}\right) \tag{2.9}$$

2.1.3 Discussion of the turbulence energy equation Although Eq. (2.9) is simpler than Eqs. (2.6), (2.7), and (2.8), it already exhibits the most disturbing feature common to these equations, namely that it contains terms involving unknown correlations of the velocity and pressure and therefore cannot be solved. Even so, the turbulence energy equation (2.9) has been extensively used by investigators who have measured its various terms. The physical meaning of these terms is easily recognized and identified as convection, generation diffusion, and dissipation, respectively.

The term *diffusion* is somewhat misleading. In fact, the term describes energy convection by the turbulent eddies or energy transfer as a result of pressure fluctuations. Yet it will be shown in Sec. 3 that the term describes transport of turbulence energy and not generation or decay of energy, and as such it has been termed turbulence diffusion. It is also common to add to this term that part of the viscous terms that contains the second spatial derivative of the energy. Unfortunately, it is difficult either to estimate this turbulence energy diffusion or to measure it. We shall refer to this question again, but at present it is preferable to discuss the generation and decay of the energy.

The generation term poses very few problems, since it contains products of the velocity gradients and the Reynolds stresses only. As these quantities are the main objects of many experiments and calculations, they are usually available. The physical meaning of the term is easily understood as well. The mean flow performs work on the turbulent eddies, and thus energy is transferred from the mean motion to the fluctuating motion.

The existence of an energy cascade, by which energy is transferred from the mean motion to the large eddies and then from the large eddies to the small eddies, has been suggested by Townsend (1956). At the end of this cascade viscous resistance counteracts the turbulent fluctuations, and all the turbulence energy is eventually dissipated into thermal energy. The last term in the turbulence energy equation describes this process. Although this process is a result of viscous forces, it is not dependent on the magnitude of the viscosity if the Reynolds number of turbulence is sufficiently high. Because of this cascade the large eddy motion is nearly nonviscous and the dissipation is basically a small eddy phenomenon. As such, it is not directly related to the geometry of the case in question, and tends to isotropy.

2.1.4 Some scale equations There are two ways to estimate the magnitude of the dissipation theoretically.

1. Dimensional analysis, coupled with the isotropy assumption leads to the following expression:

$$\epsilon = c_1 \frac{k^{3/2}}{l} \tag{2.10}$$

where c_1 is a universal function of the Reynolds number of turbulence, tending to a constant value at a high Reynolds number.

2. Algebraic manipulation of the Navier-Stokes equation yields a transport equation for the dissipation. Such derivations have been described by a number of authors (e.g., Harlow, 1968; Nakayama, 1968; Jones, 1972; Launder, 1972), and the resulting equations is given as

$$\begin{aligned}\frac{D\epsilon}{Dt} = \nu \Bigg[& -2\left(\overline{\frac{\partial u_n}{\partial x_l}}\,\overline{\frac{\partial u_m}{\partial x_l}} + \overline{\frac{\partial u_l}{\partial x_n}}\,\overline{\frac{\partial u_l}{\partial x_m}} + u_n \overline{\frac{\partial u_m}{\partial x_l}\frac{\partial}{\partial x_l}}\right)\frac{\partial \bar{U}_m}{\partial x_n} \\ & - 2\,\overline{\frac{\partial u_l}{\partial x_m}}\,\overline{\frac{\partial u_n}{\partial x_m}}\,\overline{\frac{\partial u_l}{\partial x_n}} - \frac{\partial}{\partial x_n}\left(u_n \overline{\frac{\partial u_l}{\partial x_m}}\,\overline{\frac{\partial u_l}{\partial x_m}} + 2\,\overline{\frac{\partial u_n}{\partial x_m}}\,\overline{\frac{\partial p}{\partial x_m}}\right) \\ & - 2\nu\left(\overline{\frac{\partial^2 u_l}{\partial x_n \partial x_m}}\right)^2 + \frac{\partial^2}{\partial x_l^2}\,\epsilon\Bigg]\end{aligned} \tag{2.11}$$

It is easily seen that Eq. (2.10) allows ϵ and l to be interchanged. Thus it is possible either to solve Eq. (2.11) and use ϵ as a scale parameter of the turbulence or to solve the equation for l, which is otbained by a volume integration of Eq. (2.7). These equations may be solved if all the unknown correlations are replaced by functions of $\bar{U}_i$, k, and l or ϵ, as will be shown in Sec. 2.2.2. With any of these equations a turbulence energy equation is still required, and thus we may use either an energy-length scale model or an energy-dissipation model. Some other possibilities are available as well.

Another scale equations may be obtained by deriving a fluctuating-vorticity-squared ($\overline{\omega^2}$) equation from the Navier-Stokes equation (see Spalding, 1969; Saffman, 1970) as:

$$\frac{D}{Dt}\,\overline{\omega^2} = -2\left(\overline{\omega_i u_k}\,\frac{\partial \bar{\Omega}_i}{\partial x_k} - \overline{\omega_i \omega_k}\,\frac{\partial \bar{U}_i}{\partial x_k} - \overline{\omega_i\,\frac{\partial u_i}{\partial x_k}}\cdot \bar{\Omega}_k\right) - \frac{\partial}{\partial x}\,\overline{u_k \omega^2} + 2\,\overline{\omega_i \omega_k\,\frac{\partial u_i}{\partial x_k}} + \nu\,\frac{\partial^2}{\partial x_k{}^2}\,\overline{\omega^2} - 2\nu\,\overline{\frac{\partial \omega_i}{\partial x_k}}\;\overline{\frac{\partial \omega_i}{\partial x_k}} \tag{2.12}$$

This quantity $\overline{\omega^2}$ may be assumed to be related to the length scale by

$$1 = c_2\sqrt{\frac{k}{\omega^2}} \tag{2.13}$$

Thus an energy-vorticity model may be defined.

The last model, which is not in practical use but is interesting because of its relative simplicity, is due to Kolmogorov (1942) who postulated a turbulent frequency (f) equation, and formulated an energy-frequency model. In this case the length scale is given by

$$1 = c_3\sqrt{\frac{k}{f}} \tag{2.14}$$

We have now reviewed four two-variable models. Apart from the Kolmogorov model, each of them has been used with a fair amount of success, and therefore it is interesting to compare them and study their relative advantages. To do so it is necessary to formulate the model equations. This task together with the comparison will be carried out in the following section.

2.2 Model equations

We shall now present the results of the modeling of the transport equations for the turbulence variables that were presented in the previous section. The details of the derivation of the model equations are rather long, and will not be given here. For the purpose of the presentation each equation will be divided into four terms: Convection = Diffusion + Generation − Decay. The comparison will be performed in two stages. First, the original equations as proposed by the authors will be given. A turbulence energy equation is incorporated in all the models. However, the second model-equation governs either the length scale, the dissipation, the turbulence vorticity, or the frequency. Therefore, in the second stage these equations will be reduced to a common denominator by proper transformations, and will be presented as length scale equations. The length scale was chosen as the common denominator, owing to its simple physical meaning, and because the roles of the terms of its governing equation are more transparent.

2.2.1 The turbulence energy equation As already mentioned, all energy equations are similar with the exception of those obtained by contraction of the Reynolds stress model equations. In the latter case the turbulence diffusion terms are different from those used in "original" two-variable models. If the dissipation is assumed to be governed by Eq. (2.11) and the Reynolds stresses by Eq. (2.1) the only unknown term in the energy Eq. (2.9) is the diffusion term. This term contains two parts: the triple velocity correlation $-\frac{1}{2}\,\partial/\partial x_n\,\overline{u_n u_i u_i}$ which represents energy transport by the turbulent eddies, and the pressure-velocity correlation $-\,\partial/x_n\,\overline{pu_n}$ which represents transfer of energy from one point to other points by means of the pressure fluctuations. This latter process is not yet properly understood as its measurement is subject to certain difficulties. The triple correlation may be measured, but theory can furnish very little help in its modeling. Thus the only possible course is to recognize that neither phenomenon generates or absorbs turbulence energy, acting only as agents which alter the spatial distribution of the energy. This process is similar in nature to laminar diffusion, and may therefore be modeled as a second order diffusion expression, with a diffusion coefficient that is proportional to the eddy viscosity. Therefore we postulate

$$-\frac{\partial}{\partial x_n}\left(\frac{1}{2}\,\overline{u_n u_i u_i} + \overline{u_n p}\right) = \frac{\partial}{\partial x_n}\left(\frac{\nu_t}{\sigma_k}\,\frac{\partial}{\partial x_n}\,k\right) \tag{2.15}$$

Rotta (1969) suggested that the length scale gradients also contribute to the energy diffusion, however, his suggestion was not supported by later research.

The general form of the turbulence energy model equation is then

$$\frac{D}{Dt}\,k = \frac{\partial}{\partial x_n}\left(\frac{\nu_t}{\sigma_k}\,\frac{\partial}{\partial x_n}\,k\right) + G - \epsilon \tag{2.16}$$

where σ_k is a constant, ν_t and ϵ are given by Eqs. (2.2) and (2.10), and G is given by

$$G = -\overline{u_i u_j}\, \frac{\partial \bar{U}_i}{\partial x_j} = \nu_t \left(\frac{\partial \bar{U}_i}{\partial x_j} + \frac{\partial \bar{U}_j}{\partial x_i} \right)^2 \tag{2.17}$$

The values of the constants c_μ, σ_k, and c_1 are so chosen as to agree with experimental data. As the model equations specify only the variations of l, but not its absolute value, there is one redundant constant in the equation to be determined arbitrarily. We choose $c_\mu = 1$. When this is done we get the values for the other constants given in Table 1.

It is interesting to note that the contraction of the Reynolds stress model equations does not necessarily reproduce the energy model equation, owing to the complexity of the diffusion term as was discussed by Hirt (1969). Thus the model proposed by Daly and Harlow (1970) yields the following energy diffusion term:

$$\frac{\alpha}{\sqrt{2}} \frac{\partial}{\partial x_n} \left(lk^{-1/2}\, \overline{u_n u_m}\, \frac{\partial}{\partial x_m} k \right) + \frac{\theta}{\sqrt{2}} \frac{\partial}{\partial x_n} \left(lk^{1/2} \frac{\partial}{\partial x_m} \overline{u_n u_m} \right)$$

while the model of Launder et al. (1973) results in

$$\frac{c_s}{c_1} \frac{\partial}{\partial x_n} \left(lk^{-1/2}\, \overline{u_n u_m}\, \frac{\partial}{\partial x_m} k \right) + \frac{c_s}{c_1} \frac{\partial}{\partial x_n} \left(lk^{-1/2}\, \overline{u_i u_m}\, \frac{\partial}{\partial x_m} \overline{u_i u_n} \right)$$

which are different from Eq. (2.15). Here α, θ, and c_1 are empirical constants. This subject will be discussed in Sec. 3.

In conclusion, it may be stated that the convection and generation terms of the turbulence energy equation can be precisely calculated. The modeling of the dissipation by Eq. (2.10) has not been seriously objected to, and most authors find the value of the constant $c_1 = 0.09$ in good agreement with experimental data. However, the turbulence diffusion term still raises some questions. Advanced studies of the Reynolds stress equations call into question validity of Eq. (2.15). Even when this equation is accepted, the values of the recommended constant σ_k vary between 1. and 2.5. Though the authors use $\sigma_k =$

Table 1 Coefficients of the energy equation

Reference	σ_k	c_1	Remarks and applications
Ng & Spalding (1972)	2	0.1	Boundary layers
Rodi & Spalding (1970)	1	0.09	Free shear flows
Glushko (1972)	2.5	$1.584\, \alpha^2$ $0.2 < \alpha < 0.8$	Flat plate boundary layer
Wilcox & Alber (1972)	2	0.09	Boundary layers
Jones & Launder (1972)	1	0.09	Laminarization in boundary layers
Harlow & Nakayama (1968)	1	0.2	

1.5 ÷ 2, which produce good results for some cases, they feel that the study of turbulence diffusion of the turbulence requires further work.

2.2.2 The length scale equation We shall now describe the second model equation. In fact, the second model equation was originally formulated by different authors in terms of the length-energy product, length squared, dissipation turbulence vorticity, or turbulence frequency. However, in most of these models the turbulence diffusion must be defined, and therefore a length scale may be defined as well. Thus the second model variable may be considered to be given by

$$\phi = \phi(k, l)$$

where the functional form of ϕ may be so chosen as to simplify the ϕ-equation. All the second variable equations have the general form:

$$\frac{D\phi}{Dt} = T + S - D \tag{2.18}$$

where T, S, and D stand for the turbulence diffusion, source, and decay of ϕ respectively. These equations will now be summarized after being slightly simplified and reorganized. The models considered are listed in Table 2 together with the definitions of ϕ and the turbulence viscosity. The terms used in these models for T, S, and D are listed in Table 3.

Some of the models are much simpler than the others. Kolmogorov's and Glushko's models are very simple, and the Jones and Launder, and Wilcox and Alber models are still not very complex. On the other hand, these models have not been extensively tested. The models which have been extensively tested, such as those of Ng and Spalding, and Rodi and Spalding, are more complex, and some later tests by Launder et al. (1973) conclude that even these models are not sufficiently accurate. Thus it may be necessary to use even more complex models in the future.

To make possible a better comparison between the models, the energy-length product $L = kl$ was chosen as a general dependent variable and substituted in all the model equations. Then, the general length equation is written as:

$$\begin{aligned} \frac{D}{Dt} L = {} & \frac{\partial}{\partial x_n}\left(\frac{\nu_t}{\sigma_L}\,\frac{\partial L}{\partial x_n}\right) + \frac{\partial}{\partial x_n}\left(\frac{L}{K}\,\frac{\nu_t}{\sigma_{L1}}\,\frac{\partial k}{\partial x_n}\right) \\ & + \frac{L}{k^2}\,\frac{\nu_t}{\sigma_{L2}}\left(\frac{\partial k}{\partial x_n}\right)^2 + \frac{\nu_t}{\sigma_L 3_k}\left(\frac{\partial k}{\partial x_n}\,\frac{\partial L}{\partial x_n}\right) \\ & + \frac{\nu_t}{\sigma_L 3L}\left(\frac{\partial L}{\partial x_n}\right)^2 + c_g 16 - c_d 1\epsilon + R \end{aligned} \tag{2.19}$$

Table 2 Second model variable

Reference	ϕ variable	ϕ definition	Turbulence viscosity	Application
Kolmogorov (1942)	Frequency	$f = c\,\frac{k^{1/2}}{l}$	$\nu_t = \frac{k}{f} = k^{1/2} l$	
Rotta (1968)	Length-energy product	$L = kl$		Duct flow
Rodi & Spalding (1970)	Length-energy product	$L = kl$	$\nu_t = k^{1/2} l$	Free shear flows
Ng & Spalding (1972)	Length-energy product	$L = kl$	$\nu_t = k^{1/2} l$	Boundary layers
Wolfshtein (1970)	Length-energy product	$L = kl$	$\nu_t = 0.22 k^{1/2} l$	One-dimensional flows
Glushko (1972)	Length-squared	$\lambda = \frac{1}{2} l^2$	$\nu_t = \alpha l k^{1/2}$ $0.2 < \alpha < 0.8$	High turbulence version of a low turbulence model applied to transitional flat plate boundary layer
Harlow & Nakayama (1968)	Dissipation	ϵ	$\nu_t = 0.2\,\frac{k^2}{\epsilon} = 1.41 k^{1/2} l$	
Jones & Launder (1972)	Dissipation	ϵ	$\nu_t = 0.09\,\frac{k^2}{\epsilon}$	High turbulence version of a low turbulence model applied to laminarizing boundary layer
Wilcox & Alber (1972)	Vorticity squared	Ω^2	$\nu_t = \frac{k}{\Omega} = k^{1/2} l$	Boundary layers

Table 3 Second-variable equation

Reference	ϕ	T	S	D
Kolmogorov (1942)	f	$\frac{\partial}{\partial x_n}\left(\frac{\nu_t}{\sigma\omega}\frac{\partial}{\partial x_n}f\right)$	–	$\frac{\lambda}{11}f^2$
Rotta (1968)	L	$\frac{\partial}{\partial y}\left(l\frac{\nu_t}{\sigma_{L1}}\frac{\partial k}{\partial y}\right)+\frac{\partial}{\partial y}\left(k\frac{\nu_t}{\sigma_L}\frac{\partial l}{\partial y}\right)$	$C_{g1}l\tau\frac{\partial \bar{U}}{\partial y}$	$C_d k^{3/2}$
Ng & Spalding (1972)	L	$\frac{\partial}{\partial y}\left(l\frac{\nu_t}{2}\frac{\partial k}{\partial y}\right)+\frac{\partial}{\partial y}\left(k\frac{\nu_t}{1.2}\frac{\partial l}{\partial y}\right)$	$0.84l\nu_t\left(\frac{\partial \bar{U}}{\partial y}\right)^2$	$0.055k^{3/2}\left(1+22\frac{l^4}{y^4}\right)$
Rodi & Spalding (1970)	L	$-\frac{\partial}{\partial y}\left(l\frac{\nu_t}{0.43}\frac{\partial k}{\partial y}\right)+\frac{\partial}{\partial y}\left(k\nu_t\frac{\partial l}{\partial y}\right)$	$l\nu_t\left(\frac{\partial \bar{U}}{\partial y}\right)^2$	$0.057k^{3/2}$
Wolfshtein (1970)	L	$\frac{\partial}{\partial x_n}\left(\frac{\nu_t}{40}\frac{\partial L}{\partial x_n}\right)-5\frac{l^2}{k^{1/2}}\left(\frac{\partial k}{\partial x_n}\right)^2$	$0.788l\nu_t\left(\frac{\partial \bar{U}}{\partial y}\right)^2$	$0.36k^{3/2}$
Glushko (1972)	λ	–	$0.24\nu_t\frac{\lambda}{k}\left(\frac{\partial \bar{U}}{\partial y}\right)^2$	$0.314\nu_t\left(l-\phi\frac{\lambda}{y^2}\right)$
Harlow & Nakayama (1972)	ϵ	$\frac{\partial}{\partial x_n}\left(\nu_t\frac{\partial \epsilon}{\partial x_n}\right)+\frac{\nu_t}{l^2}\frac{\partial}{\partial x_n}\left(\frac{\nu_t}{10}\frac{\partial k}{\partial x_n}\right)+\frac{\partial}{\partial x_n}\left(\frac{\nu_t^2}{10l^2}\frac{\partial}{\partial x_n}\frac{\bar{P}}{\rho}\right)$	$0.2\frac{\nu_t}{l^2}\left(\frac{\partial \bar{U}_i}{\partial x_j}+\frac{\partial \bar{U}_j}{\partial x_i}\right)\frac{\partial \bar{U}_i}{\partial x_j}$	$0.1\frac{\nu_t}{l^2}\epsilon$
Jones & Launder (1972)	ϵ	$\frac{\partial}{\partial y}\left(\frac{\nu_t}{1.3}\frac{\partial \epsilon}{\partial y}\right)$	$1.45\frac{\epsilon}{k}\nu_t\left(\frac{\partial \bar{U}}{\partial y}\right)^2$	$2\frac{\epsilon^2}{k}$
Wilcox & Alber (1972)	Ω^2	$\frac{\partial}{\partial x_n}\left(\frac{\nu_t}{1.733}\frac{\partial \Omega^2}{\partial x_n}\right)$	$0.707\Omega\left(\frac{\partial \bar{U}_i}{\partial x_j}\right)^2$	$0.18\Omega^3$

Table 4 Parameters of the unified length equation

Reference	σ_L	σ_{L1}	σ_{L2}	σ_{L3}	σ_{L4}	C_g	C_d	R
Kolmogorov (1942)	*	*	*	*	*	1.5	1.21	
Ng & Spalding (1972)	1.2	−3	∞	∞	∞	0.84	0.55	$220\,(l/y)^4\,k^{3/2}$
Rodi & Spalding (1970)	0.3	−1.77	∞	∞	∞	1.	0.634	
Wolfshtein (1970)	*	*	*	∞	∞	0.788	0.8	
Glushko (1972)	∞	2.5	∞	∞	∞	$1.1 \div 4.4$	1.2	$0.157\alpha\,k^{3/2}\,\phi(l^2/y^2)$† $0.2 < \alpha < 0.8$
Daly & Harlow (1970)	*	*	*	*	*	1.7	*	
Jones & Launder (1972)	0.52	1.73	−0.766	0.306	−0.65	1.05	0.5	
Launder et al. (1973)	*	*	*	*	*	1.05	0.6	
Wilcox & Alber (1972)	1.733	0.928	−0.577	−0.290	−0.577	1.146	0.5	

†ϕ is a function of l/y, where $\phi(0) = 0$, $\phi(1) = 1$, and its distribution is adjusted to describe boundary layers.

Here σ_L, σ_{L1}, σ_{L3}, σ_{L4}, c_g, and c_d are universal functions of Re_t, which assume a constant value at high Re_t, and R is an algebraic function of k, l, and $\bar{U}_1$. The values of these functions are presented in Table 4.

In Table 4 an infinite σ means that the corresponding diffusion term is missing. The asterisks stand for cases where the corresponding coefficients cannot be specified owing to entirely different formulation. Bearing in mind that the constants given by Harlow and Nakayama, Daly and Harlow, and Kolmogorov have not been extensively tested, we see that the constant c_g is approximately $0.8 \div 1.1$, while c_d is likely to be between 0.5 and 0.8 (although the value of 1.2 has been recommended as well). Unfortunately, it is impossible to place any reasonable limits on the diffusion coefficients. Furthermore, in some cases it was necessary to use additional terms, denoted by R, in order to obtain agreement with the experimental data, and it is conceivable that the need for this correction results from the use of inaccurate expressions for the diffusion.

2.3 Conclusions

To conclude this section, the following recommendations can be made:

1. Different formulations of the energy equation appear to converge to a common form. Solutions of this equation, when the length scale is correctly specified, usually agree reasonably well with experimental data.
2. Formulations of the length scale equation do not appear to have converged towards a common form as yet. However, we may note the increasing similarity between the more recent models such as, say, those of Jones and Launder, and Wilcox and Alber.
3. Turbulence diffusion remains the weakest point in two-variable modeling. This applies in particular to the length scale equation, but the turbulence energy equation requires improvement as well.

3 REYNOLDS STRESS MODELS

3.1 Introduction

As successful as two-variable models may be, their validity can be contested on a number of grounds. The linear relation between the Reynolds stress and the mean rate of strain is not always borne out by experimental data. Therefore, the assumption of a scalar turbulence viscosity cannot be expected to be universally valid. Typical examples include turbulent wall jets, flow in annuli, and plane ducts with different roughnesses on both sides. Furthermore, some flow fields require intimate knowledge of the Reynolds stress components. We may recall cases of energy reversal (where the production of turbulence energy is negative), secondary motion induced by geometry and roughness effects, and severe nonisotropy in the production of Reynolds stresses, such as, say, on the axis of free jets or in turbulent stagnation flow. Analysis of such cases requires the solution of the Reynolds stress equations (2.8). Moreover, it is expected that the use of the Reynolds stress models will increase the universality of the prediction methods, as turbulence is described in terms of more parameters.

Reynolds stress models constitute a new branch of turbulence modeling. Only a few solutions are available, and development of the models is still progressing rapidly. Therefore we shall describe these models in more detail than the two-variable models. In particular, we shall focus our attention on the diffusion, dissipation, and redistribution terms of the Reynolds stress equations. The convection and the generation terms do not pose modeling problems and will not be discussed here.

In addition to the Reynolds stress equation, a length scale equation is required. However, this equation is usually similar to the scale equation of the two-variable models. Therefore the discussion of this equation in the present section is limited to the features derived from a knowledge of all the components of the Reynolds stress.

3.2 Turbulence diffusion

3.2.1 The triple correlation This section is concerned with the turbulence diffusion process, namely with the triple correlation term

$$T_{ij} = -\frac{\partial}{\partial x_m}\overline{u_m u_i u_j}$$

(The contribution of the pressure-velocity correlation to the diffusion is dealt with in Sec. 3.3.) It is generally accepted that this term describes a conservative process of turbulence diffusion. To see this, the Gauss theorem may be used, written as

$$\iiint\limits_{v} T_{ij}\,dv = -\iint\limits_{S}\overline{u_m u_i u_j}\,n_m\,ds \tag{3.1}$$

where n_i is the ith component of a unit vector normal to the surface element ds, and v is an arbitrary volume surrounded by the surface S. When a flow field is bounded by laminar regions, or by homogeneous isotropic regions, the surface integral vanishes, and therefore the volume integral vanishes as well. Thus T_{ij} represents a process in which the quantity $\overline{u_i u_j}$ is transferred from one part of the flow field to others without any production or losses.

As T_{ij} vanishes in the homogeneous field, it may be assumed to depend on the $\overline{u_i u_j}$ gradients. Assuming a linear dependence, a sixth order turbulence diffusion tensor may be defined to yield the most general linear formula:

$$-\overline{u_m u_i u_j} = \Gamma_{\text{mijrst}} \frac{\partial}{\partial x_r} \overline{u_s u_t} \tag{3.2}$$

This equation is rather complicated; however, it is possible to generalize all the published proposals for this term by the introduction of a second order turbulence diffusion tensor Γ_{ml}.

We commence by noting some general properties of the diffusion tensor coefficient. The three indices in $\overline{u_m u_i u_j}$ are interchangeable, and every model should retain this property. Furthermore, it is usually assumed that the turbulence diffusion tensor is symmetrical. It is also hard to conceive how any component of Γ_{ml} can be multiplied by gradients in any direction other than l or m (but the scalar Γ_{ll} may multiply any $\overline{u_i u_j}$ gradient). These three limitations yield the following general form:

$$\begin{aligned}
-\overline{u_m u_i u_j} = {} & \alpha_1 \left[\Gamma_{ll} \frac{\partial}{\partial x_n} \left(\overline{u_n u_m}\, \delta_{ij} + \overline{u_n u_i}\, \delta_{mj} + \overline{u_n u_j}\, \delta_{mi} \right) \right] \\
& + \alpha_2 \left[\Gamma_{nl} \frac{\partial}{\partial x_l} \left(\overline{u_n u_m}\, \delta_{ij} + \overline{u_n u_i}\, \delta_{mj} + \overline{u_n u_j}\, \delta_{mi} \right) \right] \\
& + \alpha_3 \left[\Gamma_{ll} \left(\frac{\partial}{\partial x_m} \overline{u_i u_j} + \frac{\partial}{\partial x_i} \overline{u_m u_j} + \frac{\partial}{\partial x_j} \overline{u_m u_i} \right) \right] \\
& + \alpha_4 \left\{ \Gamma_{ll} \left[\frac{\partial}{\partial x_m} (2k)\delta_{ij} + \frac{\partial}{\partial x_i} (2k)\delta_{mj} + \frac{\partial}{\partial x_j} (2k)\delta_{mi} \right] \right\} \\
& + \alpha_5 \left(\Gamma_{ml} \frac{\partial}{\partial x_l} \overline{u_i u_j} + \Gamma_{il} \frac{\partial}{\partial x_l} \overline{u_m u_j} + \Gamma_{jl} \frac{\partial}{\partial x_l} \overline{u_m u_i} \right) \\
& + \alpha_6 \left[\Gamma_{ml} \frac{\partial}{\partial x_l} (2k)\delta_{ij} + \Gamma_{il} \frac{\partial}{\partial x_l} (2k)\delta_{mj} + \Gamma_{jl} \frac{\partial}{\partial x_l} (2k)\delta_{mi} \right]
\end{aligned} \tag{3.3}$$

where $\alpha_1, \alpha_2, \alpha_3, \alpha_4, \alpha_5$, and α_6 are scalar coefficients.

It may be recalled that the turbulent diffusion of the turbulence energy as given by Eq. (2.15) is related to the energy gradients only, and not to the Reynolds stress gradients. The diffusion coefficients of Eq. (2.15) were scalars, but now they may be regarded as components of an eddy diffusion tensor as well. Then, if Eq. (3.3) is to contract to a series of products of diffusion scalars or tensors with the energy gradients, the α-coefficients should satisfy the conditions

$$5\alpha_1 + 2\alpha_3 = 0 \tag{3.4}$$

$$5\alpha_2 + 2\alpha_5 = 0 \tag{3.5}$$

It is not clear whether conditions (3.4) and (3.5) must be generally satisfied. However, if we use the mixing length approach to estimate the transport of the turbulence energy as a conserved scalar, we find that it is preferable to satisfy these conditions.

3.2.2 Outline of existing models The available approximations to T_{ij} will now be described as special cases of Eq. (3.3). The simplest approximation is obtained by taking a diffusion scalar:

$$\Gamma_{ij} = c_t k^{1/2} l \delta_{ij} \tag{3.6}$$

$$\alpha_3 = 1 \quad \text{and} \quad \alpha_1 = \alpha_2 = \alpha_4 = \alpha_5 = \alpha_6 = 0 \tag{3.7}$$

This is the Donaldson and Rosenbaum (1968) formula:

$$T_{ij} = \frac{\partial}{\partial x_m}\left[c_t k^{1/2} l \left(\frac{\partial}{\partial x_m}\overline{u_i u_j} + \frac{\partial}{\partial x_j}\overline{u_m u_i} + \frac{\partial}{\partial x_i}\overline{u_m u_j}\right)\right] \tag{3.8}$$

which satisfies the symmetry of $\overline{u_m u_i u_j}$ and Eq. (3.5), but does not satisfy (3.4).

The simpler expression

$$T_{ij} = \frac{\partial}{\partial x_m}\left(c_t k^{1/2} l \frac{\partial}{\partial x_m}\overline{u_i u_j}\right) \tag{3.9}$$

contracts directly to (2.15), but does not satisfy the symmetry condition. It was used by Donaldson (1969) to predict boundary layer transition and by Naot et al. (1974) to predict fully developed flow in a square duct.

A diffusion tensor may be defined by

$$\Gamma_{ij} = c_t k^{1/2} l \frac{\overline{u_i u_j}}{\frac{2}{3}k} \tag{3.10}$$

Hirt (1969) has shown that this equation results from the same reasoning as the Prandtl mixing length hypothesis. Hirt (1969) and Hanjalic and Launder (1971) assumed $\alpha_5 = 1, \alpha_1 = \alpha_2 = \alpha_3 = \alpha_4 = \alpha_6 = 0$, and obtained:

$$T_{ij} = \frac{\partial}{\partial x_m}\left[\gamma\left(\frac{\overline{u_m u_l}}{\frac{2}{3}k}\frac{\partial\overline{u_i u_j}}{\partial x_l} + \frac{\overline{u_i u_l}}{\frac{2}{3}k}\frac{\partial\overline{u_j u_m}}{\partial x_l} + \frac{\overline{u_j u_l}}{\frac{2}{3}k}\frac{\partial\overline{u_i u_m}}{\partial x_l}\right)\right] \tag{3.11}$$

where $\gamma = c_t k^{1/2} l$ (3.12)

Again, the symmetry condition and Eq. (3.4) are satisfied, but Eq. (3.5) and the reduction to (2.15) are not. This expression was applied by Hirt to channel flow and by Hanjalic and Launder to asymmetric channel flow, wall jets, boundary layers, plane jets, and plane mixing layers.

Daly and Harlow (1970) regarded the contraction to (2.15) as more important than the symmetry of $\overline{u_m u_i u_j}$ and shortened Eq. (3.11) to

$$T_{ij} = \frac{\partial}{\partial x_m}\left(\gamma \frac{\overline{u_m u_l}}{\frac{2}{3}k} \frac{\partial}{\partial x_l} \overline{u_i u_j}\right) \tag{3.13}$$

An attempt to satisfy the symmetry condition and Eqs. (3.4) and (3.5) simultaneously led Harlow and Hirt (1969) to the following proposal:

$$\begin{aligned} T_{ij} = \frac{\partial}{\partial x_m}\Bigg(\gamma\Bigg\{\Bigg(&\frac{\overline{u_m u_l}}{\frac{2}{3}k}\frac{\partial \overline{u_i u_j}}{\partial x_l} + \frac{\overline{u_i u_l}}{\frac{2}{3}k}\frac{\partial \overline{u_j u_m}}{\partial x_l} + \frac{\overline{u_j u_l}}{\frac{2}{3}k}\frac{\partial \overline{u_i u_m}}{\partial x_l}\Bigg) \\ &- \frac{2}{5}\Bigg[\Bigg(\frac{\overline{u_n u_l}}{\frac{2}{3}k}\frac{\partial \overline{u_n u_m}}{\partial x_l} - 2\frac{\overline{u_m u_l}}{\frac{2}{3}k}\frac{\partial k}{\partial x_l}\Bigg)\delta_{ij} \\ &+ \Bigg(\frac{\overline{u_n u_l}}{\frac{2}{3}k}\frac{\partial \overline{u_n u_i}}{\partial x_l} - 2\frac{\overline{u_i u_l}}{\frac{2}{3}k}\frac{\partial k}{\partial x_l}\Bigg)\delta_{jm} \\ &+ \Bigg(\frac{\overline{u_n u_l}}{\frac{2}{3}k}\frac{\partial \overline{u_n u_j}}{\partial x_l} - 2\frac{\overline{u_j u_l}}{\frac{2}{3}k}\frac{\partial k}{\partial x_l}\Bigg)\delta_{im}\Bigg]\Bigg\}\Bigg) \end{aligned} \tag{3.14}$$

with $\alpha_2 = -\frac{2}{5}$; $\alpha_5 = 1$; $\alpha_6 = \frac{2}{5}$, and $\alpha_1 = \alpha_3 = \alpha_4 = 0$. However, these authors did not use (3.14), preferring the simpler Eq. (3.13).

3.2.3 Discussion It is impossible at present time to specify the value of the α coefficients. Yet it seems clear that the expressions used in the past are not necessarily the most general ones, and many other expressions are possible. This is equally true of a diffusion scalar and a diffusion tensor.

All the expressions which are derived from Eq. (3.2) for the turbulent diffusion contain second order derivatives of $\overline{u_i u_j}$. As these are the higher order derivative of the Reynolds stresses in their model equations, the use of these suggestions yields elliptic equations for the Reynolds stresses. These equations are similar to the transport equations of energy and matter and may be solved by existing procedures (see Runchal et al. 1969, or Naot et al., 1974). The physical significance of the ellipticity of the equations is that disturbances propagate in all directions in the flow field. Yet the transport coefficient Γ_{mijrst} (Eq. [3.2]) is not yet sufficiently understood, and the directional properties of this quantity deserve more attention, as will be discussed in Sec. 4.

Finally, most available models assume ellipticity of the transport equations for $\overline{u_i u_j}$. However, if algebraic rather than differential representation is used for $\overline{u_m u_i u_j}$ the transport equations become hyperbolic (Bradshaw et al., 1967).

3.3 Pressure-velocity interaction

In this section we shall study the pressure interaction term $P_{ij}{}^t$, defined by

$$P_{ij}{}^t = -\frac{1}{\rho}\left(\overline{u_i \frac{\partial p}{\partial x_j}} + \overline{u_j \frac{\partial p}{\partial x_i}}\right) \tag{3.15}$$

$P_{ij}{}^t$ may be divided into two parts: pressure transport τ_{ij} and redistribution ϕ_{ij}. These are defined by

$$\tau_{ij} = -\frac{1}{\rho}\left(\frac{\partial}{\partial x_i}\overline{pu_j} + \frac{\partial}{\partial x_j}\overline{pu_i}\right) \tag{3.16}$$

and

$$\phi_{ij} = \frac{1}{\rho}\left(\overline{p\frac{\partial u_i}{\partial x_j}} + \overline{p\frac{\partial u_j}{\partial x_i}}\right) \tag{3.17}$$

The contraction of ϕ_{ij} vanishes in an incompressible flow owing to the continuity equation. Thus the action of ϕ_{ij} results in changes in the individual components of $\overline{u_i u_j}$ without changing the energy level. Therefore it has been termed "redistribution." In fact, this section is mainly concerned with the redistribution, ϕ_{ij}. However before considering it, we shall refer to the pressure transport τ_{ij}.

3.3.1 The pressure transport The meaning of this term will be better understood if we consider the following integral identity, obtained by use of the Gauss theorem,

$$\iiint\limits_{v} \tau_{ij}\, dv = -\frac{1}{\rho} \iint\limits_{s} (\overline{pu_i} n_j + \overline{pu_j} n_i)\, ds \tag{3.18}$$

where n_i is the component of a unit vector normal to the surface S enveloping the arbitrary volume V. The integrand of (3.18) vanishes on those parts of "S" that are parallel to i and j and on those parts of "S" where the flow is either laminar or homogeneous-isotropic. Thus, τ_{ij} changes the spatial distribution of $\overline{u_i u_j}$, but does not change its total amount in the entire flow field. Such interaction is not simple model, and most researchers add τ_{ij} to the T_{ij} terms, and model them together.

We note the attempt made by Harlow and Hirt (1969) to treat the pressure as a conserved scalar, and thus to obtain

$$\tau_{ij} \propto \frac{\partial}{\partial x_i}\left(\frac{\gamma}{\rho}\,\frac{\overline{u_i u_m}}{\frac{2}{3}k}\,\frac{\partial \bar{p}}{\partial x_m}\right) + \frac{\partial}{\partial x_j}\left(\frac{\gamma}{\rho}\,\frac{\overline{u_i u_m}}{\frac{2}{3}k}\,\frac{\partial \bar{p}}{\partial x_m}\right) \tag{3.19}$$

This model was abandoned owing to the negative diffusion predicted by it in favor of the suggestion made by Hirt (1969) and Daly and Harlow (1970), namely

$$\tau_{ij} \propto \frac{\partial}{\partial x_i}\left(\gamma \frac{\partial}{\partial x_m}\, \overline{u_j u_m}\right) + \frac{\partial}{\partial x_j}\left(\gamma \frac{\partial}{\partial x_m}\, \overline{u_i u_m}\right) \tag{3.20}$$

Here γ is the turbulence diffusion given by Eq. (3.12). This proposal is commonly used by other investigators as well.

3.3.2 Exact equations for the redistribution The analytic approach to the specification of the total pressure interaction term, $P_{ij}{}^t$, was initiated by Chou (1945) and Rotta (1951). It is possible to develop ϕ_{ij} and τ_{ij} individually, yet it is simpler to develop the total interaction $P_{ij}{}^t$. In the present discussion we consider homogeneous turbulence (not necessarily isotropic). In this case the results presented below contribute only to the redistribution ϕ_{ij}.

By taking the divergence of the Navier-Stokes equations, a Poisson equation for the fluctuating pressure is obtained:

$$\frac{1}{\rho}\,\nabla^2 p = -\left[2\,\frac{\partial \bar{U}_m}{\partial x_l}\,\frac{\partial u_l}{\partial x_m} + \frac{\partial^2}{\partial x_m \partial x_l}\,(u_m u_l - \overline{u_m u_l})\right] \tag{3.21}$$

The solution of this equation is:

$$\frac{1}{\rho}p(\mathbf{x}) = \frac{1}{4\pi}\iiint_v \left[2\frac{\partial \bar{U}_m}{\partial r_l}\frac{\partial u_l}{\partial r_m} + \frac{\partial^2}{\partial r_m \partial r_l}(u_m u_l - \overline{u_m u_l})\right]\Bigg|_{\substack{\text{calculated}\\ \text{at } \mathbf{x}+\mathbf{r}}} \frac{dv}{r} + S$$

where $\mathbf{x}$ is a position vector, $\mathbf{r}$ is the vector separating the running point of integration from $\mathbf{x}$, v is an arbitrary volume surrounding $\mathbf{x}$, and S is a surface integral. We can use $p(\mathbf{x})$ and correlate its derivatives statistically with the velocity fluctuations at $\mathbf{x}$. At this stage we neglect the surface integral since all the two point correlations may be assumed to vanish on S. Thus we obtain

$$-P_{ij} = \frac{1}{2\pi}\iiint_v \frac{\partial}{\partial r_k}\overline{u_i(\mathbf{x})u_l(\mathbf{x}+\mathbf{r})}\frac{\partial^2}{\partial r_j \partial r_l}\bar{U}_k(\mathbf{x}+\mathbf{r})\frac{dv}{r} \qquad \text{A}$$

$$+\frac{1}{2\pi}\iiint_v \frac{\partial^2}{\partial r_k \partial r_j}\overline{u_i(\mathbf{x})u_l(\mathbf{x}+\mathbf{r})}\frac{\partial}{\partial r_l}\bar{U}_k(\mathbf{x}+\mathbf{r})\frac{dv}{r} \qquad \text{B}$$

$$+\frac{1}{4\pi}\iiint_v \frac{\partial^3}{\partial r_k \partial r_j \partial r_l}\overline{u_i(\mathbf{x})u_l u_k(\mathbf{x}+\mathbf{r})}\frac{dv}{r} \qquad \text{C}$$

$$+\frac{1}{2\pi}\iiint_v \frac{\partial}{\partial r_k}\overline{u_j(\mathbf{x})u_l(\mathbf{x}+\mathbf{r})}\frac{\partial^2}{\partial r_j \partial r_l}\bar{U}_k(\mathbf{x}+\mathbf{r})\frac{dv}{r} \qquad \text{A}'$$

$$+\frac{1}{2\pi}\iiint_v \frac{\partial^2}{\partial r_k \partial r_i}\overline{u_j(\mathbf{x})u_l(\mathbf{x}+\mathbf{r})}\frac{\partial}{\partial r_l}\bar{U}_k(\mathbf{x}+\mathbf{r})\frac{dv}{r} \qquad \text{B}'$$

$$+\frac{1}{4\pi}\iiint_v \frac{\partial^3}{\partial r_k \partial r_i \partial r_l}\overline{u_j(\mathbf{x})u_l u_k(\mathbf{x}+\mathbf{r})}\frac{dv}{r} \qquad \text{C}' \qquad (3.23)$$

The total pressure interaction may now be divided into two parts:

$$P_{ij}{}^t = P_{ij} + P_{ij}^* \qquad (3.24)$$

where $P_{ij} = A + A' + B + B'$

$P_{ij}^* = C + C'$

The term P_{ij} will be referred to as the forced interaction since it exists only in the presence of mean velocity gradients, and hence is forced by the mean velocity field. P_{ij}^* is not explicitly dependent on the mean velocity field, and is referred to as the free redistribution.

3.3.3 The free redistribution The free redistribution, P_{ij}^*, vanishes in homogeneous and isotropic turbulence. When turbulence is homogeneous but non-isotropic, P_{ij}^* is finite. Rotta (1951) postulated a linear relation between P_{ij}^* and the deviatoric Reynolds stress $(\overline{u_i u_j} - \frac{2}{3} k\, \delta_{ij})$. An exact evaluation of this term has not yet been made. (The isotropic triple velocity correlation could have been substituted in Eq. [3.23C], but this yields zero value for this term.) The exact form of $P^*{}_{ij}$ does not suggest an explicit dependence on the mean rate of strain. However, our experience with homogeneous nonisotropic shear flow far from walls (Rose, 1966) suggests that P_{ij}^* is, indeed, related to the rate of strain tensor as well. Thus we assume

$$P_{ij}^* = \Lambda^* \frac{\epsilon}{k} (\overline{u_i u_j} - \frac{2}{3} k\delta_{ij}) + \gamma^* \cdot \frac{4}{3} k\bar{E}_{ij} \tag{3.25}$$

$$\text{where } \bar{E}_{ij} = \frac{1}{2}\left(\frac{\partial \bar{u}_i}{\partial x_j} + \frac{\partial \bar{u}_j}{\partial x_i}\right) \tag{3.25a}$$

3.3.4 The forced redistribution To estimate the forced redistribution, P_{ij}, we must calculate the integrals A and B in Eq. (3.23). For this purpose we require a specification of the two-point velocity correlation and of the mean velocity gradient at the point $(\mathbf{x} + \mathbf{r})$. The latter is obtained by a Taylor series:

$$\frac{\partial}{\partial r_l} \bar{U}_k(\mathbf{x} + \mathbf{r}) = \frac{\partial}{\partial x_l} \bar{U}_k(\mathbf{x}) + \frac{\partial^2}{\partial x_l \partial x_m} \bar{U}_k(\mathbf{x}) r_m + \cdots$$

When this expression is substituted in Eq. (3.23) we obtain

$$P_{ij} = (a_{kj}{}^{li} + a_{ki}{}^{lj}) \frac{\partial \bar{U}_k}{\partial x_l} \tag{3.26}$$

$$\text{where } a_{kj}{}^{li} = -\frac{1}{2\pi} \iiint_v \frac{\partial^2}{\partial r_k \partial r_j} \overline{u_i(\mathbf{x}) u_l(\mathbf{x} + \mathbf{r})} \frac{dv}{r} \tag{3.26a}$$

Then p_{ij} may be obtained by either of two possible methods:

1. selecting an implicit model for $\overline{u_i(\mathbf{x}) u_l(\mathbf{x} + \mathbf{r})}$, or
2. selecting a formula for the coefficients $a_{ki}{}^{li}$.

An attempt to use the first method was already reported by Chou (1945). Rotta (1951) used the homogeneous isotropic two-point correlation tensor:

$$\overline{u_i(\mathbf{x})u_l(\mathbf{x}+\mathbf{r})} = \frac{2}{3}\,k(\mathbf{x})\left\{[f(r)-g(r)]\,\frac{r_i r_l}{r^2} + g(r)\delta_{ij}\right\} \tag{3.27}$$

where f and g are correlation functions related to one another by the continuity equation. The result is

$$P_{ij} = 0.6\,\frac{4}{3}\,k\bar{E}_{ij} \tag{3.28}$$

where $\bar{E}_{ij}$ is given by (3.25a). However, Rotta (1951) had to modify this expression in order to obtain agreement with channel-flow data. Naot et al. (1970) have shown that this equation, which tends to produce similarity between the Reynolds stresses and the mean rate of strain tensor, is not realistic when experimental data on grid generated turbulence is examined. Therefore, it is necessary to use another correlation model.

A quasi-isotropic, homogeneous two-point correlation tensor was developed by assuming that : (a) the two point velocity correlation is linearly related to the Reynolds stresses, and (b) the quasi-isotropic two point correlation functions have the same symmetry properties as the isotropic ones. These assumptions yield

$$\begin{aligned}
\overline{u_i(\mathbf{x})u_l(\mathbf{x}+\mathbf{r})} &= \overline{u_i u_l}(\mathbf{x})G(r) \\
&\quad + u_m u_n(\mathbf{x})\,\frac{r_n r_m}{r^2}\,\frac{r_i r_l}{r^2}\,[F(r)+G(r)-2J(r)-K(r)] \\
&\quad + \overline{u_m u_n}(\mathbf{x})\,\frac{r_n r_m}{r^2}\,\delta_{il}K(r) \\
&\quad + \left[u_i u_m(\mathbf{x})\,\frac{r_m r_l}{r^2} + \overline{u_m u_l}(\mathbf{x})\,\frac{r_i r_m}{r^2}\right][J(r)-G(r)] \\
&\quad + \frac{2}{3}\,k(\mathbf{x})\,\frac{r_i r_l}{r^2}\,[f(r)-g(r)+G(r)-F(r)+K(r)] \\
&\quad + \frac{2}{3}\,k(\mathbf{x})\delta_{il}\,[g(r)-G(r)-K(r)]
\end{aligned} \tag{3.29}$$

Here F, G, J, and K are new correlation functions related to one another by the continuity equation (see Naot et al., 1973). Using Eq. (3.29), Naot et al. (1973) found

$$P_{ij} = -\alpha\left(\pi_{ij} - \frac{1}{3}\pi_{ll}\delta_{ij}\right) + \beta\left(D_{ij} - \frac{1}{3}D_{ll}\delta_{ij}\right) + \gamma\,\frac{4}{3}\,k\bar{E}_{ij} \tag{3.30}$$

$$\text{where } \pi_{ij} \equiv -\left(\overline{u_i u_l}\,\frac{\partial \bar{U}_j}{\partial x_l} + \overline{u_j u_l}\,\frac{\partial \bar{U}_i}{\partial x_l}\right) \tag{3.30a}$$

and

$$D_{ij} \equiv -\left(\overline{u_i u_l}\,\frac{\partial \bar{U}_l}{\partial x_j} + \overline{u_j u_l}\,\frac{\partial \bar{U}_l}{\partial x_i}\right) \tag{3.30b}$$

The parameters α, β, γ are related to a single macro-scale ϕ by

$$\alpha = \frac{76 - 8\phi}{105} \qquad \beta = \frac{22 + 64\phi}{105} \qquad \text{and} \qquad \gamma = \frac{9 + 72\phi}{105} \tag{3.30c}$$

The macro-scale ϕ depends on the function $K(r)$ as:

$$\phi = \int_0^\infty \frac{1}{r} K(r)\,dr \tag{3.30d}$$

It is possible to eliminate ϕ, and to relate two parameters to the third one by

$$\alpha = 0.75 - 0.125\,\beta \tag{3.30e}$$

and

$$\gamma = -0.15 + 1.125\,\beta \tag{3.30f}$$

In fact, both Eqs. (3.28) and (3.30) can be regarded as special cases of Eq. (3.26). By assuming that p_{ij} is a linear function of $\overline{u_i u_j}$ and $\partial \bar{U}_i/\partial x_j$, and incorporating certain kinematic restrictions, P_{ij} can be obtained from (3.26). This possibility was outlined by Rotta (1951) and pursued by Hanjalic and Launder (1971), Hassid (1973), and Launder et al. (1973). They substituted

$$\overline{u_i(\mathbf{x})u_l(\mathbf{x}+\mathbf{r})} = \overline{u_n u_m}\,(\mathbf{x})F_{\text{nmil}}(\mathbf{r}) \tag{3.31}$$

into Eq. (3.26a) and obtained

$$\begin{aligned} a_{kj}{}^{li} = {} & \alpha_1\,\overline{u_i u_l}\,\delta_{jk} + \alpha_2\,\overline{u_i u_j}\,\delta_{lk} + \alpha_3\,\overline{u_i u_k}\,\delta_{lj} \\ & + \alpha_4\,\overline{u_l u_j}\,\delta_{ik} + \alpha_5\,\overline{u_l u_k}\,\delta_{ij} + \alpha_6\,\overline{u_j u_k}\,\delta_{il} \\ & + \alpha_7\,\delta_{il}\delta_{jk} + \alpha_8\,\delta_{ij}\delta_{lk} + \alpha_9\,\delta_{ik}\delta_{lj} \end{aligned} \tag{3.32}$$

By virtue of Eq. (3.26a) the $a_{kj}{}^{li}$ coefficients must satisfy

$$a_{kj}{}^{li} = a_{jk}{}^{li} \tag{3.33a}$$

Furthermore, the continuity equation implies

$$a_{kl}{}^{li} = 0 \tag{3.33b}$$

and it can be shown, by application of the Gauss theorem, that

$$a_{jj}{}^{li} = 2\overline{u_l u_i} \tag{3.33c}$$

In a homogeneous field

$$\overline{u_i(\mathbf{x})u_l(\mathbf{x}+\mathbf{r})} = \overline{u_l(\mathbf{x})u_i(\mathbf{x}-\mathbf{r})}$$

and therefore it can be shown that

$$a_{kj}{}^{li} = a_{kj}{}^{il} \tag{3.33d}$$

Restrictions (3.33) are sufficient to reduce the number of free coefficients α_1 in Eq. (3.32) to one, and Eq. (3.30) is again obtained. This is not surprising since the quasi-isotropic two-point correlation model satisfied also all the above-mentioned restrictions. This generalization shows also that regardless of the number of independent correlation functions, the P_{ij} term depends only on one of these functions, which is $K(r)$ in the quasi-isotropic case.

3.3.5 Discussion Equation (3.30) contains three terms. The term $-\alpha(\pi_{ij} - \frac{1}{3}\pi_{ll}\delta_{ij})$, with $\alpha > 0$, describes the isotropization of the Reynolds stress. The first part $-\alpha\pi_{ij}$ reduces the production of all the components, while the second part $\alpha\pi_{ll}\,\delta_{ij}/3$ describes equal distribution of the total production π_{ll} among the diagonal components, enforcing isotropy. Naot et al. (1970) have shown that grid-generated turbulence under acceleration, homogeneous shear, and plane strain can be predicted by means of this term alone with $\alpha \sim 0.80$. The same authors (1974) have also calculated fully developed square duct flow using the full formula with $\alpha \sim 0.75$. A similar value was also used by Launder et al. (1973) to cover a wide range of boundary layer flows. Thus it is quite evident that about three-quarters of the production of the Reynolds stresses is reduced and redistributed isotropically. Yet, the forced pressure interaction contributes also to the production of nonisotropy by means of the second and third terms of Eq. (3.30).

To understand the role of the second term we have to add it to the former:

$$-\alpha\left(\pi_{ij} - \frac{1}{3}\,\pi_{ll}\delta_{ij}\right) + \beta\left(D_{ij} - \frac{1}{3}\,D_{ll}\delta_{ij}\right) = \frac{1}{3}\,(\alpha - \beta)\pi_{ll}\delta_{ij}$$
$$+ (\alpha - \beta)\,[\overline{u_i u_l}\bar{E}_{lj} + \overline{u_j u_l}\,\bar{E}_{li}$$
$$+ (\alpha + \beta)(\overline{u_i u_l}\,\overline{\omega_{li}})] \tag{3.34}$$

where ω_{lj} is the mean vorticity

$$\overline{\omega_{lj}} = \frac{1}{2}\left(\frac{\partial \bar{U}_j}{\partial x_l} - \frac{\partial \bar{U}_l}{\partial x_j}\right) \tag{3.34a}$$

The last two terms of (3.34), expressing interactions with the mean rate of strain and with the mean vorticity, were proposed by Reynolds (1970), as well. With β equal to zero, both interactions are equally accounted for. Though the recommended value of β is small ($\beta \sim -\ 0.046$), this term was found to be essential for realistic predictions of channel flow. It is this term which reduces the fluctuations normal to the wall in favor of the fluctuations parallel to it. Thus this term controls the secondary motion in noncircular ducts. Since β is negative the interaction with the mean rate of strain is preferred by this term.

In the limiting case of isotropic turbulence Eq. (3.30) reduces to Eq. (3.28). This is caused by the third term $\gamma\,\frac{4}{3}\,k\,\bar{E}_{ij}$ as $\alpha - \beta + \gamma = 0.6$ (independent of ϕ) and since π_{ij} and D_{ij} transform to $\gamma\,\frac{4}{3}\,k\,\bar{E}_{ij}$ in this limiting case. This mode of interaction produces proportionality between the Reynolds stresses and the mean rate of strain tensor. However, in shear flows γ appears to be small and the above mentioned proportionality is not very significant.

Finally, it is possible to use more terms in the Taylor series expansion for the mean velocity gradients mentioned in the beginning of this section, and obtain higher-order approximation. If we also assume that the two point correlation is symmetric, namely $\overline{u_i(\mathbf{x})\ u_l(\mathbf{x}+\mathbf{r})} = \overline{u_i(\mathbf{x})\ u_l(\mathbf{x}-\mathbf{r})}$, the result obtained will not include the second order terms. Such an expression, including third order derivatives of the mean velocity, was already proposed by Rotta (1951) but was not used later. Application of the homogeneity restriction resulted in an expression for the forced redistribution. However, if this restriction will be removed, the described methods will result in additional terms which express forced transport. It may be noted that Eq. (3.32) for $a_{kj}{}^{li}$ is not the only starting point. Hanjalic and Launder (1971) proposed a nonlinear addition to this equation which was abandoned later by Launder et al. (1973).

3.4 Dissipation

In the present section the dissipation term $L_{ij} = -\,\nu\overline{(\partial u_i/\partial x_l)(\partial u_j/\partial x_l)}$ will be discussed. The term L_{ij} derives from the viscosity term of the Navier-Stokes equation. The latter has yielded also another term, $\nu(\partial^2/\partial x_l^2)\,\overline{u_i u_j}$, which was neglected since the laminar viscosity ν is generally much smaller than the turbulence diffusion γ. Obviously, the diagonal components of L_{ij} are negative and express decay of Reynolds stresses. Furthermore, in homogeneous turbulence the contraction $L_{ii}/2$ has the thermodynamic meaning of the work dissipated into thermal energy. Usually it is postulated that the off-diagonal terms of L_{ij} describe decay of the Reynolds stresses as well, and L_{ij} is usually referred to as dissipation. We shall now study the relation between the components L_{ij} and the energy loss defined by

$$\epsilon = \frac{1}{2} L_{ii} \tag{3.35}$$

3.4.1 The decay of the Reynolds stresses To approximate the L_{ij} term we restrict the analysis to homogeneous flows where it can be shown that

$$L_{ij} = -2\nu \frac{\partial u_i}{\partial x_m} \frac{\partial u_j}{\partial x_m} = 2\nu \frac{\partial^2}{\partial r_m\, \partial r_m} \overline{u_i(\mathbf{x})u_j(\mathbf{x}+\mathbf{r})}\Big|_{r=0} \tag{3.36}$$

The L_{ij} term is obtained by substituting in Eq. (3.36) an implicit formula for the two-point correlation tensor. Using Eq. (3.29) and the relations enforced by the continuity equation, the quasi-isotropic approximation is obtained

$$L_{ij} = -\frac{\epsilon}{k}\left[\overline{u_i u_j}\,\Lambda_0 + (1-\Lambda_0)\frac{2}{3}k\delta_{ij}\right] \tag{3.37}$$

where ϵ is given by

$$\epsilon = \left(\frac{10\nu}{\lambda_g^{\,2}}\right) \tag{3.37a}$$

Here Λ_0 is a function of the microscales λ_g, λ_G, and λ_K as follows:

$$\Lambda_0 = \frac{44}{50}\left(\frac{\lambda_g}{\lambda_G}\right)^2\left[1 - \frac{4}{11}\left(\frac{\lambda_G}{\lambda_K}\right)^2\right] \tag{3.37b}$$

Two modes of decay may be identified in Eq. (3.37). The first one, $\overline{u_i u_j}\,\Lambda_0$, reduces each component of the Reynolds stresses in proportion to its own value. The second, $(1-\Lambda_0)\frac{2}{3}k\,\delta_{ij}$, reduces the diagonal components equally, but does not affect the off-diagonal components. The relative importance of these two modes of decay is governed by the coefficient Λ_0. To estimate the value of Λ_0 by Eq. (3.37b) we must recall that λ_g, λ_G, and λ_K are the microscales of the correlation functions g, G, and K. The correlation function g describes the isotropic part of the two point correlation, and G and K describe the nonisotropic part. Therefore, in a nonisotropic turbulent flow, where the energy spectra are self-similar, that is to say, both large and small eddies are nonisotropic to the same extent, we should expect λ_g to be equal to λ_G and $\Lambda_0 \sim 1$. Grid generated turbulence and low Reynolds number turbulence are such situations. However, when the small eddies are isotropic and large ones are nonisotropic, as in high Reynolds number flow, we expect λ_g to be much smaller than λ_G and $\Lambda_0 \ll 1$.

In practice, there is no need to specify the value of Λ_0, since both the decay L_{ij} and the free redistribution always exist together and can be expressed by a single formula:

$$L_{ij} + P_{ij}^* = -\frac{\epsilon}{k}\,\overline{u_i u_j}\,\Lambda + (1-\Lambda)\,\frac{2}{3}\,k\delta_{ij} + \gamma^*\,\frac{4}{3}\,k\bar{E}_{ij} \tag{3.38}$$

where $\Lambda = \Lambda_0 + \Lambda^*$ is the single parameter.

3.4.2 The energy decay rate To complete the closure we have to specify the energy loss, ϵ, and the length scale, l, used to express the turbulence diffusion γ. It is suggested that these quantities are related, and therefore it suffices to specify either ϵ or l. The relations between ϵ and l are known in the following two cases:

1. At low Reynolds number of turbulence Eq. (3.37a) is applicable. It is assumed also that l is proportional to λ_g.
2. At high Reynolds number the total energy decay is not dependent on the viscosity. Then dimensional considerations lead to $\epsilon \sim k^{3/2}/L$.

Rotta (1951) suggested a formula that exhibits the correct behavior in the two limiting cases:

$$\epsilon = \frac{k^{3/2}}{l}\left[c_1 + c_D\left(\frac{\nu}{k^{1/2}\,l}\right)\right] \tag{3.39}$$

where c_D and c_1 are numerical coefficients. The second term is neglected at high Reynolds number and Eq. (3.39) then transforms to Eq. (2.10).

The dissipation function ϵ may be calculated from its governing Eq. (2.11). However, when the Reynolds stresses are available a better model for this equation may be used. Hirt (1969), and Hanjalic and Launder (1971) suggested

$$\frac{D\epsilon}{Dt} = c_\epsilon\,\frac{\partial}{\partial x_n}\left(\gamma\,\frac{\overline{u_n u_m}}{\frac{2}{3}k}\,\frac{\partial \epsilon}{\partial x_m}\right) + c_{\epsilon 1}\left(\frac{\pi_{ll}}{2k}\right)\epsilon - c_{\epsilon 2}\,\frac{\epsilon^2}{k} \tag{3.40}$$

where the three terms in the right-hand side represent turbulence diffusion, generation, and decay, respectively.

Substitution of the quasi-isotropic model Eq. (3.29) in Eq. (2.11) may simplify its generation term:

$$-2\nu\left[\left(\overline{\frac{\partial u_n}{\partial x_l}}\,\overline{\frac{\partial u_m}{\partial x_l}}+\overline{\frac{\partial u_l}{\partial x_m}}\,\overline{\frac{\partial u_l}{\partial x_n}}\right)\frac{\partial \bar{U}_m}{\partial x_n}+\overline{u_n\frac{\partial u_m}{\partial x_l}}\,\frac{\partial^2 \bar{U}_m}{\partial x_l \partial x_n}\right] \tag{3.41}$$

The results are as follows:

$$-2\nu\left(\overline{\frac{\partial u_n}{\partial x_l}}\,\overline{\frac{\partial u_m}{\partial x_l}}\right)\frac{\partial \bar{U}_m}{\partial x_n}=L_{nm}\frac{\partial \bar{U}_m}{\partial x_n}=\Lambda_0\left(\frac{\pi_{ll}}{2k}\right)\epsilon \tag{3.42a}$$

$$-2\left(\overline{\frac{\partial u_l}{\partial x_m}}\,\overline{\frac{\partial u_l}{\partial x_n}}\right)\frac{\partial \bar{U}_m}{\partial x_n}=-2\nu\frac{\partial^2}{\partial r_n\,\partial r_m}Q_{ll}(r=0)\frac{\partial \bar{U}_m}{\partial x_n}=\Lambda_1\left(\frac{\pi_{ll}}{2k}\right)\epsilon \tag{3.42b}$$

and

$$-2\nu\,\overline{u_n\frac{\partial u_m}{\partial x_l}}=\frac{\partial}{\partial r_l}Q_{nm}(r=0)=0 \tag{3.42c}$$

$$\text{where } \Lambda_1=\frac{16}{50}\left(\frac{\lambda_g}{\lambda_G}\right)^2\left[1-\frac{11}{4}\left(\frac{\lambda_G}{\lambda_K}\right)^2\right] \tag{3.42d}$$

Equations (3.42a–d) suggest that Λ_0 and Λ_1 contribute to $c_{\epsilon 1}$. However at high Reynolds number $\lambda_g \ll \lambda_G$. Thus Λ_0 and Λ_1 are expected to be small and the generation term (3.41) becomes negligible. It is plausible (Launder et al., 1973) to assume that the term associated with $c_{\epsilon 1}$ also represent pressure velocity interaction.

3.5 Discussion of the Reynolds stress models

In Sec. 3.5.1 we shall summarize some Reynolds stress models. In Sec. 3.5.2 we shall briefly discuss the empirical coefficients used in these models.

3.5.1 The model equations The general form of the Reynolds stress transport equations is

$$\frac{D}{Dt}\overline{u_i u_j}=\text{diffusion}+\text{generation}+\text{redistribution}+\text{dissipation} \tag{3.45}$$

The terms of this equation should be substituted by the following expressions:

1. Diffusion is given by either (3.8), (3.9), (3.11), (3.13) or (3.14).
2. Generation is given by π_{ij} of Eq. (3.30a).
3. Redistribution is the sum of Eq. (3.25) and Eq. (3.30).
4. Dissipation is given by Eq. (3.37).

It may be noted that the sum of Eqs. (3.25) and (3.37) is given in Eq. (3.38). In addition to Eq. (3.45), a length scale equation is usually required. This can be either one of the length scale equations given in Sec. 2, or the dissipation equation (3.40).

To enable selection of a proper model, Table 5, summarizing the models used by Daly and Harlow (1970), Launder et al. (1973), and Naot et al. (1974), was prepared.

3.5.2 The empirical coefficients It is a consequence of the quasi-isotropic model that the coefficients ϕ (i.e., α, β, γ) and Λ_0 are not universal and may vary, even within a given flow field. It is therefore encouraging to see that the difference between the ϕ values recommended by Launder et al. and Naot et al. is not very large.

The Λ-values appear different but this is somewaht deceptive, since only the sum $\Lambda_0 + \Lambda^*$ appears in the equation. Therefore we can optimize this sum by comparison with experimental data, but we cannot separate it into its two components. This sum seems to vary between the value of 0.5 near walls and 1.5 away from walls. Experimental data give some support to this conclusion. The plane duct data of Clark (1966) and Comte-Bellot and Craya (1965) yields $\Lambda_0 + \Lambda^* = 1.25$ in the duct centre. The Comte-Bellot and Corrsin (1966) measurements of the decay of free grid-turbulence correspond to $\Lambda_0 + \Lambda^* = 1.11$, and Rotta (1951) suggested $\Lambda_0 + \Lambda^* = 1.4$. Apparently Launder et al. did not have to lower $\Lambda_0 + \Lambda^*$ near walls, but this was compensated by a special wall proximity function.

Similar considerations apply also to the sum $\gamma + \gamma^*$. However, here the experimental data allow determination of each constituent separately, since γ is related through the function ϕ to α and β, which can be individually determined. Launder et al. (1973) chose $\gamma^* = 0$, and obtained agreement with the experimental data. Yet this may be attributed to their special wall proximity function. Such a function will not be effective in the calculation of a homogeneous shear flow remote from walls (Rose, 1966 and Champange et al., 1970). This type of flow may be considered as a laboratory realization of quasi-isotropic turbulence. These experiments yield the following coefficients:

$$\alpha = 0.755 \qquad \beta = -0.0379 \qquad \gamma + \gamma^* = -0.0201 \qquad \Lambda_0 + \Lambda^* = 1.138$$

Table 5 Reynolds stresses models

Reference	Diffusion	Free redistribution	Forced redistribution	Dissipation	Length scale and dissipation	Application
Daly & Harlow (1970)	Eq. (3.13) $\gamma = c_t \frac{k^2}{\epsilon}$ $c_t = 1.33$	Eq. (3.25) $\Lambda^* = \frac{1}{2}\ 1 - \frac{\pi_{II}}{\epsilon}$ $\gamma^* = 0$	Eq. (3.30) $\alpha = 0$ $\beta = 0$ $\gamma = 0$	Eq. (3.37) $\Lambda_0 = 1$	Eq. (3.40) modified $c_\epsilon = 1$ $c_{\epsilon_1} = 0.8$ $c_{\epsilon_2} = 1$	Plane channel flow
Launder, Reece, & Rodi (1973)	Eq. (3.11) $\gamma = c_t \frac{k^2}{\epsilon}$ $c_t = 0.073$ Eq. (3.13) $\gamma = c_t \frac{k^2}{\epsilon}$ $c_t = 0.133$	Eq. (3.25) $\Lambda^* = 1.5$ $\gamma^* = 0$	Eq. (3.30) $\alpha = 0.764$ $\beta = -0.109$ $\gamma = -0.273$ $(\phi = -0.523)$	Eq. (3.37) $\Lambda_0 = 0$	Eq. (3.40) $c_\epsilon = 0.1/c_t$ $c_{\epsilon_1} = 1.45$ $c_{\epsilon_2} = 1.90$	Wide range of boundary layers and free shear flows
Naot, Shavit, & Wolfshtein (1974)	Eq. (3.9) $\gamma = c_t \frac{k^2}{\epsilon}$ $c_t = 0.077$	Eq. (3.25) $\Lambda^* = 0.465 \div 1.250$ near walls — channel center $\gamma^* = 0.1415$	Eq. (3.30) $\alpha = 0.756$ $\beta = -0.0465$ $\gamma = -0.202$ $(\phi = -0.42)$	Eq. (3.37) $\Lambda_0 = 0$	$1 = y\ 0 < y < 0.27H$ $1 = 0.27H \quad 0.27H < y < H$ H = half channel width $\epsilon = 0.224 \frac{k^{1/2}}{l}$	Plane and square duct flows. Grid generated turbulence

The value of ϕ corresponding to α and β is $\phi = -0.406$, and thus γ is -0.193. On this basis $\gamma^* = 0.173$. The authors believe that a finite γ^* is necessary in Eq. (3.25).

4 SPECIAL PROBLEMS

We shall now briefly discuss a number of problems, namely, improvements in the diffusion term by means of a length scale tensor, wall effects, wall boundary conditions, and some numerical problems.

4.1 Diffusion tensor

Various researchers recommend different formulae for the diffusion term. Even the value for the diffusion used with similar formulae has not yet been completely finalized. The purpose of the present section is to discuss some improvements in the calculation of this term. In particular, we shall consider the use of an eddy diffusion tensor. The general diffusion tensor given in Eq. (3.2) is a sixth order tensor. However, in Eq. (3.3) we proposed to use a second order diffusion tensor Γ_{ij}, and to this tensor we shall confine our attention.

The diffusion tensor has already been used (Daly and Harlow, 1970; Launder et al., 1973). The components of the eddy diffusion were assumed to be proportional to the Reynolds stresses. However, diffusion depends on the length scale of the phenomenon. Therefore we expect the diffusion coefficients to be related to a length scale tensor, rather than to the Reynolds stresses, and propose

$$\Gamma_{ij} = c_t l_{ij} k^{1/2} \tag{4.1}$$

Here c_t is a constant, and l_{ij} is a length scale tensor, yet to be defined.

4.1.1 Length scale tensor The length scale tensor must satisfy the following requirements:

1. The length scale should be a weighted average of the two point correlation Q_{ij}.
2. The directional properties of l_{ij} should be determined by the separation radii.
3. Contraction of l_{ij} should reproduce the scalar length scale, l, given by Eq. (2.5).

The above requirements are satisfied by the definition:

$$2kl_{ij} = \int_v Q_{ll} \frac{r_i r_j}{r^2} \frac{dv}{r^2} \tag{4.2}$$

To understand the meaning of l_{ij} we consider the diagonal components $i = j$. For these, it can be shown that l_{ij} is an average of Q_{ll} weighted so that the points near the axis $x_i = x_j$ weigh more heavily than those that are remote from it. The tensor l_{ij} is symmetrical, and as such possesses three principle axes. These may be regarded as the axes of the eddy. Obviously the contraction of the length scale tensor is:

$$l_{ij} = 1 \tag{4.3}$$

where l is given by Eq. (2.5).

4.1.2 Quasi-isotropic length scale tensor A simple way to gain insight into l_{ij} is to substitute the quasi-isotropic model for Q_{ll} in Eq. (4.2). Q_{ll} is given by:

$$Q_{ll} = \frac{2}{3} k(\mathrm{x})[\chi(r) - \psi(r)] + \overline{u_s u_t}(\mathrm{x}) \frac{r_s r_t}{r^2} \psi(r) \tag{4.4}$$

where $\chi = f + 2g$ and $\psi = F + 2K - G$. Proper integration of Eq. (4.4) yields the length scale tensor in terms of the macro-scales of the quasi-isotropic correlation functions. These are related due to the continuity equation by

$$L_f = 2L_g \qquad \text{and} \qquad L_F = \frac{7}{2} L_G - L_K \tag{4.5}$$

Thus we obtain

$$l_{ij} = \frac{16\pi}{9} L_g \left[(1 - \lambda)\delta_{ij} + \lambda \frac{\overline{u_i u_j}}{\frac{2}{3} k} \right] \tag{4.6}$$

$$\text{where } \lambda = \frac{5L_G + 2L_K}{20L_g} \tag{4.7}$$

Introduction of the contraction:

$$l = l_{ii} = \frac{16\pi}{3} L_g \tag{4.8}$$

yields

$$l_{ij} = \frac{1}{3} l \left[(1 - \lambda)\delta_{ij} + \frac{\overline{u_i u_j}}{\frac{2}{3} k} \lambda \right] \tag{4.9}$$

We note that l_{ij} and therefore also Γ_{ij} have two terms. $(1 - \lambda)\delta_{ij}$ is isotropic and generates an eddy diffusion scalar. The term $3\lambda u_i u_j / 2k$ yields the nonisotropic part of the diffusion. When $\lambda = 0$, expressions of the type suggested by Donaldson and Rosenbaum (1968) are obtained. When $\lambda = 1$, the expressions take the form proposed by Harlow and Hirt (1969). To obtain an approximation for λ from the duct flow data of Comte-Bellot and Craya (1965), we follow Naot et al. (1970) and Naot (1972), who estimated

$$1.4 < \frac{L_G}{L_g} < 2 \quad \text{and} \quad \frac{L_K}{L_G} \sim 0.44 \tag{4.10}$$

Therefore

$$0.3 < \lambda < 0.4 \tag{4.11}$$

Consequently, both the scalar and tensorial parts of Γ_{ij} should be considered.

4.1.3 Discussion Equation (4.9) is the principal result of the present section. It suggests that l_{ij} is a linear combination of isotropic and nonisotropic parts, where the non-isotropic part is proportional to the Reynolds stress. This linear form is a consequence of the quasi-isotropic model. As the quasi-isotropic model is based on the assumption that the mean-rate-of-strain tensor has no influence on the two-point correlations, this tensor could not have appeared in Eq. (4.9). However, it is conceivable that the length scale tensor is dependent on the rate-of-strain tensor as well, and this point requires some further investigation. In the future, differential equations for l_{ij} will have to be formulated either by direct integration of the Q_{ll} equation or by calculating the G, g, and K functions of the quasi-isotropic model.

4.2 Wall boundary conditions

The models and model equations which were presented in this paper are not applicable to low turbulence levels and strong nonisotropy. Regions adjacent to solid walls are an important example of such situations. It is impossible to ignore these regions because important boundary conditions are specified (or required) on the walls. Two main approaches to this problem have been suggested.

1. To formulate special versions of the model equations, applicable near solid walls.
2. To match the solution for the region away from the wall with an available distribution of the variables near the wall.

These two methods will be discussed below.

4.2.1 Wall layer models Many investigators proposed special corrections to the model equations, either by specifying coefficients to the terms in the equations or by adding some additional terms to them. The wall region is characterized by (a) low turbulence level, (b) high rate of strain, and (c) the very presence of the wall itself. Therefore it is difficult to formulate universal wall models. For these reasons the second approach to the problem, using special distributions, seems preferable at the present time. Furthermore, even if wall models are available, it seems likely that greater economy of computational effort will be achieved by the second approach.

4.2.2 Wall distributions If the distribution of the flow variables in the neighborhood of the wall is available it is not necessary to apply the general model equations to the wall layer. Instead, the solution of the model equations in the flow field is matched to the wall distributions by a proper matching procedure in a region where both the field solution and the wall distributions are valid. Wall distributions can be obtained by a number of methods:

1. Using experimental data.
2. Using a theoretical solution applicable to the matching region. In such procedures it is usually necessary to specify a number of integration constants by comparison with experimental data. A typical example is the application of local equilibrium solutions to the so-called logarithmic boundary layers.
3. Obtaining a one dimensional solution of the wall model equations.

At the present time the second method seems to yield the best results.

4.3 Numerical problems

In most models the model equations have the form of regular transport equations which can be solved by standard numerical procedures. However, the algebraic source terms of these equations are by far larger than those encountered in simple convection problems. Therefore the numerical methods require special care. Often the sources become so large that the differential terms represent but a small correction to the algebraic solution obtained by equalizing the source terms to zero. Unfortunately, the source terms are very strongly coupled to one another. Therefore it is often necessary to solve all the equations simultaneously. This procedure may involve the difficult task of inverting medium size matrices (Naot et al., 1974, had to invert a 9×9 matrix). The problem may sometimes be simplified by splitting the variables into groups and obtaining simultaneous solution by matrix inversion only in the groups. If these matrices are small (up to 4×4) they may be inverted analytically and at a low computational cost.

NOTATION

$a_{jk}{}^{li}$	coefficient of forced redistribution
$c_1, c_2, c_3, c_\mu, c_t, c_g, c_d$	numerical coefficients
D_{ij}	part of forced redistribution
$\bar{E}_{ij}$	mean rate of strain
f	frequency
F, f, G, g	correlation functions
G	energy generation (Sec. 2)
J, K	correlation functions
l	length scale
l_{ij}	length scale tensor
L	energy-length product
$\bar{P}, p$	mean and fluctuating pressure
$P_{ij}{}^t$	total pressure interaction
P_{ij}, P_{ij}^*	forced and free redistribution
Q_{ij}	two-point double-velocity correlation tensor
Re_t	Reynolds number of turbulence
T_{ij}	turbulence diffusion
$\bar{U}_i, u_i$	mean and fluctuating velocity
$\overline{u_i u_j}$	Reynolds stresses

Greek symbols

α, β, γ	numerical constants of forced redistribution
α_1	numerical constants of turbulent transport, also of redistribution
ϵ	energy dissipation
$\Lambda, \Lambda^*, \Lambda_0, \Lambda_1, \lambda, \gamma^*$	parameters of turbulence
$\Gamma_{\mathrm{mijrst}}, \Gamma_{ml}, \gamma$	diffusion tensor and diffusion scalar
ν, ν_t	molecular and turbulence viscosity
π_{ij}	production of Reynolds stresses
ρ	density
$\sigma_k, \sigma_L, \sigma_{L2}, \sigma_{L3}, \sigma_{L4}, \sigma_{L1}$	constants
τ_{ij}	pressure transport
ϕ_{ij}	total distribution
ϕ	macro-scale

REFERENCES

P. Bradshaw, D. H. Ferris, and N. P. Atwell, 1967, Calculations of boundary-layer development using the turbulent energy equation. *J. Fluid Mech.*, vol. 28, part 3, pp. 593–616.

F. H. Champange, V. G. Harris, and S. Corrsin, 1970, Experiments on nearly homogeneous turbulent shear flow. *J. Fluid Mech.*, vol. 41, 1, pp. 81–139.

P. Y. Chou, 1945, On velocity correlations and the solution of the equations of turbulent fluctuations. *Quart. App. Math.*, vol. III. no. 1, pp. 38–54.

J. A. Clark, 1966, Study of incompressible turbulent boundary layers in a two-dimensional wind-tunnel. *Ph.D. Thesis,* Queen's University of Belfast.

G. Comte-Bellot and S. Corrsin, 1966, The use of a contraction to improve the isotropy of grid generated turbulence. *J. Fluid Mech.*, vol. 25, no. 4, pp. 657–682.

G. Comte-Bellot and P. de A. Craya, 1965, Ecoulement turbulent entre deux parois parrallels. *Pub. Sci. et Tech. du Ministers de l'Air,* no. 419.

B. J. Daly and F. H. Harlow, 1970, Transport equations in turbulence. *Phys. of Fluids,* vol. 13, no. 11, pp. 2634–2649.

C. du.P. Donaldson, 1969, A computer study of an analytical model of boundary-layer transition. *AIAA Journal,* vol. 7, no. 2, pp. 271–278.

C. du.P. Donaldson and H. Rosenbaum, 1968, Calculation of turbulent shear flows through closure of the Reynolds equations by invariant modelling. *Aero Res. Assoc. of Princeton,* Report 127.

G. S. Glushko, 1965, Turbulent boundary layer on a flat plate in an incompressible fluid. *Izv. Akad. Nauk,* SSSR, Mekh. No. 4, p. 13.

G. S. Glushko, 1972, Differential equation for the scale of turbulence and calculation of the turbulent boundary layer on a flat plate. *Fluid Mech.*, Soviet Research, vol. 1, no. 2, pp. 155–162.

K. Hanjalic and B. E. Launder, 1971, A Reynolds stress model of turbulence and its application to asymmetric boundary layers. *Imperial College, London,* TM/TN/A/8.

K. Hanjalic and B. E. Launder, 1972, A Reynolds stress model of turbulence and its application to thin shear flows. *J. Fluid Mech.*, vol. 52, p. 609.

F. H. Harlow and C. W. Hirt, 1969, Generalized transport theory of anisotropic turbulence. *Los Alamos Scientific Lab.*, LA–4086, VC–34 Physics, TTD–4500.

F. H. Harlow and P. I. Nakayama, 1967, Turbulence transport equations. *Physics of Fluid,* vol. 10, no. 11, pp. 2323–2332.

F. H. Harlow and P. I. Nakayama, 1968, Transport of turbulence energy decay rate, *Los Alamos Scientific Lab. Rep.* LA 3854.

S. Hassid, 1973, On the modelling of the redistribution terms of the Reynolds stress equation. *Technion, Faculty of Civil Eng., Report* 202.

J. O. Hinze, 1959, *Turbulence.* McGraw-Hill, New York.

C. W. Hirt, 1969, Generalized turbulence transport equations. *Los Alamos Scientific Lab.*, 1–13.

W. P. Jones and B. E. Launder, 1972, The prediction of laminarisation with a two equation model of turbulence. *Int. J. Heat Mass Transfer,* vol. 15, p. 301.

A. N. Kolmogorov, 1942, Equation of turbulence motion of an incompressible fluid (in Russian). *Izv. Akad SSR Ser. Phys.*, vol. VII, no. 1–2.

B. E. Launder, A. Morse, W. Rodi, and D. B. Spalding, (1972), The prediction of free shear flows—a comparison of the performance of six turbulence models. *Free turbulent shear flows, NASA SP*–321, pp. 361–426.

B. E. Launder, G. J. Reece, and W. Rodi, 1973, Progress in the development of a Reynolds stress turbulence closure. *Imperial College, London,* HTS/73/31.

B. E. Launder and D. B. Spalding, 1972, *Mathematical models of turbulence.* Academic Press, New York.

D. Naot, 1972, The influence of the pressure fluctuations on the redistribution of the velocity correlation in turbulent flow. *Ph.D. Thesis,* Technion, Haifa, Israel.

D. Naot, A. Shavit, and M. Wolfshtein, 1970, Interaction between components of the turbulent velocity correlation tensor due to pressure fluctuations. *Israel J. Tech.*, vol. 8, no. 3, pp. 259–269.

D. Naot, A. Shavit, and M. Wolfshtein, 1971, On the two point double-velocity correlation tensor. *Israel J. Tech.*, vol. 9, no. 1-2, pp. 83-91.

D. Naot, A. Shavit, and M. Wolfshtein, 1973, Two-point correlation model and the redistribution of Reynolds stress. *Phys. of Fluids*, vol. 16, no. 6, pp. 738-743.

D. Naot, A. Shavit, and M. Wolfshtein, 1974, Numerical calculation of Reynolds stresses in a square duct with secondary motion. *Warme und Stoffubertragung.*

K. G. Ng and D. B. Spalding, 1972, Turbulence model for boundary layer near walls. *Phys. of Fluids*, vol. 15, no. 1, pp. 20-30.

L. Prandtl, 1945, Uber ein neues Formelsystem fur die ausgebildete Turbulenz. *Nachrichten von der Akad. der Wissenschaften in Gottingen*, pp. 6-19, Van der Loeck & Ruprecht, Gottingen.

W. C. Reynolds, 1970, Computation of turbulent flow–state-of-the-art–1970. *Stanford University, California, Report* MD-27.

W. Rodi and D. B. Spalding, 1970, A two-parameter model of turbulence and its application to free jets. *Warme und Stoffubertragung*, no. 3, p. 85.

W. G. Rose, 1966, Result of an attempt to generate a homogeneous turbulent shear flow. *J. Fluid Mech.*, vol. 25, no. 1, pp. 97-120.

J. Rotta, 1951, Statistical theory of non-homogeneous turbulence. *Imperial College*, TWF/TN/39 and TWF/TN/38, London, 1968. Translated from: *Physik*, vol. 129, pp. 547-572, and *Physik*, vol. 131, pp. 51-77.

J. C. Rotta, 1969, Uber eine methode zur berechnung turbulenter Scherstromungsfelder, *Aerodynamische Versuchsanstalt, Gottingen*, 69A14.

A. K. Runchal, D. B. Spalding, and M. Wolfshtein, 1969, Numerical solution of the elliptic equations for transport of vorticity, heat and matter in two dimensional flow. *Phys. of Fluids. Supp. S* 11-21.

P. G. Saffman, 1970, A model for inhomogeneous turbulent flow. *Proc. Roy. Soc. London*, vol. A317, pp. 417-433.

D. B. Spalding, 1967, Heat transfer from turbulent separated flow. *J. Fluid Mech.*, vol. 27, no. 1, pp. 97-109.

A. A. Townsend, 1956, *The structure of turbulent shear flow.* Cambridge Univ. Press.

D. C. Wilcox and I. E. Alber, 1972, A turbulence model for high speed flows. *Proc. of 23rd Heat Transfer and Fluid Mechanics Institute*, 14-16 June 1972, Stanford Univ. Press, Northridge, California.

M. Wolfshtein, 1969, The velocity and temperature distribution in one-dimensional flow with turbulence augmentation and pressure gradients. *Int. J. Heat & Mass Transfer*, vol. 12, no. 301-318.

M. Wolfshtein, 1970, On the length-scale-of-turbulence equation. *Israel J. Tech.*, vol. 8, no. 1-2, pp. 87-99.

2
Turbulent Heat and Mass Transfer at Interfaces: Transport Models and Mechanisms

S. Sideman and W. V. Pinczewski
Department of Chemical Engineering
Technion–Israel Institute of Technology
Haifa, Israel

CONTENTS

1 INTRODUCTION

2.1 Subject treatment

Practically all engineering processes involving fluids depend on the interaction of a fluid with a phase boundary. Fluid friction in conduits and over extended surfaces, heat and mass transfer to fluids in evaporation, distillation, and gas absorption are some of the processes involving the transport of momentum, heat and mass from a phase boundary to a fluid in turbulent flow.

An understanding of the turbulent transport mechanism and the subsequent ability to predict the relevant transport rates are essential to the development of rational design procedures for the various processes. Because of the highly complex and, as yet, not fully understood nature of the turbulent transport mechanism, the prediction of transport rates necessarily involves the formulation of conceptual models which embody many simplifying assumptions. These are required to reduce the problem to a form amenable to mathematical treatment and to rationalize the observed variation of the dependence of the transport coefficient on the molecular diffusivity i.e. $K \sim D^n$ where n ranges from unity to zero.

The various models proposed in the literature constitute the framework of present day predictive theory and may be broadly divided into three general classes:

1. models based on the film theory
2. classical models based on the concept of an eddy or turbulent diffusivity
3. models based on Danckwerts' (1951) surface renewal concept

The film theory, usually related to Whitman and Lewis (1924), assumes that a thin, nearly stagnant layer of fluid is formed at the interface between the fluid and the transport surface. All the transport gradients concentrate in this so called "film." The turbulent mixing in the region outside the film is assumed to maintain the temperature or concentration at a constant value of the bulk. Under these steady state conditions the transport rate depends on the first power of the diffusivity. The film theory represents a gross oversimplification of the actual phenomenon for most practical cases and the fictitious nature of a constant thickness "film," where most of the transfer resistance lies, is by now well established. Nevertheless, the film concept is a simple and convenient tool for numerous practical transport applications. The applicability of this simple intuitive concept to many engineering problems is undoubtedly due to its inherent relationship to the "boundary layer" concept. Indeed, it can easily be shown that for a laminar flow over a flat plate the concentration thickness boundary layer thickness δ_D is 1.5 times the constant "film" thickness l_1. Similarly, $\delta_T = 1.5\, l_1$.

The eddy diffusivity class of models provides a more realistic representation of the actual turbulent transport process. In principle, the approach consists of obtaining solutions for the time-averaged Navier-Stokes equations for the turbulent flow field. The virtue of this approach is that the predictions for the turbulent flow field are based on the time-averaged properties of turbulence rather than on the detailed highly complex small-scale character of the actual turbulence mechanism. However, the process of time averaging the Navier-Stokes equations results in the appearance of statistical correlations involving fluctuating velocities, temperatures and concentrations in the resultant time-averaged conservations equations. As the conservation equations provide no information regarding the relative magnitude or nature of these fluctuation terms, it becomes necessary to simulate their effects by "turbulence models" which are formulated on plausible arguments regarding the nature of the turbulence process. These models are generally based on the concept of an eddy or turbulent transport coefficient which allows to express the unknown turbulent fluctuation correlations in terms of the mean flow field variables.

The various empirical constants introduced in the formulation of the turbulence models must be evaluated through comparison with experimental data. Since almost all the available reliable data come from momentum transport studies, the turbulence models have been developed for this case. The turbulent transport coefficients for heat and mass transport are then inferred from the corresponding coefficient for momentum transport by arguments based on the similarity of the three processes. This usually leads to the formulation of models for the turbulent Prandtl or Schmidt numbers which relate the ratio of the turbulent transport of momentum to that of heat and mass, respectively. With the introduction of a model of turbulence and a model for the turbulent Prandtl or Schmidt number the time-averaged conservation equations may be solved, either analytically or by the application of available numerical techniques, for the mean flow field variables of interest.

The surface renewal class of turbulent transport models is based on Danckwerts' (1951) idea that the replacement of fluid elements at interphase is a stochastic process motivated by the turbulent field. Danckwerts visualized that distinct eddies intermittently move from the fully turbulent region to the transport surface where they replace "older" surface elements. For the relatively short residence times of the eddies in the viscous region at the interface, heat and mass transfer to the stagnant eddies at the surface is calculated by utilizing Higbie's (1935) shallow penetration concept, which allows the transport process to be treated as an impulsive transient transfer into a semi-infinite element. Practically none of the earlier transport models in this category include the convection effects on the diffusional process; this effect is represented only in the nature of the intermittent interruptions of the transport region.

Recent hydrodynamic studies of the flow field in the immediate vicinity of a solid boundary indicate the existence of a relatively well-ordered, large scale, periodic motion and suggest that the viscous wall region in turbulent flow consists of a mosaic of periodically replaced local elements. This has lent strong support to the surface renewal concept and Danckwerts' original model has been subsequently modified to relate the renewal rate to the many flow parameters and to account for the existence of a micro-flow fluid within the individual eddies. Further modifications relate to the fact that not all the penetrating eddies reach all the way to the transport surface.

Although it may first appear that there is little in common between the approach based on the classical concept of an eddy transport coefficient and that based on the concept of surface renewal, the two concepts are by no means mutually exclusive. Rather, we may consider the former to represent a macro-scale study of the phenomena while the latter is a micro-scale analysis limited to the viscous transport region. In this sense, these two approaches are complementary. While in the classical approach solutions are obtained for the time-averaged transport equations, the solutions to the non-averaged laminar transport equations obtained by the renewal-penetration concept are quantitatively averaged. Furthermore, knowledge of the turbulent flow characteristics in the proper geometry is required for a reliable determination of the dependence of eddies flow and renewal rates at or near the transport surface.

1.2 Scope

This study is an attempt to review some existing models for the transport mechanisms between a turbulent fluid and an interface in a manner which highlights the various difficulties involved, the concepts developed to overcome them, and the ensuing limitations. In view of the limited knowledge and understanding of the turbulence flow characteristics at the transport boundary, the problem of turbulent transport at a phase boundary is by no means "closed." Organizing the different concepts of the various schools of thought under one roof may hopefully stimulate new "ideas."

The development of concepts, rather than the design of apparatus, is the guiding line of this study. The material contained here should provide the reader with the second tool, the first being common sense, towards the understanding of the fundamentals of many systems. Some of the calculated results were related to experimental data, but space does not allow to include the numerous applications to which some of the models have been used to provide practical solutions. Transport in fixed and fluidized beds and drag reducing systems are just a few examples of these possible applications.

The study concentrates mainly on transfer at solid-fluid interfaces, though some models associated with "free," fluid-liquid, interfaces are included, when appropriate to demonstrate a particular line of reasoning or an interesting approach. As such, this work may be considered an extension of Scriven's (1968, 1969) excellent lecture series on flow and transfer at fluid interfaces. Other useful works include Sherwood's (1959) review on mass, heat and momentum transfer between phases. Harriott's (1962) review on mass transfer to interfaces and Ruckenstein's (1967) study of physical models for mass or heat transfer processes.

1.3 Notations

Following Prandtl, the turbulent models were historically developed in relation to heat transfer problems. Following Danckwerts, the surface renewal approach remained, until quite recently, within the realm of chemical engineering, and most of the studies relate to mass transfer with or without chemical reactions. Thus, for the convenience of the reader the sections dealing with eddy-diffusivity based models are presented with the standard heat transfer notation. Uniformity would dictate these notations to continue thought. This, however, will present undue difficulties for extending the models to account for chemical reaction. Following Chung, Fan, and Hwang (1971), we therefore present the sections dealing with the penetration-surface-renewal based models with the "general" notations. These are compared in Table 1 with those most commonly used in the literature.

It is perhaps helpful to note the different units of the thermal conductivity in Fourier's law as compared with the units of the molecular diffusivity in Fick's

Table 1 Analogous Notation

Heat	T	q_{ave}	q_w	$\alpha = \dfrac{k}{\rho C_p}$	h	h_s	$\text{Nu} = \dfrac{hL}{k}$	$\text{Pr} = \dfrac{\nu}{\alpha}$
Mass	c	N_{ave}	$N_{y=0}$	D	K	k_s	$\text{Sh} = \dfrac{KL}{D}$	$\text{Sc} = \dfrac{\nu}{D}$
General	P	Ψ	Ψ_i	κ	H	h_s	$\text{Nu} = \dfrac{HL}{\kappa}$	$\text{Pr} = \dfrac{\nu}{\kappa}$

law. Accordingly, mass transfer studies, in terms of c and K, are transformed to heat transfer notation by substituting $c = \rho C_p T$ and $K = h/\rho C_p$. It must, however, be emphasized that this simple similarity between Fourier's and Fick's laws holds only for low concentrations of the diffusing component under discussion. For a detailed analysis of the limitations of the similarities of the temperation-concentrations fields, the reader is referred to Stewart's (1962, 1963) and Fulford and Pei's (1969) papers. It is nevertheless noteworthy that, as shown by these authors and others (Acrivos, 1960; and Bird et al., 1958, the simple analogous relationships between momentum, heat and mass transfer can be extended to high mass flux (i.e. flow through the interface) or heat transfer with mass transfer (i.e. condensation) by adopting appropriate notations which reduce the x-momentum equation, the diffusion and energy equations to a (dimensionless) form similar to the classical equations for the laminar regime. However, these similarity solutions are acceptable for transient, time dependent sublimation; localized mass transfer which occurs on surfaces consisting of combinations of permeable and impermeable materials, etc. (Libby, 1966).

2 MODELS BASED ON THE EDDY DIFFUSIVITY CONCEPT

2.1 Basic equations and mathematical formulation

2.1.1 Introduction In order to arrive at satisfactory general solutions for the problem of predicting heat and mass transfer rates in turbulent flow fields, it is necessary to solve a system of partial differential equations which embody the implications of the conservation laws of momentum, energy, and mass transport. This is required since the equations of energy and mass transport are coupled to the equation for momentum transport via the velocity components of the flow field. Because of the complex nature of the turbulence mechanism, a solution to the transport equations is presently achieved by introducing a number of simplifying assumptions and some rather huresic hypotheses regarding the nature of the statistical functions describing the turbulent fluctuations of the flow field.

The formal analysis of turbulent transport processes is generally restricted to the case of two-dimensional, quasi-steady (steady with respect to the temporal mean values of the velocities and pressure) flows. This restriction is not as severe as it may first appear as the general case of three dimensional axi-symmetrical flows (of which the planar flow is a limiting case) reduces to two dimensional systems. These include most flows of practerial importance, such as flows in tubes, in ducts, over walls, and free flows. Furthermore, it is assumed that the only significant shear stresses, heat fluxes, and diffusional fluxes are caused by gradients in the direction normal to the streaming axis, i.e. the boundary layer approximations (Persen, 1972). These assumptions lead to the truncation of the general elleptic equations for momentum, heat and mass transport to parabolic form for which fast and stable numerical finite-difference techniques of solution are now readily available (Patankar and Spalding, 1970).

In the present section, we redevelop the working partial-differential equations governing the rates of momentum, heat, and mass transport in a turbulent flow field in a fashion which will highlight the various assumptions made and review the hypotheses proposed for the characterization of the turbulent fluctuations and subsequent closure of the turbulent transport problem.

2.1.2 Conservation equations for a turbulent flow field The present formulation of the relevant partial-differential equations will, for reasons of simplicity, be restricted to the rectangular cartesian coordinate system. For the corresponding formulation in a general curvilinear orthogonal coordinate system the reader is referred to the monographs of Townsend (1956) and Goldstein (1938).

For a steady flow, as defined above, the Navier-Stokes equations for an incompressible laminar flow with constant physical properties may be written as

$$\rho u \frac{\partial u}{\partial x} + \rho v \frac{\partial u}{\partial y} + \rho w \frac{\partial u}{\partial z} = -\frac{\partial P}{\partial x} + \frac{\partial}{\partial x}\left(\mu \frac{\partial u}{\partial x}\right) + \frac{\partial}{\partial y}\left(\mu \frac{\partial u}{\partial y}\right) + \frac{\partial}{\partial z}\left(\mu \frac{\partial u}{\partial z}\right) \tag{2.1}$$

$$\rho u \frac{\partial v}{\partial x} + \rho v \frac{\partial v}{\partial y} + \rho w \frac{\partial v}{\partial z} = -\frac{\partial P}{\partial y} + \frac{\partial}{\partial x}\left(\mu \frac{\partial v}{\partial x}\right) + \frac{\partial}{\partial y}\left(\mu \frac{\partial v}{\partial y}\right) + \frac{\partial}{\partial z}\left(\mu \frac{\partial v}{\partial z}\right) \tag{2.2}$$

$$\rho u \frac{\partial w}{\partial x} + \rho v \frac{\partial w}{\partial y} + \rho w \frac{\partial w}{\partial z} = -\frac{\partial P}{\partial z} + \frac{\partial}{\partial x}\left(\mu \frac{\partial w}{\partial x}\right) + \frac{\partial}{\partial y}\left(\mu \frac{\partial w}{\partial y}\right) + \frac{\partial}{\partial z}\left(\mu \frac{\partial w}{\partial z}\right) \tag{2.3}$$

$$\frac{\partial u}{\partial x} + \frac{\partial v}{\partial y} + \frac{\partial w}{\partial z} = 0 \tag{2.4}$$

where u, v, and w are, respectively the x, y, and z components of velocity, p is the pressure, and ρ and μ denote the fluid density and viscosity, respectively.

The corresponding equations for convective heat and mass transport are

$$\rho C_p u \frac{\partial T}{\partial x} + \rho C_p v \frac{\partial T}{\partial y} + \rho C_p w \frac{\partial T}{\partial z} = \frac{\partial}{\partial x}\left(k \frac{\partial T}{\partial x}\right) + \frac{\partial}{\partial y}\left(k \frac{\partial T}{\partial y}\right) + \frac{\partial}{\partial z}\left(k \frac{\partial T}{\partial z}\right) \tag{2.5}$$

$$u \frac{\partial c}{\partial x} + v \frac{\partial c}{\partial y} + w \frac{\partial c}{\partial z} = \frac{\partial}{\partial x}\left(D \frac{\partial c}{\partial x}\right) + \frac{\partial}{\partial y}\left(D \frac{\partial c}{\partial y}\right) + \frac{\partial}{\partial z}\left(D \frac{\partial c}{\partial z}\right) \tag{2.6}$$

where T and c are the temperature and concentration, k and D are the coefficients of molecular heat and mass diffusivity and C_p is the specific heat at constant pressure.

In writing the above equations, the following additional assumptions are implied:

1. the effect of body forces is negligibly small when compared with viscosity and inertia forces.
2. flow velocities and fluid viscosity are sufficiently small so that viscous dissipation may be neglected.
3. diffusional mass fluxes are sufficiently small, and the induced velocities have a negligible influence on the flow field velocities (i.e. relatively small diffusional fluxes).

The restriction of constant physical properties requires, among other things, that temperature differences be kept small. Conditions (1) and (2) are satisfied for most turbulent fluid flows. Condition (3) necessitates low diffusional fluxes and this is generally satisfied for most fluids with low molecular diffusivities, i.e., large Schmidt numbers.

For laminar flows, Eq. (2.1) to Eq. (2.6) form a closed system which together with the relevant boundary conditions may, in principle at least, be solved for the velocity, temperature, and concentration distributions.

The turbulent flow field, however, differs from the laminar flow field in that the velocity components are not steady, but fluctuate randomly with respect to space and time about some mean value. These random fluctuations are responsible for the diffusiveness which is one of the principle characteristics of turbulent motion. Reynolds (1874) suggested that the turbulent flow variables (velocity, pressure, temperature, and concentration) may each be expressed as a simple sum of a time averaged mean value (denoted by a subscript bar) and a fluctuating component (denoted by a prime), the time average of which converges to zero for sufficiently large times. Thus, we write

$$\begin{aligned} u &= \bar{u} + u' & p &= \bar{p} + p' \\ v &= \bar{v} + v' & c &= \bar{c} + c' \\ w &= \bar{w} + w' & T &= \bar{T} + T' \\ \bar{u}' &= \bar{v}' = \bar{w}' = \bar{p}' = \bar{c}' = \bar{T}' = 0 \end{aligned} \tag{2.7}$$

It is now accepted that the complex details of the turbulent motion of ordinary fluids are described by the Navier-Stokes equations (Townsend, 1965). Thus, the conservation equations for the turbulent flow field are obtained by substituting Eq. (2.7) into Eqs. (2.1)-(2.6). Integrating the resultant expressions with respect to time yields the following equations for the mean motion.

Momentum transport:

$$\rho\bar{u}\frac{\partial\bar{u}}{\partial x} + \rho\bar{v}\frac{\partial\bar{u}}{\partial y} + \rho\bar{w}\frac{\partial\bar{u}}{\partial z} = -\frac{\partial\bar{p}}{\partial x} + \frac{\partial}{\partial x}\left(\mu\frac{\partial\bar{u}}{\partial x} - \rho\overline{u'u'}\right) + \frac{\partial}{\partial y}\left(\mu\frac{\partial\bar{u}}{\partial y} - \rho\overline{u'v'}\right) + \frac{\partial}{\partial z}\left(\mu\frac{\partial\bar{u}}{\partial z} - \rho\overline{u'w'}\right) \tag{2.8}$$

$$\rho\bar{u}\frac{\partial\bar{v}}{\partial x} + \rho\bar{v}\frac{\partial\bar{v}}{\partial y} + \rho\bar{w}\frac{\partial\bar{v}}{\partial z} = -\frac{\partial\bar{p}}{\partial y} + \frac{\partial}{\partial x}\left(\mu\frac{\partial\bar{v}}{\partial x} - \rho\overline{u'v'}\right) + \frac{\partial}{\partial y}\left(\mu\frac{\partial\bar{v}}{\partial y} - \rho\overline{v'v'}\right) + \frac{\partial}{\partial z}\left(\mu\frac{\partial\bar{v}}{\partial z} - \rho\overline{v'w'}\right) \tag{2.9}$$

$$\rho\bar{u}\frac{\partial\bar{w}}{\partial x} + \rho\bar{v}\frac{\partial\bar{w}}{\partial y} + \rho\bar{w}\frac{\partial\bar{w}}{\partial z} = -\frac{\partial\bar{p}}{\partial z} + \frac{\partial}{\partial x}\left(\mu\frac{\partial\bar{w}}{\partial x} - \rho\overline{u'w'}\right) + \frac{\partial}{\partial y}\left(\mu\frac{\partial\bar{w}}{\partial y} - \rho\overline{v'w'}\right) + \frac{\partial}{\partial z}\left(\mu\frac{\partial\bar{w}}{\partial z} - \rho\overline{w'w'}\right) \tag{2.10}$$

$$\frac{\partial\bar{u}}{\partial x} + \frac{\partial\bar{v}}{\partial y} + \frac{\partial\bar{w}}{\partial z} = 0 \tag{2.11}$$

Heat transport:

$$\rho C_p \bar{u} \frac{\partial \bar{T}}{\partial x} + \rho C_p \bar{v} \frac{\partial \bar{T}}{\partial y} + \rho C_p \bar{w} \frac{\partial \bar{T}}{\partial z} = \frac{\partial}{\partial x}\left(k \frac{\partial \bar{T}}{\partial x} - \rho C_p \overline{T'u'}\right) + \frac{\partial}{y}\left(k \frac{\partial \bar{T}}{\partial y} - \rho C_p \overline{T'v'}\right) + \frac{\partial}{\partial z}\left(k \frac{\partial \bar{T}}{\partial z} - \rho C_p \overline{T'w'}\right) \quad (2.12)$$

Mass transport:

$$\bar{u} \frac{\partial \bar{c}}{\partial x} + \bar{v} \frac{\partial \bar{c}}{\partial y} + \bar{w} \frac{\partial \bar{c}}{\partial z} = \frac{\partial}{\partial x}\left(D \frac{\partial \bar{c}}{\partial x} - \overline{c'u'}\right) + \frac{\partial}{\partial y}\left(D \frac{\partial \bar{c}}{\partial y} - \overline{c'v'}\right) + \frac{\partial}{\partial z}\left(D \frac{\partial \bar{c}}{\partial z} - \overline{c'w'}\right) \quad (2.13)$$

The above set of equations are referred to as the Reynolds equations. Comparison of Eqs. (2.8)–(2.13) with the corresponding expressions for laminar flow, Eqs. (2.1)–(2.6), shows that the Reynolds equations differ from the corresponding laminar equations only by the addition of the "Reynolds transport terms" of the form $\overline{u'v'}$, $\overline{T'v'}$, $\overline{c'v'}$, etc., to the bracketed expressions on the right hand side. These Reynolds transport terms may be interpreted as the mean rates of transport of momentum, heat, and mass across the corresponding surfaces by the turbulent velocity fluctuations.

As in the case of laminar flow, the determination of the mean flow variables in a turbulent flow field requires the solution of the system of Reynolds equations. However, unlike the closed laminar flow system, Eqs. (2.8)–(2.13) for turbulent flow present only six Reynolds equations for eighteen unknowns: the mean pressure, three mean velocity components, one mean temperature, one mean concentration, and twelve Reynolds transport terms. Clearly, closure of the mean turbulent problem requires additional information regarding the spatial dependence between the Reynolds transport terms and the mean flow field. The determination of this dependence constitutes the central problem in turbulence.

The problem of closure is generally overcome by the introduction of various simplifying assumptions and hypothesis in the form of "turbulence models" of varying degrees of complexity which relate the Reynolds stresses to the mean flow field variables. An extensive survey of the effectiveness of these models for the general problem of a turbulent flow field closure has been recently presented by Mellor and Herring (1973). Before discussing the problem of closure in detail we consider the simplifications of the Reynolds equations associated with the restriction to two-dimensional flow.

2.1.3 Simplification of the turbulent transport equations—the boundary layer approximation Many practical flows, particularly flow in pipes and ducts, and flows over solid surfaces, may be characterized by

1. a single predominant direction of flow parallel to the interface
2. changes in shear stresses, heat fluxes, and diffusional fluxes are significant only in the direction perpendicular to the direction of motion.

Under these stipulations, the major changes in velocity, temperature, and concentration occur in a region close to the surface, i.e., in a relatively thin boundary layer. For purely two-dimensional flow (i.e., $\bar{w} = 0$, and independence of all quantities from the variable z) result in the following approximations (Persen, 1972).

$$
\begin{gathered}
\left|\frac{\partial}{\partial y}\left(\mu \frac{\partial \bar{u}}{\partial y} - \rho\overline{u'v'}\right)\right| \gg \left|\frac{\partial}{\partial x}\left(\mu \frac{\partial \bar{u}}{\partial x} - \rho\overline{u'u'}\right)\right| \\
\left|\frac{\partial}{\partial y}\left(\mu \frac{\partial \bar{u}}{\partial y} - \rho\overline{u'v'}\right)\right| \gg \left|\frac{\partial}{\partial x}\left(\mu \frac{\partial \bar{v}}{\partial x} - \rho\overline{u'v'}\right)\right| \\
\left|\frac{\partial}{\partial x}\left(\mu \frac{\partial \bar{v}}{\partial x} - \rho\overline{v'v'}\right)\right| \ll \left|\frac{\partial}{\partial x}\left(\mu \frac{\partial \bar{v}}{\partial y} - \rho\overline{v'v'}\right)\right| \\
\left|\frac{\partial \bar{u}}{\partial y}\right| \gg \left|\frac{\partial \bar{u}}{\partial x}\right| \\
\left|\frac{\partial \bar{v}}{\partial y}\right| \gg \left|\frac{\partial \bar{v}}{\partial x}\right|
\end{gathered}
\tag{2.14}
$$

and

$$
\begin{gathered}
\left|\frac{\partial}{\partial y}\right|\left(k \frac{\partial \bar{T}}{\partial y} - \rho C_p\overline{T'v'}\right)\right| \gg \left|\frac{\partial}{\partial x}\left(k \frac{\partial \bar{T}}{\partial x} - \rho C_p\overline{T'u'}\right)\right| \\
\left|\frac{\partial}{\partial y}\left(D \frac{\partial \bar{c}}{\partial y} - \overline{c'v'}\right)\right| \gg \left|\frac{\partial}{\partial x}\left(D \frac{\partial \bar{c}}{\partial x} - \overline{c'u'}\right)\right|
\end{gathered}
\tag{2.15}
$$

Consequently, Eqs. (2.8)-(2.13) reduce to the following set of equations for the turbulent flow field in the vicinity of a solid surface:

Momentum transport:

$$
\rho\bar{u}\frac{\partial \bar{u}}{\partial x} + \rho\bar{v}\frac{\partial \bar{u}}{\partial y} = -\frac{\partial \bar{p}}{\partial x} + \frac{\partial}{\partial y}\left(\mu \frac{\partial \bar{u}}{\partial y} - \rho\overline{u'v'}\right) \qquad \frac{\partial \bar{u}}{\partial x} + \frac{\partial \bar{v}}{\partial y} = 0
\tag{2.16}
$$

Heat transport:

$$
\rho C_p\bar{u}\frac{\partial \bar{T}}{\partial x} + \rho C_p\bar{v}\frac{\partial \bar{T}}{\partial y} = \frac{\partial}{\partial y}\left(k \frac{\partial \bar{T}}{\partial y} - \rho C_p\overline{T'v'}\right)
\tag{2.17}
$$

Mass transport:

$$\bar{u}\,\frac{\partial \bar{c}}{\partial x} + \bar{v}\,\frac{\partial \bar{c}}{\partial y} = \frac{\partial}{\partial y}\left(D\,\frac{\partial \bar{c}}{\partial y} - \overline{c'v'}\right) \tag{2.18}$$

Equations (2.16) to (2.18) may be solved together with the relevant boundary conditions, provided sufficient information is available to determine the correct functional dependence between the fluctuating terms $\overline{u'v'}$, $\overline{c'v'}$, and $\overline{T'v'}$, and the mean flow variables $\bar{u}$, $\bar{v}$, $\overline{T}$, and $\bar{c}$. The turbulent transport problem is essentially reduced to determining these functional relationships.

The experimental difficulties in obtaining precise movement of velocities, temperatures, and concentrations in the region of interest close to the wall are formidable and the associated uncertainties in the measurements (Blom, 1970) have hindered the development of purely empirical correlations for the fluctuation terms.

Further, since detailed experimental measurements pertaining to turbulent transport processes are generally limited to the velocity field and its fluctuations, turbulence models have been developed mainly for the case of momentum transport in Section 2.2. The corresponding expressions for heat and mass transport inferred on the basis of heuristic arguments regarding the similarity of the three transport processes, are discussed in Sections 2.4 and 2.5.

2.2 Turbulence models

2.2.1 Introduction The process of time-averaging of the conservation equations for a turbulent flow field must be augmented by statistical correlations involving the fluctuating velocities, temperatures, and concentrations, e.g., the Reynolds transport terms. The conservation equations themselves provide no information regarding the magnitudes of these statistically correlated terms. It is therefore necessary to simulate their effects by introducing "turbulence models" which relate the unknown quantities in terms of the more readily measured mean flow variables, and consequently affect the closure of the system of Reynolds equations.

The models of turbulence, of varying complexity, proposed in the literature have recently been the subject of an excellent survey in the monograph of Launder and Spalding (1972).

2.2.2 Boussinesq's hypothesis The simplest models of turbulence are based on Boussinesq's (1877) suggestion that the turbulent transport terms may be expressed in a form analagous to their laminar counterparts, i.e., as the product of a mean variable gradient and a quantity termed the turbulent, or eddy, transport coefficient. Thus

$$\sigma_t = -\rho\overline{u'v'} = \mu_t \frac{\partial \bar{u}}{\partial y}$$

$$q_t = -\rho C_p \overline{T'v'} = k_t \frac{\partial \bar{T}}{\partial y} \tag{2.19}$$

$$N_t = -\overline{c'v'} = D_t \frac{\partial \bar{c}}{\partial y}$$

where μ_t, k_t, and D_t are the turbulent viscosity, conductivity, and diffusivity respectively, analogues of the molecular transport coefficients μ, k, and D. Unlike the molecular coefficients which are functions of the fluid alone, the turbulent coefficients are largely determined by the local structure of the turbulence and consequently their value will generally vary from point to point in the flow field. The turbulent transport coefficients are generally many magnitudes greater than their molecular counterparts, except in the very thin region adjacent to a solid surface, or in the case of liquid metals where the molecular conductivity is extremely high.

Substitution of Eq. (2.19) into the Reynolds Eqs. (2.16)–(2.18) yields the following expressions for the shear stress, and the heat and mass fluxes in the turbulent flow field.

$$\frac{\sigma}{\rho} = (\nu + \nu_t)\frac{d\bar{u}}{dy} = \nu_e \frac{d\bar{u}}{dy}$$

$$\frac{q}{\rho C_p} = (\alpha + \alpha_t)\frac{d\bar{T}}{dy} = \alpha_e \frac{d\bar{T}}{dy} \tag{2.20}$$

$$N = (D + D_t)\frac{d\bar{c}}{dy} = D_e \frac{d\bar{c}}{dy}$$

where ν_e, α_e, and D_e denote the "effective" transport coefficients. Although Boussinesq's hypothesis does not in itself constitute a complete model for turbulence, its importance lies in the fact that it provides a relationship between the apriori unknown Reynolds transport terms and the mean flow field variables. To complete the model it is necessary to specify the turbulent transport coefficients ν_t, α_t, and D_t in terms of known quantities. The simplest expressions were developed in Prandtl's (1925) momentum transport, and Taylor's (1929) vorticity transport theories.

2.2.3 Prandtl mixing-length hypothesis A detailed treatment of mixing-length hypothesis may be found in standard texts (Goldstein, 1938; Hinze, 1959; Schlicting, 1960). Prandtl (1925) drew an analogy between the kinetic theory of gases (i.e., $\mu = \frac{1}{3}\rho L_0 V_m$ where L_0 is the molecular mean free path and V_m is the molecular velocity) and postulated that the turbulent viscosity, μ_t, may be expressed as a product of the fluid density, a characteristic "mixing length" l_m, and a characteristic random fluctuation velocity, V_t, i.e.,

$$\mu_t = \rho l_m V_t \tag{2.21}$$

The analogy, however, is not complete since the influence of a wall may be expected to limit both the extent of the eddies motion as well as the size of the eddies as the wall is approached. Consequently, l_m and V_t must be expected to vary both with position in the flow field and the mean flow velocity.

Prandtl further proposed that the magnitude of the fluctuation velocity, V_t, is proportional to the product of the mixing-length, l_m, and the mean velocity gradient. Thus

$$V_t = l_m \left| \frac{d\bar{u}}{dy} \right| \tag{2.22}$$

Combining Eqs. (2.3) and (2.4) gives the mixing-length relationship

$$\mu_t = \rho l_m^2 \left| \frac{d\bar{u}}{dy} \right| \tag{2.23}$$

An equivalent alternative derivation assumes that the fluctuation velocity components are of the same order, i.e., $u' \sim v'$. As in Eq. (2.22), we write $u' = l_m \, d\bar{u}/dy$, thus $\rho u' v' = \rho l_m^2 \, |d\bar{u}/dy| d\bar{u}/dy$ from which Eq. (2.23) follows.

Before Eq. (2.23) may be utilized to predict turbulent velocity profiles, it is first necessary to specify the mixing-length, l_m. Prandtl suggested that for flows in the vicinity of a wall the mixing-length was directly proportional to the distance y from the wall.

$$l_m = \kappa y \tag{2.24}$$

where κ is a constant to be determined. From comparison with experimental measurements, Coles (1956) recommended that $\kappa = 0.4$.

For turbulent flows other than flows in the vicinity of a wall, the distribution of the mixing length is found to be different from that predicted by Eq. (2.24). Examples of such turbulent flows are tabulated in Table 2. Table 2 shows that although the simple mixing-length concept goes a long way in generalizing a considerable amount of experimental data for a wide variety of turbulent flows, it still lacks complete universality since it is necessary to specify the distribution of l_m apriori for different flow configurations.

Von Karman (1930) attempted to generalize the mixing-length concept by postulating that l_m is determined primarily by the local nature

Table 2 Distribution of Mixing-Length in Turbulent Flows

Type of flow	Distribution of l_m	Author
External boundary layer	$l_m = \kappa y$ for $\frac{y}{\delta} \leqslant \frac{\lambda}{\kappa}$ $l_m = \lambda\delta$ for $\frac{\lambda}{\delta} > \frac{\lambda}{\kappa}$ $\kappa = 0.41$; $\lambda = 0.09$ δ = boundary layer thickness	Escuider (1966)
Pipe flow	$l_m = R\left[0.14 - 0.08\left(1 - \frac{y}{R}\right)^2 - 0.06\left(1 - \frac{y}{R}\right)^4\right]$ $\left(l_m \to 0.4y \text{ as } \frac{y}{R} \to 0\right)$ R = tube radius	Nikuradse (1932)
Free turbulent flows in stagnant surroundings	$l_m = 0.098$ for Plane jet $l_m = 0.1258$ for Fan jet $l_m = 0.0758$ for Round jet $\delta = \frac{1}{2}$ jet width	Launder and Spalding (1972)

of the velocity profile. Based on dimensional reasoning, Von Karman suggested that

$$l_m \sim \left| \frac{d\bar{u}}{dy} \middle/ \frac{d^2\bar{u}}{dy^2} \right| \tag{2.25}$$

Equation (2.25) would appear to be a desirable extension of the original mixing-length concept since it removes the necessity to prescribe the mixing-length profile apriori. However, in practice it is found that the relationship for l_m which is predicted by Eq. (2.25) is not in agreement with measurements except in the inner region of boundary layer flow where it reduces to the original Prandtl hypothesis, $l_m \sim y$, Eq. (2.24). Equation (2.25) is known as the Von Karman similarity hypothesis, and although of limited utility, it suggests the direction followed by subsequent workers.

2.2.4 Limitations of the mixing-length hypothesis Although the simple mixing-length concept has been extremely successful in predicting a wide range of boundary layer flows (Patankar and Spalding, 1970), it is nevertheless subject to a number of important limitations. Since the turbulent viscosity is defined as the product of a mixing-length and a mean velocity gradient, it follows that the theory predicts a zero Reynolds shear stress in the center of a conduit where the velocity gradient is zero, and at the edge of the boundary layer. This prediction is in contradiction to experimental evidence (Hinze, 1959) which shows that both the longitudinal and transverse velocity fluctuations differ from zero in the center of a conduit. This objection has been overcome by a number of workers (Doshi and Gill, 1970, 1971; Gill and Scher, 1961) including Prandtl himself (see Schlicting 1960). The generalization proposed by Doshi and Gill is interesting in that it includes both the Prandtl mixing-length concept and the Von Karman similarity hypothesis as special cases.

More fundamental objections to the general validity of Prandtl's hypothesis concern the dependence of the predicted turbulent transport coefficient on the local conditions in the flow field, as well as the basic assumption of a dominant transport by a gradient type diffusion process, as per Boussinesq's hypothesis. The former of these objections is supported by observations that the turbulent kinetic energy at a point in the flow field may depend as much on transport processes from remote parts of the flow as on local conditions of production and dissipation (Townsend, 1961). The objection to a dependence on the mean velocity gradient is supported by the observation that, in the region of maximum shear, the distribution of the turbulent shear stress is almost uniform (Hinze, 1959). Moreover, a gradient-type transport process is consistent only with a uniform small scale random structure of turbulence. Although this was previously assumed to be the case for the region near a solid boundary, recent experimental evidence (see Section 3.1) suggests that the turbulence in this region is characterized by a regular, highly repetitive, large scale structure. Further discussions regarding the inapplicability of a purely gradient-type transport process are given by Schubauer and Tchen (1959) and Townsend (1956). Townsend concludes that although the concept of transport by means of a gradient-type diffusion is useful in that it yields a successful description of mean momentum distribution, it may have little bearing on the actual physical processes taking place.

The above general criticisms notwithstanding, Townsend (1961) has demonstrated that, for the most part, these objections are unimportant for the special case of flows near a rigid boundary where the balance of turbulent kinetic energy is virtually unaffected by the nature of the flow in adjacent regions, i.e., in equilibrium layers. These equilibrium layers possess an essential universality of structure which is manifested by a simple dependence of mean velocity gradient on the Reynolds stress and the distance from the wall, thus apparently validating the simple Prandtl hypothesis.

An outline of models of turbulence which were developed to overcome some of the above mentioned deficiencies of the simple mixing-length hypothesis follows. These models are necessarily more complex than the simple Prandtl hypothesis, but for the special case of equilibrium layers, i.e., flows near a solid wall, the resultant predictions are only marginally superior to those of the simple mixing length concept (Launder and Spalding, 1972), and some of these models indeed reduce directly to the simple mixing length formulae. However, for more complicated turbulent flows, such as flows with circulation, the simple mixing-length hypothesis is inadequate, and more complicated models have been demonstrated to give good predictions (Gosman et al., 1969).

2.2.5 Multi-equation models of turbulence

One-equation models of turbulence The major limitation of the simple mixing-length concept is its failure to account for the nonlocal nature of turbulence. A model to overcome this deficiency was proposed by Prandtl (1945) and Kolmogorov (1942). Instead of relating the turbulent velocity fluctuation to the mean velocity gradient, as in Eq. (2.22), Prandtl proposed that the velocity fluctuation is proportional to the time-averaged turbulence kinetic energy, e, i.e.,

$$V_t \sim e^{1/2} \tag{2.26}$$

where $e = \frac{1}{2}(\bar{u}'^2 + \bar{v}'^2 + \bar{w}'^2)$

Equation (2.26) overcomes one of the objections to the simple mixing-length concept since the turbulence energy at the center line of a conduit is not zero, although the velocity gradient is zero. Combining Eq. (2.21) $\mu_t = \rho l_m V_t$ with Eq. (2.26) gives

$$\mu_t = \rho l_m e^{1/2} \tag{2.27}$$

Equation (2.27) is known as the Prandtl-Kolmogorov formulae for the turbulent viscosity.

For two-dimensional boundary layer flows at high Reynolds numbers, the distribution of turbulent kinetic energy is given by (Townsend, 1956):

$$\frac{\rho De}{Dt} = -\frac{\partial}{\partial y}(\overline{\rho v'e'} + \overline{v'p'}) - \rho u'v' \frac{\partial \bar{u}}{\partial y} - \mu \sum_{i,j}\left(\frac{\partial \bar{u}_i}{\partial x_j}\right)^2 \tag{2.28}$$

convection diffusion production dissipation

Equation (2.28) is derived from the Navier-Stokes equations in a manner similar to that used in deriving the Reynolds equations for turbulent flow (i.e., multiplying the momentum equation for each coordinate direction by its corresponding fluctuating velocity to form the terms u'^2 etc., then time-averaging and summing the three equations). It is immediately evident that the specification of e by Eq. (2.26) overcomes the second objection to the simple mixing-length concept since the nonlocal nature of turbulence is accounted for by the convection and diffusion terms in Eq. (2.28).

The solution of Eq. (2.28) for the turbulence energy, e, requires the introduction of a number of simplifying assumptions regarding the unknown turbulent fluctuation correlation terms. Prandtl (1945) assumed that

1. the diffusive nature of the turbulence may be modeled by a gradient-type diffusion process, i.e.,

$$-(\rho\overline{v'e'} + \overline{v'p'}) = \frac{\mu_t}{\mathrm{Pr}_{t,e}} \frac{\partial e}{\partial y} \tag{2.29}$$

where $\mathrm{Pr}_{t,e}$ is the turbulent Schmidt or Prandtl number for the transfer of turbulent kinetic energy.

2. the Boussinesq hypothesis holds:

$$-\rho\overline{u'v'} = \mu_t \frac{\partial \bar{u}}{\partial y} \tag{2.30}$$

3. the dissipation rate at high Reynolds number is controlled by the processes which transfer energy from the larger energy containing eddies to the smaller eddies which dissipate this energy. This process depends only on ρ, e, and l_c (i.e., local isotropy) where l_c is a characteristic length-scale of the turbulent motion, and, from dimensional considerations

$$-\mu \sum_{i,j} \left(\frac{\partial \bar{u}_i}{\partial x_j}\right)^2 = \frac{C_D \rho e^{3/2}}{l_c} \tag{2.31}$$

where C_D is a constant which remains to be determined.

Based on the above assumptions, Eq. (2.28) becomes

$$\rho \frac{De}{Dt} = \frac{\partial}{\partial y}\left(\frac{\mu_t}{\sigma_e} \frac{\partial e}{\partial y}\right) + \mu_t \left(\frac{\partial \bar{u}}{\partial y}\right)^2 - \frac{C_D \rho e^{3/2}}{l_c} \tag{2.32}$$

Provided that numerical values are assigned to the constants σ_e and C_D (Launder and Spalding, [1972] suggest the values $\sigma_e = 1$ and $C_D = 0.09$), and that the length-scale distribution is specified apriori, Eq. (2.32) may be solved together with the Reynolds Eqs. (2.16)-(2.18), and the relevant boundary conditions for the mean flow variables which are of interest.

For the special case of turbulence in the vicinity of a wall, the convection and diffusion terms are negligibly small compared to the production and dissipation terms. This condition is referred to as "local-equilibrium" turbulence. Equation (2.32) reduces to

$$\mu_t \left(\frac{\partial \bar{u}}{\partial y}\right)^2 = \frac{C_D \rho e^{3/2}}{l_c} \tag{2.33}$$

Combining Eqs. (2.33) and (2.27) yields

$$\mu_t = \rho \frac{l_c^2}{C_D^{1/2}} \left(\frac{\partial \bar{u}}{\partial y}\right) \tag{2.34}$$

which is identical with the Prandtl mixing-length hypothesis, Eq. (2.23), when

$$l_c = C_D^{1/4} l_m \tag{2.35}$$

Although Prandtl's (1945) model overcomes some of the limitations of the simple mixing-length hypothesis, it still suffers from the restrictions that,

1. precise distributions of l_c can rarely be made apriori for flows where convection and diffusion of energy are important. For flows where these terms are unimportant, the model reduces to the simple mixing-length hypothesis.
2. the concept of a gradient-type diffusion is retained.

Several other one-equation models of turbulence have been developed. Bradshaw, Ferris, and Atwell (1967) follow Prandtl in assuming that the conserved quantity is the turbulence energy. Abandoning the concept of a gradient-type diffusion process, they assume that the turbulent shear stress, σ_t, is directly proportional to the turbulence energy, consistent with experimental evidence in many external wall boundary layers. Also, the diffusive flux for the turbulence energy is assumed to be proportional to the product of the energy itself and a large eddy diffusive velocity, which is taken as σ_0/ρ. Although the model has the advantage that it frees the Reynolds stress from the local velocity gradient, it still suffers from the limitation that the turbulence length-scale distribution must still be specified apriori. Ng and Spalding (1972) and Sivasegaram and Whitelaw (1971) conclude that the model offers no real advantages over Prandtl's formulation.

Nee and Kovasznay (1968) proposed that the kinetic turbulent viscosity, ν_t $(= \mu_t/\rho)$, may be determined directly by means of a transport equation which is obtained in a manner similar to that used in obtaining Eq. (2.28). As for the former two models, the length scale must be specified apriori, and like Prandtl's model it reduces to the simple mixing-length hypothesis for equilibrium-layers.

In practice, the above one-equation models are only marginally superior to the simple mixing-length hypothesis (Launder and Spalding, 1972), since transport effects on the turbulence length-scale are unaccounted for. These effects are accounted for in the two-equation models discussed below.

Two-equation models of turbulence As in the one-equation models, the Prandtl-Kolmogorov formulae, Eq. (2.27)

$$\mu_t = \rho l_c \, e^{1/2}$$

is used for the turbulent viscosity. However, for the two-equation models, both the turbulence energy, e, and the length-scale, l_c, are determined by way of the transfer equations.

The first such model was proposed by Kolmogorov (1942) and later elaborated upon by a number of workers (Hanjalic, 1970, 1972; Harlow and Nakayama, 1968; Jones and Launder, 1972; Ng and Spalding, 1972; Rodi and Spalding, 1970; Rotta, 1968, 1971; Spalding, 1967, 1969, 1971). Launder and Spalding (1972) show that the resulting equations may be expressed in terms of a general functional variable Z which contains the turbulence length-scale and has the form of $e^{1/2}/l_c, e^{3/2}/l_c, l_c, el_c$, or e/l_c^2.

The equation for the conservation of Z is given as

$$\rho \frac{DZ}{Dt} = \frac{\partial}{\partial y}\left(\frac{\mu_t}{\mathrm{Pr}_{t,Z}} \frac{\partial Z}{\partial y}\right) + Z\left[c_1 \frac{\mu_t}{e}\left(\frac{\partial \bar{u}}{\partial y}\right)^2 - c_2 \frac{\rho^2 e}{\mu_t}\right] + S_Z \qquad (2.36)$$

where $\mathrm{Pr}_{t,Z}$, c_1, and c_2 are constants to be determined from experimental data. Launder and Spalding (1972) recommend the values $\mathrm{Pr}_{t,Z} = 1, c_1 = 1.45, c_2 = 0.17$. S_Z is a secondary source or sink term which differs from model to model. Equation (2.36) is obtained from the Navier-Stokes equations in a manner exactly analagous to that used in deriving Eqs. (2.16) and (2.32).

The functions $Z = el_c$, $Z = e/l_c^2$, and $Z = e^{3/2} l_c$ have been extensively tested by Spalding's group at Imperial College (Launder and Spalding, 1972) for a wide variety of turbulent flows, including flows with circulation. The models provide predictions of about the same level of accuracy for high Reynolds number boundary layer flows. The $Z = e$ model of Rodi and Spalding (1970) gives good predictions for free boundary layers, and $Z = e/l_c^2$ (Spalding, 1969) successfully predicts a number of turbulent recirculating flows (Gosman et al., 1969).

Multi-equation models of turbulence Although the above two-equation models possess a high degree of universality, they are still generally restricted to cases where the Boussinesq turbulent viscosity concept adequately represents the generated turbulent stress.

The above criticism can be overcome by discarding the concept of an effective viscosity and formulating transport equations for the unknown turbulent stresses themselves. However, these equations, obtained in a manner analogous to Eq. (2.28) contain further unknown velocity correlations, including triple velocity products (e.g., $\overline{u'v'v'}$). These correlations may either be modelled in terms of known quantities (in a manner analogous to the way in which the effective viscosity concept was used to model the double velocity correlation $\overline{u'v'}$) or they may themselves be made the dependent variables of yet another set of transport equations. These equations will, in turn, contain unknown correlations of yet higher order (e.g., $\overline{u'v'v'v'}$). These unknowns may also be made the dependent variables of yet another set of differential equations and the preceeding process may be carried on indefinitely, with the model becoming increasingly complex as the number of equations increases. However, at some stage, closure of the resulting system of equations must be achieved by modeling the unknown quantities in terms of known mean flow field variables.

Models of the above type have been proposed by Chou (1945), Rotta (1951), Davidov (1961), Daly and Harlow (1970), Hanjalic (1970), Hanjalic and Launder (1972), and Kolavandin and Vatutin (1969), among others. The models vary in complexity with the treatment of Kolavandin requiring the simultaneous solution of a maximum of 28 equations.

Summary An excellent review of the above models is given in the monograph of Launder and Spalding (1972). For equilibrium boundary layer flows the simple mixing-length hypothesis is sufficiently accurate for most purposes. The more elaborate models become necessary only for the more complex cases of flows with circulation. The one-equation models are suitable for axisymmetric flows in pipes or annuli where the simple Prandtl hypothesis fails because of the existence of a zero velocity gradient in the flow field. However, even here, the simple mixing-length hypothesis has been modified to overcome this limitation (see Section 2.2.3).

2.3 Law of the wall

2.3.1 Introduction Numerous experimental investigations (Coles, 1956; Hinze, 1959; Schlicting, 1960) of the velocity field in the vicinity of a solid boundary in external flows over flat plates and internal tube flows have shown that in this region the distribution of velocity follows the simple relationship

$$u^+ = u^+(y^+) \tag{2.37}$$

where $u^+ = \dfrac{u}{u^*} \qquad y^+ = y\,\dfrac{u^*}{\nu} \qquad \text{and} \qquad u^* = \sqrt{\dfrac{\sigma_w}{\rho}}$

This is the universal "law of the wall" for boundary layer flows. The flow region described by Eq. (2.37) is usually considered to consist of two composite layers called the inner and outer regions. The existence of the two regions is due to the different response to shear and pressure gradient by the fluid near the wall. The thickness of the inner region is approximately 10–20% of the total boundary layer thickness, and it is this region which is of most importance in heat and mass transfer calculations since it constitutes the major part of the resistence to transfer.

The bulk of the inner region is characterized by a nearly constant shearing stress which is equal to the value of the shear stress at the wall, i.e.,

$$\sigma \cong \sigma_w \tag{2.38}$$

and the flow is dominated by turbulent mixing, so that

$$\nu_t \gg \nu \qquad \text{and} \qquad \frac{\nu_t}{\nu} \gg 1 \tag{2.39}$$

In practice the shearing stress cannot remain strictly constant. The shear stress in pipe flows varies over the cross-section because of the pressure gradient necessary to overcome wall friction. However, because the width of the inner region is very small compared with the channel width, the stress variation in this zone is small and $\sigma \cong \sigma_w$.

This distribution of the shear stress for a turbulent flow field, given by Eq. (2.20), may be rewritten in dimensionless form as

$$\frac{\sigma}{\sigma_w} = \left(1 + \frac{\nu_t}{\nu}\right)\frac{du^+}{dy^+} = \epsilon^+\,\frac{du^+}{dy^+} \tag{2.40}$$

where ϵ^+ is a dimensionless effective viscosity defined as

$$\epsilon^+ = 1 + \frac{\nu_t}{\nu}$$

For the inner or constant shear stress region, Eq. (2.40) simplifies to

$$\epsilon^+ = \frac{dy^+}{du^+} \tag{2.41}$$

Equation (2.41) constitutes a very important result since it relates the effective viscosity to the velocity distribution.

In view of Eq. (2.39), Eq. (2.41) reduces simply to

$$\frac{\nu_t}{\nu} \cong \frac{dy^+}{du^+} \tag{2.42}$$

Utilizing the Prandtl mixing-length expression, Eqs. (2.23) and (2.24), Eq. (2.42) reduces to

$$\kappa y^+ = \frac{dy^+}{du^+} \tag{2.43}$$

which may be integrated to yield the velocity distribution

$$u^+ = \frac{1}{\kappa} \ln (y^+) + G \tag{2.44}$$

where κ and G are constants. The values of κ and G recommended by Coles (1955) on the basis of extensive experimental data are

$$\kappa = 0.4, G = 5.1 \tag{2.45}$$

Equation (2.44) is the "law of the wall" expression attributed to Prandtl, and is compared with experimental data in Figure 1. Reference to Figure 1 shows that the equation is in satisfactory agreement with the experimental data for the most part of the inner region of the boundary layer, i.e., the region where the Reynolds stress is of considerably greater magnitude than the viscous stress. However, as the wall is approached and the magnitude of the turbulent viscosity diminishes (see the mixing-length hypothesis, Eq. [2.24]) and becomes quantitatively comparable with the molecular viscosity, the data deviates from Eq. (2.44). In this region of deviation, commonly called the *viscous sublayer,* viscous effects are significant and a special treatment is required to describe the simultaneous action of both viscous and turbulent forces. The exact description of this region is of great importance since this region is characterized by very steep gradients of velocity, temperature, and concentration, particularly at high Prandtl or Schmidt numbers.

Many expressions have been proposed to describe the distribution of velocity (and consequently the turbulent viscosity) in the viscous sublayer. For a comprehensive review of these studies the reader is referred to the works of Hinze (1959), Jayatilleke (1969), Kestin and Richardson (1963), and Schlicting

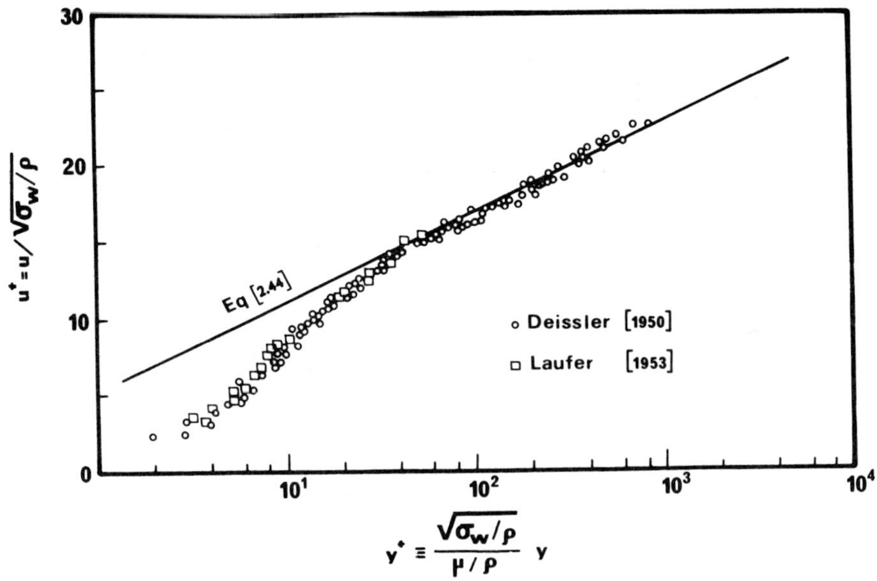

Figure 1 Generalized velocity distribution for adiabatic turbulent flow.

(1960). The more commonly used expressions due to Prandtl-Taylor (Deissler, 1955; Rannie, 1956; Reichardt, 1951; Schlicting, 1960; Spalding, 1961; Van Driest, 1956; and Von Karman, 1939) are listed in Table 3 along with the corresponding expression for ϵ^+.

An expression which is currently enjoying widespread popularity in the literature is that due to Van Driest (1956).

2.3.2 The Van Driest modification of the Prandtl mixing-length hypothesis

Van Driest (1956) considered that the turbulent motion in the viscous sublayer in the immediate vicinity of a wall is similar to the motion induced by the oscillatory motion of a plate in an infinite fluid. The oscillatory plate problem was solved by Stokes (Schlicting, 1960), who showed that the motion diminishes with distance from the plate by a factor of exp $(-y/A)$, where A is a constant which depends on both the frequency of oscillation of the plate and the fluid kinematic viscosity. Assuming that the plate motion in the viscous sublayer is analagous to the fluid executing oscillations relative to a fixed plate, Van Driest suggested that the damping factor had the form $[(1 - \exp(-y/A)]$, where A is a damping length constant which is interpreted as the distance from the wall beyond which viscous effects are negligible.*

*Based on their experimental data, Shlanchyauskas et al. (1974) propose that the damping factor take the form exp $[-(1-y^+/23)/0.4]$.

Table 3 Some Analytical Forms Proposed for the Universal Velocity Profile

Formula for $u^+(y^+)$	$\epsilon^+ = \frac{dy^+}{du^+}$	Range	Author
$u^+ = y^+$ $u^+ = 2.5 \ln y^+ + 5.5$	$\epsilon^+ = 1$ $\epsilon^+ = 0.4y^+$	$0 \leqslant y^+ \leqslant 11.5$ $y^+ > 11.5$	Prandtl (1928) Taylor (1916)
$u^+ = y^+$ $u^+ = 5 \ln y^+ - 3.05$ $u^+ = 2.5 \ln y^+ + 5.5$	$\epsilon^+ = 1$ $\epsilon^+ = 0.2y^+$ $\epsilon^+ = 0.4y^+$	$0 \leqslant y^+ \leqslant 5$ $5 \leqslant y^+ \leqslant 30$ $y^+ > 30$	Von Karman (1939)
$u^+ = y^+$ $u^+ = 5 \ln y^+ - 3.05$ $u^+ = 2.5 \ln y^+ + 5.5$	$\epsilon^+ = 1$ $\epsilon^+ = 0.2y^+\left(\frac{1 - y^+}{R^+}\right)$ R = tube radius $\epsilon^+ = 0.4y^+\left(\frac{1 - y^+}{R^+}\right)$	$0 \leqslant y^+ \leqslant 5$ $5 \leqslant y^+ \leqslant 30$ $y^+ > 30$	Martinelli (1947)
$u^+ = 2.5 \ln (1 + 0.4y^+) + 7.8\left[1 - \exp\left(-\frac{y^+}{11}\right) + -\left(\frac{y^+}{11}\right)\exp(-0.33y^+)\right]$	$\frac{1}{\epsilon^+} = \frac{1}{1 + 0.4y^+} + 7.8\left[\left(\frac{1}{11}\right)\exp\left(-\frac{y^+}{11}\right) + -\left(\frac{1}{11}\right)\exp-(0.33y^+) + 0.03y^+\exp(-0.33y^+)\right]$	all y^+	Reichardt (1951)
$u^+ = \frac{14.5}{3}\left[\frac{1}{2}\ln\frac{(1 + y^+/14.5)^2}{1 - (y^+/14.5) + (y^+/14.5)^2} + \sqrt{3}\tan^{-1}\frac{(2y^+/14.5) - 1}{\sqrt{3}} + \frac{\pi\sqrt{3}}{6}\right]$ $u^+ = 4.77 + 5 \ln (0.2y^+ + 0.041)$ $u^+ = 2.5 \ln y^+ + 5.5$	$\epsilon^+ = 1 + \left(\frac{y^+}{14.5}\right)^3$ $\epsilon^+ = 1 + 0.2y^+ - 0.959$ $\epsilon^+ = 0.4y^+$	$0 < y^+ < 5$ $5 < y^+ < 30$ $y^+ > 30$	Lin, Moulton and Putnam (1953)

$\dfrac{du^+}{dy^+} = \dfrac{1}{1 + n^2u^+y^+[1 - \exp(-n^2u^+y^+)]}$ $n = 0.124$ $u^+ = 2.78 \ln y^+ + 3.8$	$\epsilon^+ = 1 + n^2u^+y^+[1 - \exp(-n^2u^+y^+)]$ $\epsilon^+ = \dfrac{y}{2.78}$	$0 \leqslant y^+ \leqslant 26$ $y^+ > 26$	Deissler (1955)
$u^+ = 14.53 \tanh\left(\dfrac{y^+}{14.53}\right)$ $u^+ = 2.5 \ln y^+ + 5.5$	$\epsilon^+ = 1 + \sinh^2\left(\dfrac{y^+}{14.53}\right)$ $\epsilon^+ = 0.4y^+$	$0 \leqslant y^+ \leqslant 27.5$ $y^+ > 27.5$	Rannie (1956)
$\dfrac{dy^+}{du^+} = \dfrac{2}{1 + \{1 + 4K^2y^{+2}[1 - \exp(-y^+/A^+]^2\}^{1/2}}$ $K = 0.4 \quad A^+ = 26$	$\epsilon^+ = 1 + \left\{1 + 4K^2y^{+2}\left[1 - \exp\left(-\dfrac{y^+}{A^+}\right)\right]^2\right\}^{1/2}$	all y^+	Van Driest (1956)
$y^+ = u^+ + A\left[\exp Bu^+ - 1 - Bu^+ - \dfrac{1}{2}(Bu^+)^2 - \dfrac{1}{6}(Bu^+)^3 - \dfrac{1}{24}(Bu^+)^4\right]$ (last term in u^{+4} may be omitted)	$\epsilon^+ = 1 + AB\left[\exp Bu^+ - 1 - Bu^+ - \dfrac{1}{2}(Bu^+)^2 - \dfrac{1}{6}(Bu^+)^3\right]$ (last term in u^{+3} may be omitted) $A = 0.1108$ $B = 0.4$	all y^+	Spalding (1961)
$u^+ = y^+ - 1.04 \times 10^{-4}y^{+4} + 3.03 + 10^{-6}y^{+5}$	$\epsilon_+ = 1 + \dfrac{4.16 \times 10^{-4}y^{+3} - 15.15 \times 10^{-6}y^{+4}}{1 - 4.16 \times 10^{-4}y^{+3} + 15.15 \times 10^{-6}y^{+4}}$	$0 \leqslant y^+ \leqslant 20$	Wasan and Wilke (1963)

Accordingly, the fluctuating velocity component, u' is given by

$$u' = u'_0 \left[1 - \exp\left(-\frac{y}{A}\right)\right] \tag{2.46}$$

where u'_0 is the velocity fluctuation which remains unaffected by the pressure of the wall.

Following Prandtl's original hypothesis that $u' \sim v'$ yields

$$v' = v'_0 \left[1 - \exp\left(-\frac{y}{A}\right)\right] \tag{2.47}$$

The Reynolds shear stress is then

$$-\rho\overline{u'v'} = -\rho\overline{u'_0 v'_0}\left[1 - \exp\left(-\frac{y}{A}\right)\right]^2 \tag{2.48}$$

Introducing Prandtl's mixing-length concept,

$$-\rho\overline{u'_0 v'_0} = \rho(\kappa y)^2\left(\frac{d\bar{u}}{dy}\right)^2 \tag{2.49}$$

and combining Eq. (2.48) and Eq. (2.49) yields

$$-\rho\overline{u'v'} = \rho(\kappa y)^2\left[1 - \exp\left(-\frac{y}{A}\right)\right]^2\left(\frac{d\bar{u}}{dy}\right)^2 \tag{2.50}$$

Comparison of Eq. (2.50) with the definition of the turbulent viscosity, Eq. (2.19), yields

$$\mu_t = \rho(\kappa y)^2\left[1 - \exp\left(-\frac{y}{A}\right)\right]^2 \frac{d\bar{u}}{dy} \tag{2.51}$$

or

$$\frac{\mu_t}{\mu} = \frac{\nu_t}{\nu} = (\kappa y^+)^2\left[1 - \exp\left(-\frac{y^+}{A^+}\right)\right]^2 \frac{du^+}{dy^+} \tag{2.52}$$

Comparison of Eq. (2.52) with the mixing-length hypothesis, Eq. (2.23), indicates that both expressions are equivalent if we write

$$l_m = \kappa y \left[1 - \exp\left(-\frac{y}{A}\right)\right] \tag{2.53}$$

This expression is commonly referred to as the Van Driest modification of the Prandtl mixing-length. At large values of y^+, away from the wall, the exponential term in Eq. (2.53) becomes negligible and the expression reduces to $l_m = \kappa y$ i.e., the simple mixing-length concept. Utilizing the value of $\kappa = 0.4$ and comparing Eq. (2.52) with experimental data at high Reynolds number for smooth wall, Van Driest empirically determined that $A^+ = 26$. Note that A^+ can be interpreted as the distance from the wall beyond which the viscous effects are negligible.

Combining Eqs. (2.52) and (2.42) gives

$$\frac{du^+}{dy^+} = \frac{2}{1 + \{1 + 4(\kappa y^+)^2 \left[1 - \exp\left(-y^+/A^+\right)\right]^2\}^{1/2}} \tag{2.54}$$

which may be integrated to yield the velocity profile. The calculated profile is compared in Figure 2 with the profiles obtained by utilizing the expressions listed in Table 3.

The Van Driest expression has been modified by a number of workers to account for such effects as a non zero pressure gradient and finite mass transport at the wall, and these modifications have been successfully applied to a wide range of transpired turbulent boundary layer flows (Launder and Priddin, 1973). The amenability of the Van Driest expression to these modifications has contributed in small part to the status enjoyed by the expression in the present literature.

2.3.3 Limiting form of the law of the wall, $y^+ \to 0$ Comparison of the various expressions for the velocity distribution in the wall region given in Figure 2 shows that the differences between the expressions are of the same order as the scatter in the experimental data shown in Figure 1. Thus, this comparison cannot discriminate between the various expressions.

However, predictions of heat and mass transfer rates based on the above expressions (see section 2.6.3) exhibit considerable disagreement at high Prandtl and Schmidt numbers. Since the region of interest at high Prandtl and Schmidt numbers is very close to the wall (i.e., $y^+ < 1$ for Pr or Sc $> 10^4$), the important difference between the various expressions is restricted to the behavior of the predicted ν_t as $y^+ \to 0$. This is the region where the uncertainties associated with experimental measurements are greatest. Inspection of Table 1 shows that the

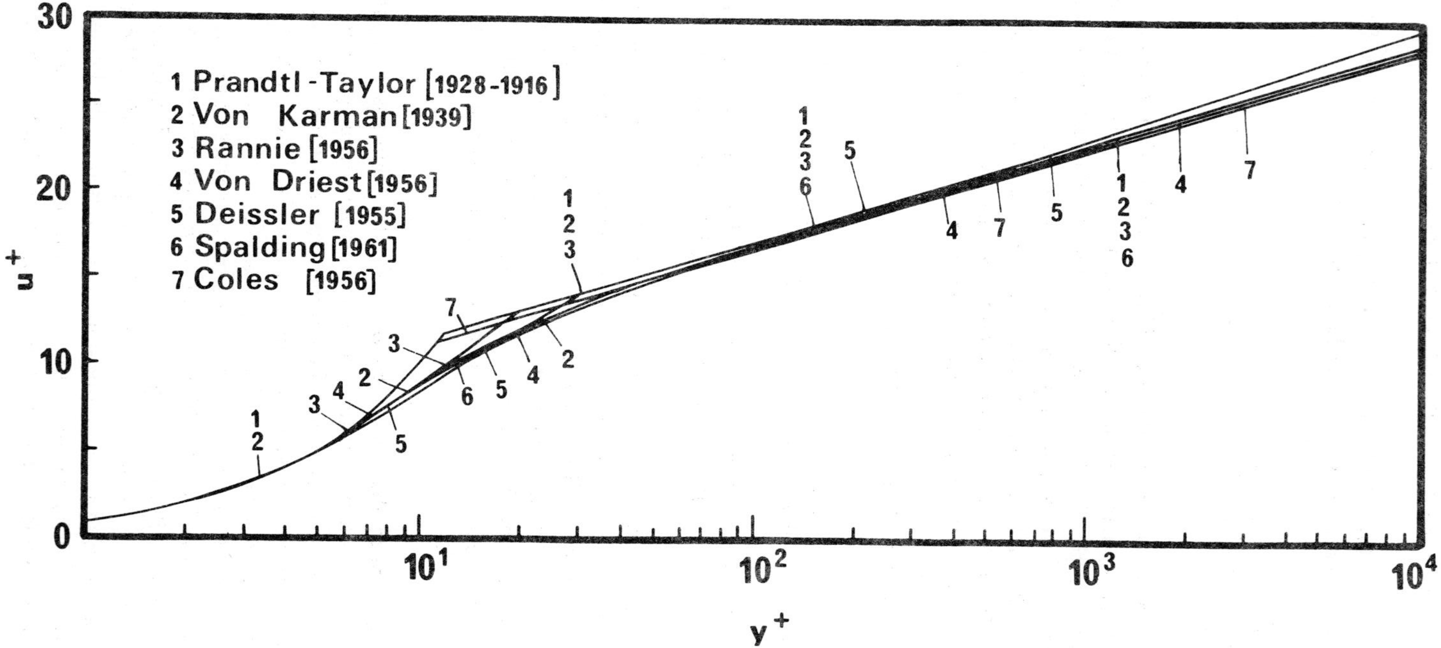

Figure 2 Various analytic forms proposed for the universal velocity profile. (Kestin and Richardson, 1963).

expressions of Van Driest and Deissler result in $\nu_t \sim (y^+)^4$ as $y^+ \to 0$, while Reichardt's expression leads to $\nu_t \sim (y^+)^3$ as $y^+ \to 0$. The expression due to Spalding is flexible, with ν_t related either to $(y^+)^3$ or $(y^+)^4$, depending on whether the last term in $(u^+)^3$ is omitted or retained. The presence or absence of this term has no significant effect on the "goodness of fit" of the expression to the experimental data (Spalding, 1961).

Several attempts have been made to provide a theoretical basis for either a $(y^+)^3$ or $(y^+)^4$ variation of ν_t as $y^+ \to 0$. Clearly, the velocity fluctuations must satisfy the equation of continuity as well as the no slip condition at the wall. Reichardt (1951) and Elrod (1957) argued that ν_t must vary with a minimum power of $(y^+)^3$ if the streamwise variation in the time-mean values of turbulence quantities is to be maintained, whereas a minimum power of $(y^+)^4$ is required if the streamwise variation no longer exists. In an erratum to his earlier work, Elrod (1960) states that while the conclusion regarding $(y^+)^3$ as the minimum rate of variation as $y^+ \to 0$ remain firm, it is not yet possible to draw firm conclusions regarding the $(y^+)^4$ variation. Townsend (1956), Notter and Sleicher (1969, 1971), and Wasan, Tien, and Wilke (1963) conclude that $\nu_t \sim (y^+)^3$ as $y^+ \to 0$. On the other hand, Ohji (1967) has argued that for wall turbulence which is homogeneous in the x- and z- directions, $\nu_t \sim (y^+)^4$ as $y^+ \to 0$.

In view of the inconclusive nature of both the theoretical arguments and the experimental velocity data (a good example of the latter is the work of Sherwood, Smith, and Fowles, 1968) it would seem that the only meaningful discrimination between the various expressions in Table 1 can be based on the heat and mass transport rates at high Pr and Sc. A comparison of this type is made in section 2.6, where it is concluded that as $y^+ \to 0$ $\nu_t \sim (y^+)^3$ and not $(y^+)^4$. This conclusion is supported by the investigations of Notter and Sleicher (1971), Hubbard and Lightfoot (1966), and Sirkar and Hanratty (1969). In an earlier study Son and Hanratty (1967) suggested that $\nu_t \sim (y^+)^4$ as $y^+ \to 0$. However, their argument has been reexamined by Hughmark (1968) and by Hubbard (1968) who conclude that the variation is with $(y^+)^3$.

It is important to remember that the conclusion that $\nu_t \sim (y^+)^3$ as $y^+ \to 0$ is subject to the assumption that Pr_t is either unity or a constant. In spite of the considerable indirect evidence which supports this relationship, (see Section 2.6.3), the evidence is by no means conclusive.

2.4 The similarity between heat, mass, and momentum transfer

Inspection of the partial-differential equations for the conservation of momentum, energy, and mass in a turbulent flow field, Eqs. (2.8)–(2.13) and Eqs. (2.16)–(2.18) show that the equations are similar, provided that the pressure gradient $\partial P/\partial x$ in the momentum equation, is equal to zero. The solutions of these equations will then also be similar if the corresponding boundary conditions are similar. This similarity plays an important part in the theory of heat and mass transfer as it suggests the possibility that heat and mass transfer rates may be predicted from the rate of momentum transport.

For strict similarity to exist between the processes of momentum, heat, and mass transport, the following conditions must be satisfied;

1. the velocity field must be strictly two-dimensional. Since momentum is a vector quantity a general description of this quantity requires three equations (conservation in x, y, and z directions, respectively). In contrast to this, heat and mass are scalar quantities requiring only one conservation equation each. Thus, a comparison between the three processes can only be made for the transport of heat, mass and but one conponent of momentum transport.
2. the pressure gradient, dp/dx, must be zero. The condition of zero pressure gradient is satisfied for flows over flat surfaces at zero incidence (e.g., flat plate) and may be considered to be sufficiently satisfied for flows over surfaces of gentle curvature.
3. the viscous dissipation and the diffusive mass flux must be negligibly small, as is implied by Eqs. (2.5) and (2.6). These conditions have been discussed in Section 2.1.2.
4. fluid physical properties ρ, μ, k, and D must be constant. This condition is made necessary by the generally different functional dependences of these properties on the temperature.
5. the spectral fluctuations of the fluctuating quantities u', T', and c' must be similar. This condition implies that the Reynolds transport terms, $\overline{u'v'}$, $\overline{T'v'}$, and $\overline{c'v'}$ must be described by similar functions.

It is evident from the above stipulations that the strict validity of the three-way similarity between the processes of heat, mass, and momentum transport is quite limited. Nevertheless, a large number of useful "analogies" have been proposed (Section 2.6). It is also evident that the similarity between heat and low-flux mass transfer may be expected to be somewhat less restricted. This expectation is supported by considerable experimental evidence (Chilton and Colburn, 1934; Sherwood and Pigford, 1952) and suggests that correlations developed for heat transfer can be applied to mass transfer studies simply by replacing the Prandtl number by the Schmidt number.

Whether or not there exists a strict similarity between the processes of heat, mass, and momentum transfer, an apriori specification of the relationship between the turbulent kinematic viscosity, ν_t, and the turbulent thermal diffusivity, α_t, reduces the problem of closure of the set of Reynolds partial differential equations to one of adequately specifying ν_t. This relationship is usually expressed in terms of a turbulent Prandtl number, Pr_t, where

$$\mathrm{Pr}_t = \frac{\nu_t}{\alpha_t} \tag{2.55}$$

The importance of the concept of a turbulent Prandtl number lies in the fact that while there exists sufficiently precise measurements of the turbulent velocity field from which ν_t may be evaluated to some degree of accuracy, the experimental difficulties in making corresponding uncertainty in the calculated values of α_t and D_t. Until these experimental difficulties are overcome it remains necessary to estimate α_t and D_t by making some plausible assumptions regarding the similarity of the heat, mass, and momentum transport processes. These assumptions generally take the form of assumptions concerning the value of the turbulent Prandtl (or Schmidt) number.

2.5 Turbulent Prandtl (Schmidt) numbers

2.5.1 Introduction As the complex nature of the turbulent transport process is not yet well understood, the development of models for the prediction of Pr_t requires the introduction of many tenuous assumptions regarding the behavior of turbulence. The validity of these assumptions may be verified only indirectly by comparing the predicted Pr_t values with the measured ones.

There are basically two routes by which Pr_t may be evaluated experimentally.

1. Utilizing experimental time-averaged velocity and temperature (or concentration) profiles together with the integrated Reynolds equations. The latter, Eqs. (2.16) to (2.18), are solved directly for ν_t, α_t, and D_t. For fully developed turbulent flow over a flat plate at zero incidence (section 2.3.1).

$$\frac{\sigma}{\sigma_w} = \left(1 + \frac{\nu_t}{\nu}\right)\frac{du^+}{dy^+} \tag{2.56}$$

$$\frac{q}{q_w} = \left(1 + \frac{\alpha_t}{\alpha}\right)\frac{dT^+}{dy^+} \tag{2.57}$$

   Thus, if the momentum and heat flux variation with the distance from the wall are known, ν_t, α_t, hence Pr_t can be evaluated.
2. Direct measurement of the Reynolds transport terms ($\overline{u'v'}$, $\overline{T'v'}$): From the definition of the eddy diffusivity (Eq. [2.19]) we may write

$$\frac{\nu_t}{\nu} = -\frac{\overline{u'v'}}{du^+/dy^+} \qquad \frac{\alpha_t}{\nu} = -\frac{\overline{T'v'}}{du^+/dy^+} \tag{2.58}$$

or

$$\mathrm{Pr}_t = \frac{\overline{u'v'}}{\overline{T'v'}} \cdot \frac{dT^+}{dy^+} \cdot \frac{dy^+}{du^+} \tag{2.59}$$

Thus, Pr_t can be evaluated from the experimentally determined velocity and temperature fields, and the turbulent fluctuating terms $\overline{u'v'}$ and $\overline{T'v'}$.

The characteristic feature of all the reported experimentally determined Pr_t is the great scatter of the data. The results of different investigations lack in agreement and are often conflicting. This situation arises directly from the necessity to differentiate the two measured profiles and determine the shear stress and heat flux profiles for the calculation of Pr_t. The differentiation procedure greatly amplifies the uncertainties associated with the experimentally measured velocity and temperature profiles. These difficulties rise from the inaccessibility of probes to the thin region adjacent to the wall in which the velocity gradient and temperature gradient concentrate, particularly at high molecular Prandtl numbers. For the case of liquid metals there is the additional difficulty of correctly interpreting hot wire measurements where the molecular conduitivity is large. (Hochreiter and Sesonske, 1974; and Bremhorst and Bullock, 1973).

2.5.2 Experimentally determined values of the turbulent Prandtl numbers

Experimentally determined turbulent Prandtl numbers have been surveyed by Kestin and Richardson (1963) and more recently by Blom and de Veres (1968). The results of these surveys together with the more recent data of Blom (1970), Simpson et al. (1970), Sleicher et al. (1973), and Quarmby and coworkers (1969, 1972, 1974) are summarized in Figure 3. Although the results are greatly scattered, various investigators have concluded, on the basis of their own results, that Pr_t is varyingly affected by the molecular Prandtl number, the flow Reynolds number, and the distance from the wall. The suggested trends of Pr_t with these parameters for the case of liquid metals ($\mathrm{Pr} < 0.1$) differ from those for ordinary fluids ($\mathrm{Pr} > 0.7$). For liquid metals Pr_t is greater than unity and decreases with increasing Reynolds number and the distance away from the wall (Carr and Balzhiser, 1967). This effect is opposite to that for air (Sleicher, 1958), where Pr_t is less than one, and increases towards unity with increase in Re and distance from the wall.

Although the above trends have provided a basis for the formulation of models for the prediction of Pr_t to be considered below, the picture is by no means conclusive, and many studies have produced results which exhibit a different behavior. Quarmby and Quirk (1972, 1974) have concluded, on the basis of an extensive study of turbulent flow in a plain circular tube with Schmidt numbers varying from 0.7 to 1200 and Reynolds numbers ranging from 5230 to 23550, that Pr_t is a simple function of the nondimensional distance from the wall (radius), independent of both the Prandtl number and Reynolds

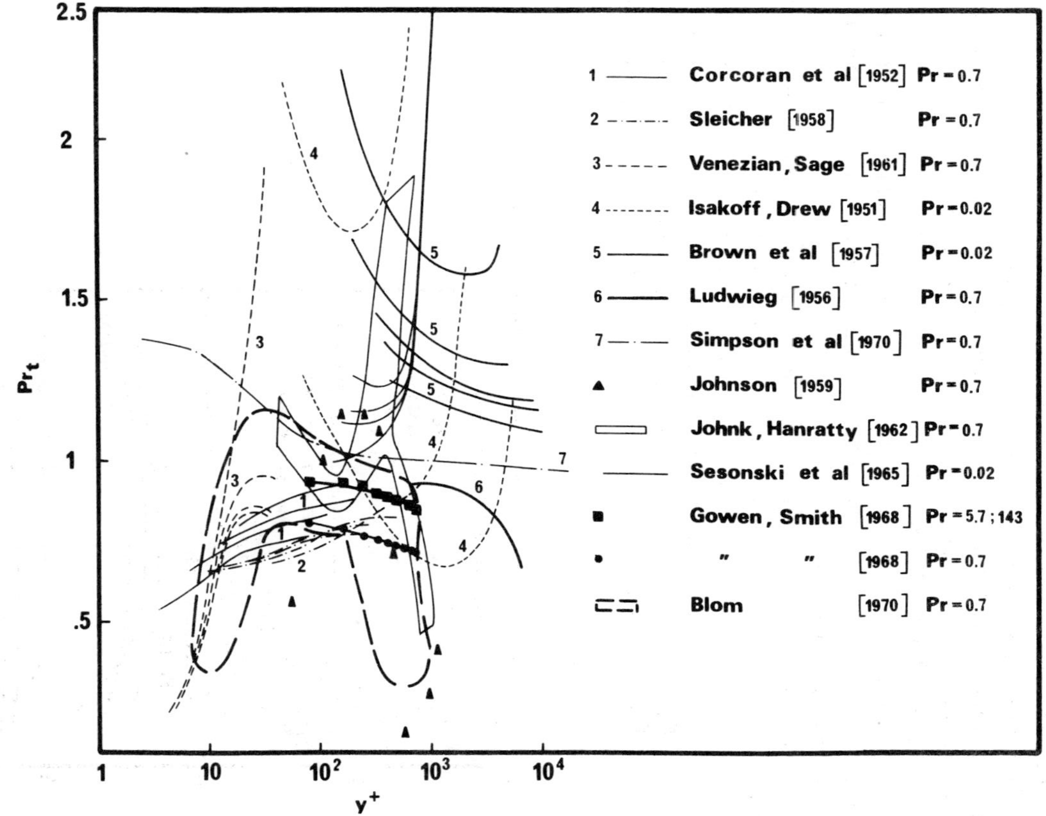

Figure 3 Summary of published Pr_t values. (After Blom, 1970).

number. They found that Pr_t varies smoothly from 0.5 at the wall to unity at the tube center. Brienkworth and Smith (1969) (Pr = 6) and Eckelman and Hanratty (1972) (Pr = 0.7) suggest, on the other hand, that Pr_t is constant and equal to approximately unity over the whole of the boundary layer. Yet a different trend was reported by Simpson et al. (1970) who found that for boundary layer flows of air (Pr = 0.77), the local value of Pr_t near the wall is greater than unity, displaying a maximum value of approximately 1.4 at the wall and decreasing to approximately unity in the outer region. Finally, the often quoted results of Ludwig (1956) for air show Pr_t to vary from about unity at the wall and decrease smoothly to about 0.6 at the pipe center. This result is consistent, in a way, with the constant value of $Pr_t = 0.75$ suggested by Shlanchyauskas et al. (1974) for $Pr > 1$.

In view of the contradictory experimental data it is significant to note the analysis of Deissler (1963) based on the statistical nature of the turbulent process. Deissler suggests that Pr_t approaches unity as the velocity gradient increases, regardless of the molecular Prandtl number of the fluid. This is indeed consistent with the observed trends of the effect of Re on Pr_t for both liquid metals and ordinary fluids depicted in Figure 4. Deissler also suggests that the

Figure 4 Ratio of radial eddy diffusivities of heat and mass to that of momentum.

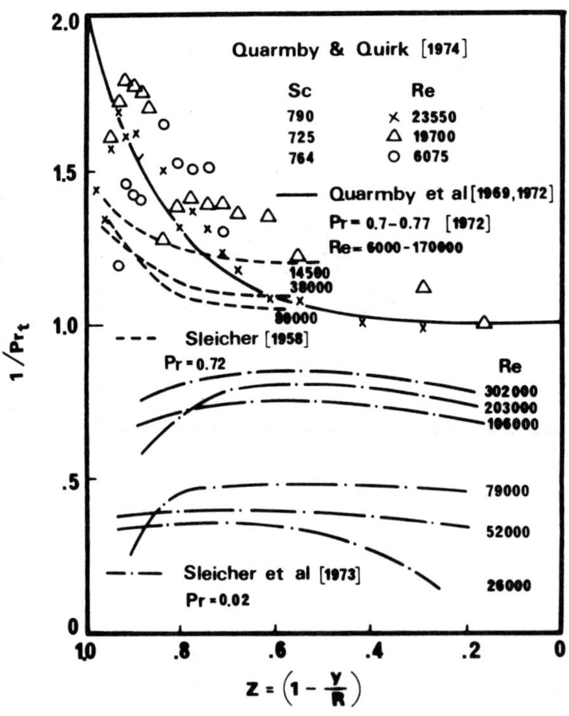

effect of Pr is much greater at low values of Pr (i.e., liquid metals) than at higher ones. This is consistent with the results of both Sliecher (1958) and Quarmby and Quirk (1974), if one considers that the Reynolds number effect reported by Sliecher for air lies within the experimental error band of Quarmby and Quirk's data.

A complete analysis of the available experimental data is beyond the scope of the present work. However, the above comments are sufficient to indicate the general inconsistencies and lack of agreement which are characteristic of much of the published work. These inconsistencies, and low accuracy of the available measurements, have hindered the development of a satisfactory empirical theory. This in turn has necessitated the formulation of semi-empirical theories in an attempt to both rationalize the existing data and to provide a starting point for heat and mass transfer calculations.

2.5.3 Models for Pr_t

Reynolds analogy The first model for the prediction Pr_t was proposed by Reynolds in 1874. Reynolds argued that in turbulent flow both heat and momentum are transported as a result of the same eddy motion. The implementation of this argument is that the eddy diffusivities of heat and momentum transport are equal, i.e., $\nu_t = \alpha_t$, or that

$$Pr_t = 1 \tag{2.60}$$

Equation (2.60) is widely referred to as the Reynolds analogy. It is the simplest of the models proposed for the estimation of Pr_t, and is the most widely used. Its applicability to practice is evidenced by its long standing. It has, in fact, been shown (Tien, 1959) to be exact under certain idealized conditions. Chi and Spalding (1966) have suggested the value 0.9, while Kader and Yaglom (1972) recommend a value of 0.85. These slight modifications proved effective in predicting heat and mass transfer rates for a wide variety of flows (Launder and Spalding, 1972; Goldman and Marchello, 1969).

The experimental observations discussed above would suggest that Reynolds analogy is an oversimplification of the general phenomenon. Not surprisingly, predictions based on the Reynold's analogy show significant deviations from experimental observations at low Prandtl numbers encountered with liquid metals (Dwyer, 1963). This prompted various workers to suggest modifications of the basic Reynolds analogy. Most of these modifications involve corrections of the basic mixing-length concept of Prandtl to account for the heat lost by eddies during their flight times. A compilation of various models for Pr_t is given in Table 4.

Models based on heat losses from a spherical eddy One of the first modifications of the Reynolds analogy was proposed by Jenkins (1951). As in

Table 4 Models for the Turbulent Prandtl Number

Equation for Pr_t	Author
$\mathrm{Pr}_t = \dfrac{1}{\mathrm{Pr}} \left\{ \dfrac{1 - 90\left(\dfrac{\nu_t}{\nu}\right)\left\{\sum_{n=1}^{\infty} \dfrac{1}{n^6}\left[1 - \exp\left(-\dfrac{n^2\pi^2}{\nu_t/\nu}\right)\right]\right\}}{1 - 90\left(\mathrm{Pr}\,\dfrac{\nu_t}{\nu}\right)\left\{\sum_{n=1}^{\infty} \dfrac{1}{n^6}\left[1 - \exp\left(-\dfrac{n^2\pi^2}{\mathrm{Pr}\,\nu_t/\nu}\right)\right]\right\}} \right\}$	Jenkins (1951)
$\dfrac{1}{\mathrm{Pr}_t} = n\,\mathrm{Re}\,\mathrm{Pr}\,[1 - \exp(-n\,\mathrm{Re}\,\mathrm{Pr})]$ n = experimental constant = 0.000153	Diessler (1952)
$\dfrac{1}{\mathrm{Pr}_t} = 0.032\,\mathrm{Pr}\,\mathrm{Re}\left(\dfrac{1}{8}f\right)^{1/2} \dfrac{Z}{1-Z}\left\{1 - \exp\left[-1/0.032\,\mathrm{Pr}\,\mathrm{Re}\left(\dfrac{1}{8}f\right)^{1/2}\dfrac{Z}{1-Z}\right]\right\}$ $Z = \left(1 - \dfrac{y}{R}\right)$ f = fanning friction factor R = pipe radius	Voskrenskii (1956)
$\dfrac{1}{\mathrm{Pr}_t} = \dfrac{6}{\pi^2}\left[\exp\left(-\dfrac{b}{\mathrm{Pr}}\right) + 0.25\exp\left(-\dfrac{4b}{\mathrm{Pr}}\right)\right]$ b = experimental constant	Lykoudis and Toulukian (1958)
$\dfrac{1}{\mathrm{Pr}_t} = \dfrac{1 + 135\,\mathrm{Re}^{-0.45}\exp(-\eta^{0.25})}{2 + 57\,\mathrm{Re}^{-0.46}\,\mathrm{Pr}^{0.58}\exp(-\eta^{0.25})}$ $0.6 < \mathrm{Pr} < 15$ $\dfrac{1}{\mathrm{Pr}_t} = \dfrac{1 + 135\,\mathrm{Re}^{-0.45}\exp(-\eta^{0.25})}{1 + 380\,\mathrm{Pe}^{-0.58}\exp(-\eta^{0.25})}$ $\mathrm{Pr} < 0.6$ $\eta = \dfrac{y}{R}$ R = pipe radius	Azer and Chao (1960)
$\dfrac{1}{\mathrm{Pr}_t} = 1.5\,\dfrac{C_4\,\dfrac{\nu_t}{\nu}\,\mathrm{Pr}}{1 + C_3\left(\dfrac{\nu_t}{\nu}\right)^{1/2}\mathrm{Pr}^{1/3}}\left[1 - \exp\left(\dfrac{1 + C_3\left(\dfrac{\nu_t}{\nu}\right)^{1/2}\mathrm{Pr}^{1/3}}{C_4\,\dfrac{\nu_t}{\nu}\,\mathrm{Pr}}\right)\right]$ C_3, C_4 = experimental constants	Mizushina and Sasano (1961)
$\dfrac{1}{\mathrm{Pr}_t} = 0.0612\,\mathrm{Pr}\left(\dfrac{\nu_t}{\nu}\right)\left\{1.1 - \exp\left[-\dfrac{\pi^2}{\mathrm{Pr}}\left(\dfrac{\nu_t}{\nu}\right)\right]\right\}$	Buyco (1961)
$\dfrac{1}{\mathrm{Pr}_t} = \mathrm{Pr}\,\mathrm{Re}\left(\dfrac{1}{8}f\right)^{1/2} Z\left\{1 - \exp\left[-\dfrac{1}{Z}\,\mathrm{Pr}\,\mathrm{Re}\left(\dfrac{1}{8}f\right)^{1/2}\right]\right\}$ $Z = \left(1 - \dfrac{y}{R}\right)$ f = fanning friction factor R = pipe radius	Kokorev and Ryaposov (1962)

Table 4 Models for the Turbulent Prandtl Number (*Continued*)

Equation for Pr_t	Author
$\left(\frac{1}{\mathrm{Pr}_t}\right)_{\text{average}} = 1 - \frac{1.82}{\mathrm{Pr}\left(\frac{\nu_t}{\nu}\right)_{\text{maximum}}}$	Dwyer (1963)
$\mathrm{Pr}_t = 1.0$ low turbulence intensity $\mathrm{Pr}_t = \sqrt{\mathrm{Pr}}$ high turbulence intensity	Marchello and Toor (1963)
$\mathrm{Pr}_t = \frac{2 + 6\,\mathrm{Pr}}{9\,\mathrm{Pr}}$ low turbulence intensity $\mathrm{Pr}_t = \frac{2 + 9\,\mathrm{Pr}}{3 + 9\,\mathrm{Pr}}$ high turbulence intensity	Tyldesley and Silver (1968)
$\left(\frac{1}{\mathrm{Pr}_t}\right)^2 = \frac{1 - \frac{r_e}{l_m}\mathrm{Pe}^* \sum_{k=1}^{\infty} \epsilon_k \left[1 - \exp\left(-k^2\pi^2 \frac{l_m}{r_e \mathrm{Pe}^*}\right)\right]}{1 - \pi^2 \sum_{k=1}^{\infty} k^{\epsilon} k \left(-k^2\pi^2 \frac{l_m}{r_e \mathrm{Pe}^*}\right) + \frac{r_e}{l_m}\mathrm{Pe}^* \sum_{k=1}^{\infty} \epsilon_k \left[1 - \exp\left(-k^2 \frac{\pi^2 l_m}{r_e \mathrm{Pe}^*}\right)\right]}$ $\epsilon_k = \frac{3(-1)^{k+1}}{(k\pi)^3} \frac{1}{k\pi} \frac{\int_0^{k\pi} \frac{\sin t}{t}\,dt}{\int_0^{k\pi} \frac{\sin^2 t}{t^2}\,dt}$ $\mathrm{Pe}^* = r_e^*\,\mathrm{Pr}$ = eddy Peclet number $r_e^* = \frac{v^* r_e}{\nu}$ = eddy Reynolds number r_e = eddy radius	Senechal (1968)
$\frac{1}{\mathrm{Pr}_t} = F\left[1 - \exp\left(-\frac{1}{F}\right)\right]$ $F = \frac{0.0166\left(\frac{\nu_t}{\nu}\right)^{1/2} \mathrm{Pr}}{1 + 0.200\left(\frac{\nu_t}{\nu}\right)^{1/2} \mathrm{Pr}^{1/3}}$ $y^+ > 20$ $F = \frac{0.0344\left(\frac{\nu_t}{\nu}\right)^{1/3} \mathrm{Pr}}{1 + 0.240\left(\frac{\nu_t}{\nu}\right)^{1/3} \mathrm{Pr}^{1/3}}$ $y^+ < 20$	Mizushina, Ito, Ogino, and Muramoto (1969)
$\mathrm{Pr}_t = \mathrm{Pr} \frac{\exp\left(y^+ \sqrt{\frac{f}{2}}\right) - 1}{\exp\left(\sqrt{\mathrm{Pr}}\, y^+ \sqrt{\frac{f}{2}}\right) - 1}$	Thomas and Rajagopal (1973)

Table 4 Models for the Turbulent Prandtl Number (*Continued*)

Equation for Pr_t	Author
$\mathrm{Pr}_t = \dfrac{C_3}{C_1\,\mathrm{Pr}}\left\{\dfrac{1-\exp\left[-\dfrac{C_4}{(\nu t/\nu)}\right]}{1-\exp\left[-\dfrac{C_2}{\mathrm{Pr}(\nu_t/\nu)}\right]}\right\}$ C_1, C_2, C_3, C_4 = experimental constants	Wassel and Cotton (1973)
$\mathrm{Pr}_t = \dfrac{1-\exp(-y^+/A^+)}{1-\exp(-y^+/B^+)}$ $A^+ = 26$ $B^+ = B^+(\mathrm{Pr})$ (Eq. [2.74])	Cebeci (1973)

the Prandtl mixing-length concept, eddies are assumed to move with constant velocity and temperature, so that the temperature fluctuation T' is given by

$$T' = l_H \frac{d\overline{T}}{dy} \tag{2.61}$$

and

$$\alpha_t = l_H v' \tag{2.62}$$

where v' is the eddy velocity and l_H is the mixing length relevant to the process of heat transfer. Jenkins suggested that the moving eddies lose energy and momentum by thermal conductivity and molecular viscosity. The amount of heat transferred when the eddy mixes with the fluid as some particular level is smaller than its original heat content. This results in a decrease of the value of l_H from that suggested by the simple mixing-length theory where the heat loss is assumed to be zero, i.e.,

$$\alpha_t = l_H v' \frac{T_f - T_i}{dT/dy} \tag{2.63}$$

where T_i is the initial temperature of the eddy and T_f the temperature on mixing. The temperature of the moving eddy is estimated by assuming it to be a solid sphere of a radius equal to the mixing length, with a surface temperature which varies linearly with time. The average temperature of the sphere at time t is given by (Carslaw and Jaeger, 1958).

$$\bar{T} - T_i = \frac{d\bar{T}}{dy}\left\{\left(t - \frac{l_H^2}{15\alpha}\right) + \frac{6l_H^2}{\pi^4}\sum_{n=1}^{\infty}\frac{1}{n^6}\left[1 - \exp\left(-n^2\frac{\pi^2\alpha}{\rho}\right)\right]\right\} \tag{2.64}$$

where α is the thermal diffusivity. If the eddy lifetime is defined as l_H/v', the temperature drop $(T_f - T_i)$ in Eq. (2.63) is obtained from Eq. (2.64) by substituting l_H/v' for t.

Assuming that the momentum loss during the eddy movement may be treated in a manner analogous to that for the heat loss, the Jenkins model leads to the following expression for the turbulent Prandtl number

$$\mathrm{Pr}_t = \frac{1}{\mathrm{Pr}}\,\frac{1 - 90\left(\frac{\nu_t}{\nu}\right)\left\{\sum_{n=1}^{\infty}\frac{1}{n^6}\left[-\exp\left(\frac{-n^2\pi^2}{\nu t/\nu}\right)\right]\right\}}{1 - 90\left(\mathrm{Pr}\,\frac{\nu_t}{\nu}\right)\left\{\sum_{n=1}^{\infty}\frac{1}{n^6}\left[-\exp\left(\frac{-n^2\pi^2}{\mathrm{Pr}\,\nu_{t/\nu}}\right)\right]\right\}} \tag{2.65}$$

where it has been assumed that the mixing lengths of heat and momentum transfer are equal.

Subsequent workers have utilized the spherical eddy concept to provide variations of the original Jenkins model (see Table 4). The variations generally arise from the assumptions made regarding the mechanism by which the eddies lose energy and momentum. Azer and Chao (1960), for instance, used the Pohlhausen expressions for heat and momentum transfer for the laminar boundary layer on a flat plate to estimate the heat and momentum loss from the eddy. Mizushina et al. (1969) assumed Newton's law for the drag force on the eddy particle and used the Ranz and Marshall (1952) expression to calculate the heat loss from the eddy. The models generally contain one or more constants, usually introduced to account for the inadequacies of the chosen description of the heat and momentum transport from the eddy. These constants are evaluated by comparison with experimental observations, and as such, are more flexible than the original Jenkins formulation in representing experimental data.

A comparison between the predictions of some of the above models is shown in Figure 5. The general disagreement between the different formulations is of the same order as that between different sets of experimental data. This is not unexpected since different experimental data were used to evaluate the constants in the equations in Table 4.

Model based on a modification of the Van Driest modification of the mixing-length theory A somewhat different approach to that taken above was recently reported by Cebeci (1973). His model is based on the application of the

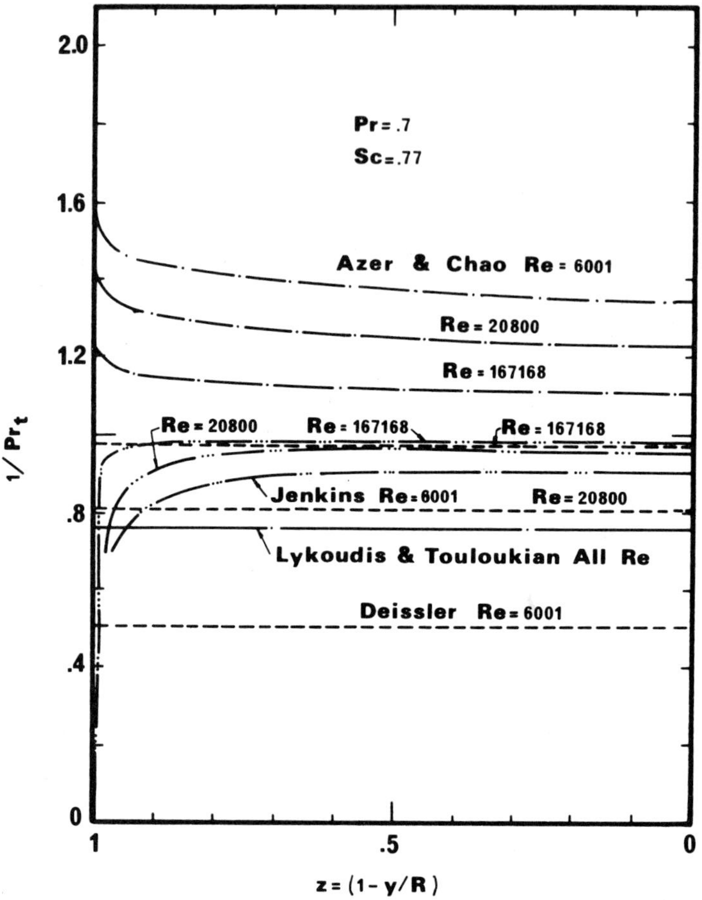

Figure 5 Ratio of mass and heat diffusivities to that of momentum. Comparison of various theories.

Van Driest (1956) modification of the mixing-length theory, which has proved extremely useful.

According to Van Driest, the turbulent velocity fluctuations are damped by the viscous forces as the wall is approached. Considering the damping action to be similar to the damping of laminar stokes type flow, an infinite plate that executes oscillations parallel to itself, the turbulent eddy viscosity is given by Eq. (2.52), rewritten here as

$$\nu_t = (\kappa y)^2 \left[1 - \exp\left(-\frac{y^+}{A^+}\right)\right]^2 \frac{d\bar{u}}{dy} \tag{2.66}$$

Adopting Van Driest's procedure to heat transfer, Cebeci writes for the turbulent temperature fluctuations

$$T' = T_0'\left[1 - \exp\left(-\frac{y^+}{B^+}\right)\right] \tag{2.67}$$

where T_0' is the temperature fluctuation unaffected by the molecular conductivity, and B^+ is the distance from the wall, beyond which the influence of molecular conductivity is negligible. Unlike A^+ (= 26), B^+ is a function of the molecular Prandtl number, the effect of the molecular conductivity is felt further from the wall as the Prandtl number decreases. Thus,

$$B^+ = B^+(\mathrm{Pr}) \tag{2.68}$$

Introducing Eqs. (2.47), (2.66), and (2.67) into the Reynolds heat flux term, gives

$$-\rho C_p \overline{T'v'} = -\rho C_p T_0' v_0' \left[1 - \exp\left(-\frac{y^+}{A^+}\right)\right]\left[1 - \exp\left(-\frac{y^+}{B^+}\right)\right] \tag{2.69}$$

According to the mixing-length concept (see Section 2.2.3)

$$-\rho C_p \overline{T'v'} = l_m l_H \frac{d\bar{u}}{dy}\frac{d\bar{T}}{dy} \tag{2.70}$$

where

$$l_H = \kappa_H y \qquad \text{and} \qquad \kappa_H = \text{constant of proportionality}$$

Combining Eq. (2.69) and Eq. (2.70) with the definition of the eddy conductivity, Eq. (2.19) gives

$$\alpha_t = \kappa_H \kappa y^2 \left[1 - \exp\left(-\frac{y^+}{A^+}\right)\right]\left[1 - \exp\left(-\frac{y^+}{B^+}\right)\frac{d\bar{u}}{dy}\right] \tag{2.71}$$

Dividing Eq. (2.66) by Eq. (2.71), and taking $\kappa/\kappa_H \cong 1$, yields (Cebeci, 1973):

$$\mathrm{Pr}_t = \frac{1 - \exp(-y^+/A^+)}{1 - \exp(-y^+/B^+)} \tag{2.72}$$

At the wall itself, $y^+ = 0$, and Eq. (2.72) reduces to

$$\mathrm{Pr}_t = \frac{B^+}{A^+} = \frac{B^+(\mathrm{Pr})}{26} \tag{2.73}$$

Before Eq. (2.72) may be used to predict values of Pr_t, the functional form of the parameter B^+ (Pr) must be specified. Cebeci (1973) suggested B^+ (Pr) may be calculated by Eq. (2.73) from experimental data for Pr_t at the wall. Following this procedure, Na and Habib (1973) proposed the expression

$$B^+ = \frac{1}{(\mathrm{Pr})^{1/2}} \sum_{i=1}^{5} C_i (\log \mathrm{Pr})^{i-1} \tag{2.74}$$

where

$$C_1 = 34.96 \qquad C_4 = 6.33$$
$$C_2 = 28.97 \qquad C_5 = -1.186$$
$$C_3 = 33.95$$

provided the best fit of the available experimental data. The predicted values of Pr_t are presented in Figure 6.

Tyldesley and Silver's model Tyldesley and Silver (1968) abandoned the concept of a mixing-length and assumed that the detailed fluid behavior may be modeled by considering it to consist of a simple ensemble of "entities" or eddies

Figure 6 Turbulent Prandtl number calculated by Cebecci's model (1973).

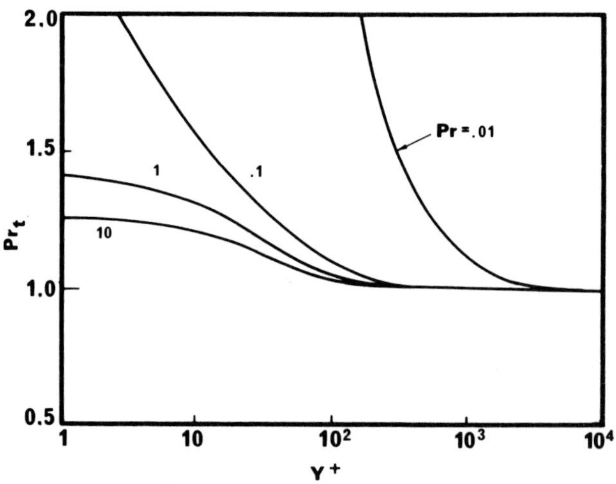

of varying size, shape, velocity, and temperature. The analysis was divided into two parts: first, valid for low turbulent intensities, i.e., low Re; second, valid for high turbulent intensities, i.e., high Re.

At low turbulent intensity levels, the transfer of momentum and energy from an entity along its random trajectory is assumed to occur by a purely molecular mechanism in Stokes flow.

The average behavior of an entity, assumed to be an elliptically shaped rigid body, is obtained by averaging the laws of conservation of momentum and energy over the ensemble. Assuming the bulk velocity is steady and unidirectional, with constant velocity and temperature gradients, and a completely random entity formation process, the analysis yields

$$\mathrm{Pr}_t = \frac{\nu_t}{\alpha_t} = \frac{2 + 6\mathrm{Pr}}{9\mathrm{Pr}} \tag{2.75}$$

For high turbulence intensities, Tyldesley and Silver suggest that the dominant mechanism of transport is no longer purely molecular, but that penetration of smaller entities into larger ones becomes increasingly important as the intensity increases. At sufficiently high turbulence intensities the penetration effect becomes dominant. The equations describing this mechanism have exactly the same form as those developed for the molecular case, with the turbulent transfer coefficients ν_t^* and α_t^* now replacing the corresponding molecular coefficients ν and α. The corresponding turbulent Prandtl number then becomes

$$\mathrm{Pr}_t = \frac{\nu_t^*}{\alpha_t^*} = \frac{2 + 9\mathrm{Pr}}{3 + 9\mathrm{Pr}} \tag{2.76}$$

The predictions of Eqs. (2.75) and (2.76), together with some experimental data, are shown in Figure 7. The most striking feature of the model is that the predicted trend for the dependence of Pr_t on Re is consistent with experimental data for both low and high Prandtl numbers. (Sleicher et al., and Sleicher, 1958, Figure 4). Figure 8 shows the effect of increasing turbulence intensity on Pr_t for both high and low Pr. The trend towards a constant value of Pr_t at high turbulence intensities is consistent with Deissler's (1963). Although successful in predicting overall trends, the model does not account for the expected variation of Pr_t with the distance from the wall.

Random mixing layer model Marchello and Toor (1963) assume that the low level of turbulence near a boundary causes random localized mixing. The fluid in the vicinity of an interface is visualized as made up of layers of some thickness L. The steady, fully developed, transport is interrupted at random times by local mixing within one layer or another. In the interval between these mixings, the transport between the freshly mixed layer and its adjacent neighbors, is by unsteady molecular diffusion.

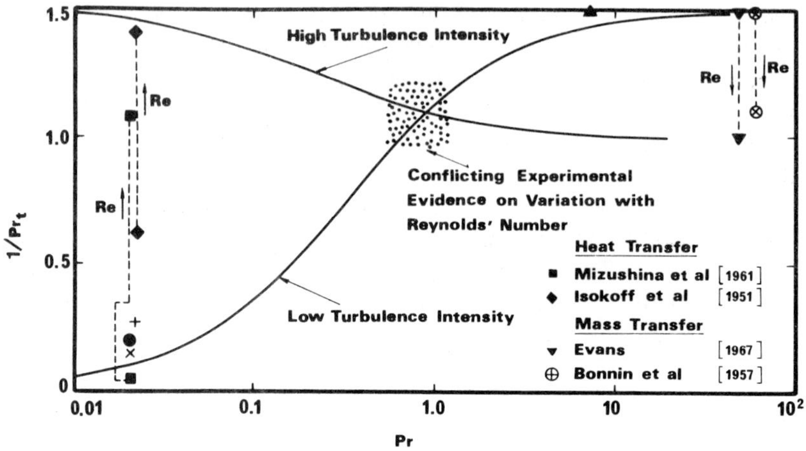

Figure 7 Predicted and experimental results for the effect of turbulence intensity on the ratios of turbulent diffusivities. (After Tyldesley and Silver, 1968).

The eddy thermal diffusivity is evaluated by equating the expression derived for the flux by unsteady diffusion through a finite thickness fluid layer (see Section 3.2.4) with Boussinesq's type expression (Eq. [2.20])

$$\frac{q}{\rho Cp} = (\alpha + \alpha_t)\,\frac{\Delta \overline{T}}{L} \tag{2.77}$$

Figure 8 Variation in diffusivity ratio with increasing turbulence intensity. (After Tyldesley and Silver, 1968).

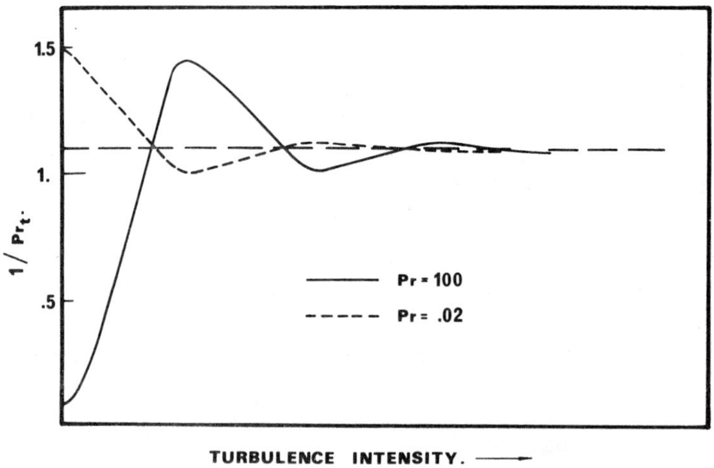

where $\Delta\bar{T}/L$ is the space average temperature gradient over the mixing layer. This procedure gives the eddy diffusivity as a function of the dimensionless group SL^2/α, or

$$\frac{\alpha_t}{\alpha} = g_1 \sqrt{\frac{SL^2}{\alpha}} \tag{2.78}$$

where S is the mean frequency of mixing. A similar procedure leads to

$$\frac{\nu_t}{\nu} = g_1 \sqrt{\frac{SL^2}{\nu}} \tag{2.79}$$

The turbulent Prandtl number is obtained by eliminating (SL^2) from Eq. (2.78) and Eq. (2.79). The limiting values of Pr_t depend on the turbulence intensity.

It is interesting that at low turbulence intensity (small SL^2/α or SL^2/ν), Eqs. (2.78) and (2.79) predict that α_t or ν_t are independent of their molecular counterparts α, or ν, and $\mathrm{Pr}_t = 1$. At high turbulence intentisy, α_t or ν_t, are proportional to $\sqrt{\alpha}$ or $\sqrt{\nu}$, and $\mathrm{Pr}_t \cong \sqrt{\mathrm{Pr}}$.

As seen by inspection of Table 4, these conclusions are essentially contrary to most other studies. For instance, Jenkins' (1951) models predicts $\mathrm{Pr}_t \to 1$ for high ν_t, and $\mathrm{Pr}_t \to 1/\mathrm{Pr}$ for low turbulent intensities. As pointed out by Marchello and Toor, the model is satisfactory at high Prandtl numbers. In this case the transfer rate is controlled by the thin thermal layer at the wall, which is affected only by low intensity turbulence, and $\mathrm{Pr}_t \cong 1$ is probably the better representation in this region.

A model based on the surface renewal concept Thomas and Rajagopal (1973) applied Danckwerts' concept of surface renewal (see Section 3.2 for a detailed discussion of this concept) to calculate the mean rate of energy transport from the wall, i.e., the Reynolds transport flux $-\rho C_p \overline{T'v'}$ at a given y-plane, and the mean temperature profile in the wall region. This information was then used to evaluate the eddy conductivity by utilizing Eq. (2.19).

Based on the surface renewal model, and using Danckwerts' random contact time distribution of the eddies at the wall, the mean temperature profile within the wall region is written as

$$\bar{T} = (T_0 - T_b) \exp\left(-\frac{y}{\sqrt{\alpha\tau}}\right) + T_b \tag{2.80}$$

where T_0 and T_b are the wall and bulk stream temperatures, respectively, and τ is the mean contact time.

The heat flux from the surface is equal to the heat accumulated in the eddy during its residence time τ. Thus the mean flux per unit time, i.e., the Reynolds heat flux, $-\rho C_p \overline{T'v'}$, is simply

$$-\rho C_p \overline{T'v'} = \frac{1}{\tau} \int_0^y \rho C_p (T - T_b)\, dy \tag{2.81}$$

Combining Eqs. (2.80) and (2.81), and carrying out the integration, gives

$$-\rho C_p \overline{T'v'} = \rho C_p (T_0 - T_b) \sqrt{\frac{\alpha}{\tau}} \left[1 - \exp\left(- \frac{y}{\sqrt{\alpha \tau}} \right) \right] \tag{2.82}$$

Differentiating Eq. (2.80) yields

$$\frac{\partial \bar{T}}{\partial y} = - \frac{1}{\sqrt{\alpha \tau}} (T_0 - T_b) \exp\left(- \frac{y}{\sqrt{\alpha \tau}} \right) \tag{2.83}$$

Introducing Eqs. (2.82) and (2.83) into the definition of the turbulent diffusivity

$$\alpha_t = - \frac{\rho C_p \overline{T'v'}}{\partial \bar{T} / \partial y}$$

yields

$$\frac{\alpha_t}{\nu} = \frac{1}{\mathrm{Pr}} \left[\exp\left(\frac{y}{\sqrt{\alpha \tau}} \right) - 1 \right] \tag{2.84}$$

An analogous formulation for momentum transfer leads to expression for ν_t / ν, i.e.,

$$\frac{\nu_t}{\nu} = \left[\exp\left(\frac{y}{\sqrt{\nu \tau}} \right) - 1 \right] \tag{2.85}$$

Dividing Eq. (2.84) by Eq. (2.85) yields

$$\mathrm{Pr}_t = \mathrm{Pr}\,\frac{\exp\left(\dfrac{y}{\sqrt{\nu\tau}}\right) - 1}{\exp\left(\dfrac{y}{\sqrt{\alpha\tau}}\right) - 1} \tag{2.86}$$

As shown below in Section 3.3.5, the mean contact time of an eddy with the transport surface can be evaluated by utilizing the surface renewal model to evaluate the shear at the wall. The result is given by

$$u_* \sqrt{\frac{\tau}{\nu}} = \sqrt{\frac{2}{f}} \tag{2.87}$$

where $u^* = \sqrt{\sigma_w/\rho}$ is the friction velocity, and f is the friction factor. Introducing τ from Eq. (2.87) into Eq. (2.86) gives

$$\mathrm{Pr}_t = \mathrm{Pr}\,\frac{\exp\left(y^+ \sqrt{\dfrac{f}{2}}\right) - 1}{\exp\left(\sqrt{\mathrm{Pr}}\,y^+ \sqrt{\dfrac{f}{2}}\right) - 1} \tag{2.88}$$

It is important to note that although the previous models discussed here, in principle at least, were valid for all Prandtl numbers; Eq. (2.88) is valid only for $\mathrm{Pr} \geqslant 1$. This restriction arises from assumption implicit in Eq. (2.80), that penetration of the thermal boundary layer is always within the "eddy", i.e., smaller than the penetration of the momentum boundary layer. Equation (2.88) has a singularity at the wall (i.e., $y^+ = 0$), and the value of Pr_t at the wall is thus indeterminant. Moreover, as $y^+ \to 0$, $\mathrm{Pr}_t \to \sqrt{\mathrm{Pr}}$ which is not consistent with the most recently reported experimental data of Quarmby and Quirk (1974).

Since the model of Thomas and Rajagopol is based on the surface renewal theory, it may be concluded that it is valid only for the conditions for which the simple renewal theory is valid. Pinczewski and Sideman (1974) have recently demonstrated that the simple renewal theory is valid only for $\mathrm{Pr} = 0(1)$. For $\mathrm{Pr} \cong 1$, Eq. (2.88) predicts that $\mathrm{Pr}_t \cong 1$, which is consistent with the bulk of published data. In view of its very limited range of validity, the model in its present form is of little predictive value.

Empirical correlations for $\mathbf{Pr}_t$ Several authors have presented purely empirical correlations for Pr_t.

Goldman and Marchello (1969) correlated their data ($0.22 < \mathrm{Sc} < 2.56$) using the expression

$$\mathrm{Pr}_t = 1 + \frac{\nu_t}{\nu}\frac{1}{\mathrm{Sc} + 2.93} - 0.254 \exp\left[-\left(\frac{\nu_t}{\nu}\right)(0.015\,\mathrm{Sc} + 0.00769)\right] \tag{2.89}$$

Equation (2.89) was reported to provide a better fit to the data than the expression due to Azer and Chao (1960).

Notter and Sleicher (1972) have constructed an empirical equation which fits their experimental data at Pr = 0.02 − 0.025 and 0.72. Their equation is

$$\frac{1}{\mathrm{Pr}_t} = \frac{0.025\,\mathrm{Pr}\left(\dfrac{\nu_t}{\nu}\right) + 90\,\mathrm{Pr}^{2/3}\left(\dfrac{\nu_t}{\nu}\right)^{1/4}}{1 + 90\,\mathrm{Pr}^{2/3}\left(\dfrac{\nu_t}{\nu}\right)^{1/4}}\left[1 + \frac{10}{35 + \left(\dfrac{\nu_t}{\nu}\right)}\right] \tag{2.90}$$

More recently, Quarmby and Quirk (1974), have found that their data for Sc = 0.7, Pr = 0.77, and 725 < Sc < 790, may be represented by a simple function of the distance from the wall

$$\frac{1}{\mathrm{Pr}_t} = 1 + 400^{(Z-1)} \tag{2.91}$$

where $Z = \left(1 - \dfrac{y}{R}\right)$

General comments In view of the inconclusive nature of the existing experimental data it is, at present, not possible to confidently recommend any particular correlation or model. Nevertheless, a few observations are instructive. As recently shown by Shlanchyauskas et al. (1974), the difference in the *calculated* amount of heat transferred, based on $\mathrm{Pr}_t = 0.5$ and $\mathrm{Pr}_t = 1.0$, decreases from 33% at Pr = 10 to 28% at Pr = 100. Hence, the sensitivity to the assumed value of Pr_t reduces as Pr increases. More important still, the effect of the variation (if it exists) of Pr_t across the boundary layer decreases as Pr increases. While at Pr = 1, the resistance to heat transfer is uniformly distributed over the thickness of the boundary layer, the primary resistance is concentrated near the wall at larger values of the Prandtl number, and the value of Pr_t at the wall determines the transfer rate. It is therefore not surprising that constant values of Pr_t usually suffice for engineering calculations, and the simple Reynolds analogy $\mathrm{Pr}_t = 1$ is chosen by many workers for lack of confidence in other values.

2.6 Simple analogy theories for convective transport

2.6.1 Introduction The basic problem of turbulent transport was formulated in Section 2.1. It was shown that closure of the resulting system of Reynolds equations requires a knowledge of the turbulent viscosity and the turbulent Prandtl (Schmidt) number. This information has been introduced above, and it is now possible to proceed with the actual solution of the turbulent transport problem. This is done in Section 2.7.

In this section we explore a class of solutions commonly referred to as the simple analogy theories. These theories start with Eq. (2.20) for the momentum, heat, and mass fluxes, assuming the flow variables to be functions of the transverse coordinate only, and the mean transverse velocity, v, to be zero (the Couette flow assumption). The analogies start by relating the fluxes to each other. Here we shall consider only momentum and energy fluxes. Analogous relationships can be written for the case of mass transfer, provided the mass flux does not affect the flow field.

The assumptions is often made that shear stress and heat flux vary linearly with respect to y, i.e.,

$$\frac{\sigma}{\sigma_w} = \frac{q}{q_w} = Z \qquad Z = \left(1 - \frac{y}{R}\right) \tag{2.92}$$

while others, following Reynolds, simply assume that

$$\frac{\sigma}{\sigma_w} = \frac{q}{q_w} = 1 \tag{2.93}$$

With reference to Eq. (2.92), we note that $q/q_w = Z$ is strictly valid for the ideal flow, or plug flow case. The relation $\sigma/\sigma_w = Z$ is valid only for fully developed flow with constant physical properties. Obviously, Eq. (2.93) is not satisfied for pipe flow. However, the introduction of either assumptions, Eq. (2.92) or Eq. (2.93), does not introduce great errors into the solutions for many practical cases. Deissler and Eian (1952) demonstrated that velocity, temperature, and concentration profiles are relatively insensitive to shear stress distribution for fully developed turbulent flow in a tube. More recently, Hatton (1964) and Webb (1971), reached the same conclusions for turbulent boundary layer and tube flow, respectively. This is consistent with the fact that in calculating heat transfer rates, the region near the wall is of the most essential consideration, and in this region both assumptions, Eq. (2.92) and Eq. (2.93) are approximately valid.

For a Couette flow, the momentum and energy equations for fully developed turbulent flow (Eqs. [2.6]-[2.8]) reduce to

$$\frac{\sigma}{\sigma_w} = \left(1 + \frac{\nu_t}{\nu}\right)\frac{du^+}{dy^+} \tag{2.94}$$

$$\frac{q}{q_w} = \left(\frac{1}{\mathrm{Pr}} + \frac{1}{\mathrm{Pr}_t}\frac{\nu_t}{\nu}\right)\frac{dT^+}{dy^+} \tag{2.95}$$

$$\text{where } T^+ = (T_w - T)\,\frac{C_p}{u^*}\,\frac{\sigma_w}{q_w} \tag{2.96}$$

Equations (2.94) and (2.95) may be integrated to yield the velocity and temperature distributions in the turbulent flow field, i.e.,

$$u^+ = \int_0^{y^+} \frac{\dfrac{\sigma}{\sigma_w}}{1 + \dfrac{\nu_t}{\nu}}\, dy^+ \tag{2.97}$$

$$T^+ = \int_0^{y^+} \frac{\dfrac{q}{q_w}}{\dfrac{1}{\mathrm{Pr}} + \dfrac{1}{\mathrm{Pr}_t}\dfrac{\nu_t}{\nu}}\, dy^+ \tag{2.98}$$

Assuming $\mathrm{Pr}_t = 1$, and $\sigma/\sigma_w = q/q_w = 1$, Eqs. (2.97) and (2.98) become

$$u^+ = \int_0^{y^+} \frac{dy^+}{1 + \dfrac{\nu_t}{\nu}} \tag{2.99}$$

$$T^+ = \int_0^{y^+} \frac{dy^+}{\dfrac{1}{\mathrm{Pr}} + \dfrac{\nu_t}{\nu}} \tag{2.100}$$

Taking the upper bound of the integral in Eq. (2.100) as $y = \delta > \delta_T$, where δ_T is the thickness of the thermal boundary layer ($T = T_b$) gives

$$T_b = \int_0^{\delta^+} \frac{dy^+}{\dfrac{1}{\mathrm{Pr}} + \dfrac{\nu_t}{\nu}} \tag{2.101}$$

Combining Eq. (2.101) with the definition of T^+ Eq. (2.98) gives

$$\frac{\rho C_p(T_w - T_b)u^*}{q_w} = \int_0^{\delta^+} \frac{dy^+}{\frac{1}{\text{Pr}} + \frac{\nu_t}{\nu}} \tag{2.102}$$

or

$$\text{Nu} = \frac{\sqrt{\frac{f}{2}}\,\text{Re Pr}}{\int_0^{\delta^+} \frac{dy^+}{\frac{1}{\text{Pr}} + \frac{\nu_t}{\nu}}} \qquad \delta^+ > \delta_m^+ \text{ or } \delta_M^+ \tag{2.103}$$

where δ_M is the thickness of the momentum boundary layer

$$q_w = h(T_w - T_b), \text{Nu} = \frac{h2R}{k} \text{ and Re} = \frac{u_b 2R}{\nu}$$

Equation (2.103) is the basic relationship for the simple analogy theories. The evaluation of the integral in Eq. (2.103) requires the introduction of an expression for the turbulent kinematic viscosity, ν_t, as a function of y^+. The substitution of the various expressions for ν_t/ν discussed in Section 2.3, lead to the various analogy theories for heat and mass transfer proposed in the literature. A detailed discussion of numerous analogies is found in the texts of Hinze (1959), Schlicting (1960), Kays (1966), Knudsen and Katz (1958), and the reviews of Sherwood (1959), and Jayatilleke (1969).

2.6.2 The Reynolds analogy The simplest relationship for the Nusselt number was proposed by Reynolds (1874), who assumed that the processes of heat and momentum transfer were completely analogous (see Section 2.4). The implication of this assumption is

$$\text{Pr} = 1 \quad \text{and} \quad \delta_M = \delta_T \tag{2.104}$$

The integral term in Eq. (2.103) is simply the velocity $u^+ = \delta_T^+ = \delta_M^+$, (Eq. [2.99]). As the velocity at δ_M is equal to the bulk velocity, u_b and $u_b/u^* = \sqrt{2/f}$, Eq. (2.103) reduces to

$$\text{Nu} = \frac{\sqrt{\frac{f}{2}}\,\text{Re}}{\frac{u_b}{u^*}} = \frac{f}{2}\,\text{Re} \tag{2.105}$$

which is the well-known Reynolds analogy.

It is apparent from the derivation of Eq. (2.105), that this analogy is quite limited and attempts have been made to extend it.

The Chilton-Colburn correlation A well known extension is the correlation of Chilton and Colburn (1934). The correlation, based on an empirical modification of Eq. (2.105) for the effect of Pr, gives

$$\mathrm{Nu} = \frac{f}{2} \, \mathrm{Re} \, \mathrm{Pr}^{1/3} \tag{2.106}$$

Sherwood and Pigford (1952) have shown Eq. (2.106) to be in good agreement with a wide range of experimental data.

Other extensions of the simple Reynolds analogy, based on various plausible rationals, are discussed below.

2.6.3 Extensions of the Reynolds analogy For Prandtl numbers other than unity, the integral in Eq. (2.103) is broken into a number of parts, depending on the number of regions the velocity field is divided. Following Rannie (1956), we break the integral into two parts and Eq. (2.103) becomes

$$\mathrm{Nu} = \frac{\sqrt{\dfrac{f}{2}} \, \mathrm{Re} \, \mathrm{Pr}}{\displaystyle\int_0^{y_1^+} \frac{dy^+}{\dfrac{1}{\mathrm{Pr}} + \dfrac{\nu_t}{\nu}} + \int_{y_1^+}^{\delta^+} \frac{dy^+}{\dfrac{\nu_t}{\nu}}} \tag{2.107}$$

y_1^+ is chosen to be sufficiently large so that unity, or $1/\mathrm{Pr}$, whichever is larger, becomes negligible compared with the term ν_t/ν. $y^+ = y_1^+$ may be interpreted as being the distance from the wall, beyond which molecular effects (α or ν) become negligible. For $\mathrm{Pr} > 1$, ν is greater than α, and y_1^+ is simply the distance from the wall where ν becomes negligible, or $\nu \ll \nu_t$. For $\mathrm{Pr} < 1$, α is greater than ν, and y_1^+ may be expected to be a function of Pr, since the effect of α will be felt further from the wall as the value of Pr decreases (α increases). The upper limit, δ^+, for the second integral will always correspond to some point in the flow field where $u = u_b$, the bulk velocity of the flow field. Thus for $\mathrm{Pr} > 1$, $\delta^+ = \delta_M^+$, while for $\mathrm{Pr} < 1$, $\delta^+ = \delta_T^+$.

From Eq. (2.99) the bulk stream velocity may be written as

$$u_b^+ = \int_0^{y_1^+} \frac{dy^+}{1 + \dfrac{\nu t}{\nu}} + \int_{y_1^+}^{\delta^+} \frac{dy^+}{\dfrac{\nu t}{\nu}} \tag{2.108}$$

Adding and subtracting Eq. (2.108) to the denominator of Eq. (2.107), and noting that $u_b^+ = \sqrt{2/f}$, we have

$$\mathrm{Nu} = \frac{\sqrt{\frac{f}{2}}\,\mathrm{Re}\,\mathrm{Pr}}{\sqrt{\frac{2}{f}} + \int_0^{y_1^+} \left(\frac{1}{\frac{1}{\mathrm{Pr}} + \frac{\nu_t}{\nu}} - \frac{1}{1 + \frac{\nu_t}{\nu}} \right) dy^+} \tag{2.109}$$

The Taylor-Prandtl analogy Taylor (1916), and Prandtl (1925), provided the simplest extension of the Reynolds analogy. They assumed that in the region adjacent to the wall ($0 < y^+ < y_1^+$) conditions were purely laminar. Performing the integration in Eq. (2.109) with ν_t/ν set to zero gives

$$\mathrm{Nu} = \frac{\sqrt{\frac{f}{2}}\,\mathrm{Re}\,\mathrm{Pr}}{\sqrt{\frac{2}{f}} + (\mathrm{Pr} - 1)u_1^+} \tag{2.110}$$

where u_1^+ corresponds to the velocity at $y^+ = y_1^+$. Prandtl (1928) chose y_1^+ empirically as 8.7. Utilizing the following relationship for the friction factor

$$\frac{f}{2} = 0.04\,\mathrm{Re}^{-1/4}$$

Equation (2.110) reduces to

$$\mathrm{Nu} = \frac{\mathrm{Re}\,\mathrm{Pr}}{25\,\mathrm{Re}^{1/4}\,[1 + 1.74\,\mathrm{Re}^{-1/8}(\mathrm{Pr} - 1)]} \tag{2.111}$$

Equation (2.111) agrees well with heat transfer data for Pr = 0(1). At very low Prandtl numbers the derivation is in serious error because the thermal boundary layer is much greater than $y^+ = 8.7$. At very high Prandtl numbers the assumption of no turbulence in the layer adjacent to the wall constitutes a considerable oversimplification, (see Section 3.3) and the agreement of Eq. (2.111) with the data is quite poor.

Von Karman's analogy Von Karman (1939) extended the Prandtl-Taylor analysis by further dividing the region $0 < y^+ < y_1^+$. Simple empirical expressions relating u^+ and y^+ in the region $0 < y^+ < 30$ were differentiated to provide the following relevant expressions for ν_t/ν (see Table 3).

$$\frac{\nu}{\nu_t} = 0 \qquad 0 < y^+ < 5$$
$$\frac{\nu}{\nu_t} = 0.2y^+ - 1 \qquad 5 < y^+ < 30 \tag{2.112}$$

Carrying out the integration indicated in Eq. (2.107) leads to

$$\mathrm{Nu} = \frac{\sqrt{\frac{f}{2}}\,\mathrm{Re\,Pr}}{\sqrt{\frac{2}{f}} + 5(\mathrm{Pr} - 1) + \ln\left[1 + \frac{5}{6}(\mathrm{Pr} - 1)\right]} \tag{2.113}$$

Von Karman found Eq. (2.113) to be in good agreement with experimental data for $0.73 < \mathrm{Pr} < 40$. The improvement over the Prandtl-Taylor treatment for $\mathrm{Pr} > 1$ is due to the allowance made for the existence of turbulence down to $y^+ = 5$, rather than $y^+ = 8.7$, as taken in the development of Eq. (2.111). The Von Karman treatment is subject to the same limitations as the earlier Prandtl-Taylor treatment for $\mathrm{Pr} \ll 1$.

Martinelli's analogy The major limitation of the treatments of Prandtl-Taylor and Von Karman of heat transfer at very low values of the Prandtl number relates to specifying the upper limit y_1^+ of the integration in Eq. (2.107). Martinelli (1947) overcame this problem by specifying the distribution of ν_t/ν over the complete flow field. For the range $0 < y^+ < 30$, the treatment is similar to that of Von Karman, Eqs. (2.112) and (2.115). In the region $y^+ > 30$, ν_t/ν was obtained by utilizing the universal logarithmic velocity profile

$$u^+ = 5.5 + 2.5 \ln y^+ \tag{2.114}$$

and the assumption that σ/σ_w and q/q_w are proportional to the tube radius, Eq. (2.92). Integration of Eq. (2.98) between the tube wall and the tube axis resulted in values of Nu which were shown to agree well with data for heat transfer to mercury ($\mathrm{Pr} = 0.024$). A comparison with other studies is presented in Figure 9.

Levich's analogy, **Pr** $\ll$ *1* An alternative approach to the problem of heat transfer at low Prandtl numbers was proposed by Levich (1962). Following Prandtl, he assumes $\nu_t = 0$ in the region $0 < y^+ < y_1^+$, and that turbulent effects predominate in the range $y_1^+ < y^+ < \delta^+$. Whereas Prandtl-Taylor assume that y_1^+ is some constant, Levich suggests that y_1^+ may be estimated by the point at which the thermal diffusivity α, and turbulent thermal diffusivity α_t are equal, or, with $\mathrm{Pr}_t = 1$, at the point where $\alpha = \nu_t$. For the fully turbulent region, Eq. (2.43) with $\kappa = 0.4$ yields

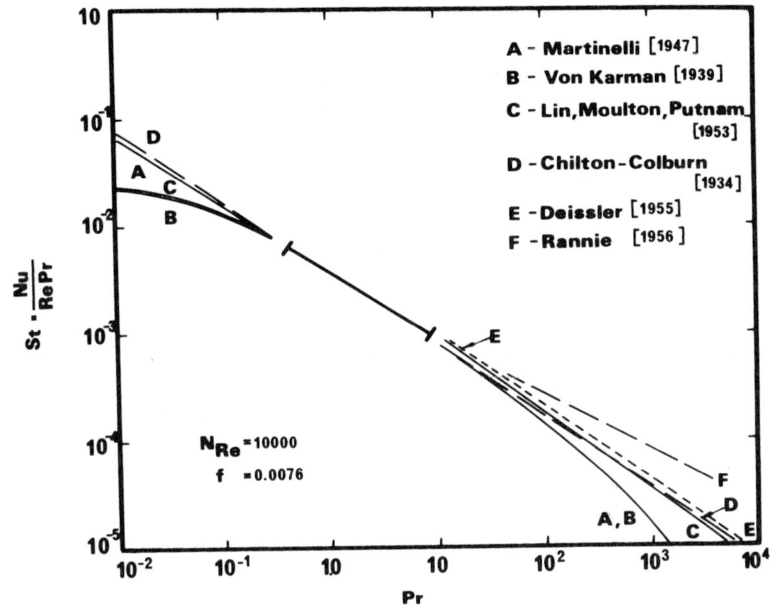

Figure 9 Comparison of Stanton numbers calculated from several analogies. (After Sherwood, 1959)

$$\frac{\nu_t}{\nu} = 0.4y^+ \tag{2.115}$$

and ν_t decreases as the wall is approached. It follows that at some point $y^+ = y_1^+$, ν_t will reach the value of α. Although the division of the flow field into two sharply defines zones at $y^+ = y_1^+$ is somewhat unrealistic, the treatment amplifies the important effect of increasing molecular influence with distance from the wall as the molecular diffusivity increases (or Pr decreases).

Introducing $\alpha = \nu_t$ at $y^+ = y_1^+$ into Eq. (2.115) yields

$$\frac{\alpha}{\nu} = 0.4y_1^+ \quad \text{or} \quad y_1^+ = \frac{1}{0.4\,\text{Pr}} \tag{2.116}$$

Integrating Eq. (2.107) with y_1^+ given by Eq. (2.116) and $\delta^+ = R^+$, the dimensionless tube radius, yields

$$\text{Nu} = \frac{0.4\sqrt{\frac{f}{2}}\,\text{Re Pr}}{\ln \text{Re} + \ln \text{Pr} + \ln\sqrt{\frac{f}{2}} + (1 + \ln 0.2)} \tag{2.117}$$

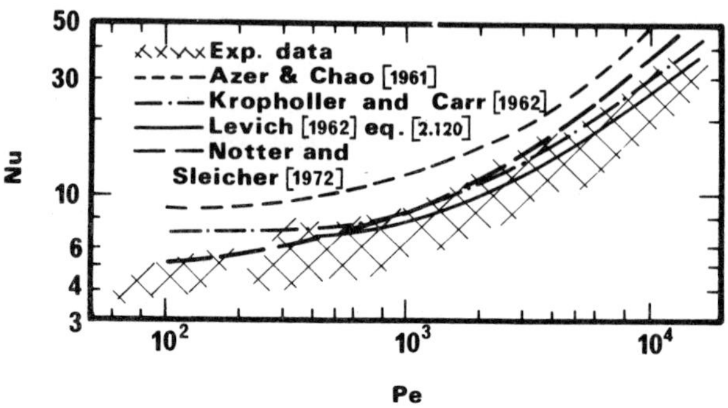

Figure 10 Comparison of analogy calculations with experimental data for heat transfer to liquid metals.

Levich shows that Eq. (2.117) is in good agreement with experimental data on heat transfer to liquid metals, Figure 10. It is noted that no arbitrary constants were introduced into the analysis. It is interesting that simple division of the flow field into two sharply defined regions describes the characteristic peculiarities of the low Pr turbulent heat transport with sufficient accuracy.

Murphee and Lin, Moulton, and Putnam's analogy Lin, Moulton, and Putnam (1953) extended the Prandtl-Taylor and Von Karman analyses to higher Prandtl numbers by assuming (following Murphee, 1932) that the turbulence persists down to the wall itself. The assumed distributions for the turbulent viscosity were

$$\frac{\nu_t}{\nu} = \left(\frac{y^+}{14.5}\right)^3 \qquad 0 < y^+ < 5 \tag{2.118}$$

$$\frac{\nu_t}{\nu} = \frac{y^+}{5} - 0.949 \qquad 5 < y^+ < 33 \tag{2.119}$$

Substituting Eqs. (2.118) and (2.119) into Eq. (2.107), and performing the required integration, results in an expression of the form

$$\mathrm{Nu} = \frac{\sqrt{\dfrac{f}{2}}\,\mathrm{Re}\,\mathrm{Pr}}{\sqrt{\dfrac{2}{f}} + F(\mathrm{Pr})} \tag{2.120}$$

where $F(\mathrm{Pr})$ is a rather complicated function of the Prandtl number. Equation (2.120) was found to agree well with data over a wide range of Prandtl and Schmidt numbers (0.5 to 3,200).

Levich's model—$\Pr \gg 1$ Noting that absolute values of D are usually much smaller than α, Levich (1962) approached the problem mass transport at high Schmidt numbers by dividing the flow field into three zones, Figure 11.

1. a fully turbulent zone,

$$\delta_0 < y < y_b \qquad \text{where} \qquad \frac{\nu_t}{\nu} = 0.4y^+ \tag{2.121}$$

2. a damped turbulent zone,

$$\delta_1 < y < \delta_0 \qquad \text{where} \qquad \frac{\nu_t}{\nu} = b(y^+)^3 \tag{2.122}$$

3. a laminar wall zone,

$$0 < y < \delta_1 \qquad \text{where} \qquad \frac{\nu_t}{\nu} = 0 \tag{2.123}$$

The fully turbulent zone was assumed to extend to some level, $y = \delta_0$, where the turbulent viscosity, ν_t, was equal to the value of its molecular counterpart, ν. From Eq. (2.115), or Eq. (2.121), it follows that

$$y^+ = \delta_0^+ = \frac{1}{0.4} \tag{2.124}$$

Figure 11 Levich's (1962) model for mass transfer at high Schmidt number.

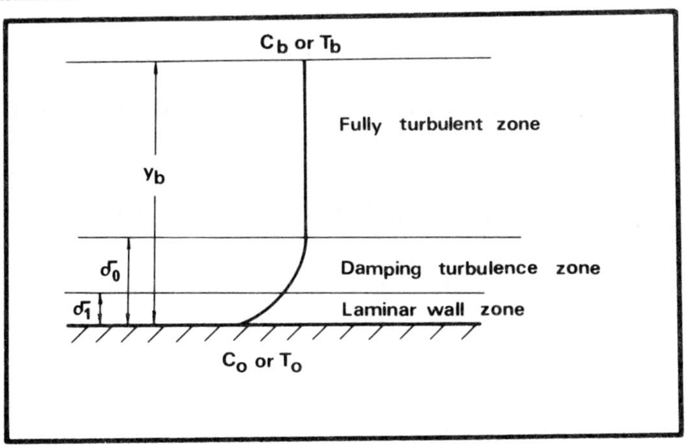

The laminar wall zone was assumed to extend to some distance, $y = \delta_1$, where the turbulent diffusivity, D_t (or ν_t, once we assume that $\mathrm{Sc}_t = 1$) is equal to the molecular diffusivity, D. From Eq. (2.122) we have, by substituting $\nu_t = D$ at $y^+ = \delta^+$, that

$$\delta_1^+ = \frac{1}{b\,\mathrm{Pr}^{1/3}} \tag{2.125}$$

The relevant Sherwood number may now be calculated by integrating Eq. (2.103) in three parts between the wall and some characteristic distance, y_b, which is located within the turbulent zone, say the center of a tube.

$$\mathrm{Nu} = \frac{\sqrt{\dfrac{f}{2}}\ \mathrm{Re\,Pr}}{\displaystyle\int_0^{\delta_1^+} \frac{dy^+}{\dfrac{1}{\mathrm{Pr}}} + \int_{\delta_1^+}^{\delta_0^+} \frac{dy^+}{b(y^+)^3} + \int_{\delta_0^+}^{y_b^+} \frac{dy^+}{0.4y^+}} \tag{2.126}$$

Integrating Eq. (2.126) within the limits given by Eqs. (2.124) and (2.125) for δ_1^+ and δ_0^+, results in

$$\mathrm{Nu} = \frac{0.4\sqrt{\dfrac{f}{2}}\ \mathrm{Re\,Pr}}{0.4\left(\dfrac{b^2+2}{2b}\right)\mathrm{Pr}^{2/3} + \ln \mathrm{Re} + \ln\left(0.4\sqrt{\dfrac{f}{2}}\right) - \dfrac{0.032}{b}} \tag{2.127}$$

where Nu stands for the Sherwood number ($= Ky_b/D$), $\mathrm{Re} = y_b U_0/\nu$ and b is a constant which may be evaluated from a comparison of Eq. (2.127) with experimental data.

For $\mathrm{Sc} \gg 1$, the last three terms in the denominator of Eq. (2.127) are small when compared with the leading term, and the equation reduces to

$$\mathrm{Nu} = \left(\frac{2b}{b^2+2}\right)\sqrt{\frac{f}{2}}\ \mathrm{Re\,Pr}^{1/3} \tag{2.128}$$

which is identical to the high Schmidt number asymptote of the treatment of Lin, Moulton, and Putnam (1953) (see below).

Analogies based on power law expressions for ν_t Reichardt (1951), Deissler (1955), Rannie (1956), Van Driest (1956), Spalding (1961), Wassan and Wilke (1964), Rasmussan and Karamcheti (1965), Son and Hanratty (1967),

Mizushina, Ito, and Ogino (1970), Notter and Sliecher (1972), Hanna and Sandall (1972), and others, have proposed power law expressions for ν_t for the region in the immediate proximity of a solid surface. These expressions may be used to write additional analogy expressions in the exact manner shown above. A comparison of some of the above analogies is given in Figure 9. It is evident that for Pr = 0(1) there is little difference between the theories. A considerable disagreement exists at $\mathrm{Pr} \ll 1$ and $\mathrm{Pr} \gg 1$. As noted previously, the disagreement is due to the difficulty of correctly specifying the upper limit of integration in Eq. (2.107).

The nature of the variation of ν_t with y^+ in the wall region is important in determining the dependence of the transport rate on the molecular diffusivity. At very high Prandtl numbers the bulk of the temperature change is restricted to a region very close to the wall, where u^+ is very nearly equal to y^+. The various expressions for the variation of ν_t with y^+ in this region (Table 3), may be expanded by setting $u^+ = y^+$ to a series of increasing powers of y^+. Since for $\mathrm{Pr} \gg 1$, the region of interest is very close to the wall ($y^+ \to 0$), we may neglect all but the lowest power term in y^+ in the expression for ν_t we obtain

$$\frac{\nu_t}{\nu} = b(y^+)^n \tag{2.129}$$

where b is a numerical constant which generally differs for different investigators.

The corresponding expression for the Nusselt number is obtained by substitution of Eq. (2.129) into Eq. (2.103), i.e.,

$$\mathrm{Nu} = \frac{\sqrt{\dfrac{f}{2}}\ \mathrm{Re\,Pr}}{\displaystyle\int_0^\infty \frac{dy^+}{\dfrac{1}{\mathrm{Pr}} + b(y^+)^n}} \tag{2.130}$$

We note that the upper limit on the integral in Eq. (2.103), δ^+, has been replaced by $y^+ = \infty$. This is possible because T^+ is nearly constant except in the region very close to the wall and, as a consequence, the limit $y^+ = \delta^+$ coincides with the limit $y^+ = \infty$. Performing the integration in Eq. (2.130) results in the expression

$$\mathrm{Nu} = \frac{n}{\pi}\, b^{1/n} \sin\left(\frac{\pi}{n}\right) \sqrt{\frac{f}{2}}\ \mathrm{Re\,Pr}^{1/n} \tag{2.131}$$

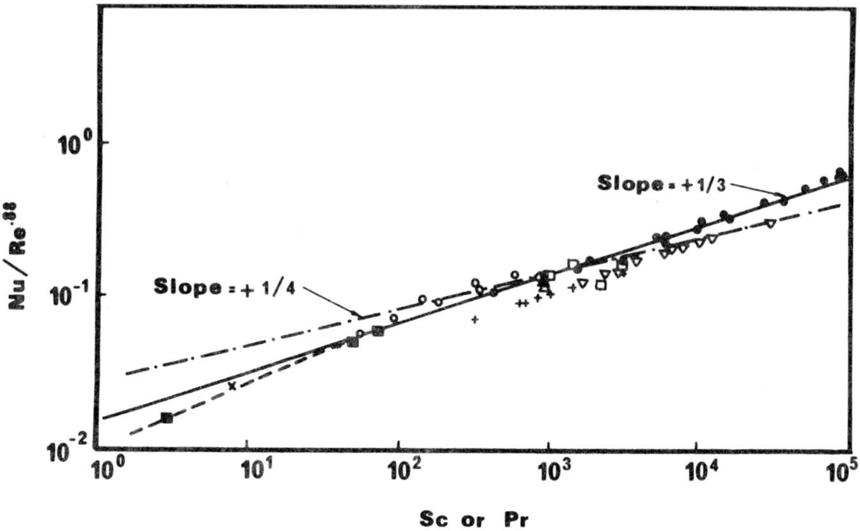

Figure 12 Effect of Schmidt or Prandtl number on transfer rate. Notation: ○ Harriott and Hamilton; △ Meyerink and Friedlander; □ Friend and Metzner; ■ Malina and Sparrow; X Allen and Eckett; ▲ Hubbard and Lightfoot; + Lin, Denton, Gaskill, and Putnam; □ Linton and Sherwood. (After Notter and Sleicher, 1971).

The experimental data plotted in Figure 12 suggest that $n = 3$ in Eq. (2.131). This is consistent with the data of Hubbard and Lightfoot (1966) and the conclusions of Notter and Sleicher (1971).

Friend and Metzner's correlation The above extensions of the simple Reynolds analogy may be classified as "theoretical" in that they arise from the analytical integration of Eqs. (2.103), or (2.109), with relevant assumptions regarding the distribution of ν_t in the turbulent flow field. Friend and Metzner (1958) rearranged Eq. (2.109) to read

$$\mathrm{Nu} = \frac{\sqrt{\dfrac{f}{2}}\ \mathrm{Re\,Pr}}{1 + \sqrt{\dfrac{f}{2}}\,(\mathrm{Pr} - 1)\displaystyle\int_0^{y_1^+} \frac{dy^+}{\left(1 + \mathrm{Pr}\,\dfrac{\nu_t}{\nu}\right)\left(1 + \dfrac{\nu_t}{\nu}\right)}} \tag{2.132}$$

Using experimental data of ten investigations, including their own, Friend and Metzner empirically determined the integral to be represented by the function $11.8\ \mathrm{Pr}^{-1/3}$. Equation (2.132) then becomes

$$\mathrm{Nu} = \frac{\sqrt{f/2}\,\mathrm{Re\,Pr}}{1 + 11.8\sqrt{f/2}(\mathrm{Pr} - 1)/\mathrm{Pr}^{1/3}} \tag{2.133}$$

(In the original treatment of Friend and Metzner [1958], the unity term in the denominator of Eq. [2.133] is given as 1.20. The difference arises as a correction for the ratios of the mean to maximum velocities and temperatures, which are not exactly unity. This correction is very small for $\mathrm{Pr} \gg 1$ and has been ignored in the present formulation.)

Equation (2.133) was shown to correlate experimental data for $0.7 < \mathrm{Pr} < 3.000$ with a standard deviation of 9.4%. At very large values of Pr, the equation reduces to $\mathrm{Nu}\ \alpha\ \mathrm{Pr}^{1/3}$, which is consistent with the conclusions in *Analogies based on power law expressions for* ν_t.

2.6.4 Analogy calculations with Pr_t other than unity The simple analogy theories discussed above are all based on the assumption that $\mathrm{Pr}_t = 1$. This assumption may be relaxed and the calculations performed with Pr_t given by the various models and empirical correlations discussed in Section 2.5. Such calculations have been performed by many workers (see Table 4), mainly for the case of low Prandtl heat transport. Some of the results have been demonstrated to be in good agreement with experimental data, Figure 10.

A significant aspect of Figure 10 is that calculations based on $\mathrm{Pr}_t = 1$ (Levich, 1962; Kropholler and Carr, 1962), and those of Azer and Chao (1962), and Notter and Sleicher (1972), based on more elaborate models for Pr_t, agree within the scatter of the experimental data. Moreover, it may be noted that calculations for $\mathrm{Pr} \geqslant 1$, based on the assumption that $\mathrm{Pr}_t = 1$, are also in good agreement with experimental data. It may be concluded that the smoothing effect of the integration in Eq. (2.103) renders the final result insensitive to the assumptions regarding Pr_t. As a consequence, analogy calculations do not provide a basis for differentiation between the various propositions for the behavior of Pr_t. It would seem that the value of $\mathrm{Pr}_t = 1$ is a good first approximation for most practical applications.

2.7 Solution of the Reynolds system of equations

2.7.1 Introduction The simple analogy equations discussed in Section 2.6 are based on the simplification of the Reynolds Eqs. (2.16)–(2.18) by introduction of the Couette-flow assumption. In principle, this simplification need not be made and it is possible to solve the complete system of Reynolds equations for the relevant boundary conditions with the appropriate system closure relationships, i.e., the turbulence models discussed in Sections 2.2 and 2.3, and the turbulent Prandtl number relationships in Section 2.5.

A detailed discussion of the various techniques which have been developed to solve the system of Reynolds equations for turbulent flow is beyond the scope of the present work and is indeed unnecessary since excellent treatments of this subject have recently appeared in the monographs of Patankar and Spalding (1970), and Gosman et al. (1969). In what follows we demonstrate the manner in which the information presented in the previous sections may be combined to provide solutions to practical problems.

2.7.2 Spalding's theory Utilizing Boussinesq's hypothesis, the Reynolds Eqs. (2.16) and (2.17) for momentum and heat transport across a turbulent boundary layer over a flat plate at zero incidence, may be written as

Momentum transport:

$$\bar{u}\frac{\partial \bar{u}}{\partial x} + \bar{v}\frac{\partial \bar{u}}{\partial y} = \frac{\partial}{\partial y}\left[(\nu + \nu_t)\frac{d\bar{u}}{dy}\right] \tag{2.134}$$

Heat transport:

$$\bar{u}\frac{\partial \bar{T}}{\partial x} + \bar{v}\frac{\partial \bar{T}}{\partial y} = \frac{\partial}{\partial y}\left[(\alpha + \alpha_t)\frac{d\bar{T}}{dy}\right] \tag{2.135}$$

The introduction of an expression (experimental) for the "law of the wall" $u^+ = u^+(y^+)$, is known to hold well for the case of turbulent flow over a flat plate at zero incidence and eliminates the need for an explicit integration of the equation for momentum transport since the mean flow velocity components and the turbulent viscosity are implied by the law of the wall.

Using Von Mises transformation, Spalding (1961), reduced Eq. (2.135) to read

$$\frac{\partial \bar{T}}{\partial x^+} = \frac{1}{\epsilon^+ u^+}\frac{\partial}{\partial u^+}\left(\frac{\alpha^+}{\epsilon^+}\frac{\partial \bar{T}}{\partial u^+}\right) \tag{2.136}$$

where $x^+ = \displaystyle\int_{x_0}^{x} \frac{u^*}{\nu}\,dx$

$$\epsilon^+ = 1 + \frac{\nu_t}{\nu} = \frac{dy^+}{du^+}$$ (for the constant shear stress layer where $\sigma = \sigma_w$)

and

$$\alpha^+ = \frac{1}{\Pr} + \frac{1}{\Pr_t}\frac{\nu_t}{\nu}$$

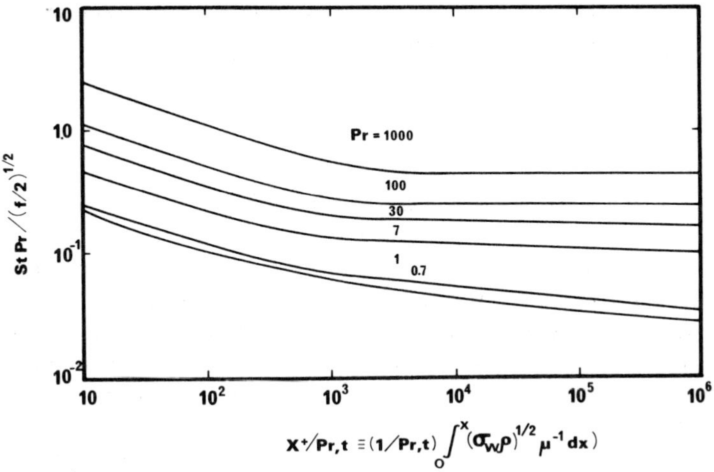

Figure 13 Numerically calculated solutions of Eq. (7.14). (Spalding, 1964).

Introducing a new variable ζ (Smith and Shah, 1962),

$$d\zeta = \frac{\epsilon^+}{\alpha^+}\, du^+$$

into Eq. (2.136) leads to

$$\frac{\partial \bar{T}}{\partial x^+} = \frac{1}{u^+\alpha^+}\frac{\partial^2 \bar{T}}{\partial \zeta^2} \tag{2.137}$$

Equation (2.137) is similar to the heat conduction equation and may be readily solved by the finite difference method (Patankar and Spalding, 1970) provided that ϵ^+ and α^+ (or ν_t/ν and Pr_t) are specified. This specification may take the form of any of the expressions presented in Sections 2.2, 2.3, and 2.5.

Numerical solutions of Eq. (2.137) have been carried out by Spalding (1964) for $\mathrm{Pr}_t = 1$ and ϵ^+ given by the Spalding expression for the law of the wall (see Table 3). The results for a number of different Prandtl numbers are plotted in Figure 13.

2.7.3 General Solutions Wassel and Cotton (1973) recently solved Eqs. (2.134) and (2.135) with a number of different models for turbulence and turbulent Prandtl number. The turbulence models used were

1. the Van Driest (1956) expression, Eq. (2.52) denoted as (D-model)
2. the single equation Nee-Kovasznay's (1968) model (Section 2.2.5), (N-K model)

3. a modification of the two-equation model of Harlow and Nakayama (1968), (Section 2.2.5), (H-N model)

The turbulent Prandtl number was given by either

1. $Pr_t = 0.9$, or
2. the Pr_t model of Wassel and Cotton (1973) (see Table 4)

The numerical procedure employed for the solution of the system of partial differential equations was a modification of the Patankar-Spalding (1967) procedure, developed by Denny and Mills (1969). This procedure involves a solution of the finite-difference versions of the relevant nonlinear parabolic partial differential equations in a step-wise manner in the downstream direction.

The results of the computations were compared with the experimental skin friction results of Weighardt (1968) and the heat transfer data of Moffat and Kays (1968). The predictions based on the three turbulence models and $Pr_t = 0.9$ are shown in Figures 14 and 15. The agreement with the experimental data is good. Figure 16 shows a comparison of the prediction based on the Van Driest expression for ϵ^+ with $Pr_t = 0.9$ and Pr_t given by the Wassel and Cotton (1973) expression. As is found with the simple analogy theories (Section 2.6),

Figure 14 Comparison of computed skin friction with Wieghardt's data. (After Wassel and Cotton, 1973).

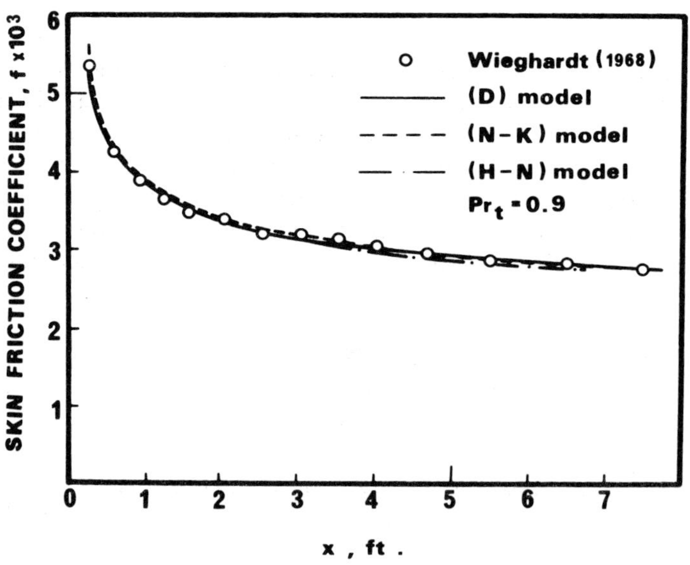

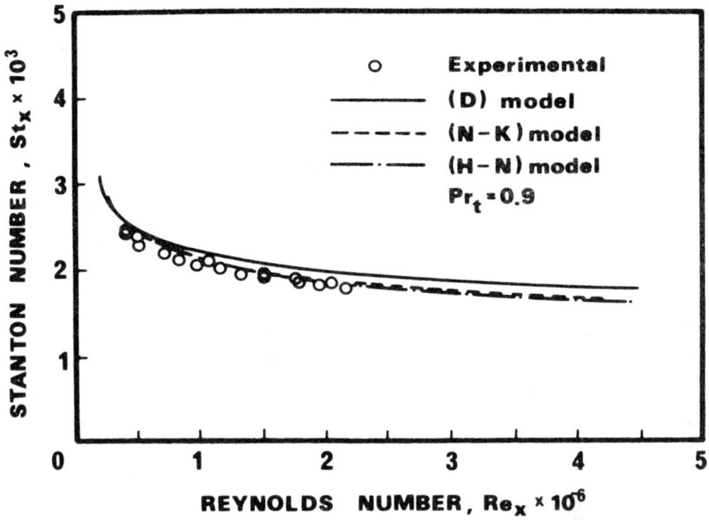

Figure 15 Comparison of computed Stanton number with Moffat and Kay's data (1968). (After Wassel and Cotton, 1973).

the prediction is relatively insensitive to the model chosen for the value of Pr_t.

A detailed survey of similar calculations may be found in the recent monograph of Launder and Spalding (1972). More recent treatments of interest are the papers by Cebeci (1973), Na and Habib (1973), Larson and Yerazunis (1973), and Ng and Spalding (1972).

Figure 16 Comparison of computed Stanton number with Moffat and Kay's data (1968).

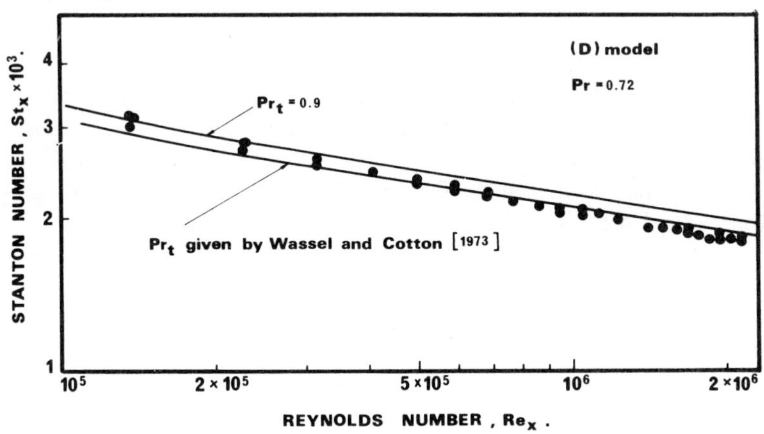

3 MODELS BASED ON THE SURFACE RENEWAL CONCEPT

3.1 Formulation and basic equations

3.1.1 Description of physical phenomena Although the application of the eddy diffusivity concept has proven useful, phenomenological models that have been formulated in terms of the turbulent diffusivities provide limited insight into the mechanism of turbulent transfer processes.

The introduction of the surface renewal concept by Danckwerts (1951) initiates a new class of transport models which seem to provide a basic and yet simple approach to the study of turbulent heat and mass transport. These models are based on the premise that fluid elements, usually denoted as "eddies," intermittently move from the turbulent core to the close vicinity of the transport surface. These eddies are assumed to maintain their identity during their residence time in the proximity of the surface, and transport occurs during the exposure time of the eddies to the transport surface. Applying a broad definition, the surface renewal approach is based on a micro scale study of the transfer characteristics of a single fluid element at the transport surface. The overall effect is then obtained by integrating the individual effects of all the eddies at the (unit) surface.

The bulk of recent hydrodynamic studies indicate the existence of a relatively well-ordered, large-scale, periodic motion in the wall region, and suggest that in turbulent flow, this consists of a mosaic of periodically replaced local elements. The recent flow visualization study of Corino and Brodkey (1969) elucidates the physical mechanisms associated with turbulent transport from a solid surface. According to this study the penetration of an eddy into the wall region (fluid inrush phase) affects a simultaneous acceleration of the low momentum fluid in the region. This acceleration is associated with a rapidly growing violent instability which results in an abrupt ejection of the accelerated fluid mass "outwards" from the wall into the turbulent core region (fluid ejection phase). Because of viscous interaction with the wall, a velocity gradient is developed in the inrushing fluid and is gradually decelerated in a progressively developing, time-dependent, viscous motion at the wall. This is the viscous sublayer development phase. The cycle is completed by a subsequent eddy which initiates the next fluid ejection phase. It is noted that the development time for the viscous sublayer is very much greater than the time required for its subsequent breakdown, i.e., the inrush and ejection phases.

Without getting involved in the question of the origin of turbulence, we note that since all the following models relate to the eddies during their contact time with the surface, it is immaterial whether the renewal is caused by the inrushing high energy eddies, or as suggested by Einstein and Li (1956), by the spontaneous eruption of the eddies from the surface due to inherent instability.

3.1.2 The basic equations It is well established that the resistance to transfer (for $\mathrm{Sc} > \mathrm{Pr} \geqslant 1$) is mainly in the periodic, fluctuating, viscous sublayer in the vicinity of the transport surface. The problem of closure thus reduces to solving the familiar set of boundary layer equations for laminar flow. Neglecting viscous dissipation and assuming constant properties the governing equations in rectangular coordinates for a single eddy are:

$$\frac{\partial u}{\partial t} + u \frac{\partial u}{\partial x} + v \frac{\partial u}{\partial y} = \nu \frac{\partial^2 u}{\partial y^2} \tag{3.1}$$

$$\frac{\partial P}{\partial t} + u \frac{\partial P}{\partial x} + v \frac{\partial P}{\partial y} = \kappa \frac{\partial^2 P}{\partial y^2} - k_n P^n \tag{3.2}$$

$$\frac{\partial u}{\partial x} + \frac{\partial v}{\partial y} = 0 \tag{3.3}$$

k_n denotes the rate constant. For mass transfer k_n is positive with respect to the reactant. For heat transfer k_n is negative if heat is generated within the eddy. $k_n = 0$ for nonreacting, or nonheat generating systems.

When $n = 0$, Eq. (3.2) represents mass transfer accompanied by zero order irreversible chemical reaction, or heat conduction with a constant rate of heat generation (or absorption).

When $n = 1$, Eq. (3.2) represents mass transfer accompanied by a first order irreversible chemical reaction, or heat conduction accompanied by heat generation (or absorption) for which the rate is proportional to the temperature of the fluid element under consideration.

Obviously, the choice of the initial and boundary conditions will determine the resulting mathematical expressions for the transport rate.

Solution of the governing equations with the appropriate boundary conditions yields the local instantaneous flux of solute to the eddy at the interface

$$\Psi(t = \theta) = \Psi_i(\theta) = \kappa \left. \frac{\partial P}{\partial y} \right|_{y=0} \tag{3.4}$$

The corresponding instantaneous transfer coefficient H_i, and Nusselt number $\mathrm{Nu}(\theta)$, are then obtained by dividing $\Psi_i(\theta)$ by the driving potential and nondimensionizing. The instantaneous transfer rate depends on the "age" of the element, i.e., the time passed since its arrival at the surface. At any given instant, the distribution of ages of the elements at the transport surface is given by $\phi(\theta)$, where $\phi(\theta)d\theta$ represents the relative portion of the area covered with fluid elements having ages between θ and $\theta + d\theta$. Since the ages can theoretically range from zero to infinity, it follows that

$$\int_0^\infty \phi(\theta)\, d\theta = 1 \tag{3.5}$$

The local instantaneous flux across any portion of the surface having eddies of age θ is given by

$$\Psi_i(\theta)\phi(\theta)\, d\theta$$

The average transfer rate per unit area, Ψ, is obtained by integrating the local fluxes over the entire surface, which corresponds to averaging the fluxes over the range of age distributions from zero to infinity:

$$\Psi = \int_0^\infty \Psi i(\theta)\phi(\theta)\, d\theta \tag{3.6}$$

The generalized average transfer coefficient H is then defined by

$$H = \frac{\Psi}{P_0 - P_b} \tag{3.7}$$

where P_0 and P_b are the interfacial and bulk concentrations, respectively.

Similarly, the time averaged Nusselt (Sherwood) number is then given by

$$\text{Nu} = \overline{\text{Nu}}(\tau) = \frac{H\bar{L}}{\kappa} = \int_0^\infty \text{Nu}(\theta)\phi(\theta)\, d\theta \tag{3.8}$$

where τ represents the mean contact time of the various eddies at the transport surface, and $\bar{L}$ is the characteristic length of the system under consideration.

In general, the various transport models to be discussed below differ by the assumptions made regarding the nature of

1. the governing equations, Eqs. (3.1) and (3.2)
2. the boundary conditions
3. the age distribution function $\phi(\theta)$, and/or
4. the mean contact time τ.

For convenience, the transport models are classified here according to the relative significance attributed by the different authors to the convective and temporal acceleration terms in Eqs. (3.1) and (3.2). Two major groups of models emerge:

1. Models based on the hypothesis that the convective terms are negligibly small in comparison with the temporal acceleration term. These models of eddies with no internal gradients are detailed in Section 3.2. (Also included in this section are models based on the Boussinesq's type flux in stagnant eddies with internal eddy turbulent fluctuations.)
2. Models based on the assumption that the convective terms predominate, either directly or indirectly. These models are presented in Section 3.3 dealing with eddies with internal velocity gradients.

Additional assumptions, within each of these two groups, give the models their special identity.

3.2 Models based on stagnant eddies with no internal velocity gradients

3.2.1 The penetration theory When a fluid comes in contact with a transport surface, a stagnant layer or "film" is assumed to form at the interface and a concentration gradient is immediately initiated at the fluid interface. Higbie (1935) visualized that the "stale" fluid at the surface is continuously replaced by fresh fluid from the bulk, and transport from the surface to the fluid element (or "film") occurs by unsteady state diffusion during the residence of the fluid element at the surface. Higbie further assumes that fluid elements which arrive at the transporting surface have identical residence times, τ, or in terms of the age distribution function,

$$\phi(\theta)\begin{cases} = \dfrac{1}{\tau} & 0 < \theta < \tau \\ = 0 & \theta > \tau \end{cases} \tag{3.9}$$

Neglecting convective terms and taking $k_n = 0$, we assume that the particular fluid element reaching the transport surface occupies a certain volume, has a finite thickness, and that is stays in place for a time period τ. Equation (3.2) reduces to

$$\frac{\partial P}{\partial t} = \kappa \frac{\partial^2 P}{\partial y^2} \tag{3.10}$$

The boundary conditions normally associated with such a problem would be specified by

$$P = P_0 \text{ at } y = 0 \qquad t > 0 \tag{3.11}$$

$$P = 0 \quad \text{at } l \geqslant y > 0 \qquad t = 0 \tag{3.12}$$

$$P = 0 \quad \text{at } y = l \qquad t = 0 \tag{3.13}$$

where P represents the reduced concentration, with respect to the bulk concentration P_b, and P_0 denotes the interfacial concentration. (Note that if $k_n \neq 0$, one may not use reduced concentrations!)

The solution of Eqs. (3.10) to (3.13) is discussed below in Section 3.2.4 dealing with the Film-Penetration model. For the special, but common case of short contact times, the boundary condition given by Eq. (3.13) may be modified to yield an immediate solution. This modification is the essence of the penetration theory.

The concept of the penetration theory was apparently first introduced formally by Higbie (1935), and later revived and extended by Danckwerts (1951). As implied by its name, the basis of this theory lies in the fact that, in unsteady-state transport, the depth of penetration into the exposed boundary depends on the time of contact. The longer the contact time, the deeper the penetration, and vice versa. Furthermore, at short contact times, if the penetrating agent does not penetrate very deeply, say up to $y = \delta$, where $\delta/l \ll 1$, it does not "know" where the other boundary is; neither does it "feel" nor is it affected by its presence. In other words, as far as this particular process is concerned, the second boundary at $y = l$ may be at infinity as far as the penetrating component "knows."

Thus, whether or not the fluid film is of an infinite thickness, the diffusing agent would not "know" the difference. This physical argument suggests that for the particular case of short contact time, the boundary condition given by Eq. (3.7) may be replaced by the mathematical expression:

$$P = 0 \quad \text{at } y = \infty \qquad t > 0 \tag{3.14}$$

which obviously implies that mathematically $l = \infty$.

It is important to note that the set of Eqs. (3.10), (3.11), (3.12), and (3.14) is also obtained when transfer to an ideal fluid is considered. Obviously, the ideal fluid is not affected by the presence of the transport boundary, whether a solid or a fluid. Thus, for a fluid flowing in the x-direction in a constant velocity u_b, we have

$$u_b \frac{\partial P}{\partial x} = \kappa \frac{\partial^2 P}{\partial y^2} \tag{3.15}$$

where conduction in the x-direction is assumed negligible compared with the convection in the same direction. Since $x/u_b = t$, Eq. (3.15) is identical with Eq. (3.10). A similar relationship is derived for transport at a gas-liquid interface, where zero tangential shear stress may be assumed and "ideal flow" is approached (Sideman, 1966).

The solution of Eqs. (3.10), (3.11), (3.12), and (3.14) for diffusion into a semi-infinite plane is given by the familiar error function form

$$\frac{P - P_b}{P_0 - P_b} = \operatorname{erfc} \frac{y}{\sqrt{4\kappa t}} = 1 - \frac{2}{\sqrt{\pi}} \int_0^{\eta} e^{-\eta^2}\, d\eta \qquad \eta \equiv \frac{y}{\sqrt{4\kappa t}} \tag{3.16}$$

By Eq. (3.4), the instantaneous flux Ψ_i is given by

$$\Psi_i(\theta) = (P_0 - P_b)\left(\frac{\kappa}{\pi\theta}\right)^{1/2} \tag{3.17}$$

Utilizing Eq. (3.9), the average flux, Eq. (3.6), is given by

$$\Psi = \frac{1}{\tau} \int_0^{\tau} \Psi_i(\theta)\, d\theta \tag{3.18}$$

By Eq. (3.7), the average transfer coefficient is

$$H = \frac{\Psi}{P_0 - P_b} = \frac{2}{\sqrt{\pi}} \left(\frac{\kappa}{\tau}\right)^{1/2} \tag{3.19}$$

In the familiar nondimensional form,

$$\mathrm{Nu} = \frac{HL}{\kappa} = \frac{2}{\sqrt{\pi}}\left(\frac{L^2}{\kappa\tau}\right) \equiv \frac{2}{\sqrt{\pi}} (\mathrm{Fo}')^{-1/2} \tag{3.20}$$

where Fo′ is the familiar Fourier modulus, defined here in terms of the length scale L, (instead of y in the classical Fourier modulus) and the mean residence time τ.

For the corresponding case of an ideal flow, Eq. (3.15), $\tau = L/u_b$ in Eqs. (3.16) to (3.20) and

$$\mathrm{Nu} = \frac{HL}{\kappa} = \frac{2}{\sqrt{\pi}}\left(\frac{Lu_b}{\kappa}\right)^{1/2} = \frac{2}{\sqrt{\pi}} \mathrm{Pe}^{1/2} \tag{3.21}$$

It can easily be shown (Sideman, 1966) that the application of the concept of a thin penetration layer is only valid for a fluid system of high Peclet numbers. A further generalization showing the equivalence of the penetration and laminar boundary layer theories for high Sc(Pr) number fluids was recently presented by Akiyama (1971).

3.2.2 Danckwerts surface renewal theory Danckwerts (1951) adopted Higbie's shallow penetration concept to evaluate $\Psi_i(\theta)$ by Eq. (3.4). However, whereas Higbie assumes the fluid elements that arrive at the transporting surface to have identical residence times, Danckwerts proposes the probability of replacement of a surface fluid element is independent of its age, and the distribution of ages is a random one.

If we define Sdt as the relative portion of the interfacial area replaced in time dt, then S denotes the average rate of renewal of fluid elements at the wall. Obviously, $S = \phi(t)$ at $t = 0$, and

$$\phi(t) = \mathrm{S}e^{-St} \tag{3.22}$$

Hence, the local rate of transfer of the solute across any portion of the film having θ age is given by

$$\Psi_i(\theta)\phi(\theta)\, d\theta = \kappa \mathrm{S}e^{-S\theta}\left(\frac{\partial P}{\partial y}\right)_{y=0} d\theta \tag{3.23}$$

and the total rate of transfer across a unit area (dm/Adt) is obtained by integrating Eq. (3.23) for all ages from zero to infinity.

Introducing Eq. (3.17) into Eq. (3.23) and integrating yields the average transfer coefficient

$$H = \frac{\Psi}{P_0 - P_b} = \int_0^\infty Se^{-S\theta}\left(\frac{\kappa}{\pi\theta}\right)^{1/2} d\theta = \sqrt{\kappa S} \tag{3.24}$$

In deriving Eq. (3.24), the mean bulk concentration P_b is assumed constant with time, and spatially uniform. The fulfillment of these conditions depends on the flow system under discussion.

Danckwerts also considered the more general model where the surface consists of different areas, a, each characterized by a different renewal frequency, S_a. Utilizing a surface-age distribution similar to Eq. (3.22) he obtained

$$H = \sqrt{\kappa}\,\frac{\Sigma a\sqrt{S_a}}{\Sigma a} \tag{3.25}$$

Except for its generality, this last model has not been useful since the additional information about the flow characteristics in the different areas is usually unavailable.

It is important to note that the above penetration-renewal models predict that the transfer coefficient is proportional to the molecular diffusivity to the half power. Equation (3.24) and Eq. (3.19) can be generalized to yield:

$$H = \lambda \sqrt{\frac{\kappa}{\tau}} \tag{3.26}$$

where $\lambda = 2/\sqrt{\pi} = 1.13$ for Higbie's uniform age distribution (constant τ), and $\lambda = 1$ for Danckwerts' random age distribution with an average renewal frequency $S = 1/\tau$, where τ now represents the mean contact time, or residence time, of the eddies at the transport surface.

The quantitative determination of the transfer coefficient requires to relate S or τ to the measurable parameters of the system under consideration. This is demonstrated below for the interesting case of heat transfer at a turbulent liquid-liquid interface.

3.2.3 The surface renewal model for a turbulent liquid-liquid interface Consider two immiscible liquids, 1 and 2, in stratified, counter- or co-current flow, which are in contact at $y = 0$. The interfacial zone of primary interest is the region near the interphase in which the temperatures differ from those of the bulks, T_{b_1} and T_{b_2}, respectively.

Following Danckwerts' random renewal concept, relatively large scale eddies are carried from the respective bulks to the interface. After staying there for a mean contact time τ, these eddies are replaced by other eddies coming from the bulk. This process continues as long as the fresh eddies have sufficient energy (or momentum) to replace the momentarily dormant eddies at the interface.

Braginski and Pavlushensko (1965) show that work W, required to replace a unit area at one side of the interface, is given by

$$W_i \sim (\rho_i^2 \sigma_0 \epsilon_i)^{0.2} \qquad i = 1 \text{ and } 2 \tag{3.27}$$

where σ_0 denotes the interfacial surface tension, and ϵ_i represents the average specific energy input per unit mass of phase i. As the high energy eddies hit the interface, their momentum tends to carry them through the interface into the other phase, simultaneously affecting local flow, and "surface renewal" on the other side of the deformed interface. The eddy is prevented from crossing the interface by the capillary pressure brought about by the deformation of the interface.

If the two liquids are independently motivated by two stirrers, the renewal frequency S is proportional to the total replacement work of the eddies on both sides of the interface, and inversely proportional to the interfacial tension. Relating the specific energy input ϵ_i to the stirring rate, N_i, in Eq. (3.27), the renewal frequency is given by (Braginski and Pavlushensko):

$$S \sim (\rho_1^{0.4} N_1^{1.8} + \rho_2^{0.4} N_2^{1.8})^{1/2} \tag{3.28}$$

Since the renewal rate in one phase is synonymous with renewal in both phases, the overall transfer coefficient U across the interface, rather than individual phase coefficients, is related to S. Figure 17 represents the experimental heat transfer data of Sideman and Uliss (1972) obtained in a stirred cell with a layer of water above a layer of Freon −113. As N_1 and N_2 go to zero, the transfer coefficient takes the value of 42 (Kcal/hr °Cm²), which is considered to represent the contribution of natural convection to the overall transfer rate. Which ρ in g/cm³ and N in rpm, the experimental correlation for the overall heat transfer coefficient in Kcal/hr °Cm² is given by (Sideman and Uliss):

$$U = 42 + 3.45(\rho_1{}^{0.4}N_1{}^{1.8} + \rho_2{}^{0.4}N_2{}^{1.8})^{1/2} \tag{3.29}$$

Figure 17 Effect of stirring in both layers on the overall heat transfer coefficient surface renewal model. (Sideman and Uliss, 1972).

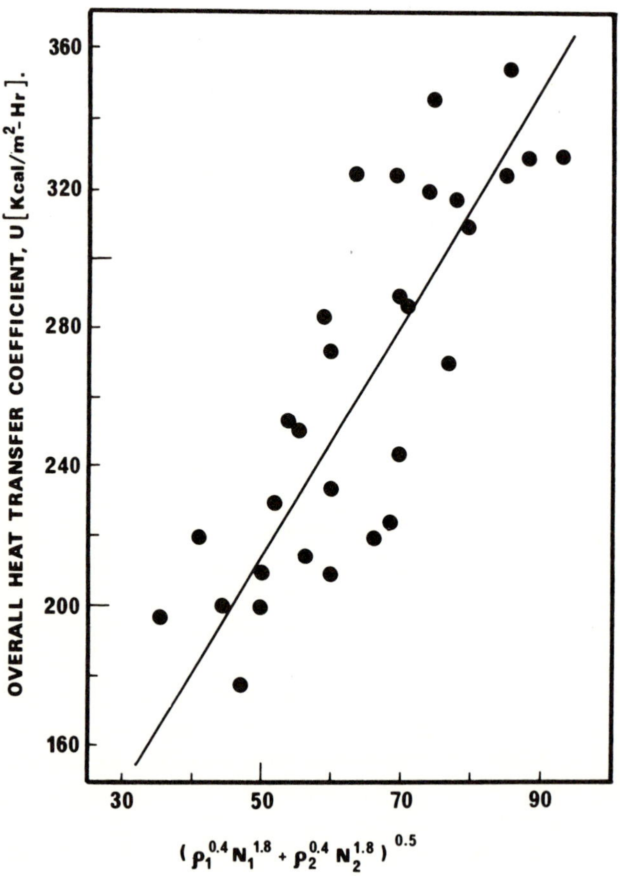

It is interesting to note that no effects of the flow direction could be distinquished, and Eq. (3.29) correlates the data obtained in both current and co-current flows. More important, the effect of the mechanical energy input (into each phase separately) on the overall transfer rate is additive.

3.2.4 The film-penetration theory The dependence of H on $\kappa^{0.5}$, predicted by the penetration-renewal theory, fails to explain a large body of data which shows that $H \sim \kappa > 0.5$. The film-penetration model attempts to extend the range of this dependence. Adopting the idea that interfacial transport is, in essence, restricted to periodically replaced elements of fluid of finite thickness l at the interface, Dobbins (1958), and Toor and Marchello (1958), solved the set of Eqs. (3.10) to (3.13). The solution can be obtained by utilizing the Laplace transform (i.e., multiplying by e^{-st}, where s is a positive number, and integrating with respect to t between the limits 0 and ∞) or by the usual Fourier analysis. The former technique yields, with P reduced with respect to P_b (or $P_b = 0$),

$$\frac{P}{P_0} = \sum_{n=0}^{\infty} \operatorname{erf}\left(\frac{n+1-y/2l}{\mathrm{Fo}^{0.5}}\right) - \operatorname{erf}\left(\frac{n+y/2l}{\mathrm{Fo}^{0.5}}\right) \tag{3.30}$$

where $\mathrm{Fo} = (\kappa t/l^2)$ is the Fourier modulus. Utilizing the Fourier analysis yields

$$\frac{P}{P_0} = 1 - \frac{y}{l} - \frac{2}{\pi} \sum_{n=1}^{\infty} \frac{1}{n} \sin \frac{n\pi y}{l} \, e^{-(\pi^2 n^2 \mathrm{Fo})} \tag{3.31}$$

Equations (3.30) and (3.31) give identical results, but Eq. (3.30) converges more rapidly than Eq. (3.31) for small values of $\mathrm{Fo} = \kappa t/l^2$. Figure 18 represents the concentrations in the film as a function of Fo (or time). It is noted that the transfer across the film reaches a steady state linear profile as $\mathrm{Fo} \geqslant 0.6$. As shown below, this parametric behavior approaches the classical film theory at large values of Fo, and Higbie's penetration, or Danckwerts' renewal theory, at low values of Fo.

The mean rate of transfer across a surface consisting of a multiplicity of elements with different surface ages is given by Eq. (3.6). Explicit expressions for transfer coefficients depend on the assumed age distribution function $\phi(\theta)$, and the choice of either Eq. (3.30) or Eq. (3.31) to evaluate $\Psi_i(\theta)$.

Identical residence times; constant age distribution of surface elements If the residence times of all elements at the surface are identical, Higbie's age distribution function, $\phi(\theta) = 1/\tau$ applies. Utilizing Eq. (3.30) to evaluate Ψ_i, Eq. (3.6) gives (Toor and Marchello, 1958):

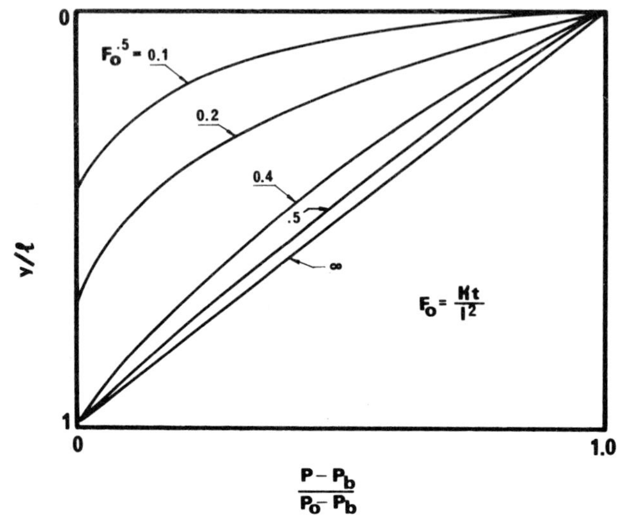

Figure 18 Transfer across a fluid element of finite thickness.

$$H = 2\sqrt{\frac{\kappa}{\pi\tau}}\left(1 + 2\sqrt{\pi}\sum_{n=1}^{\infty} \operatorname{ierfc}\frac{n}{\mathrm{Fo}_\tau^{1/2}}\right) \tag{3.32}$$

where $\mathrm{Fo}_\tau = (\kappa\tau/l^2)$. Similarly, utilizing Eq. (3.31) suitable for large exposure times, yields

$$H = \frac{\kappa}{l}\left[1 + \frac{1}{3\mathrm{Fo}_\tau} - \sum_{n=1}^{\infty}\frac{1}{n^2}\exp\left(-n^2\pi^2\mathrm{Fo}_\tau\right)\right] \tag{3.33}$$

Inspection of Eq. (3.32) shows that for $\mathrm{Fo}_\tau \ll 1$ ($\tau \ll l^2/\kappa$, i.e., small penetration depths and short contact times) Eq. (3.23) reduces to

$$H = \frac{2}{\sqrt{\pi}}\left(\frac{\kappa}{\tau}\right)^{1/2} \tag{3.34}$$

which is identical with the penetration theory transfer coefficient, Eq. (3.19). Conversely, if $\mathrm{Fo}_\tau \gg 1$ or $\tau \gg l^2/\kappa$ synonymous with large contact times, Eq. (3.33) reduces to

$$H = \frac{\kappa}{l} \tag{3.35}$$

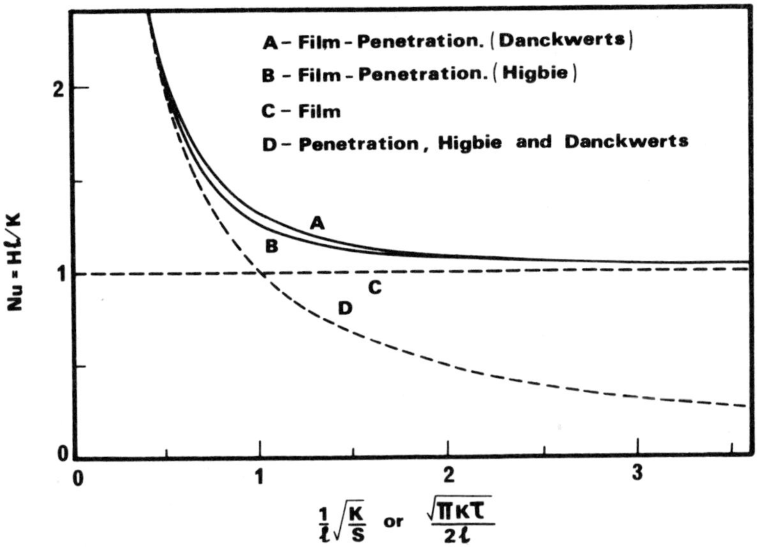

Figure 19 Mean transfer rates as a function of time. (Toor and Marchello, 1958).

which is identical with the classical film theory transfer coefficient. This is demonstrated in curve B, Figure 19.

Random age distribution of surface elements If the fluid elements at the transport surface are assumed to be randomly replaced, Danckwerts' age distribution function, Eq. (3.22), can be used in Eq. (3.6). Utilizing Eq. (3.30) to evaluate $\Psi_i(\theta)$, the average transfer coefficients for rapid replacements is given by (Dobbins, 1958):

$$H = \sqrt{\kappa S} \coth \sqrt{\frac{Sl^2}{\kappa}} \tag{3.36}$$

As shown in the next section, *Evaluation of the surface renewal frequency and the "film" thickness,* Eq. (3.36) reduces to Danckwerts' surface renewal expression $H = \sqrt{\kappa S}$, Eq. (3.24), for large ($>$ 2.5) values of (Sl^2/κ) and to the classical film-model expression $H = \kappa/l$ for values of $Sl^2/\kappa < 0.6$.

A fast converging expression for long contact times is given by (Toor and Marchello, 1958):

$$H = \frac{\kappa}{l}\left[1 + 2\sum_{n=1}^{\infty}\left(1 + n^2\pi^2\frac{\kappa}{Sl^2}\right)^{-1}\right] \tag{3.37}$$

Thus, the film-penetration theory shows that for a given system, high replacement frequencies (large S) leads to the penetration-renewal theory where $H \sim \kappa^{0.5}$. Conversely, when the surface is replaced at a low rate, the fluid elements are completely penetrated and $H \sim \kappa$. This is demonstrated in Figure 19, curve A.

When operating at identical flow conditions, S and l are presumably fixed by the hydrodynamics of the system. Changing fluids will show that transfer to fluids with low Schmidt number (say high κ) would follow the film model whereas the transfer to high Schmidt number fluids (low κ) will be controlled by the renewal mechanism and $H \sim \kappa^{1/2}$. As shown by Toor and Marchello (1958), this conclusion is consistent with experimental evidence supplied by Lin et al. (1951, 1953), Diessler (1954), Hanratty (1956), and Johnson and Huang (1956). It is nevertheless wrong (see *Analogies based on power law expressions for* v_t in Section 2.6.3).

Evaluation of the surface renewal frequency and the "film" thickness Toor and Marchello (1958) attempted to evaluate S and l for turbulent pipe flow by utilizing the mass transfer data of Linton and Sherwood (1950), and Lin et al. (1951), and assuming

$$l = l^{+}(2R)\left(\frac{f}{2}\right)^{-\frac{1}{2}} \text{Re}$$

where $f/2 = 0.023\ \text{Re}^{-0.2}$, R denotes the tube radius, and $l^{+} = 5$. This pioneering attempt is limited by the use of an arbitrary constant to evaluate S.

Brusset et al. (1973) recently evaluated S and l by utilizing mass transfer data obtained from evaporation studies in wetted-wall columns into a turbulent ($2{,}000 < \text{Re} < 20{,}000$) gas. The values of S and l were solved from a set of equations obtained by writing Eq. (3.36) for each value of the Reynolds number. Comparison with different columns led to the following empirical correlations:

$$S \cong 2 \times 10^{-5} \nu R^{-2} \text{Re}^{1.8} \tag{3.38}$$

and

$$l \cong 24\ \text{Re}^{-0.7} \tag{3.39}$$

The constants were averaged from four sets of data (Jackson et al., 1950; McCarter et al., 1959; Vivian et al., 1968; and Gilliland and Sherwood, 1934). Substituting Eqs. (3.38) and (3.39) into Eq. (3.36) yields

$$\text{Nu} \cong 8.9 \times 10^{-3} \text{Sc}^{\frac{1}{2}} \text{Re}^{0.9} \coth\left(0.214\, \text{Sc}^{\frac{1}{2}} \text{Re}^{0.2}\right) \tag{3.40}$$

It is clear by inspection, that for $\mathrm{Sc} \geqslant 1$ and $\mathrm{Re} \geqslant 10^4$, the coth term approaches unity. This implies that the value of l is practically unimportant for turbulent systems. It is noteworthy, however, that Brusset et al. obtained a simple theoretical relationship essentially similar to Eq. (3.38) by utilizing $l^+ = 30$, rather than the $l^+ = 5$ suggested by Toor and Marchello.

3.2.5 A renewal model for eddies with internal eddy diffusivity King (1966) presented a modification of Danckwerts basic model; he proposed that small scale fluctuations within the eddy augment molecular transport rate to the eddy during its residency at the transport surface.

Utilizing the same assumption led to Eq. (3.10), the transfer to an eddy was described by

$$\frac{\partial P}{\partial t} = \frac{\partial}{\partial y}\left[(\kappa + E)\frac{\partial P}{\partial y}\right] \tag{3.41}$$

where $E = ay^n$ (3.42)

is interpreted to represent an eddy diffusivity which accounts for the enhanced transport rate due to turbulent fluctuation within the eddy, and a and n are taken to be independent of t.

Equation (3.41) is solved for penetration type boundary conditions, Eqs. (3.11), (3.12), and (3.13), rewritten here as

$$\begin{aligned} P &= P_b \quad \text{at} \quad t = 0, \quad y > 0 \\ P &= P_0 \quad \text{at} \quad t > 0, \quad y = 0 \\ P &= P_b \quad \text{at} \quad t > 0, \quad y \to \infty \end{aligned} \tag{3.43}$$

Although the model incorporates the concept of an eddy diffusivity, it is important to appreciate that E in Eq. (3.41) and D_t, (or ν_t), introduced in Section 2.2, are not equivalent. The difference between the two quantities lies in the fact that while ν_t refers to the overall time averaged motion of the eddies, E refers specifically to the fluctuations within the eddy, excluding the motion induced by the arrival and departure of fluid elements from the surface (which is accounted for in terms of the renewal concept). The two eddy diffusivities become equivalent only when the effect of eddy renewal on the velocity field at the interface is negligible with respect to the contribution due to the motion within the eddy, i.e., when the renewal rate is negligibly small, or the eddy residence time is large. Under these conditions, $\partial P/\partial t \cong 0$, and Eq. (3.41) reduces to Eq. (2.20). The solution (see Section 2.6.3) is given by (King, 1966):

$$H = H_i = \frac{n}{\pi} a^{1/n} \sin\left(\frac{\pi}{n}\right) \kappa^{1-1/n} \tag{3.44}$$

For small residence times the temporal acceleration term is not negligible, solution of Eqs. (3.41) and (3.43) depend on the values of a and n.

For low a and $n > 0, E \cong 0$, and Eq. (3.41) reduces to Eq. (3.10), yielding the classical penetration solution, Eq. (3.17):

$$H_i = \frac{\Psi_i}{P_0 - P_b} = \sqrt{\frac{\kappa}{\pi t}} \tag{3.45}$$

or

$$\Psi^* = \frac{1}{\sqrt{\pi\tau^*}} \tag{3.46}$$

Where the dimensionless transfer coefficient and surface age are given by:

$$\Psi^* = \frac{H_i}{a^{1/n}\kappa^{1-1/n}}$$
$$\tau^* = \frac{a^{2/n}t}{\kappa^{2/(n-1)}} \tag{3.47}$$

For $n = 0, E = a$, the solution is analagous to the penetration solution

$$H_i = \sqrt{\frac{\kappa + a}{\pi t}} \tag{3.48}$$

or

$$\Psi_* = \sqrt{\frac{2}{\pi\tau_*}} \tag{3.49}$$

For $n = 1$, the solution is given by Carslaw and Jaeger (1959), and may be written in the form

$$\Psi_* = \frac{2}{\pi^2}\int_0^\infty \frac{\exp\left(-\dfrac{\tau^*\xi^2}{4}\right) d\xi}{\xi[J_0{}^2(\xi) + Y_0{}^2(\xi)]} \tag{3.50}$$

where J_0, Y_0, are Bessel functions.

For $n \to \infty$, the solution is interpreted in the following manner: E is zero at the wall ($y = 0$), and increases exponentially with distance from the wall; it follows that at some distance from near the wall, say $y = \lambda$, we have $E = \kappa$.

Provided that a be infinitesimal (i.e., low intensity turbulence) we may write that

$$\begin{aligned} E &\ll \kappa \quad \text{where} \quad y < \lambda \\ E &\gg \kappa \quad \text{where} \quad y > \lambda \end{aligned} \tag{3.51}$$

For small contact times, the penetration of the concentration wave is within the region $0 < y < \lambda$. As a consequence of Eq. (3.51), solution for this region is given simply by Eq. (3.45) or Eq. (3.46), consistent with the penetration theory. For large contact times the solution once again is given by Eq. (3.44) which tends to the result

$$H = \frac{n}{\pi} a^{1/n} \kappa \tag{3.52}$$

where $n/\pi \; a^{1/n}$ is finite as $n \to \infty$, provided that a is infinitesimally small.

Equation (3.52) is analogous to the simple film-theory result. Thus for $n \to \infty$, the model reduces to the penetration theory at small contact times, and the film-theory at large contact times. This result is similar to the film-penetration theories of Dobbins (1958), and Toor and Marchello (1958). The various results are compared in Figure 20.

Figure 20 King's (1966) model for transfer.

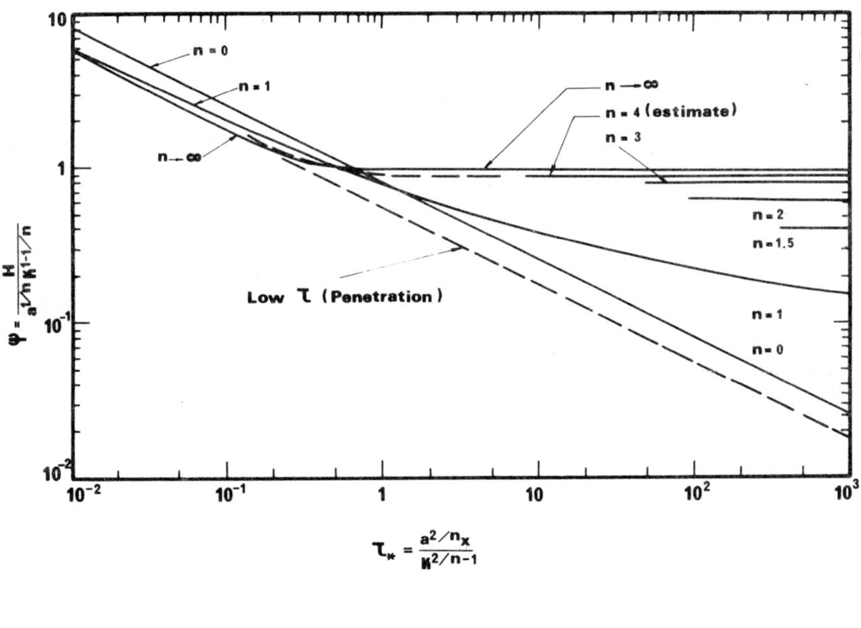

The parameters a, n, and τ^* must be evaluated by relating the model to experimental data. King (1966) assumed that for liquid phase gas absorption across a free interface in a turbulent agitated vessel, the mass transfer rate is given by Eq. (3.44), i.e., the high τ^* asymptotic solution. Comparing Eq. (3.44) with the correlation proposed by Calderbank and Moo-Young (1961) he concluded that

$$n = 4$$
$$a = 0.006 \frac{\epsilon}{\mu^2} \tag{3.53}$$

where ϵ is the energy dissipation rate per unit liquid mass.

Although King did not apply this model to the case of transport from a solid boundary, it is nevertheless interesting to follow up the ramifications of such a treatment. As seen in Section 2.6, the transfer rate to a wall is given by Eq. (3.44) with $n = 3$. This implies, according to King's treatment, the transport at a solid surface is characterized by "long contact times" and not by a penetration-renewal type mechanism. This conclusion is inconsistent with the model proposed by Pinczewski and Sideman (1974), where it is suggested that a penetration type mechanism is indeed operative.

3.2.6 Additive resistance models: interfacial resistance The additivity of resistances is probably most familiar when related to the classical Ohm's law: the flux is inversely proportional to the total resistance. Whitman and Lewis (1924) two film-theory represents one of the simplest cases in which individual resistances at two sides of an interface are combined in series.

Danckwerts (1951) considered a two-layered model, where the transfer in the wall-layer is characterized by the simple surface renewal mechanism, and the transport in the layer immediately above it is carried by eddy turbulence. The overall transfer rate is thus given by connecting the resistances of these two layers in series. An essentially similar approach was reported by Hughmark (1969). Here, we limit the discussion to the effect of the interfacial resistance on the overall transfer rate.

There is ample evidence to suggest an interfacial resistance to mass transfer may exist, particularly where liquid surfaces are modified by the presence of surface active agents. As pointed out by Coughlin (1967), who studied the surface resistance at a vapor-liquid interface, the application of Danckwerts' random contact time distribution function conveniently yields that resistances to transfer at the interface are algebraically additive.

Consider a fluid element (eddy) reaching the transport surface. For short contact times the eddy is treated as a semi-infinite extent. The interfacial resistance h_s is expressed in terms of the interfacial equilibrium (or pure) concentration P_0^*, i.e.,

$$-\kappa \frac{\partial P}{\partial y} = h_s(P_0^* - P) \qquad t > 0 \;\; y = 0 \tag{3.54}$$

while the other boundary conditions are given by Eqs. (3.12) and (3.14). The solution of Eq. (3.10), with the above boundary conditions, gives the instantaneous surface concentration P as

$$\frac{P}{P_0^*} = 1 - \exp\left(\frac{h_s^{\,2} t}{\kappa}\right) \operatorname{erfc}\,(h_s\sqrt{t/\kappa}) \tag{3.55}$$

Utilizing Danckwerts' age distribution function, the overall transfer coefficient is given by

$$H = \frac{\Psi}{P_0^*} = \frac{\displaystyle\int_0^\infty \Psi_i(\theta)\phi(\theta)\,dt}{P_0^*} = \frac{Sl\,[\Psi_i(\theta)]}{P_0^*} = \frac{1}{(1/h_s) + \sqrt{1/\kappa S}} \tag{3.56}$$

$$\text{where } \Psi_i(t) = h_s P_0^* \exp\left(\frac{h_s^{\,2} t}{\kappa}\right) \operatorname{erfc}\,(h_s\sqrt{t/\kappa}) \tag{3.57}$$

Perlmutter (1961) suggested a model based on an exponential decay of the non-equilibrium interfacial condition, yielding

$$h_s = \lambda_1 \sqrt{\kappa\tau}$$

where λ_1 is a measure of the rate of equilibration approach at the interface. Analysis of Eq. (3.56) indicates that accurate experimental verification, and determination of h_s, requires systems with relatively small τ (or a correspondingly smaller λ_1).

As deviation from equilibrium is rather unreasonable at solid-fluid interfaces, the interfacial resistance has been attributed to the effect of reduced effective diffusivity in a narrow region adjacent to the interface. The consequences of this approach are discussed by Perlmutter (1961).

3.2.7 Renewal models based on the gamma distribution function So far we have seen two distribution functions: Higbie's uniform distribution, and Danckwerts' random distribution. We now introduce the gamma distribution function, which has some interesting and useful properties. These are discussed in the next section.

Some properties of the gamma function It is evident by comparisons of Eqs. (3.10) and (3.15) that Higbie's postulate that all eddies have identical residence time τ at the transport surface is equivalent to the constant velocity plug flow case. On the other hand, Danckwerts' random age distribution function, Eq. (3.22), is based on the idea that all eddies have equal chances to be replaced, regardless of age, which is characteristic of a "well mixed vessel."

Zwietering (1959) derived a relation between the age distribution function $\phi(t)$ (at an instant in time), and the residence time distribution function $F(\theta)$ (after time $t = \theta$), in a continuous flow system. In terms of the random eddy surface renewal model $F(\theta) = \int_0^\infty f(t)dt$ where $f(t)$, the frequency function of residence time, is defined as the probability that an eddy chosen randomly from those leaving the interface has been in contact with it for a time which is between t and $t + dt$. The relationship is given by

$$\phi(t) = \frac{1}{\tau}\,[1 - F(t)] \tag{3.58}$$

where τ, the average residence time, or lifetime, of the eddies in the "vessel" is obtained either by dividing the volume V by volumetric flow rate Q, or by the formal relationship:

$$\tau = \int_0^\infty tf(t)\,dt = \frac{1}{S} \tag{3.59}$$

As noted by Kovasy (1968), the relationship $\tau = 1/S$ between the rate of surface renewal S, and the mean lifetime of the eddies τ, is valid for any arbitrary lifetime distribution.

Thus, if we view the interface region as a thin "mixed vessel," we can identify Danckwerts' age-distribution function (actually it is the eddy age function, but for ease of comparison and future reference we follow Danckwerts' original definition), Eq. (3.22), with the residence time frequency function of a single vessel, i.e.,

$$f(\theta) = \frac{1}{\tau}\exp\left(-\frac{\theta}{\tau}\right) \tag{3.60}$$

As noted by Perlmutter (1961), Danckwerts' age-distribution function, now identified with the well mixed vessel characterized by Eq. (3.60), leads to the unreasonable conclusion that the most likely eddy residence time is zero. Perlmutter suggested that correct distribution function lies somewhere between Higbie's plug flow, and Danckwerts' well mixed vessel, and proposed the following residence time distribution function for a system of m-vessels:

$$f(\theta) = \frac{m^m \theta^{m-1}}{(m-1)!\tau^m} \exp\left(-\frac{m\theta}{\tau}\right) \qquad m > 0 \tag{3.61}$$

Utilizing Eq. (3.58), the gamma form for the age-distribution function is obtained (Koppel, 1966):

$$\phi(\theta) = \frac{1}{\tau} e - \frac{m\theta}{\tau}\left[\sum_{j=0}^{m-1}\left(\frac{m\theta}{\tau}\right)^j \frac{1}{j!}\right] \quad m = 1, 2, \ldots \tag{3.62}$$

where $\tau = 1/S$ represents the average contact time of the elements with the transport surface. For the special case of $m = 1/2$, Eqs. (3.58) and (3.61) yield

$$\phi(\theta) = \frac{1}{\tau} \operatorname{erfc} \sqrt{\frac{\theta}{\tau}} \qquad m = +\frac{1}{2} \tag{3.63}$$

The age-distributions are plotted in Figure 21 for different values of the system parameter m. It is noteworthy that the integration leading from Eq. (3.61) to Eq. (3.62) leads to much smaller differences in the age-distribution functions than those generated by the residence time distributions, Eq. (3.61). Since the age-distribution has a fixed intercept, and must be monotonously decreasing, the choice of shapes is quite limited.

As $m \to 0$, the age-distribution of Figure 21 will approach a pulse function around the age of zero, but with a long tail at high values of age. Thus, the effect of decreasing m is to increase the proportion of old elements at any given instant of time.

For $m = 1$, Eq. (3.61) reduces to Danckwerts' distribution function, Eq. (3.22), whereas introducing $m = \infty$ yields Higbie's uniform residence time model. Intermediate values of m represent successive degrees of compromise

Figure 21 Age distributions resulting from gamma residence time distribution. (Koppel et al., 1966).

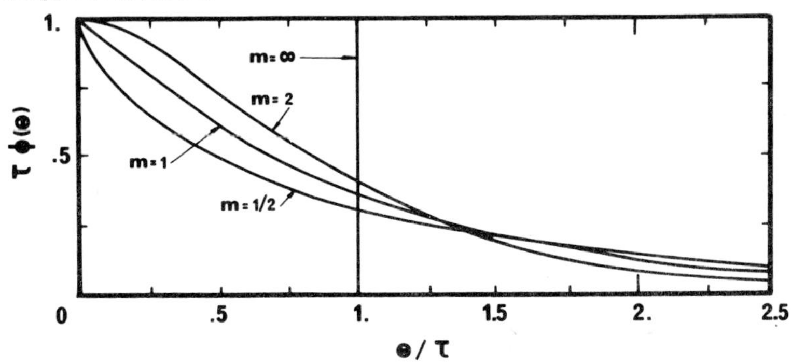

between these two extremes. Physically, this model visualizes adjacent well mixed regions, all a part of the interfacial zone, with mass transferred between the regions by molecular diffusion. Mathematically, this distribution offers convenient operational properties for calculating the average transport coefficient.

For a piecewise continuous function $f(\theta)$

$$\int_0^\infty e^{-P\theta}\,\theta^j f(\theta)\,d\theta \equiv L\,[\theta^j f(\theta)] = (-1)^j \frac{d^i}{dP^i}\,\{L\,[f(\theta)]\} = (-1)^j \frac{d^j f(P)}{dP^j} \tag{3.64}$$

where $f(P)$ is the Laplace transform of the function $f(\theta)$.

Introducing Eq. (3.62) into the expression for the average flux Ψ, and utilizing Eq. (3.64), yields:

$$\Psi = \int_0^\infty \phi(\theta)\Psi_i(\theta)\,d\theta$$

$$= \int_0^\infty \text{Se}^{-mS\theta}\left[\sum_{j=1}^{m-1}\frac{(mS\theta)^j}{j!}\right]\Psi_i(\theta)\,d\theta$$

$$= S\sum_{j=0}^{m-1}\frac{-(mS)^j}{j!}\left\{\frac{d^j}{dpj}\,L\,[\Psi_i(\theta)]\right\}_{P=mS}$$

$$\equiv SD_m\,[g(P)] = SG(S) \tag{3.65}$$

The transform operator D_m is defined (Chung et al., 1971) for positive integers of m, by

$$D_m\,[g(P)] \equiv \sum_{j=0}^{m-1}\left\{\frac{(-mS)^j}{j!}\left[\frac{d^j}{dP^j}\,g(P)\right]\right\}_{P=mS} = G(S) \text{ or } G\left(\frac{1}{\tau}\right) \tag{3.66}$$

and

$$\Psi_i(\theta) = L^{-1}\,[g(P)] \tag{3.67}$$

$G(S)$ is a transform function of $g(P)$ in the S (or $1/\tau$) domain. Equation (3.66) represents a function $g(P)$ transformed to $G(S)$ under the operator D_m. The sufficient condition for the existence of $G(S)$ is that $g(P)$ is a function which is differential to the $(m-1)$th order with respect to P.

The physical meaning of Eq. (3.65) is instructive. We note that $g(P)$ is the Laplace transform of the instantaneous flux, $\Psi_i(\theta)$. Thus, Eq. (3.65) represents a particularly convenient property of the gamma distribution: one can calculate the time average transfer coefficient directly from the Laplace transform of the instantaneous coefficient, *without inversion.* Since the instantaneous rate coefficient is often obtained by utilizing the Laplace transform to solve the partial differential equations, it is quite convenient not to have to invert the transform to obtain the average rate coefficient.

The convenient operational properties of the gamma distribution function (which for integer values of m is identical with Perlmutter's (1961) multiple capacitance model) were utilized by Koppel et al. (1966) and later extended by Chung, Fan, and Hwang (1971) to present a generalized approach to generate the transfer rate expressions for processes with and without chemical reaction or heat generation.

Derivation of models and discussions Following Chung et al. (1971), some of the analytical solutions for the transport coefficients based on the multiple capacitance are left in the form of the D_m transform for the sake of brevity. The evaluation of the D_m transform is straightforward and requires only a knowledge of differentiation.

In the case of heat transfer between a solid boundary and a turbulent fluid, or stirred fluid without heat generation, or in the case of mass transfer between a gas and a liquid without chemical reaction, ($k_n = 0$), the equation defining the transport process at the interface is given by

$$\frac{\partial P}{\partial \theta} = \kappa \frac{\partial^2 P}{\partial y^2} \tag{3.10}$$

The initial and boundary conditions are given by

$$P = P_b \qquad t = 0 \quad y \geqslant 0 \tag{3.68}$$

$$P = P_b \qquad t \geqslant 0 \quad y \geqslant l \tag{3.69}$$

$$-\kappa \frac{\partial P}{\partial y} = h_s(P_0 - P) \qquad t > 0 \quad y = 0 \tag{3.70}$$

Equation (3.68) implies that the elements arrive at the interface at the bulk concentration. Equation (3.69) implies that the transfer is contained within a finite thickness, l. (This B.C. was the basis for the film-penetration theory.) The diffusing component at $y > l$ is immediately carried into the bulk due to highly turbulent transfer. Equation (3.70) defines the interfacial resistance h_s in terms of the equilibrium concentration (or wall temperature).

The instantaneous transfer rate of the fluid element at the interface is given by Eq. (3.4), and the generalized local transport coefficient H is given by Eq. (3.7).

Solving the set of Eqs. (3.10) with (3.68) to Eq. (3.70) for $\Psi_i(\theta)$, and introducing the multiple capacitance contact time distribution, Eq. (3.62) into Eq. (3.7), yields the following general solution for the transport coefficient:

$$H_m = \kappa S\gamma \sum_{j=0}^{m-1} \frac{(-mS)^j}{j!} \left[\frac{d^j}{dP^j} \frac{1}{P\left(1 + \gamma \sqrt{\frac{\kappa}{P}} \tanh \sqrt{\frac{P}{\kappa}}\, l\right)} \right]_{P=mS} \tag{3.71}$$

where $\gamma = h_s/\kappa$.

Introducing the linear operator D_m defined by Eq. (3.66), the above expression may be written as

$$H_m = \kappa S\gamma D_m \left[\frac{1}{P\left(1 + \gamma \sqrt{\frac{\kappa}{P}} \tanh \sqrt{\frac{P}{\kappa}}\, l\right)} \right] \tag{3.72}$$

If the surface resistance is neglected, which implies a constant surface temperature for heat transfer, or a constantly saturated surface concentration for mass transfer, Eq. (3.72) reduces to

$$H_{m\gamma} = \lim_{\gamma \to \infty} H_m = SD_m \left(\sqrt{\frac{\kappa}{P}} \coth \sqrt{\frac{P}{\kappa}}\, l \right) \tag{3.73}$$

If the time of contact is short, the depth of penetration is very small in comparison with the thickness of the fluid element, l. Taking the limiting value of $l \to \infty$, Eq. (3.71) reduces to form:

$$H_{ml} = \lim_{l \to \infty} H_m = \kappa S\gamma D_m \left[\frac{1}{P\left(1 + \gamma \sqrt{\frac{\kappa}{P}}\right)} \right] \tag{3.74}$$

If both l and γ approach infinity, Eq. (3.71) becomes

$$H_{ml\gamma} = \lim_{\substack{l \to \infty \\ \gamma \to \infty}} H_m = \sqrt{\kappa S}\, D_m\, [\bar{P}^{1/2}] = 2 \sqrt{\frac{\kappa S}{\pi m}} \frac{m + 1/2}{(m-1)!} \tag{3.75}$$

The above equation is identical to the mass transfer coefficient proposed by Koppel et al. (1966) by using a gamma residence time distribution.

When m equals unity, Eq. (3.62) reduces to Eq. (3.22), Danckwerts' random contact time distribution. Equation (3.71) reduces to Danckwerts' model with interfacial resistance (see H_{1l} in Table 5). Similarly, when m equals unity, Eq. (3.73) yields the expression by Toor and Marchello:

$$H_{1\gamma} = \sqrt{\kappa S} \coth \sqrt{\frac{S}{\kappa}}\, l \tag{3.76}$$

It is very difficult to obtain a solution directly from Eq. (3.71) by letting m approach infinity. However, it has been noted that when m approaches infinity, the multiple capacitance distribution function reduces to that of Higbie's uniform contact time distribution. Replacing the former by the latter, the solution for the transport coefficient in the case that $m \to \infty$ is obtained as follows

$$H_{\infty} = \frac{\kappa}{l}\left[\frac{\gamma l}{1+\gamma l} + 2\,\frac{\gamma^2 l^4}{\kappa T}\sum_{n=1}^{\infty}\frac{1-e^{-\beta_n^2 \kappa T/l^2}}{\beta_n^2(\gamma l + \gamma^2 l^2 + \beta_n^2)}\right] \tag{3.77}$$

where β_n are the consecutive roots of the equation

$$\beta \cot \beta + \gamma l = 0$$

When the mean residence time of fluid elements on the transport surface is very short, the depth of penetration becomes much less than the thickness of the fluid element. Letting l approach infinity, Eq. (3.77) reduces to the same expression of mass transfer coefficient suggested by Perlmutter (1961). This limiting case (see $H_{\infty l}$ in Table 5) may also be approximated by the following expression according to Penny and Bell (1967):

$$H_{\infty l} = \frac{1}{\dfrac{1}{h_s} + \dfrac{1}{2\sqrt{\dfrac{\kappa}{\pi T}}}} \tag{3.78}$$

This approximation is essentially based on Danckwerts' rule of additive resistance.

When the mean residence time approaches infinity, i.e., when the surface is no longer renewed and the fluid elements form a stagnant film on the interface,

Table 5 Local Transport Coefficients for the Systems Without Chemical Reaction or Heat Generation of Various Models (*Chung, Fan, and Hwang, 1971*)

Interface boundary conditions	Boundary extent conditions	Age distribution function	Local transport coefficient	Author
With* surface resistance	Finite† extent	Multiple capacitance	$H_m = \kappa s \gamma D_m \left[\dfrac{1}{P\left(1 + \gamma \sqrt{\dfrac{\kappa}{P}} \tanh \sqrt{\dfrac{P}{\kappa}}\, l\right)} \right]$ $\quad D_m[g(p)] = \sum_{j=0}^{m-1} \dfrac{(-ms)^j}{j!} \left[\dfrac{d_j}{dp^j} g(p) \right]_{P=ms}$	
		Random distribution	$H_1 = \dfrac{\kappa \gamma}{1 + \gamma \sqrt{\dfrac{\kappa}{s}} \tanh \sqrt{\dfrac{s}{\kappa}}\, l}$	
		Uniform distribution	$H_\infty = \dfrac{\kappa}{l} \left[\dfrac{\gamma l}{1 + \gamma l} + \dfrac{2\gamma l^3}{\kappa \tau} \sum_{n=1}^{\infty} \dfrac{(1 - e^{-\beta_n^2 n \tau / l^2})}{\beta_n(\gamma l + \gamma^2 l^2 + \beta_a^2)} \right]$ $\quad \gamma l = \dfrac{h_s l}{\kappa}$ $\quad \beta_n$ are the roots of Eq. $\beta \cot \beta + \gamma l = 0$	
	Infinite‡ extent	Multiple capacitance distribution	$H_{ml} = \kappa s \gamma D_m \left[\dfrac{1}{P\left(1 + \gamma \sqrt{\dfrac{\kappa}{p}}\right)} \right]$	
		Random distribution	$H_{1l} = \dfrac{1}{\dfrac{1}{\kappa \gamma} + \dfrac{1}{\sqrt{\kappa s}}}$	Danckwerts (1951)
		Uniform distribution	$H_{\infty l} = \dfrac{1}{\tau \gamma} \left(e^{\kappa \gamma^2 \tau} \operatorname{erfc} \gamma \sqrt{\kappa \tau} + 2\gamma \sqrt{\dfrac{\kappa \tau}{\pi}} - 1 \right)$	Perlmutter (1961)

Without** surface resistance	Finite extent	Multiple capacitance distribution	$H_{m\gamma} = sD_m\left(\sqrt{\frac{\kappa}{p}} \coth \sqrt{\frac{p}{\kappa}}\, l\right)$	Dobbins (1956)
		Random distribution	$H_{1\gamma} = \sqrt{\kappa s} \coth \sqrt{\frac{s}{\kappa}}\, l$	Toor & Marchello (1958)
		Uniform distribution	$H_{\infty\gamma} = 2\sqrt{\frac{\kappa}{\tau\pi}}\left(1 + 2\sqrt{\pi} \sum_{n=1}^{\infty} i \operatorname{erfc} \frac{nl}{\sqrt{\kappa\tau}}\right)$	Dobbins (1956)
			$= \frac{\kappa}{l}\left\{1 + \frac{2l^2}{\pi^2\kappa\tau}\left[\frac{\pi^2}{6} - \sum_{n=1}^{\infty} \frac{\exp(-\kappa\tau n^2\pi^2/l^2)}{n^2}\right]\right\}$	Toor & Marchello (1958)
	Infinite extent	Multiple capacitance distribution	$H_{m\gamma l} = \sqrt{\kappa s}\, D_m(p^{-1/2}) = \sqrt{\frac{\kappa s}{\pi m}}\, \frac{2\Gamma\left(m + \frac{1}{2}\right)}{(m-1)!}$	Koppel et al. (1966)
		Random distribution	$H_{1\gamma l} = \sqrt{\kappa s}$	Danckwerts (1951)
		Uniform distribution	$H_{\infty\gamma l} = \frac{2}{\sqrt{\pi}} \sqrt{\frac{\kappa}{\tau}}$	Higbie (1935)

*h_s or γ is finite
**h_s or γ approaches infinity
†l is finite
‡l approaches infinite

Eq. (3.77) yields the following express transfer coefficient based on the film theory:

$$\lim_{\tau \to \infty} H_\infty = \frac{1}{\dfrac{1}{h_s} + \dfrac{l}{\kappa}} \tag{3.79}$$

When the interfacial resistance is neglected, Eq. (3.77) yields the expression based on the film theory, as derived by Dobbins (1958), and Toor and Marchello (1958), based on the film-penetration concept.

Expressions for the transport coefficient corresponding to different age-distribution functions and different boundary conditions for the transport systems without chemical reaction or heat generation are summarized in Table 5.

Some general conclusions can be deducted by considering the above equations from a mathematical viewpoint:

1. The effect of surface resistance tends to decrease the transport coefficient.
2. The increase in the thickness of fluid elements tends to decrease the transport coefficient. The result, based on the simple penetration concept ($l \to \infty$), gives a lower bound if other variables such as τ, κ, and γ are fixed.
3. Only the simple penetration model without interfacial resistance predicts that the transfer coefficient is proportional to the one-half power of molecular (or thermal) diffusivity. The assumption of finite thickness of fluid elements gives rise to a higher power than 0.5. The interfacial resistance gives rise to a lower power than 0.5.
4. The transport coefficient increases as the number of "capacitors," m, increases. This point is easily demonstrated quantitatively by utilizing the penetration concept to evaluate the local instantaneous flux, $\Psi_i(\theta)$ by Eq. (3.17).

Figure 22 demonstrates the relatively small effect of the various forms of the gamma function on the average transfer coefficient calculated by Eq. (3.7), or on $\overline{\text{Nu}}\ \text{Fo}_\tau^{1/2}$.

The maximum difference between the transfer rates, based on $m = 1$ (well mixed) and $m = \infty$ (plug flow), is only 13%, which is certainly within the range of experimental errors. Perlmutter (1961) concluded that it does not matter much whether the interfacial zone acts as a single, well-mixed region, or a plug flow region, provided that all the fluid in this region is in motion. This conclusion is not surprising in view of the equivalence of the penetration-potential flow (Sideman, 1966), and penetration-boundary layer theories (Akiyama, 1972). For shallow mass penetration depths, and/or high Schmidt (Prandtl) numbers, the average mass transfer coefficient will be practically identical regardless of the form of $\phi(t)$. Similar conclusions were obtained by Hanratty (1956), utilizing exponential forms for the age-distribution function.

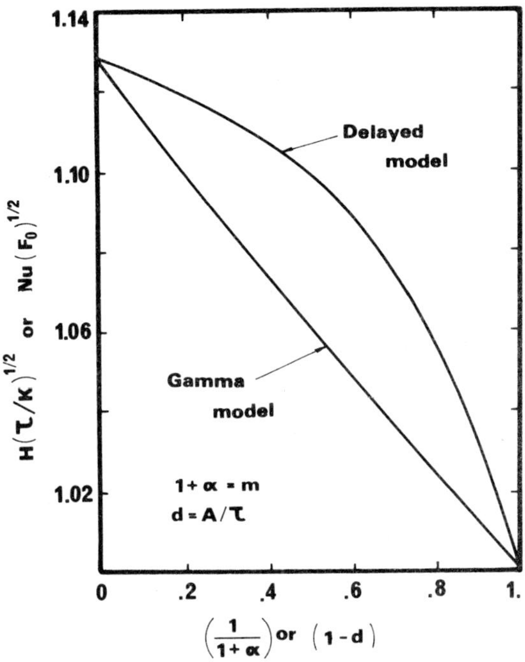

Figure 22 Average Nusselt number from gamma and delayed exponential distributions. (Koppel et al., 1966).

The relative insensitivity of the age-distribution function to the residence time distribution function, as well as the insensitivity exhibited by the average transfer coefficient to the shape of the age-distribution function, do not allow for a clear understanding of the surface renewal processes. It has, however, practical advantages as it allows for calculating the average transfer coefficient from the average residence time τ only, without recourse to the shape of the age-distribution. This, in a sense, is the inherent advantage of utilizing the surface renewal concept for prediction and analysis of various transport processes.

Methods to estimate the mean residence times of elements at a transporting surface are presented by Koppel et al. (1966, part II).

3.2.8 Models based on functions incorporating dead time Imperfect mixing of the interfacial fluid may possibly occur in the presence of local laminar flow. Lynn et al. (1955) noted this phenomenon in mass transfer over a string of spheres, and similar effects are to be expected when operating with packed (fixed) beds containing regions of laminar flow.

Perlmutter (1961) suggested that the presence of stagnant, unreplaced, fluid elements at the transporting surface would alter the residence time distribution, and introduced the concept of a delayed exponential distribution.

$$\phi(\theta) = \begin{cases} \dfrac{1}{\tau + A} & 0 \leqslant \theta \leqslant A \\[2ex] \dfrac{1}{\tau + A} \exp\left(-\dfrac{(\theta - A)}{\tau}\right) & \theta \geqslant A \end{cases} \tag{3.80}$$

As seen from Eq. (3.80), the dead time A and the average residence time of the undelayed system are assumed to be additive. This approach is severely criticized by Koppel et al. (1966), who claimed the average residence time must remain constant regardless of the amount of dead time. The rationale is that average residence time is a fundamental property of the time scale of the transport process. The suggested age-distribution function is

$$\phi(\theta) = \frac{1}{\tau}\left\{ u(\theta - A) \exp\left(-\frac{\theta - A}{\tau - A}\right) + [u(\theta) - u(\theta - A)] \right\} \tag{3.81}$$

where $u(\theta)$ is a unit step function. Equation (3.81) is presented in Figure 23 for various values of the dead time fraction, $d(= A/\tau)$. It is noted that for larger amounts of dead time, there is a smaller fraction of old elements present at the surface, as must be true in order to maintain a constant average residence time. As $A/\tau \to 1$, Eq. (3.81) reduces to an impulse located at $\theta = \tau$, i.e., to Higbie's uniform age-distribution.

Perlmutter's calculated transfer rates were based on the well-mixed residence time, "corrected" for the dead time, and showed great effects of the

Figure 23 Age distributions resulting from delayed exponential distributions. (Koppel et al., 1966).

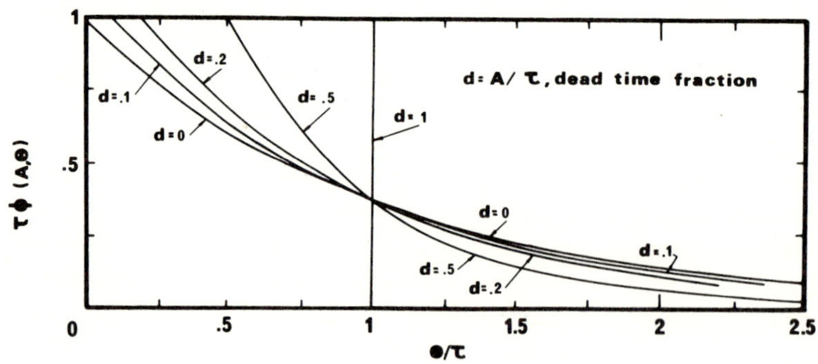

dead time. For instance, the mass transfer rate will be reduced by some 50% if $A/\tau = 4$. At high (> 10) values of A/τ, his average transfer coefficient reaches the asymptotic value of

$$H = \frac{2}{\sqrt{\pi}}\left(\frac{\alpha}{A}\right)^{1/2} \tag{3.82}$$

the transfer coefficient is a function of the dead time only. Comparison of Eq. (3.82) with Eq. (3.19) suggests, according to Perlmutter, that Higbie's model is equivalent to an all dead time distribution.

The averaged (over τ) transfer coefficient, based on Koppel's finite, dead time inclusive, residence time is given by:

$$H = \frac{2}{\sqrt{\pi}}\left(d\,\frac{\kappa}{\tau}\right)^{1/2} + \left[(1-d)\,\frac{\kappa}{\tau}\right]^{1/2} \exp\left(\frac{d}{1-d}\right) \operatorname{erfc}\left(\frac{d}{1-d}\right)^{1/2} \tag{3.83}$$

where $d \equiv A/\tau$. As $d \to 1$, Eq. (3.83) reduces to Eq. (3.19). This conclusion is, of course, consistent with Higbie's original concept of an uninterrupted, continuous film with a uniform contact time.

Figure 22 also represents the dependence of the transfer coefficient (or Nu $\mathrm{Fo}_\tau^{1/2}$) on the dead time fraction as given by Eq. (3.83). The variation indicated is only 13% over all values of m and d considered. Figure 22 demonstrates the limited effect of the dead time, which is rather small compared with Perlmutter's results based on allowing $A/\tau > 1$. Obviously, Perlmutter's results represent the effect of changing the average residence time, rather than the dead time fraction.

3.2.9 Models for transport systems accompanied by zero order reaction When molecular diffusion, or conduction, through fluid elements in contact with the transport surface is accompanied by a zero order chemical reaction (or constant heat generation), the material or energy balance equation for the surface elements may be obtained from Eq. (3.2) by letting n be zero. This gives, again neglecting convective terms,

$$\frac{\partial P}{\partial \theta} = \frac{\partial^2 P}{\partial y^2} - k_0 \tag{3.84}$$

Equation (3.84) is subject to the boundary conditions specified by Eqs. (3.68) to (3.70). Again, the multiple capacitance contact time distribution function is employed. With the same approach employed in Section 3.2.7, we obtain the general expression for the transport coefficient (Chung et al., 1971):

$$H_m{}^0 = \frac{S\gamma}{P_0 - P_b} D_m \left[\frac{\dfrac{k_0}{P\sqrt{P/\kappa}}\left(1 - \operatorname{sech}\sqrt{\dfrac{P}{\kappa}}\, l\right) + \dfrac{P_0 - P_b}{\sqrt{P/\kappa}}}{\sqrt{P/\kappa} + \tanh\sqrt{P/\kappa}\, l} \right] \tag{3.85}$$

When $k_0 = 0$, the above equation becomes identical to the transfer coefficient given by Eq. (3.71). When m is unity, and the interfacial resistance is negligibly small, Eq. (3.85) yields

$$H_1{}^0 = \lim_{\gamma\to\infty} H_1{}^0$$

$$= \sqrt{S\kappa}\coth\sqrt{\frac{S}{\kappa}}\, l + \frac{\sqrt{\kappa/S}\, k_0}{(P_0 - P_b)}\left(\coth\sqrt{\frac{\kappa}{S}}\, l - \operatorname{csch}\sqrt{\frac{S}{\kappa}}\, l\right) \tag{3.86}$$

A different expression for the transfer coefficient in a transport system involving a zero order reaction may be found in the work of Onda and Sada (1966). They obtained the following result based on the film theory in the absence of interfacial resistance:

$$H = \frac{\kappa}{l}\left[1 + \frac{k_0 l^2}{2(P_0 - P_b)\kappa}\right] \tag{3.87}$$

Equation (3.87) represents a limiting case of Eq. (3.86) when the rate of surface renewal becomes extremely slow.

If m approaches infinity, the general expression for the transport coefficient, Eq. (3.85), becomes rather complicated. It is given by

$$H_\infty{}^0 = \frac{\kappa}{l}\left\{\frac{\gamma l}{1 + \gamma l} + \frac{2\gamma^2 l^4}{\kappa\tau}\sum_{n=1}^{\infty}\frac{1 - e^{-\kappa\beta_n{}^2\tau/l^2}}{\beta_n{}^2[(1+\gamma l)\gamma l + \beta_n{}^2]}\right.$$

$$+ \frac{k_0 l^2}{2\kappa(P_0 - P_b)}\left[\frac{\gamma l}{1 + \gamma l} + \frac{4\gamma^2 l^4}{\kappa\tau}\sum_{n=1}^{\infty}(\sec\beta_n - 1)\right.$$

$$\left.\left.\cdot\frac{1 - e^{-\kappa\beta_n{}^2\tau/l^2}}{\beta_n{}^4[(1+\gamma l)\gamma l + \beta_n{}^2]}\right]\right\} \tag{3.88}$$

If the surface resistance is neglected, Eq. (3.88) reduces to

$$H_\infty{}^0 = \lim_{\gamma\to\infty} H_\infty{}^0$$

$$= \frac{\kappa}{l}\left[1 + \frac{2l^2}{\kappa\tau\pi}\left(\frac{\pi^2}{6} - \sum_{n=1}^{\infty}\frac{1}{n^2}\,e^{-\kappa\pi^2 n^2\tau/l^2}\right)\right]$$

$$+ \frac{k_0 l}{2(P_0 - P_b)} - \frac{4k_0 l^3}{\pi^4\kappa\tau(P_0 - P_b)}\left(\frac{\pi^4}{96} - \sum_{n=1}^{\infty}\frac{e^{-(2n-1)\pi^2\kappa\tau/l^2}}{(2n-1)^4}\right) \tag{3.89a}$$

$$= \frac{2k_0}{(P_0 - P_b)}\sum_{n=1}^{\infty}(-1)^n$$

$$\cdot\left(\frac{4}{3}\frac{\kappa\tau}{\pi}\,e^{-n^2l^2/4\kappa\tau} + \frac{1}{3}\frac{n^2l^2}{\sqrt{\kappa\tau}}\,\text{ierfc}\,\frac{nl}{\sqrt{2\kappa\tau}} - nl\,\text{erfc}\,\frac{nl}{\sqrt{2\kappa\tau}}\right)$$

$$+ \frac{k_0}{P_0 - P_b}\frac{4}{3}\sqrt{\frac{\kappa\tau}{\pi}} + 2\sqrt{\frac{\kappa}{\pi\tau}}\left(1 + 2\sqrt{\pi}\sum_{n=1}^{\infty}\text{ierfc}\,\frac{nl}{\sqrt{\kappa\tau}}\right) \tag{3.89b}$$

If the thickness of the fluid elements is assumed to be infinity, Eq. (3.89) reduces to the identical expression presented by Astarita (1967) for the mass transfer coefficient with zero order reaction (see $H_\infty^0 \gamma l$ in Table 6).

Expressions for the transport coefficients for the system with zero order reaction under various boundary conditions and age distributions are presented in Table 6. All of these equations may be considered as the special limiting cases of Eq. (3.85).

The effect of chemical reaction on mass transfer rates has been of considerable interest. This effect may be discussed in terms of the ratio of the chemical mass transfer coefficient to the physical mass transfer coefficient and is called the reaction coefficient. The reaction coefficient may be readily obtained by taking the ratios of the corresponding transport coefficients from Tables 5 and 6.

The following observations can be made in examining all of the expressions for the transport coefficients derived above (Chung et al., 1971):

1. All of the expressions for the chemical transport coefficients are reduced to the corresponding equations for the physical transport coefficients when k_0 vanishes.
2. The reaction coefficients are always greater than 1 when k_0 is negative. In other words, the effect of the zero order irreversible chemical reaction tends to increase the mass transfer coefficients of the reactant. Similarly,

Table 6 Local Transport Coefficient for the System of Zero Order Irreversible Chemical Reaction of Various Models (*Chung, Fan, and Hwang, 1971*)

Interface boundary conditions	Boundary extent conditions	Age distribution function	Local transport coefficient	Reference
With surface resistance	Finite extent	Multiple capacitance distribution	$H_m^0 = \dfrac{s\gamma}{P_0 - P_b} D_m \left[\dfrac{-\dfrac{k_0}{P}\sqrt{\dfrac{\kappa}{P}}\left(\operatorname{sech}\sqrt{\dfrac{P}{\kappa}}\, l - 1\right) + (P_0 - P_b)\sqrt{\dfrac{\kappa}{P}}}{\sqrt{\dfrac{P}{\kappa}} + \gamma \tanh \sqrt{\dfrac{P}{\kappa}}\, l} \right]$	
		Random distribution	$H_1^0 = \dfrac{\gamma\kappa}{P_0 - P_b} \left[\dfrac{\dfrac{-\kappa l_0^2}{s}\left(\operatorname{sech}\sqrt{\dfrac{s}{\kappa}}\, l - 1\right) + P_0 - P_b}{1 + \gamma\sqrt{\dfrac{\kappa}{s}} \tanh \sqrt{\dfrac{s}{\kappa}}\, l} \right]$	
		Uniform distribution	$H_\infty^0 = \dfrac{\kappa}{l}\left\{ \dfrac{\gamma l}{1+\gamma l} + \dfrac{2\gamma^2 l^4}{\kappa\tau} \sum_{n=1}^{\infty} \dfrac{1 - \exp\left(-\dfrac{\kappa\beta_n^2\tau}{l^2}\right)}{\beta_n^2(\gamma^2 l + \gamma l + \beta_n^2)} + \dfrac{k_0 l^2}{2\kappa(P_0 - P_b)}\left(\dfrac{\gamma l}{1+\gamma l} + \dfrac{4\gamma^2 l^4}{\kappa\tau} \sum_{n=1}^{\infty} (\sec\beta_n - 1)\, \dfrac{1 - e^{-\kappa\beta_n^2\tau/l^2}}{\beta_n^4[(1+\gamma l)\gamma l + \beta_n^2]} \right) \right\}$	
	Infinite extent	Multiple capacitance distribution	$H_{ml}^0 = \dfrac{s}{(P_0 - P_b)} D_m \left\{ \dfrac{\sqrt{\dfrac{\kappa}{P}}\left[(P_0 - P_b) + \dfrac{k_0}{P}\right]}{1 + \dfrac{1}{\gamma}\sqrt{\dfrac{P}{\kappa}}} \right\}$	
		Random distribution	$H_{1l}^0 = \dfrac{\gamma\kappa}{(P_0 - P_b)} \left[\dfrac{(P_0 - P_b) + \dfrac{k_0}{s}}{1 + \gamma\sqrt{\dfrac{\kappa}{s}}} \right]$	

		Uniform distribution	$H_{\infty l}{}^0 = \frac{1}{\gamma\tau}\left(e^{\kappa\gamma^2\tau}\operatorname{erfc}\gamma\sqrt{\kappa\tau} + 2\gamma\sqrt{\frac{\kappa\tau}{\pi}} - 1\right) + \frac{k_0}{\gamma(P_0 - P_b)}\left[\frac{1}{\kappa\gamma^2\tau}\left(e^{\kappa\tau\gamma^2}\operatorname{erfc}\gamma\sqrt{\kappa\tau} + 2\gamma\sqrt{\frac{\kappa\tau}{\pi}} - 1\right) + \frac{4}{3}\gamma\sqrt{\frac{\kappa\tau}{\pi}} - 1\right]$	
Without surface resistance	Finite extent	Multiple capacitance distribution	$H_{m\gamma}{}^0 = \frac{s}{P_0 - P_b} D_m \left[\frac{k_0}{P}\sqrt{\frac{\kappa}{P}}\left(\coth\sqrt{\frac{P}{\kappa}}\, l - \operatorname{csch}\sqrt{\frac{P}{\kappa}}\, l\right) + (P_0 - P_b)\sqrt{\frac{\kappa}{P}}\coth\sqrt{\frac{P}{\kappa}}\, l\right]$	
		Random distribution	$H_{1\gamma}{}^0 = \sqrt{s\kappa}\coth\sqrt{\frac{s}{\kappa}}\, l + \sqrt{\frac{\kappa}{s}}\frac{k_0}{(P_0 - P_b)}\left(\coth\sqrt{\frac{s}{\kappa}}\, l - \operatorname{csch}\sqrt{\frac{s}{\kappa}}\, l\right)$	
		Uniform distribution	$H_{\infty\gamma}{}^0 = \frac{\kappa}{l}\left\{1 + \frac{2l^2}{\kappa\tau\pi}\left[\frac{\pi^2}{6} - \sum_{n=1}^{\infty} - \frac{1}{n^2}\exp\left(-\frac{\kappa\pi^2 n^2\tau}{l^2}\right)\right]\right\} + \frac{k_0 l}{2(P_0 - P_b)} - \frac{4k_0 l^3}{\pi^4\kappa\tau(P_0 - P_b)}\left[\frac{\pi^4}{96} - \sum_{n=1}^{\infty}\frac{1}{(2n-1)^4}\exp\left(-\frac{(2n-1)^2\pi^2\kappa\tau}{l^2}\right)\right]$	
	Infinite extent	Multiple capacitance distribution	$H^0_{ml\gamma} = 2\sqrt{\frac{\kappa s}{\pi m}}\left[\frac{\Gamma\left(m + \frac{1}{2}\right)}{(m-1)!} + \frac{2k_0\Gamma\left(m + \frac{3}{2}\right)}{3(P_0 - P_b)sm!}\right]$	
		Random distribution	$H^0_{1\gamma l} = \sqrt{\kappa s}\left[1 + \frac{k_0}{s(P_0 - P_b)}\right]$	
		Uniform distribution	$H^0_{\infty\gamma l} = 2\sqrt{\frac{\kappa}{\pi\tau}}\left[1 + \frac{2}{3}\frac{\tau k_0}{(P_0 - P_b)}\right]$	Astarita (1967)

the effect of constant rate heat generation within the fluid elements tends to decrease the heat transfer coefficient.

3. As in the case of transport systems without reactions, the transport coefficients in systems accompanied by zero order reactions depend on the mean residence time of surface fluid elements, the average thickness of fluid elements, the surface resistance, and the properties of the fluid. Furthermore, when the thickness of the fluid elements are assumed to be infinite, and the surface resistance is neglected, the chemical transfer coefficient is proportional to $\kappa^{1/2}$, no matter what contact time distribution is employed.
4. Unlike the case of transport without chemical reaction, the transport coefficient for zero order reaction always depends on the driving force of the system under consideration. More specifically, the transport coefficient for zero order reaction decreases when the potential difference increases.

Other expressions for transfer coefficients with first order reactions ($n = 1$) are given by Chung et al. (1971).

3.2.10 The random distance rejuvenation model The models discussed so far are based on the assumption that fluid elements which influence the transfer rate come in and sweep all stale elements down to the transport surface. However, due to the no slip condition at a solid (or highly viscous fluid) surface, this assumption is oversimplified. In fact, any eddy which comes close enough to the surface to sweep away some of the mass accumulated in the concentration boundary layer will favorably affect the transfer rate, since the concentration gradient will be momentarily steeper near the fresh fluid.

Harriott (1962) suggested a model which incorporates the spatial distributions of the sweeping distances (from the interface), and the age distribution of the eddies at the surface. The incoming fresh eddies are visualized to arrive at random times, at a uniform solute concentration, within a random distance l from the surface, and sweep away the material from l to infinity. In the interval between the arrival of new eddies, molecular transport occurs in the wall region.

Mathematically, the process can be described by the unsteady diffusion Eq. (3.10), utilizing the penetration concept to formulate the spatial boundary conditions, Eqs. (3.11) and (3.14). The initial condition, however, is given by $P = f(y)$ at $t = 0, y > 0$. The mean transfer rate is then said to be a function of the instantaneous molecular transport within the wall region, and the mean frequency of rejuvenation, $S = 1/\tau$. The results, correlated in terms of the Nusselt number Hl/κ as a function of the reciprocal of the Fourier number $\tau\kappa/l^2$, depend on the assumptions made regarding the spatial and temporal characteristics of the system.

1. For a constant approach distance l, and a constant contact time τ, the numerically calculated average transfer coefficient for high Fourier numbers agree exactly with Eq. (3.20), with l representing the dimensional length parameter. For $Fo_\tau < 0.1$, the Nusselt number approaches the value of 1.0, consistent with the film theory. This is due to the fact that at low Fourier numbers, eddies are frequently replaced, and concentration at $y = l$ is practically maintained at (zero) bulk concentration. For the range of $10 < Fo_\tau < 0.1$, the following approximation (within 5% error) holds:

$$\frac{1}{H} = \frac{l}{\kappa} + \frac{\sqrt{\pi}}{2}\sqrt{\frac{\tau}{\kappa}} \tag{3.90}$$

   Equation (3.90) represents another example of the simple law of additive resistances and is applicable, within an error of 5%, to the case of transfer through a thin film of negligible capacity into a semi-infinite medium.
2. Constant approach distance l and variable contact time lead, at low values of l, (or high Fo_τ numbers), to τ-averaged transfer coefficients similar to those obtained by the "well mixed" Danckwerts' random age model, Eq. (3.24). Consequently, the transfer coefficients in this case are 13% lower than those exhibited in the above constant time case. The difference between the results for cases (1) and (2) diminish as the Fourier number decreases, and the resistance beyond l become negligible due to correspondingly higher eddy replacement rates.
3. Variable approach distances and constant (or variable) contact times gave transfer rates identical with case (1) for $Fo_\tau > 1$, but showed large effects of changing l in the range of $0.1 < Fo_\tau < 1$. This is demonstrated in Figure 24. The selection of the approach distances is based on the gamma distribution function, which for integer values of m is identical with Eq. (3.62), where δ/l replaces θ/τ, and l denotes the mean approach distance. With reference to Figure 24, it is noted that the differences between constant and variable l are larger with $m = 1$ (Danckwerts' random distribution function), than with $m = 2$ (2-capacitance model), since the former distribution is most strongly steered toward small approach distances. The difference also increases with decreasing diffusivities, i.e., increasing the Prandtl (Schmidt) numbers. This is due to the fact that the thickness of the concentration boundary layer is proportional to $\kappa^{-1/3}$, which means that if κ is low, the eddies have to come closer to the transport surface to have any effect. On the other hand, the time to establish steady state gradients following an impulse change is proportional to κ^{-1}, which means that relatively infrequent eddies can have a large effect at high Prandtl (Schmidt) numbers.

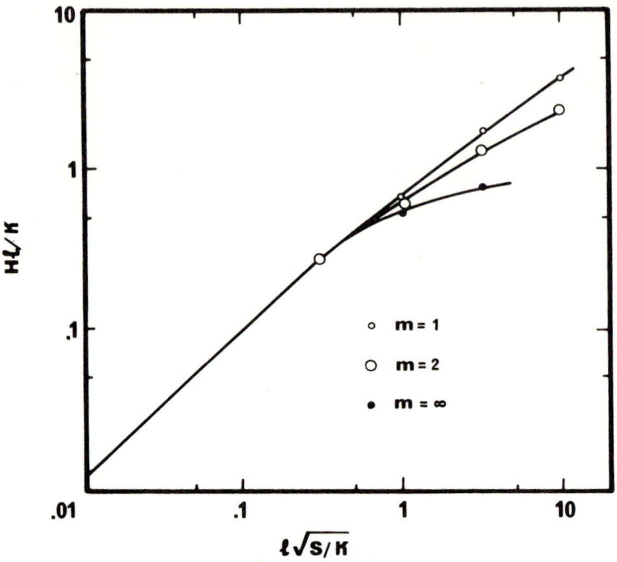

Figure 24 Average transfer coefficients for uniform residence time and constant ($m = \infty$) and random ($m = 1, 2$) approach distance distributions.

The detailed computational techniques are given by Harriott (1962) for turbulent mass transfer, and by Thomas et al. (1971) for temperature profiles in turbulent pipe flow. A formulation as a stochastic process was suggested by Dukler (1972) and extended by Thomas and Rajagopal (1974).

Quantitative evaluation of model parameters Practical utilization of Harriott's model requires assumptions regarding the nature of age distribution, and distribution of the approach distances. In view of the insensitivity of the average transport rate to the form of the age-distribution function, the choice of this function is of secondary importance. However, the assumptions pertaining to the approach distribution are of prime importance as the choice of the system parameter m dictates the functional dependence of the transport coefficient on the molecular diffusivity, from κ^{-1} to $\kappa^{-1/2}$. The absolute values of the coefficients depend on the choice of the system parameter m, the mean approach distance l, and the eddy renewal frequency or average residence time, $\tau = 1/S$.

Thomas and Fan (1971) specified these parameters and established the validity and usefulness of Harriott's surface rejuvenation model on a quantitative basis. The value of $m = 1$ leads, at large values of the abscissa (small Fourier or large Peclet number), to $\mathrm{Nu} \sim \mathrm{Sc}^{1/3}$ for large Sc (Pr) numbers, consistent with experimental evidence. (Hubbard and Lightfoot, 1966; Dilman, 1968). The value of the dimensionless approach distance was taken as $l^+ = 5$, based on the visual studies of Popovich and Hummel (1967). The mean rejuvenation frequency $S = 1/\tau$ was evaluated numerically by analogy to Harriott's model for transient momentum transfer (see Section 3.3.5 below). Since the calculation procedure for local mean wall shear stress is identical with Harriott's calculations for mean mass transfer flux, the results are identical, provided the correlating parameters are properly substituted. Hence,

$$\mathrm{Nu} = \frac{lH}{\kappa}$$

and

$$X = \frac{1}{\sqrt{\mathrm{Fo}_\tau}} \equiv \frac{l}{\sqrt{\kappa\tau}}$$

are replaced by utilizing the classical analogy, by

$$l^+ \sqrt{\frac{f}{2}}$$

and

$$x = \frac{l}{\sqrt{\nu\tau}}$$

respectively. Recall that

$$l^+ \equiv \frac{lu^*}{\nu} = \frac{lu}{\nu}\sqrt{\frac{f}{2}}$$

Figure 24 is now replotted to represent the momentum as well as the (heat) mass transfer fluxes as a function of the approach distance and the rejuvenation frequency $S = 1/\tau$.

As already mentioned with reference to Figure 24, the effect of the rejuvenation is significant only at $x > 0.5$. For smaller values of x, the penetration or surface renewal models suffice to predict the values of S or τ (see Section 3.3.5).

Figure 25 represents a reasonably useful practical tool for evaluating the transfer coefficient with turbulent transport at fluid-solid interfaces. Based on a knowledge of the friction factor associated with a given flow situation, Figure 25 relates the momentum transfer parameter $l^+ \sqrt{f/2}$ to x and then, for a given Prandtl number, $X = (\mathrm{Fo}_\tau)^{-1/2} = x\,\mathrm{Pr}^{1/2}$ is related to the mass (heat) transfer

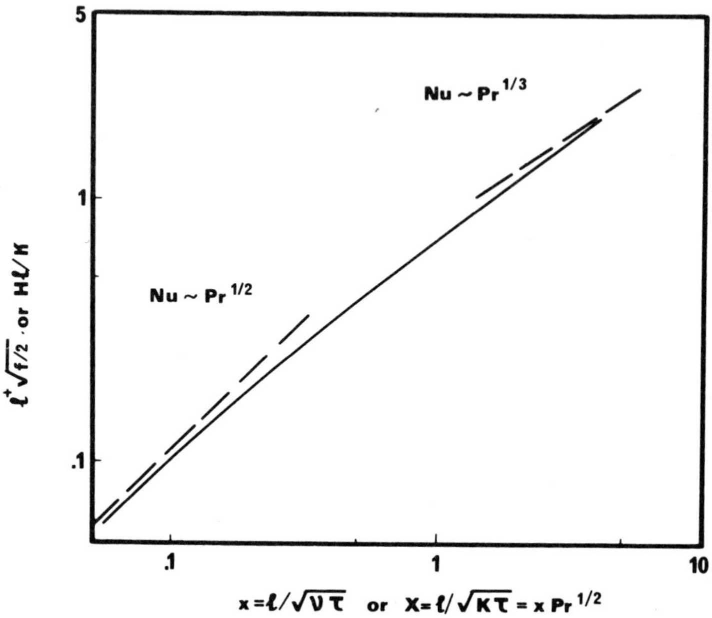

Figure 25 Calculations for heat (mass) and momentum transfer based on surface rejuvenation model ($m = 1$) and uniform residence times.

Nusselt parameter. Thus, for a given Reynolds number $l^+ \sqrt{f/2}$ is known, x is read and one can plot the dependence of the Nusselt number on the Prandtl (Schmidt) number. As demonstrated by Thomas and Fan (1971), the rejuvenation mechanism affects the mass (heat) transfer rates only for Pr(Sc) > 5, i.e., for fluids in which the major resistance is concentrated near the transport surface. For 1 < Pr(Sc) < 5, Danckwerts' basic penetration surface model holds and Nu ~ $\mathrm{Pr}^{1/2}$.

3.3 Models including convective effects

3.3.1 Introduction The limitation inherent in all models discussed in Section 3.2 is due to the assumption that convective effects inside the eddies are negligibly small in comparison with the temporal acceleration term. Fluid elements at, or near the transport surface were regarded as "solid" elements without a micro-hydrodynamic structure. By neglecting this micro-hydrodynamic characteristic, the renewal-penetration models have, in a sense, been divorced from the hydrodynamics that gave them birth. Here, we address ourselves to models which account for the convective effects, either directly through the convective-diffusion equation, or indirectly through the renewal mechanism near the transport surface. Solution of the unsteady convective-diffusion equation, Eq. (3.2), in which $u = u(x, y, t)$, and $v = v(x, y, t)$, together with the momentum equation, Eq. (3.1), requires numerical computation and is considered outside the scope of this study. Here, we apply ourselves to various models or theories brought about by physical rationale and intuition, as well as assumptions required to facilitate trackable solutions.

At a solid interface, all relative velocity vanishes, and tangential (parallel) and normal components of the velocity field near the interface are proportional to the normal distance from it. At a free interface, tangential viscous shear is usually negligible, and the tangential velocity component is practically y-independent close to the interface. However, y-dependent normal velocity is effective, and the rate of convection to or from the interface is directly proportional to the distance from it. The diffusional process is thus affected by convective motions, particularly in the vicinity of fluid interfaces. As shown by Scriven (1968, 1969), convection affects diffusion by tilting and sharpening or dulling concentration gradients, even when flow is perpendicular to the overall concentration difference.

Recognizing the difficulties associated with characterizing the turbulent chaos at a fluid interface, Chan (1964) and Scriven (1969) suggested the flow field can be approximated by a mixed population of relatively simple, practically laminar, small scale, microflow elements. The lifetime of these flows are related to the time-scales of agitating turbulence. However, unless a reasonable assumption attributing the rate dependence to a single type of the microflow elements can be made, this approach is restricted; similar to Danckwerts' general model expressed by Eq. (3.25), by the lack of information regarding the population dynamics of local flows at the interface.

3.3.2 The "generalized" convective-diffusional penetration theory Maintaining the shallow penetration depth concept, Ruckenstein (1968, 1971) assumed for a fluid-liquid interface, that tangential velocity along the transport surface is independent of the distance from the interface, y. Thus, $u \cong u(x, t)$, and by the continuity equation, $v = -y(\partial u/\partial x) = y, v_1, (x,t)$. The rate of mass transport in each fluid element may then be obtained by solving the following convective-diffusion equation:

$$\frac{\partial P}{\partial t} + u(x,t)\frac{\partial P}{\partial x} + yv_1(x,t)\frac{\partial P}{\partial y} = D\frac{\partial^2 P}{\partial y^2} \tag{3.91}$$

for the following initial and boundary conditions:

$$\begin{aligned} P &= P_b \quad \text{at} \quad t = 0 \\ P &= P_0 \quad \text{at} \quad y = 0 \\ P &= P_b \quad \text{at} \quad y \to \infty \\ P &= P_b \quad \text{at} \quad x = 0 \end{aligned} \tag{3.92}$$

Introducing a new similarity variable

$$\eta = \frac{y}{\delta(x, t)}$$

transforms Eq. (3.91) into two relatively simple equations: (1) an ordinary differential equation for the concentrations

$$\frac{d^2P}{d\eta^2} + a\eta \frac{dP}{d\eta} = 0 \tag{3.93}$$

and (2) a first order equation with partial derivative for $\delta(x, t)$, the thickness of the boundary layer

$$\frac{\partial(\delta^2)}{\partial t} + u\frac{\partial(\delta^2)}{\partial x} + 2\frac{\partial u}{\partial x}(\delta^2) = 2a\kappa \tag{3.94}$$

where a is an arbitrary (but positive) numerical constant, assigned the value of 2. A different technique for solving Eq. (3.91), based on constructions of the time dependent terms, was also suggested by Ruckenstein (1970).

Ruckenstein (1967, 1968, 1970, 1971) applied Eq. (3.91) and presented solutions to a number of problems involving surface waves and "almost" periodic fluctuations approximating turbulence fluctuations. The solutions require that velocities be apriori specified. In all cases, the calculated transfer coefficient dependence on the diffusivity is identical with renewal theories, i.e., $H \sim \sqrt{\kappa S}$, although the renewal rate, S, has a somewhat more complicated meaning because of the spatial dependence of the concentration along the x-coordinate.

Solutions of the convective-diffusion equation thus require stipulating assumptions which allow simplification or neglect of different terms. Scriven and Pigford (1959) applied the steady state form of Eq. (3.91) to study transfer characteristics of laminar jets. The solution, obtained by applying the two-parameter similarity parameter $\eta = y/\delta(x)$, shows that the transfer rate increases if the normal velocity is toward the interface. Conversely, a normal velocity component away from the interface reduces the interfacial transfer. It is to be noted that the effect of positive and negative velocity gradients are in general, not identical; the effect of flow toward the interface is stronger than the reverse (away) flow effect on the transfer rate.

A somewhat modified form of Eq. (3.91) was applied by Scriven and his coworkers to solve the problem of unsteady irrotational flow associated with his stagnation flow microelement model. The transfer characteristics are qualitatively similar to those obtained with the laminar jet. For the somewhat simpler case of steady irrotational stagnation flows, they found that as long as the exposure time is no longer than the time it takes fluid particles to move 20% closer to, or further from, the interface, the simple penetration theory is a fair approximation. At very long exposure times, the flux in stagnation flow asymptotically approaches a constant value, which is characteristic of the "film" theory. Conversely, the flux in reverse stagnation flow asymptotically approaches zero quite rapidly, and the total amount transferred approaches a finite asymptote. This is similar to the penetration model, though in the latter model the total amount transferred increases without reaching a finite value. Like most other models, practical application of this model awaits experimental correlation of the flow intensity, and the quantitative description of dynamics of fluid elements at the interface.

Similarly, transformations which permit solving the time-dependent, convective-diffusion equation for time-dependent, and time-independent velocity fields, and for deforming or fixed interfaces, have been proposed by Ruckenstein (1968, 1969). A solution procedure, based on introducing so-called "transfer coordinates," has also been suggested by Ruckenstein (1971), to be applicable for cases where u is also a function of y.

3.3.3 The surface stretch model The surface stretch approach extends the stagnant fluid penetration model to systems in which the area of the transport surface is a function of time, as would be the case at a free interface in turbulent flow, or in transfer to forming and/or oscillating drops. Clearly, the penetration-stretch model is related to the penetration-renewal concept. Here, the surface as well as the fluid elements at the surface are constantly or periodically renewed. We consider an interface which is formed suddenly between two quiescent fluid phases of uniform concentrations and the convection resulting from the sudden shrinkage or expansion of the interfacial surface area. Neglecting the tangential convection and diffusion terms, Eq. (3.91) now reduces to (Angelo et al., 1966):

$$\frac{\partial P}{\partial t} + v(y, t)\,\frac{\partial P}{\partial y} = \kappa\,\frac{\partial^2 P}{\partial_y 2} \tag{3.95}$$

As shown by Scriven (1969), the rate of convection away from, or towards, the free interface is proportional to the distance away from it, and the proportionality factor is the rate of interface dilation by surface inflation and surface stretch. For constant mass density and relatively small mass fluxes, normal local fluid velocity near the interface is given by

$$v = -y\left(\frac{\partial \ln A}{\partial t}\right) \tag{3.96}$$

where A stands for the time-independent transfer surface. The corresponding initial and boundary conditions are identical with those associated with the classical penetration theory, Eqs. (3.11), (3.12), and (3.14). Utilizing the similarity solution, the time averaged coefficient is given by

$$H = \frac{2}{\sqrt{\pi}} \frac{t_0}{t} \sqrt{\frac{\kappa}{t_0}} \int_0^{t/t_0} \left[\frac{A(t/t_0)}{A_0}\right]^2 d\left(\frac{t}{t_0}\right) \tag{3.97}$$

where A_0 and t_0 are characteristic (constant) reference area and time, respectively, and must be chosen separately for each special situation.

For the simple case of transfer to a drop forming at a submerged nozzle, $t_0 = t_F$ or formation time,

$$\frac{A(t/t_F)}{A_0} = 1 + \beta\left(\frac{t}{t_F}\right) \qquad \beta = \text{const} \tag{3.98}$$

and

$$\begin{aligned} H &= \frac{2}{\sqrt{\pi}}\left(\frac{t_F}{t}\right)\sqrt{\frac{\kappa}{t_F}}\left[\frac{t}{t_F} + \beta\left(\frac{t}{t_F}\right)^2 + \frac{\beta^2}{3}\left(\frac{t}{t_F}\right)^3\right]^{1/2} \qquad t < t_F \\ &= \frac{2}{\sqrt{\pi}}\left(\frac{\kappa}{t_F}\right)^{1/2}\left(1 + \beta + \frac{\beta^2}{3}\right)^{1/2} \qquad t = t_F \end{aligned} \tag{3.99}$$

Typical values for drop formation are $A_0 = 50\ \text{mm}^2$, $\beta \cong 2.5$ (equivalent to $1.25\ \text{mm}^2/\text{millisec}$), and $t_F \cong 115$ millisec. A similar expression was reported by Beek and Kramers (1962), and a somewhat different one was given by Groothuis and Kramers (1955), for the continual formation of a completely fresh surface. The latter yields transfer rates only 6% higher than the surface-stretch model. Comparison of Eq. (3.99) with the simple penetration model yields $H_{\text{stretch}}/H_{\text{penet}} = 2.67$, indicating the large contribution of the stretch mechanism (Angelo, 1966).

A compilation of the various expressions for the instantaneous transfer coefficient is presented in Table 7. Note that all these studies relate to a constant concentration of the body of the forming drops. A modification accounting for possible variation of the drop concentration was recently presented by Walia (1973).

For oscillating drops, where $A = A_0\ [1 + \bar{\epsilon}\sin^2(t/t_0)]$, the average drop side mass transfer coefficient is given by (Angelo et al., 1966):

$$H = \frac{2}{\sqrt{\pi}}\ [\kappa\omega(1 + \bar{\epsilon}_0)]^{1/2} \tag{3.100}$$

where $t_0 = 1/\pi\omega$, and $t/t_0 = \pi$, for a single cycle.

Table 7 Equations Proposed in Literature for the Instantaneous Transfer Coefficient During Drop Formation (*Walia, 1973*)

Investigators	Mechanism	$H_i(t) = M\sqrt{\kappa/\pi t}$
Licht & Pansing (1953)	The whole area becomes older according to unsteady state diffusion theory. Only area variation with time is taken into account	$M = 1$
Heertjes et al. (1954)	The velocity of diffusion is small compared to the velocity of drop growth	4
Sawistowski & Goltz (1963)	The velocity of diffusion is small compared to the velocity of drop growth	4
Groothuis & Kramers (1955) (linear surface growth rate)	Any increase in area during formation produces really fresh surface. There is no mixing between areas of different ages	1.56
Groothuis & Kramers (1955) (linear volume growth rate)	Any increase in area during formation produces really fresh surface. There is no mixing between areas of different ages	1.78
Ilovic (1934), Baird (1959)	Interface movement influences the diffusion layer thickness, stretching it evenly around the sphere	1.53
Calderbank & Patra (1966)	Area variation with time is taken into account by the empirical equation: $A = A_f\left[1 - \left(\frac{t}{t_f}\right)^n\right]$ where $n = 1.86 \pm 0.06$	1.82
Michaels (1960) (front point)	The growing spherical bubble advances into the fluid with its centre moving. The diffusional layer is stretched by the moving interface. Maximum stretching is at the front of the sphere and least at the rear	2.08
Michaels (1960) (rear point)	The growing spherical bubble advances into the fluid with its centre moving. The diffusional layer is stretched by the moving interface. Maximum stretching is at the front of the sphere and least at the rear	0.58
Angelo et al. (1966) (linear surface growth rate)	Convectional mass due to expansion of interface is taken into account	1.35
Angelo et al. (1966) (linear volume growth rate)	Convectional mass due to expansion of interface is taken into account	1.53

$\overline{\epsilon}_0 = \overline{\epsilon} + \frac{3}{8}\,\epsilon^{-2}$ where $\overline{\epsilon}$ denotes the dimensionless amplitude factor characteristic of the particular system, and the assumed surface time relation ($\epsilon \cong 0.2$) and ω, is the oscillation frequency.

Other expressions for surface stretch rate are summarized in Angelo's et al. paper. Since two phases, *D* and *C*, are involved, the overall transfer coefficient based on phase *D* is given by

$$H_{\text{overall},D} = H_D\left(\frac{1}{1 + m\sqrt{\kappa_D/\kappa_C}}\right) \tag{3.101}$$

where m denotes the equilibrium constant, κ_D, and κ_C, are the diffusivities of solute in phase *D*, and phase *C*, respectively.

3.3.4 Roll-cell models Basic surface renewal models assume that elements of fluid are continuously renewed at the interface, but no details concerning the hydrodynamics of renewal are given. The roll cell model, schematically illustrated in Figure 26, visualizes the renewal process as hydrodynamic rotating structures superimposed on the main motion. Renewal is thus a consequence of the assumed velocity distribution, rather than just a qualitative assumption. Conceptually, this model is related to the cellular convective motion due to buoyancy forces first observed by Bernard (1900). A detailed analysis of interfacial roll-cells is given by Scriven (1968, 1969).

Eddies, or "upwellings," which flow toward the surface by turbulent forces (inertial, pressure, or viscous coupling), are deflected by the surface, flow along it and, subsequently, plunge back into the main body of the fluid. Each element of fluid which reaches the interface generates motion of opposite signs, in the $\pm x$ directions. If velocities of the upwellings are relatively small, and the frequency at which turbulent pulses infringe on the interface is sufficiently large, a quasi-steady roll cell structure may form. Fresh fluid flows near the surface, and mass (heat) is transferred by molecular diffusion. The flow pattern within the eddy differs for flow near a solid surface, where at $y = 0$ $u = 0$, and a free fluid interface where $\mu\partial u/\partial y \cong 0$. Depending upon the relative viscosities, flow patterns at liquid-liquid interfaces may vary between these two extremes. Thus,

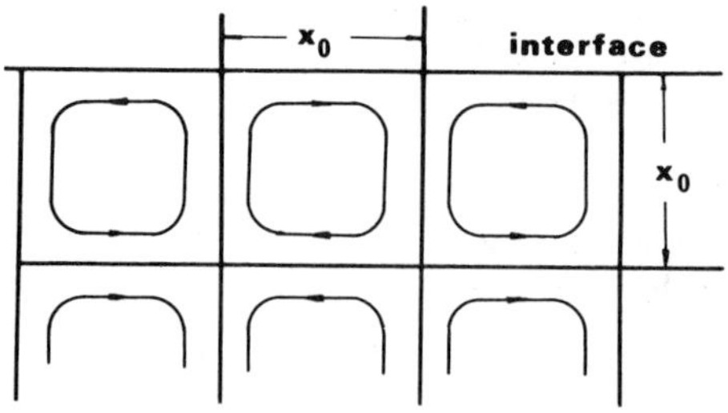

Figure 26 Roll cell model.

an element of fluid may have a translatory motion with the local average velocity of the liquid and a circulatory one in the roll cell. These characteristics are unchanged whether the roll cells are made of large scale eddies impinging on the interface, or by small scale eddies superimposed on these large scale motions.

Large scale eddy cell models Two limiting situations related to the motion inside the roll cell were considered by Ruckenstein (1971);

1. Elements of liquid keep their identity. The velocity gradient at the interface is negligible and no significant change of concentration occurs during motion along inside parts of the roll cell. The transfer coefficient can then be obtained by Higbie's model, by applying Eq. (3.15) where the contact time is taken as the time the element covers the distance X_0 along the interface. As shown experimentally by Ruckenstein and Suciu (1969), this situation may exist at small values of turbulent pulsation normal to the interface.
2. At higher pulsation rates, liquid elements inside the roll cell are completely refreshed during the inside motion, and the elements reaching the interface are fresh. In this situation the transfer mechanism is similar to the surface renewal models. Ruckenstein and Berbente (1970) assumed various expressions simulating the micro-structure velocities which account for the average x-directed velocity, and the superimposed rotating roll cell motion propagated by the average bulk liquid velocity, and solved the convective diffusion equation to obtain expressions for the transfer coefficient for some limiting cases.

For absorption into a turbulent liquid, the derived expression for the transfer coefficient is similar to Danckwerts' simple renewal model. However, the renewal rate is a function of the amplitude of the velocity oscillations and the "wave number" along the x-axis. For extraction into a liquid film, when interfacial turbulence due to the Marangoni effect is present (small amplitude of velocity fluctuation and small "wave numbers" along the x-axis, or large wave numbers along the y-axis), the transfer coefficient is related to the diffusivity by

$$H \sim \kappa^n \qquad n < 0.5 \tag{3.102}$$

Clearly, if $n \to 0$, the transfer rate becomes independent of the diffusivity, consistent with Kishinevsky's (1949, 1954) model for turbulent transport.

A detailed analysis of the quasi-steady large scale eddy cellular model at a fluid-liquid interface was presented by Fortescue and Pearson (1967). The eddy length scale is taken as 10% of the channel depth, and the intensity was taken to correspond to that of the bulk turbulent flow field ($\sqrt{\overline{u^2}} = 0.1\ u_b$). The agreement with experimental data was good.

The relatively uniform roll cells depicted in Figure 26 relate to fluid-liquid interfaces. As noted by Ruckenstein and Suciu (1969), the shapes of the cells will be quite different at a fluid-solid interface. However, if the quasi-steady condition prevails, the analysis reduces to solving the transport rate under laminar steady flow conditions along the contact length. This quasi-steady model is discussed in detail in Section 3.3.4.

Small scale eddy cell model An attempt to relate the turbulent energy spectrum to mass transfer behavior of an idealized small eddy motion near the transport surface was reported by Lamont and Scott (1970). This study is an extension of Fortescue and Pearson's (1967) analysis of transfer in gas-liquid channel flow, in which large energy containing eddies are assumed to control the mass transfer rate. Here, the whole range of eddy sizes actually present in a well developed turbulent field is considered. The small scale eddies are superimposed on the incoming eddies, Figure 27. These small scale eddies cause mixing within the very surface of large eddies, and the mass transfer rate is controlled by the motion of these predominately viscous small scale motions. This notion, at variance with Fortescue and Pearson's renewal mechanism where the transfer process is controlled by large scale energy containing eddies, is also supported by Miyauchi et al. (1970).

The model assumes that turbulent velocity fluctuations predominate over any mean relative velocity. Relating to an experimental situation of transport to bubbles (or solid particles) in turbulent flow, Lamont assumes that motions near the surface are a size scale smaller than the dimensions of the particles, so the surface may be considered nearly flat. All scales of motion smaller than the bubble are assumed to be within the equilibrium range of turbulence. Effects of surface tension, which play an important role in Levich's (1962), and the ensuing theory of Davies (1966), are assumed negligible for the small scales of motions considered in the model.

Figure 27 A small eddy superimposed on a large body. (Lamont and Scott, 1970).

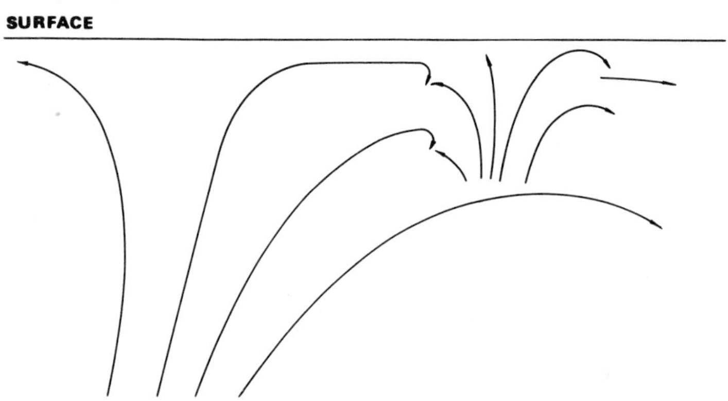

The motion, whether initiated by rotation of the element, or by a jet directed at the surface, is represented as a two dimensional roll cell of width X_0. A sinusoidal shearing motion of amplitude A exists at half height $y = X_0$ beneath the surface. This sinusoidal shearing motion is suggested by the Fourier analysis of the turbulent velocity field in which velocities are decomposed into sinusoidal shearing motions of different amplitudes and wave lengths. Solving the Navier-Stokes equation for small Reynolds numbers with this spatially periodic velocity field, and utilizing Kavaszney's (1959) spectrum for the energy spectrum in the dissipation range over the whole range of size scales (including the large inertial scales), Lamont obtained the following expressions for the average mass transfer coefficients:

$$H \cong C_1 (\mathrm{Sc})^{-2/3} (\epsilon\nu)^{1/4} \quad \text{solid surface} \tag{3.103}$$

$$H \cong C_2 (\mathrm{Sc})^{-1/2} (\epsilon\nu)^{1/4} \quad \text{fluid surface} \tag{3.104}$$

where ϵ, the rate of energy dissipation by turbulence per unit mass is related to the turbulent pipe flow Reynolds number by

$$\epsilon = 0.16\ \mathrm{Re}^{2.75} \frac{\nu^3}{2R} \quad \text{(c.g.s units)} \tag{3.105}$$

The estimated constants of proportionality, ($C_1 \cong C_2 \cong 0.4$), are larger by a factor of 2 than the experimental values of Calderbank and Moo-Young (1961) for a solid-fluid system, and Fortescue and Pearson's (1967) analysis of a gas-liquid interface. However, the analysis indicates mass transfer is not limited to any range of scales of motion, but is due to scales which extend from the smallest viscous motions well into the inertial motions, with a slight preference for the smaller scales.

Recently noted by Prasher (1973), Lamont's expression for free interface is consistent with King's (1966) model based on the eddy diffusivity concept. Comparing Danckwerts' surface renewal relations with Eq. (3.104) yields the relationship between the renewal rate and the energy of dissipation in the liquid and its kinematic viscosity, i.e.,

$$s = \frac{1}{T} \sim \left(\frac{\epsilon}{\nu}\right)^{1/2} \tag{3.106}$$

Utilizing the average velocity, $\bar{u}$, in channel flow of hydraulic radius R_H, to determine the energy dissipation rate ϵ,

$$\epsilon = \left(\frac{0.01\,\bar{u}}{R_H^{2/3}}\right)^2 \bar{u}g \quad \text{(m.k.s units)} \tag{3.107}$$

Prasher correlated mass transfer data (CO_2 to water) and obtained

$$H = 0.196\kappa^{1/2}\left(\frac{\epsilon}{\nu}\right)^{1/4} \tag{3.108}$$

thus substantiating Lamont's eddy cell model. Disagreement of the proportionality constant in Eq. (3.108), with Lamont's value of 0.4, is attributed to the use of the average velocity rather than the true unknown values at the interface.

3.3.5 Quasi-steady flow model Ruckenstein (1958, 1967) proposed a turbulent transport model which, unlike the transient surface renewal and penetration models, is based on assuming the flow within the wall region consists of a mosaic of simple, quasi-steady, boundary layer type, flow processes. Accordingly, Ruckenstein considers the flow field within individual elements in close vicinity of the wall to be analogous to steady boundary layer flow over a short flat surface of length X_0. Mathematically, temporal accelerating terms in the momentum and general convective-diffusion equations are neglected. Furthermore, for small mass fluxes (or no suction), the y-directed velocity in the boundary layer is zero, and Eqs. (3.1) and (3.2) reduce to:

$$u\frac{\partial u}{\partial x} = \nu\frac{\partial^2 u}{\partial y^2} \tag{3.109}$$

$$u\frac{\partial P}{\partial x} = \kappa\frac{\partial^2 P}{\partial y^2} \tag{3.110}$$

The corresponding boundary conditions are:

$$\begin{array}{llllll} u = u_i & P = P_b & \text{at} & x = 0 & y > 0 \\ u = 0 & P = P_0 & \text{at} & x > 0 & y = 0 \\ u = u_i & P = P_b & \text{at} & x > 0 & y = \infty \end{array} \tag{3.111}$$

where u_i is the mean velocity of the fluid elements at the first instant of renewal. Based on the standard Pohlhausen (1921) solution of the energy equation for laminar B.L. flow, Ruckenstein proposed the following expression for the mean transfer coefficient over the length X_0:

$$H = 0.666\kappa\sqrt{\frac{u_i}{\nu X_0}}\,P_r^{1/3} \tag{3.112}$$

The functional dependence of X_0 was then obtained by utilizing Blasius' solution of the momentum equation for mean shear stress (over X_0). Thus, with

$$\frac{u}{u_i} = 0.334 \sqrt{\frac{u_i}{\nu x}} \tag{3.113}$$

and

$$u_*^2 = \frac{\sigma_w}{\rho} = 0.666 u_i \sqrt{\frac{\nu u_i}{X_0}} = \frac{f}{2} u_b{}^2 \tag{3.114}$$

one obtains (Thomas and Fan, 1970):

$$X_0 = 0.443 \; \frac{u_i}{u*} \left(\frac{u_i}{u*}\right)^3 \tag{3.115}$$

Substituting Eq. (3.115) in Eq. (3.112) yields:

$$H = \left(\frac{\kappa}{\nu}\right)^{2/3} \frac{\sigma_w}{\rho} \frac{1}{u_i} \tag{3.116}$$

or

$$\mathrm{Nu} = \frac{u*}{u_i} \sqrt{\frac{f}{2}} \; \mathrm{Re}\, \mathrm{Pr}^{1/3} \tag{3.117}$$

Thomas (1971) rederived Eq. (3.117) by approximating the velocity and temperature profiles within the individual eddy at the surface ($y^+ < 5$) by

$$\frac{u}{u_i} = \frac{3}{2}\left(\frac{y}{\delta}\right) - \frac{1}{2}\left(\frac{y}{\delta}\right)^3 \tag{3.118}$$

and

$$\frac{P - P_0}{P_i - P_0} = \frac{3}{2}\left(\frac{y}{\Delta}\right) - \frac{1}{2}\left(\frac{y}{\Delta}\right)^3 \tag{3.119}$$

where from the momentum and temperature integral solutions,

$$\delta = 4.64 \frac{\nu x}{u_i} \quad \text{and} \quad \Delta = \delta \, \mathrm{Pr}^{-1/3}$$

$$\frac{\sigma_w}{\rho} = 0.647 \, u_i \sqrt{\frac{u_i \nu}{X_0}} \tag{3.120}$$

and

$$H = 0.647 \frac{P_0 - P_i}{P_0 - P_b} \kappa \sqrt{\frac{u_i}{\nu X_0}} \mathrm{Pr}^{1/3} \tag{3.121}$$

Eliminating X_0 by utilizing $u_*^2 = \sigma w/\rho = f/2\, u_b^2$ yields

$$\mathrm{Nu} = \frac{P_0 - P_i}{P_0 - P_b} \frac{u^*}{u_i} \sqrt{\frac{f}{2}} \mathrm{Re}\, \mathrm{Pr}^{1/3} \tag{3.122}$$

For fluids other than liquid metals, $P_i (= T_i)$ may be assumed to be equal to the bulk temperature P_b $(= T_b)$, and Eq. (3.122) reduces to Eq. (3.117). A generalized dimensionless temperature profile, which does not include T_i, is recommended for liquid metals (Thomas, 1971).

The value of u_i/u^* must be approximated in order to derive a quantitative expression. Hanratty (1956) and Einstein and Li (1956) assumed that fluid reaching the interface comes, in general terms, from the buffer zone rather than the main bulk core. Denoting this source by y_0^+, Hanratty suggested $y_0^+ = 30$, corresponding to $u_i/u^* = 13.5$ (based on Deissler's, 1954 experimental data). The corresponding value inferred by Einstein and Li is 15.7.

If u_i is taken as bulk velocity, u_b, substituting $u_*^2 = f/2\, u_b^2$ in Eq. (3.117) or Eq. (3.118) gives (Thomas and Fan, 1970):

$$\mathrm{Nu} = \frac{f}{2} \mathrm{Re}\, \mathrm{Pr}^{1/3} \tag{3.123}$$

It is noteworthy that Ruckenstein's (1967) derivation leads essentially only up to Eq. (3.117), and does not contain the ratio u_i/u^*. The value of u_i/u^* is an implied constant, to be evaluated by comparison with experimental mass transfer data. Utilizing Harriott and Hamilton's (1965) data for Schmidt numbers in the range of 4.3×10^{-2} to 10^5, and taking $f \sim \mathrm{Re}^{-0.2}$, Ruckenstein infers that

$$\mathrm{Nu} \cong 0.0096\, \mathrm{Re}^{0.9}\, \mathrm{Pr}^{0.33} \tag{3.124}$$

The quasi-steady model functional relationships are similar to the familiar Colburn's correlation. Significant difference between the quasi-state model, represented by Eq. (3.117), and the transient surface renewal models, characterized by $\mathrm{Nu} \sim \mathrm{Pr}^{0.5}$, lies in the predicted relationships between the Nusselt number and the Schmidt (Prandtl) number. Comparison of Eq. (3.123) with Dilman's (1968) data is presented in Figure 28.

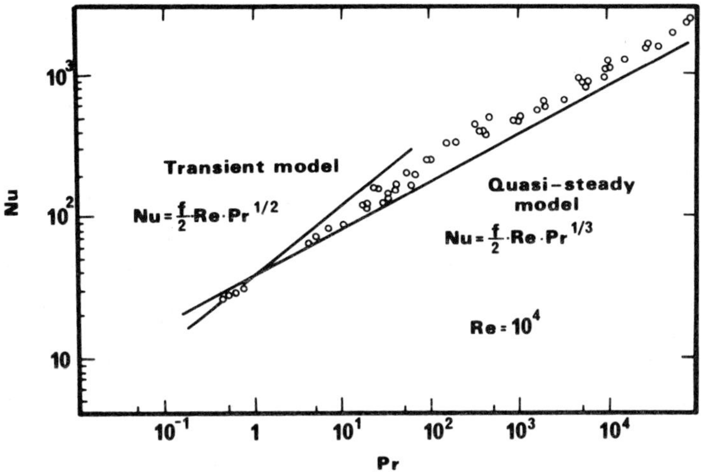

Figure 28 Comparison of quasi-steady model with Dilman's (1968) experimental data. (After Thomas and Fan, 1971).

Dilman's (1968) data is quite typical of heat and mass transfer data for turbulent tube flow, showing Nu $\sim$ Pr $^{1/3}$ for large values of the Prandtl number. However, agreements of Eq. (3.123) with this set of data, and others (Thomas, 1971), is rather unsatisfactory. This is possibly due to the fact that Ruckenstein's model, like the former basic renewal models, is based on the notion that eddies move into direct contact with the surface. As will be shown below, this restricts the above analysis to Pr $\cong$ 0(1), where the concentration or thermal resistance is not highly restricted to the wall layer region of $y^+ < 5$. Ruckenstein's quasi-steady model, which leads to the experimentally verified relationship Nu $\sim$ Pr $^{1/3}$ at high Pr numbers, does not validate Eq. (3.123), but implies that micro-hydrodynamics in the eddies at the wall surface are probably of the quasi-steady type. As shown below, the correct apriori quantitative prediction of the transfer coefficient must account for the presence of the relatively unreplenished wall layer.

Finally, accuracy of the solution for flow past a finite plate depends on the length X_0, and the solution presented here is, by necessity, limited to high Reynolds numbers. A detailed discussion of the problem of viscous incompressible flow past a finite flat plate was presented by Pao and Daugherty (1968).

3.3.6 Transient-periodic viscous sublayer models

Description of the periodic mechanism Based on Page and Towend's (1932) visual studies of dust particles in water (Pr $\cong$ 5) near a solid surface, and Lin et al. (1953) mass transfer data (Sc = 900, 400 < Re < 400), Hanratty

(1956) suggested the classical laminar sublayer "is much thinner than has commonly been assumed, or it is non-existent," as turbulent eddies seem to reach the interface, or very close to it. This apparent disagreement with Prandtl's classical laminar sublayer forms the basis for the transient-periodic-viscous sublayer models.

Einstein and Li (1956) noted that if the sublayer flow is assumed steady, quasi-laminar, it is difficult to present a physical description of transition to the turbulent part of the flow. This is due to the fact that turbulence is created only in connection with a frictional boundary or at interfaces between fluid masses of different velocities. A satisfactory model of the viscous sublayer must, therefore, explain how turbulence is created.

Einstein and Li were evidently the first to propose that the turbulent shear stress mechanism is governed by an inherent instability within the flow. The viscous sublayer undergoes regular periods of relatively slow viscous development, terminated by rapid breakdown resulting from the instability inherent in the sublayer flow. Basically similar ideas were presented by Malkus (1956), and Grant (1958), who proposed that large eddies are stress relieving motions comprising of coherent "mixing jets" generated by an instability in the turbulent shear stress mechanism.

The experiments of Kline et al. (1963) revealed a distinctive tempospatial flow pattern in the sublayer, with blending regions where fluid is periodically ejected outward from the wall in localized jet like motion. These eruptive motions exhibit a characteristic distribution over the wall which appears to scale on the sublayer thickness, and are generally consistent with the presence of a moving array of powerful, locally disruptive, instabilities located near the edge of the viscous sublayer.

Black's (1968, 1969) model is based on the notion that primary function of the instability induced eruption process is to periodically relieve stress buildup in the viscous sublayer. The model presents a detailed picture of the shear stress mechanism, the turbulence structure, and the mean motion. A schematic presentation of the model is depicted in Figure 29. This model, like others, relates to the large scale eddies which interact with the basic shear flow to affect the mean trasfer rate of mass, momentum, and energy across the layer. The eddies in this model, however, are vortex structures which are generated and maintained by a tempospatial nonlinear instability in the sublayer flow. Wall turbulence is viewed as an organized, time-dependent, viscosity governed, primary motion. It becomes unstable, and breaks down periodically in space and time to generate and maintain a system of characteristic horseshoe vortex structures, which are intimately coupled to the temporal and spatial behavior of the sublayer. The ensuing derivation leads to a description of the wall turbulence up to $y^+ = 50$, from which the time-independent law of the wall holds.

A strong support to the concept of period fluctuations near the wall is provided by excellent visual studies of Corino and Brodkey (1969). These

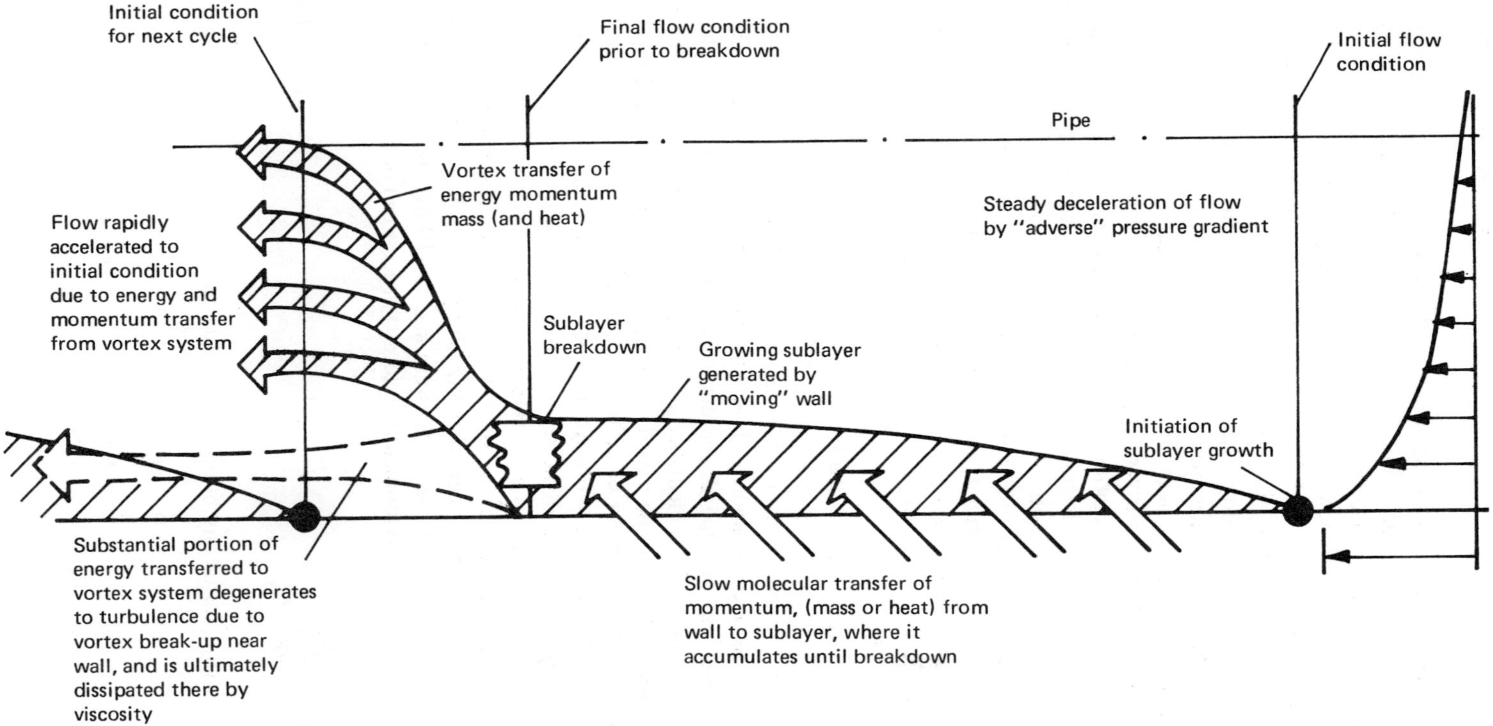

Figure 29 Cyclic development and breakdown of primary motion. (After Black, 1969).

authors attribute the onset of instability at the wall to penetrating eddies which accelerate low momentum fluid near the wall. Obviously, the transport mechanism at the wall is independent of the origin of the instability.

Classical surface renewal models assume that high energy eddies come from the turbulent core, "kick," and replace "stale" eddies from the interface, then remain there motionless. The unsteady periodic sublayer concept introduced here relates the eddy replacement mechanism to instability induced eruptions of the "old" eddies from the viscous sublayer. While Danckwerts' type eddies come to rest at, or near the surface, the eddies in the periodic sublayer model reach the interface (or very close to it) at a certain uniform axial velocity, u_i. Because of viscous interaction with the wall, a velocity gradient is developed in the rushing fluid which is gradually decelerated in a progressively developing, time-dependent, viscous motion at the wall. This process ends with a sudden breakdown and instantaneous eruption of the fluid into the main core. A new inrushing turbulent eddy starts the cycle. Clearly, the mean period of the cycle, i.e., the average contact time of the eddy with the transport surface, is the principle parameter required for a quantitative description of the rates of the transport process.

The velocity profile in the viscous sublayer Whereas an accurate description of the transient characteristics of the sublayer in turbulent flow may be quite complex, practical results can be obtained if the average cycle of eddy replacement and velocity-gradient development is very short so the shallow penetration concept can be applied. Assuming instantaneous impulsive replacement, the cycle time until now is identified only with the velocity development period.

For Sc(Pr) No $\geqslant 1$ the mass (heat) transfer rate is controlled by the fluid characteristics in the region close to the transport surface, and mathematical description of the periodic hydrodynamic phenomenon can be greatly simplified by restricting the analysis to the region close to the wall, i.e., $y^+ \leqslant 50$. Under these conditions, the "decaying" velocity profile (from some initial uniform velocity u_i) near the wall can be treated, with a large measure of confidence, as an impulsively induced phenomenon, similar to the penetration-renewal approach to heat (mass) transfer. Einstein and Li (1956) and Hanratty (1956), assumed that sublayer development (or decay) process can be described by

$$\frac{\partial u}{\partial t} = \nu \frac{\partial^2 u}{\partial y^2} \tag{3.125}$$

and the following initial and boundary conditions:

$$\begin{aligned} u &= u_i \quad \text{at} \quad y > 0 \quad t = 0 \\ u &= 0 \quad \text{at} \quad y = 0 \quad t > 0 \\ u &= u_i \quad \text{at} \quad y = \infty \quad t > 0 \end{aligned} \tag{3.126}$$

Solution of Eqs. (3.125) and (3.126) yields the well known error function relationship for the instantaneous velocity:

$$u = u_i \operatorname{erf} \eta \qquad \eta \equiv \frac{y}{2\sqrt{\nu t}} \tag{3.127}$$

This result implies a discontinuity at the wall at $t = 0$, but all time averaged values are well behaved. Introducing $y^+ = y\, u^*/\nu$, $u^+ = u/u^*$, and $t^+ = u^*\sqrt{t/\nu}$ into Eq. (3.127) yields

$$u^+ = u_i^+ \operatorname{erf}\left(\frac{y^+}{2t^+}\right) \tag{3.128}$$

The time averaged velocity distribution in the sublayer at any distance y from the surface during the average cycle period (again, neglecting the breakup and replacement time as compared to the sublayer developing time), is given by (Hanratty, 1956):

$$\bar{u}^+ = \frac{\bar{u}}{u^*} = \frac{u_i}{u^*}\int_0^1 \operatorname{erf}\left[\frac{\sqrt{\pi}}{4}\,\frac{y^+}{(u_i/u^*)\sqrt{(t/\tau)}}\right] d\left(\frac{t}{\tau}\right) \tag{3.129}$$

The time averaged shear stress, or average momentum flux of a fluid element arriving at a uniform axial velocity u_i, and staying at the wall for a time period τ is given by

$$\sigma_w = \frac{\mu}{\tau}\int_0^\tau \left.\frac{\partial u}{\partial y}\right|_{y=0} dt = \frac{2}{\sqrt{\pi}}\,\rho u_i \sqrt{\frac{\nu}{\tau}} \tag{3.130}$$

or

$$u^* = \sqrt{\sigma_{w/\rho}} = \frac{2}{\sqrt{\pi}}\left(\frac{u_i}{u^*}\right)\sqrt{\frac{\nu}{\tau}} \tag{3.131}$$

Numerical evaluation of the velocity profile near the wall thus depends on proper choice of the bounding velocity u_i. Hanratty (1956) recommended that $u_i^+ = u_i/u^* = 13.5$, corresponding to $y^+ = 30$, the experimental breakpoint between the turbulent core and buffer (and laminar) zone. Einstein and Li (1956) recommended, though indirectly, the value of $u_i^+ = 15.7$. Meek and Baer (1970) introduced the effect of the bulk flow Reynolds number on u_i^+ by including the turbulent core in their analysis. According to these authors, u_i^+

varies from 19 at Re = 2,100, to 16 at Re = 2.5×10^5. The computed time averaged velocity profiles in the sublayer are compared in Figure 30 with the experimental data of Laufer (1961), Hettler et al. (1964), and Popovich and Hummel (1967).

Since $u = u_i$ is taken at $y = \infty$ in the solution of Eq. (3.125), Eq. (3.127) to Eq. (3.130) should hold for values of y^+ much smaller than $y^+ = 30$. The discrepancy near the outer edge of Figure 30 is due to this semi-infinite boundary condition. As seen in Figure 30, the effect of the bulk flow Reynolds number on the velocity profile at the wall $y^+ < 20$ is quite small, and the simple model presented here suffices to represent the velocity profile at the wall.

A somewhat simpler way to introduce the effect of the Reynolds number is based on the assumption that fluid elements reach the surface from the main turbulent core, and u_i may be taken to represent the axial turbulent velocity of the bulk, u_b. With this assumption

$$\frac{u_i}{u^*} = \frac{u_b}{u^*} = \sqrt{\frac{2}{f}} \tag{3.132}$$

the functional dependence of the friction factor f on the Reynolds number is given by the well known correlations for turbulent flow. For Re $> 10^4$, Eq. (3.131) becomes (Thomas and Fan, 1971):

$$u^* = \lambda \sqrt{\frac{2}{f}} \sqrt{\frac{\nu}{\tau}}$$

Figure 30 Calculated time averaged velocity profiles in the sublayer compared with previously measured values. (Meek and Baer, 1970).

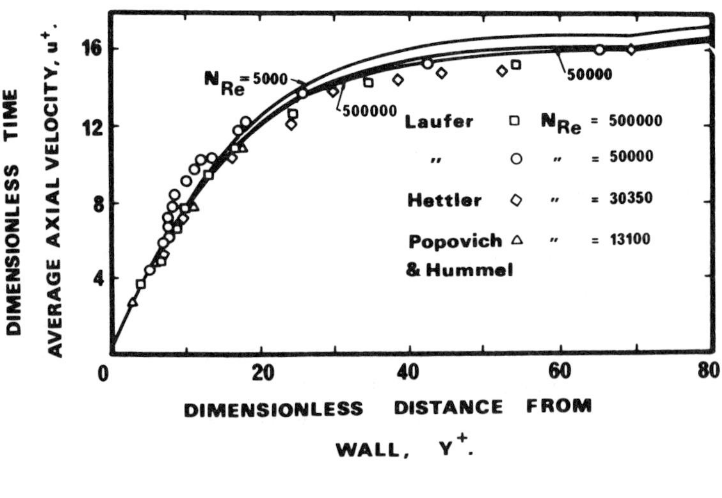

where $\lambda = 2/\sqrt{\pi}$ for Higbie's uniform time distribution, and $\lambda = 1$ for Danckwerts' random time distribution.

For fully developed turbulent pipe flow ($\mathrm{Re} < 10^4$), Gross and Thomas (1972) included the axial pressure gradient dp/dx in Eq. (3.125). The solution, utilizing Danckerts' random contact time distribution, yields

$$u^* = \sqrt{\frac{\nu}{\tau}} \frac{\dfrac{2u_i}{u^*}}{1 + \left(1 - \dfrac{16\, u_i}{\mathrm{Re}\, u^*} \sqrt{\dfrac{2}{f}}\right)^{1/2}} \tag{3.134}$$

Equation (3.134) reduces to Eq. (3.133) as the pressure gradient becomes less effective at $\mathrm{Re} > 10^4$.

All the models above assume the fluid elements reach the wall at a constant uniform approach velocity u_i, and the entire sublayer periodically becomes part of the turbulent core. Experimental evidence, however, shows that although most eddies penetrate to within a small distance from the wall, only occasional eddies actually replace the fluid down to the wall surface itself. Thomas and Fan (1971) suggested Harriott's rejuvenation model (Section 3.2.10) and computational technique be used to account for the effect, on the wall shear stress, of eddies which move to within different distances from the wall. By analogy to Harriott's formulation, the flow field during the boundary development period (between replacements), is represented by Eqs. (3.125), (3.126), and (3.127), with

$$u - u_b = F(y) \qquad \text{at} \qquad t = 0 \quad y > 0 \tag{3.135}$$

replacing $u - u_i = 0$ stipulated by Eq. (3.126a). $F(y)$ represents initial velocity profile within the wall region, and reduces to a constant (or zero) if all penetrating eddies are assumed to reach the wall as per the renewal theory. Resulting numerical calculations, correlated in terms of $l^+ \sqrt{f/2}$ vs $l/\sqrt{\nu\tau}$, are presented in Figure 25. It is, however, to be pointed out that the lower part of the curve which corresponds to Eq. (3.131) suffices for most practical hydrodynamic systems. In other words, the effect of rejuvenation is not significant for momentum transfer, but only for mass (heat) transfer in systems where $\mathrm{Sc(Pr)} > 1$.

The sublayer thickness If periodically developed viscous sublayer thickness is defined as the point where the velocity becomes 99% of the bounding velocity u_i, Eq. (3.127) yields

$$\delta = 3.64(\nu t)^{1/2} \tag{3.136}$$

and the maximum thickness, at the end of the wall contact period τ is

$$\delta_{max} = 3.64(\nu\tau)^{1/2} \tag{3.137}$$

or, in dimensionless units

$$\delta^+_{max} = 3.64(\tau^+)^{1/2} \tag{3.138}$$

where $\tau^+ = \dfrac{\tau u_*^2}{\nu}$ and $\delta^+ = \dfrac{\delta u^*}{\nu}$

By analogy to Eq. (3.136), the concentration (thermal) sublayer thickness δ_K, is given by (Meek and Baer, 1973)

$$\begin{aligned} \delta_K &= 3.64(\nu t)^{1/2}\mathrm{Sc}^{-1/2} && \text{— const. surface concentration} \\ &= 3.20(\nu t)^{1/2}\mathrm{Sc}^{-1/2} && \text{— const. surface flux} \end{aligned} \tag{3.139}$$

Clearly, $\delta_K/\delta < 1$ for $\mathrm{Sc} > 1$, and renewal models based on the shallow penetration concept hold. However, for $\mathrm{Sc(Pr)} < 1$, the concentration (thermal) wave extends beyond the growing viscous sublayer, and resistance to mass (heat) transfer in the turbulent region must be considered. (Thomas and Fan, 1971; Meek and Baer, 1973).

The thin wall-layer The assumption that the entire sublayer down to the wall is periodically replaced is an oversimplification as viscous forces will hold some of the fluid at the wall. Hydrodynamic studies of Kline et al. (1967), Popovich and Hummel (1967), and Corino and Brodkey (1969), show that although most eddies penetrate to within a small distance from the wall, only occasional eddies have sufficient energy to replace the fluid down to the wall surface itself. This results in the establishment of a thin periodic fluid layer adjacent to the wall which is replaced at a much lower frequency than most of the fluid in the sublayer.

The thickness of the thin wall-layer, δ_1, can at present only be estimated. Corino and Brodkey (1969) note that minimum distance from the wall at which ejection of fluid was observed to originate was about $y^+ = 2.5$. On the basis of dye injection experiments, Kline et al. (1967), conclude that fluid is ejected from the wall region down to $y^+ \cong 0$. However, they noted the time taken for fluid to be ejected from this level is considerably longer than for the fluid further away from the wall. Consistent with this are the observations of Popovich and Hummel (1967). Using a photoanalysis technique, they observed the existence of a broadening linear trace in the $y^+ = 1.6 \pm 0.4$ region. Based on measured temperature fluctuations at the wall of the tube, Meek and Baer (1973) suggest that the thickness of the wall-layer is of the order $y^+ \cong 1\text{-}1.5$.

Choosing a value of $\delta_1^+ = 1.5$ brought their model prediction into close agreement with Deissler's (1955) correlations up to Prandtl numbers of about 100. Katsibas and Gordon (1974) modified Hanratty's (1956) approach [Eqs. (3.125), (3.126)] to account for the thin wall-layer. By curve-fitting to Deissler's correlation they suggest $\delta_1^+ = 4.0/\mathrm{Pr}^{1/7}$. Using availabe heat and mass transfer data, Hughmark (1969) suggested the thickness of the effective laminar layer at the wall is of order $y^+ = 0.5$. Hutchinson et al. (1971) found that existence of a viscosity dominated wall-layer of thickness $y^+ = 1.25$ was consistent with experimental data for deposition of particles from turbulent gas streams.

In view of the above information, it is not unreasonable to choose the value $y^+ = 1$ for the wall-layer thickness δ_1^+. We note that variations in δ_1^+, consistent with the above range of values, do not greatly affect the predicted mass (heat) transfer results discussed below. Meek and Baer (1970, 1973) showed that persistence of this thin wall-layer does not significantly influence hydrodynamic characteristics of the locally developing boundary layer flow. However, its influence on mass (heat) transfer cannot, in general, be ignored.

For Schmidt number of order one, the concentration boundary layer extends to $y^+ \cong 30$, and the thin wall-layer will have a negligible effect on the overall transfer rate from the wall to the fluid. However, with increasing Schmidt numbers, thickness of the concentration boundary layer decreases, and an increasing proportion of the overall resistance is due to the thin wall-layer. At very high Schmidt numbers, the depth of penetration of the concentration boundary layer is smaller than the thickness of the wall-layer, and all of the resistance concentrates in the thin wall-layer. Analysis of the transport rate must be formulated accordingly. However, it is noted here, for low (0.5–10) and high ($>$ 1,000) Schmidt numbers, mass (heat) transfer rates are actually independent of the absolute value of δ_1^+.

The periodic sublayer time scales To account for the effect of the thin wall-layer, Pinczewski and Sideman (1974) suggested the penetrating eddy population be arbitrarily divided into two groups: the majority, which reach down to δ_1^+, and the minority which reach down to the wall. In order to predict transfer rates, it is first necessary to estimate the frequencies, or periodic time scales, associated with these two groups of eddies.

The mean period between turbulent penetrations (and replacement of the developing sublayer) is the principle parameter required for quantitative calculations of rates of the transport processes.

Analysis of the cyclic nature of the viscous sublayer in turbulent flow indicates the cyclic period τ scales with the wall parameter u^*, the frictional velocity, and the kinematic viscosity, ν. Thus, utilizing the universal dimensionless definition for the axis length $y^+ = y\ u^*/\nu$, and replacing y by the characteristic length $(\nu t)^{1/2}$, yields the dimensionless time $t^+ = (t/\nu)^{1/2}\ u^*$. The corresponding dimensionless mean time period is $\tau^+ \equiv \tau u_*^2/\nu$ and the time period is given by

$$\tau = \frac{\tau^+ \nu}{u_*^2} \tag{3.140}$$

Equation (3.140), which relates to all eddies approaching the wall region, can be derived directly from Eq. (3.131). Comparing Eq. (3.131) with Eq. (3.140) yields the dimensionless time constant τ^+.

$$\tau^+ = (\bar{T}^+)^2 = \lambda^2 \left(\frac{u_i}{u^*}\right)^2 \qquad \lambda = \frac{2}{\sqrt{\pi}} \text{ — uniform time period } \tau$$
$$\lambda = 1 \text{ — random time distribution, average } \tau \tag{3.141}$$

where $\bar{T}^+ = (\tau^+)^{1/2}$ corresponds to Meeks' (1972) notation, based on Eq. (3.138), which reads $\delta^+_{max} = 3.64\,\bar{T}^+$.

For a constant velocity ratio, τ^+ is independent of the bulk flow parameters. Thus, taking Hanratty's (1956) suggestion of $u_i/u^* = 13.5$ and $\lambda = 2/\sqrt{\pi}$ yields $\tau^+ = 232$ or $\bar{T}^+ = 15.2$. Meek and Baer's (1970) model yields $\bar{T}^+ = 18$ for $\mathrm{Re} > 10^4$. A compilation of available experimental data is given by Black (1969) and Meek (1972). The latter is represented in Figure 31, showing that the data range from $14 < \bar{T}^+ < 22$. However, in view of the experimental data of Corino and Brodkey (1969), Kline (1959, 1967), and Kim (1971), the value of $\bar{T}^+ = 15.6$ or $\tau^+ = 243$ is probably the best approximation. This is substantiated by the fact that for $\bar{T}^+ = 18$, the sublayer maximum thickness δ^+_{max} is of order 80, while the analysis of Black (1969), as well as Popovich and Hummel's (1967) experiments, indicates δ^+_{max} to be somewhere between 30 and 50.

The periodic time scale near the wall can formally be scaled with bulk flow parameters by taking u_i to represent bulk velocity. Introducing $u_b/u^* = \sqrt{2/f}$ in Eq. (3.141) gives

$$\tau^+ = \lambda^2 \frac{2}{f} \tag{3.142}$$

Utilizing Blasius' expression for the friction factor in turbulent flow in smooth tubes, $f/2 = 0.023\,\mathrm{Re}^{-0.2}$ yields, with $\lambda = 1$

$$\tau^+ = 43.5\,\mathrm{Re}^{0.2}$$

or (3.143)

$$\bar{T}^+ = 6.6\,\mathrm{Re}^{0.1}$$

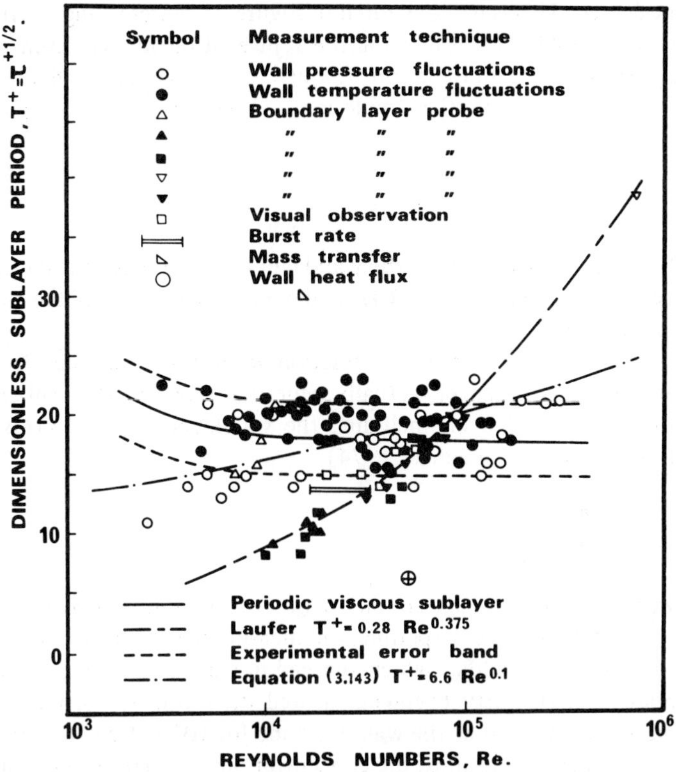

Figure 31 Dimensionless mean sublayer period versus Reynolds number. (After Meek, 1972).

In view of experimental evidence, Eq. (3.143) is probably an over-simplified approximation,* although it does seem to fit available experimental data better than Laufer and Narayanan's (1971) expression $\bar{T}^{+} = 0.28\ \mathrm{Re}^{0.375}$.

Effect of the mean axial pressure gradient on mean renewal frequency was found (Thomas and Greene, 1973) negligible for values of Reynolds number above 10^4. However, this effect is significant for deeper molecular penetration associated with low Reynolds numbers, and is helpful in correlating effects of drag reducing polymers.

It is important to note that the value of τ^+ is related to the entire sublayer growth period. However, the persistent, infrequently replaced, thin wall-layer or order $y^+ \cong 1.0$ affects the value of τ^+ only by some 2%, and as far as the hydrodynamics of the system is concerned, the thin wall-layer can usually be

*A similar approach was recently taken by Katsibas and Gordon (1974) for drag reducing solutions, utilizing Virk et al. (1967) maximum drag reduction asymptote $f/2 = 0.21\ \mathrm{Re}^{-0.55}$.

neglected. This is demonstrated in Figure 25 representing Harriott's, and Thomas and Fan's (1971) results for the rejuvenation model. Even with $\delta^+ = 5$, the hydrodynamic rejuvenation frequency is practically identical with the one derived by utilizing the simple renewal concept which, with $u_i = u_b$, is given by

$$\tau = \frac{2}{f}\,\frac{\nu}{u_*^2} \tag{3.144}$$

However, as seen below, the wall-layer effect on mass (heat) transfer rate cannot be neglected at $Sc > 0\,(10)$, and replacement rate of the thin wall-layer is of importance.

Assuming that some fraction b of the total number of eddies which approach the sublayer, actually penetrate down to the wall and displace all the fluid in the sublayer, including the wall-layer, Eq. (3.140) is modified to read (Pinczewski and Sideman, 1974)

$$\tau_1 = \frac{\tau_1^+ \nu}{u_*^2} \tag{3.145}$$

where τ_1 is the average contact period of the thin wall-layer, and $\tau_1^+ = \tau^+/b$.

Only limited experimental measurements which allow determination of the fraction of eddies which succeed in reaching the wall are presently available. Shaw and Hanratty (1964) measured the frequency of local fluctuation of the mass transfer rate at the wall of a tube for $10^4 < \mathrm{Re} < 6 \times 10^4$ and $\mathrm{Sc} = 2{,}400$. Under these conditions, the concentration boundary layer is of order $y^+ = 0(1)$. Thus, measured fluctuations are directly related to the thin wall-layer, indicating the value of b is of order 0.1. Although the measured frequencies show a slight dependence on the Reynolds number, data is limited, and the mean value of b = 0.07 is recommended (Pinczewski and Sideman, 1974). Note that a variation of b from 0.05 to 0.15 affects predicted transfer coefficients only by some 15%, which is within the normal range of experimental error. Introducing $\tau^+ = 243$ and $b = 0.07$ into Eq. (3.145) yields $\tau_1^+ = \tau^+/b \cong 3{,}470$.

The transfer rates Mean renewal frequency of the periodic sublayer can now be utilized in the quantitative solution of transient mass (heat) transfer equations. Maintaining shallow penetration depth, convective terms in the convective diffusion equation are neglected and only one dimensional molecular transport is considered. The governing equation and boundary conditions for an individual fluid element in contact with the surface are then identical with Eqs. (3.10), (3.11), (3.12), and (3.14). Utilizing the simple expression for average renewal rate, Eq. (3.144), Thomas (1971) and Thomas et al. (1971) obtained

$$\mathrm{Nu} = \frac{P_0 - P_i}{P_0 - P_b}\,\frac{u^*}{u_i}\sqrt{\frac{f}{2}}\;\mathrm{Re}\,\mathrm{Pr}^{0.5} \tag{3.146}$$

Consistent with Eq. (3.144), taking $u_i = u_b$ and $u_b/u_* = \sqrt{2/f}$ reduces Eq. (3.146) to

$$\mathrm{Nu} = \frac{P_0 - P_i}{P_0 - P_b}\frac{f}{2}\,\mathrm{Re}\,\mathrm{Pr}^{0.5} \tag{3.147}$$

The quantitative determination of the transfer rate depends on the relation between P_i and P_b.

For Sc(Pr) $\ll 1$, associated with heat transfer to liquid metals, quantitative solution of Eq. (3.147) depends on correct specification of the initial eddy concentration, P_i. Since for Sc $\ll 1$, $\delta_\kappa \gg \delta_M$, i.e., the thermal boundary layer penetrates the core region from which the eddies originate, it is likely that eddies originating in this region will have a P_i value other than P_b. Thomas (1971) proposed that P_i ($= T_i$) may be evaluated by analysis of energy and momentum transfer to eddies during their transit from the core region to the wall. The resulting expression for the Nusselt number is given as

$$\mathrm{Nu} = \frac{1}{1+\chi}\frac{f}{2}\,\mathrm{Re}\,\mathrm{Pr}^{0.5} \tag{3.148}$$

where χ is a function relating P_i ($= T_i$) to P_b ($= T_b$), taken directly from the analysis of Azer and Chao (1960) for heat and momentum loss from a spherical eddy. This analysis, discussed in Section 2.5.3, shows function χ is similar to the concept of a turbulent Prandtl number in the classical eddy diffusivity approach to turbulent transport. It is important to note that the same limitations and uncertainties inherent in the concept of Pr_t (see Section 2.5) also apply to function χ.

A plot of the temperature ratio calculated by utilizing χ is shown in Figure 32. Like analogous expressions for Pr_t, χ is only slightly dependent on the

Figure 32 Variation of the parameter $(P_0 - P_i)/(P_0 - P_b)$ with Reynolds number for various values of Pr_t.

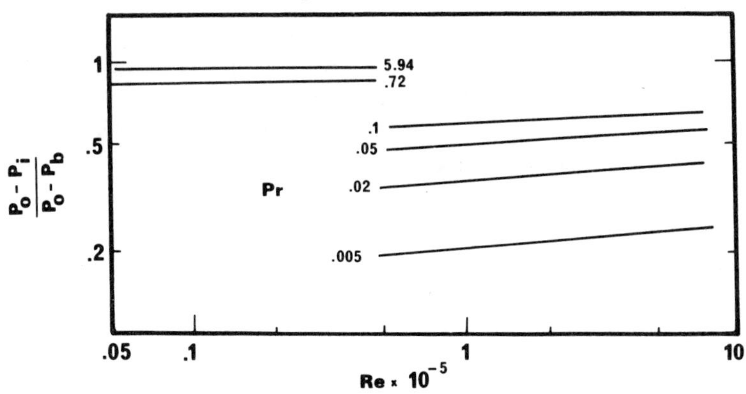

Reynolds number, it is a strong function of Pr. Equation (3.148) was demonstrated to be representative of experimental data for $\mathrm{Pe} > 5 \times 10^3$. For Peclet numbers below this value, the equation shows increasing deviation from the data. The failure of the model for $\mathrm{Pe} < 5 \times 10^3$ is most probably due to penetration of the concentration boundary layer being too great for conditions in the wall region to approach those of steady state molecular transport, as suggested by Levich (1962).

In subsequent analysis, Thomas (1971) modified his original proposal by incorporating a film adjacent to the wall through which diffusion occurred by steady molecular transport. This model is essentially similar to the earlier film-penetration model of Toor and Marchello (1958). Overall resistance to transport was evaluated by assuming the additivity of the two series resistances: the film at the wall, and the renewal mechanism, assumed to be the dominant mechanism for the region above the film. Film thickness was empirically chosen to provide a good fit to the experimental data.

In modification of the Einstein-Li (1956) model, Meek and Baer (1973) also discussed the problem of heat transfer for $\mathrm{Pr} \ll 1$. The core region in their analysis was treated using the classical eddy diffusivity approach. ν_t was obtained using the Reichardt (1951) relationship, and Pr_t was calculated from Jenkins (1951) model (see *Models based on heat losses from a spherical eddy* in Section 2.5.3). Although resultant calculations were found to be in favorable agreement with experimental data, assumption of a constant temperature for the growing thermal boundary layer is unrealistic. Agreement with experimental data is, in no small part, due to the insensitivity of calculation for $\mathrm{Pr} \ll 1$ to the assumptions regarding the nature of the wall region, provided that a correct description is given of the more important turbulent core region.

For $\mathrm{Sc}(\mathrm{Pr}) > 1$, it is quite reasonable to assume that $P_i = P_b$ and Eq. (3.147) reduces, with $u_i/u^* = \sqrt{2/f}$, to

$$\mathrm{Nu} = \frac{f}{2}\,\mathrm{Re}\,\mathrm{Pr}^{0.5} \tag{3.149}$$

Equation (3.149) is based on the assumption that eddies move into direct contact with the transport surface, and must therefore be restricted to fluids with Prandtl numbers of order one. Using Knudson and Katz (1958) suggested expression for the local friction factor

$$f = 0.0585\,\mathrm{Re}_x^{-0.2} \tag{3.150}$$

Thomas and Fan (1971) obtained

$$\mathrm{Nu}_x = 0.0292\,\mathrm{Re}_x^{0.8}\,\mathrm{Pr}^{0.5} \qquad \mathrm{Pr} = 0(1) \tag{3.151}$$

which seems to be in good agreement with air-data of Seban and Doughty (1956), and correlated experimental data for gases presented by Kays (1966) and others.

Meek and Baer (1970, 1973) modified Einstein and Li's (1956) model to include the effects of the thin wall-layer of thickness $\delta_1^+ = 1.5$, and their numerical calculations seem to agree with experimental data over the range of $1 < \text{Pr} < 200$.* Thomas and Fan (1970) extended Harriott's rejuvenation model and combined the surface renewal-penetration concept with a surface resistance of a finite thickness $\delta_1^+ = 5$. Unlike their own rejuvenation model (1971) in which they recommend the shape parameter $m = 1$, the value of $m = 0.4$ has been somewhat arbitrarily used here in the gamma distribution function. The reported range of applicability of this somewhat simplified rejuvenation model is $0.5 < \text{Pr} < 10^5$, in agreement with Dilman's (1968) data.

3.3.7 The periodic transitional sublayer model Pinczewski and Sideman (1974) suggested a periodic transitional sublayer model which accounts for infrequent penetration of eddies down to the wall and enables prediction of transfer rates over the whole range of Schmidt (Prandtl) numbers, equal to, or greater than one. The model is based on the surface renewal concept and fluid motion at the wall is assumed to be of the impulsive, semi-infinite, boundary layer type. The thin wall-layer is considered to be periodically replenished, but at a much smaller frequency than the bulk of the wall region. Unlike previously described models, no apriori assumption is made as to the steady or transient nature of the periodic flow in the sublayer. Rather, as depicted in Figure 33, flow conditions in the developing sublayer affect a transition from steady to transient state. Finally, the required proportionality constants were determined by utilizing the experimentally observed sublayer fluctuation frequencies, $1/\tau$ and

*A somewhat simpler treatment, leading to fit Deissler's (1955) correlation for $1 < \text{Pr} < 10^3$, was recently presented by Katsibas and Gordon (1974).

Figure 33 Idealized developing sublayer patch at the tube wall.

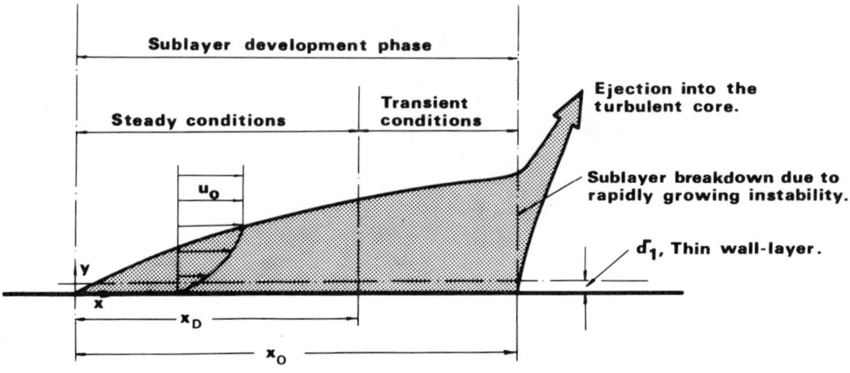

$1/\tau_1$. Treatment of the proposed model is divided according to the relative importance of the thin wall-layer.

Transfer with 0.5 $< \text{Pr} < 10$ For Pr(Sc) $= 0(1)$, effect of the thin wall-layer is negligible, and eddies may be assumed to penetrate to the wall. For $\text{Pr} \geqslant 1$, resistance to mass transfer is located in the periodically developing sublayer. For $\text{Pr} < 1$, the turbulent region outside of the developing sublayer offers an additional resistance to transfer. However, for $0.5 < \text{Pr} < 1$, this additional resistance can, to a first approximation, be assumed negligible in comparison to that of the sublayer. Validity of this assumption is verified apostiori by a comparison of the model predictions with experimental data.

The momentum and mass transport equations in the developing viscous flow are governed by Eq. (3.1) to Eq. (3.3) with $k_n = 0$, and the following initial and boundary conditions:

$$\begin{aligned}
t &= 0 \quad u = u_b \quad \text{and} \quad P = P_b \\
x &= 0 \quad u = u_b \quad \text{and} \quad P = P_b \\
y &= 0 \quad u = 0 \quad v = 0 \quad \text{and} \quad P = P_0 \\
y &\to \infty \quad u = u_b \quad \text{and} \quad P = P_b
\end{aligned} \tag{3.152}$$

consistent with the assumption that eddies arrive spontaneously, and uninterrupted, from the core region.

Utilizing the momentum integral method, the following distributions of velocity and concentration across the boundary layer are assumed:

$$\frac{u}{u_b} = 2\left(\frac{y}{\delta_M}\right) - 2\left(\frac{y}{\delta_M}\right)^3 + \left(\frac{y}{\delta_M}\right)^4 \tag{3.153}$$

$$\frac{P - P_0}{P_b - P_0} = 2\left(\frac{y}{\delta_\kappa}\right) - 2\left(\frac{y}{\delta_\kappa}\right)^3 + \left(\frac{y}{\delta_\kappa}\right)^4 \tag{3.154}$$

The infinite boundary condition in Eq. (3.152) may be replaced by the conditions

$$\begin{aligned}
y &= \delta_M \quad u = u_b \\
y &= \delta_\kappa \quad P = P_b
\end{aligned} \tag{3.152a}$$

The results for the concentration boundary layer, for $\text{Pr} \geqslant 1$, are (Rosenshtok et al., 1965)

$$\delta_\kappa = 5.72\left(\frac{\nu x}{u_b}\right)^{1/2} \text{Pr}^{-1/3} \qquad x < x_D \tag{3.155}$$

$$\delta_\kappa = 3.65(\nu t)^{1/2}\mathrm{Pr}^{-1/2} \qquad x > x_D \tag{3.156}$$

where $x_D = 0.4 u_b t \,\mathrm{Pr}^{-1/3}$ (3.157)

Equation (3.155) applies to that part of the developing flow where a steady concentration profile has already been established, and corresponds to Blasius' solution for steady state boundary layer flow. Equation (3.156) relates to that part of the flow where an unsteady concentration profile still persists and corresponds to Rayleigh's unsteady state solution.

Defining the local mass transfer coefficient by

$$H_x = -\frac{\kappa}{P_b - P_0} \cdot \left.\frac{\partial(P - P_0)}{\partial y}\right|_{y=0} \tag{3.158}$$

yields

$$\frac{H_x x}{\kappa} = 0.35\left(\frac{u_b x}{\nu}\right)^{1/2}\left(\frac{\nu}{D}\right)^{1/3} \qquad x < x_D \tag{3.159}$$

$$\frac{H_x x}{\kappa} = 0.55\left(\frac{\kappa}{t}\right)^{1/2} = 0.55\left(\frac{\kappa u_b}{x}\right)^{1/2} \qquad x > x_D \tag{3.160}$$

where time t and distance x are related to the velocity u_b at the edge of the developing boundary layer, by

$$t = \frac{x}{u_b} \tag{3.161}$$

The average mass transfer coefficient over the development length, X_0, is given by

$$H = \frac{1}{X_0}\int_0^{X_0} H_x \, dx$$

$$= \frac{1}{X_0}\int_0^{x_T} 0.35\left(\frac{u_b}{x}\right)^{1/2}(\nu\kappa^2)^{1/3}\, dx + \int_{x_T}^{X_0} 0.55\left(\frac{\kappa u_b}{x}\right)^{1/2} dx \tag{3.162}$$

where $x_T = 0.4 X_0 \mathrm{Pr}^{-1/3}$.

Defining the average development time of the sublayer flow, τ, as

$$\tau = \frac{X_0}{u_b} \tag{3.163}$$

and integrating yields

$$\text{Nu} = 0.44(2R)\left(\frac{1}{\nu\tau}\right)^{1/2}\text{Pr}^{1/2} + (1.10 - 0.70\,\text{Pr}^{-1/6})(2R)\left(\frac{1}{\kappa\tau}\right)^{1/2} \tag{3.164}$$

where, for pipe flow, $\text{Nu} \equiv H2R/\kappa$ Substituting the development time τ of the sublayer at the wall, Eq. (3.140) into Eq. (3.164) gives, with $\tau^+ = 243$ and $u_b/u^* = \sqrt{2/f}$,

$$\text{Nu} = 0.064\,\frac{f}{2}\,\text{Re}\,\text{Pr}^{1/2}(1.10 + 0.44\,\text{Pr}^{-1/3} - 0.70\,\text{Pr}^{-1/6}) \tag{3.165}$$

where $\text{Re} = u_b 2R/\nu$.

Introducing Blasius' expression for the friction factor for turbulent flow in smooth tubes $f/2 = 0.023\ \text{Re}^{-0.2}$, into Eq. (3.165) yields

$$\text{Nu} = 0.0097\,\text{Re}^{9/10}\,\text{Pr}^{1/2}(1.10 + 0.44\,\text{Pr}^{-1/3} - 0.70\,\text{Pr}^{-1/6}) \tag{3.166}$$

If we assume the complete developing boundary to be unsteady, i.e., averaging H_x given by Eq. (3.160) over the whole development length X_0, we obtain

$$\text{Nu} = 0.0107\,\text{Re}^{9/10}\,\text{Pr}^{1/2} \tag{3.167}$$

which is similar to the relationship derived from the application of the model of Einstein and Li (1956), and the expression derived by Hanratty (1956), utilizing Danckwerts' surface renewal theory. Conversely, if we assume the whole developing boundary to be in steady flow and average H_x given by Eq. (3.159) over the whole length X_0, we have

$$\text{Nu} = 0.0068\,\text{Re}^{9/10}\,\text{Pr}^{1/3} \tag{3.168}$$

The functional relationship in Eq. (3.168) is identical with Eq. (3.124) developed by Ruckenstein (1967) using his quasi-steady flow model, but the constants are different.

Equations (3.166), (3.167), and (3.168), are plotted in Figure 34 together with experimental data. The constants in Eq. (3.166) and Eq. (3.167) arise from the assumed model, utilizing Blasius' correlation, and Corino and Brodkey's observed sublayer fluctuation frequencies. It is evident from these figures that for Sc(Pr) > 1, the "transient" Eq. (3.167) overestimates the transfer coefficient, whereas the "steady" Eq. (3.168) underestimates the transfer

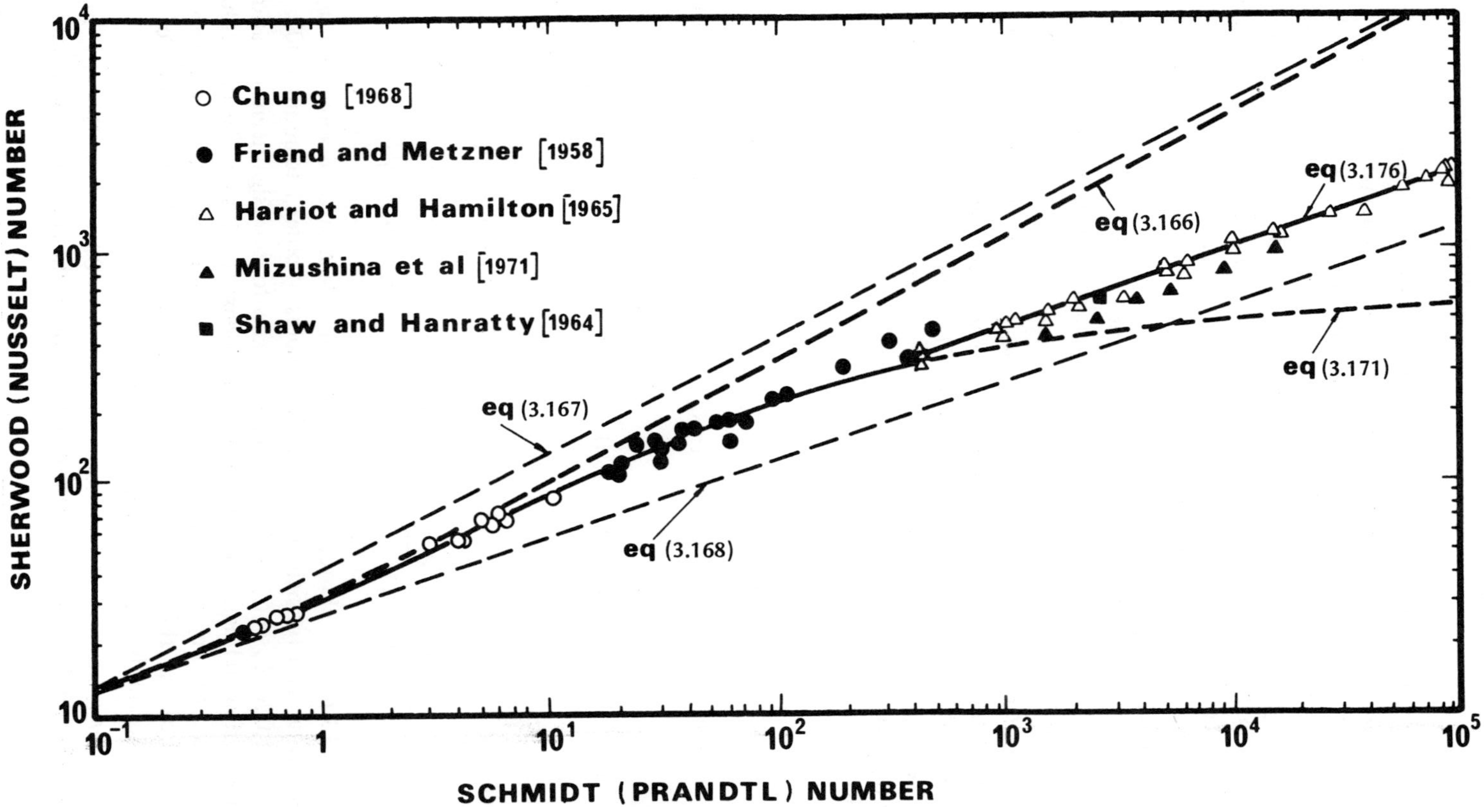

Figure 34 Comparison of the transitional model with experimental heat and mass transfer data.

coefficients. The curve representing Eq. (3.166), which lies between these two extremes, is in good agreement with experimental data in the Sc(Pr) = 0(1) range. The model suggests for Pr = 0(1), conditions in the developing sublayer are neither fully transient, nor fully steady, but rather correspond to a transient state undergoing transition to a steady state. Figure 34 shows the transitional sublayer model results in predictions superior to those derived from models based on the assumption of either fully transient (Hanratty, 1956; Einstein and Li, 1956; Meek and Baer, 1970; or fully steady conditions (Ruckenstein, 1967), in the developing sublayer.

For higher Sc(Pr) numbers, Eq. (3.166) approaches Eq. (3.167) in disagreement with the data. Disagreement is due to neglect of the effect of the thin wall-layer at higher Schmidt numbers.

Transfer with moderate Pr(Sc) ***numbers*** ($10 < \mathrm{Pr}\ 1{,}000$) At moderate Prandtl (Schmidt) numbers, thickness of the concentration boundary layer, δ_κ, is considerably greater than the thickness of the thin wall-layer, δ_1. Moreover, since renewal frequency of the wall-layer is small, accumulation in the wall-layer may be neglected, and the flux through the region $0 < y < \delta_1$ is approximated by (Ruckenstein, 1963):

$$\Psi = \frac{\kappa}{\delta_1}(P_0 - P_1) \tag{3.169}$$

The mass flux through the region $\delta_1 < y < \delta_\kappa$ is approximated by Eq. (3.166), i.e.,

$$\Psi = \frac{\kappa}{2R}\, 0.0097\, \mathrm{Re}^{9/10}\, \mathrm{Pr}^{1/2}(1.10 + 0.44\, \mathrm{Pr}^{-1/3} - 0.70\, \mathrm{Pr}^{-1/6})(P_1 - P_b) \tag{3.170}$$

Eliminating P_1 between Eq. (3.169) and Eq. (3.170), and setting $\delta_1^+ = 1$, yields

$$\mathrm{Nu} = \frac{0.0097\, \mathrm{Re}^{9/10}\, \mathrm{Pr}^{1/2}(1.10 + 0.44\, \mathrm{Pr}^{-1/3} - 0.70\, \mathrm{Pr}^{-1/6})}{1 + 0.064\, \mathrm{Pr}^{1/2}(1.10 + 0.44\, \mathrm{Pr}^{-1/3} - 0.70\, \mathrm{Pr}^{-1/6})} \tag{3.171}$$

Equation (3.171) is plotted in Figure 34, and is seen in good agreement with experimental data for Schmidt (Prandtl) numbers of order 1 to about 1,000. At Pr(Sc) > 1,000, $\delta_\kappa \to \delta_1$, and the equation begins to show significant deviations from experimental data. This behavior is not unexpected, since for Pr = 1,000, $\delta_\kappa^+ \cong 3$, i.e., the depth of penetration of the concentration boundary layer is of the order of the thickness of the thin wall-layer, and the assumption of no accumulation in the thin wall-layer is no longer valid.

Transfer at high Pr(Sc) ***numbers*** (Pr > 1,000) for sufficiently high Prandtl (Schmidt) numbers, depth of penetration of the concentration boundary layer, δ_K, is smaller than the thin fluid layer at the wall δ_1. Under these conditions, relatively low energy eddies which penetrate to $y > \delta_1$ do not affect the transfer rate, and rate of mass transfer is determined by the frequency with which higher energy eddies penetrate the sublayer; i.e., the rate of renewal of the thin wall-layer. The transfer to the fluid is specified by Eq. (3.110) and Eq. (3.111).

Solution of Eq. (3.110) and Eq. (3.111), with velocity in the wall-layer given by $u^+ = y^+$, or $u = yu_*^2/\nu$, is:

$$H_x = 0.538\left(\frac{\kappa^2 u_*^2}{x\nu}\right)^{1/3} \tag{3.172}$$

The fluid in the thin wall-layer moves for a characteristic distance, X_1 ($> X_0$), along the wall until it is disrupted by an eddy which penetrates to the wall.

The average transfer coefficient over the interval X_1 is given by

$$H = 0.807\left(\frac{\kappa^2 u_*^2}{X_1 \nu}\right)^{1/3} \tag{3.173}$$

The average velocity $\bar{u}_1$ in the wall-layer of thickness $\delta^+ = 1$, is

$$\bar{u}_1 = 0.5u^* \tag{3.174}$$

Utilizing Eq. (3.145) with $\tau^+ = 243$, and $b = 0.07$, together with Eq. (3.174), yields

$$X_1 = \tau_1 \bar{u}_1 = 1740\,\frac{\nu}{u^*} \tag{3.175}$$

Introducing Eq. (3.175) into Eq. (3.173), and utilizing Blasius' expression for the friction factor, leads to:

$$\mathrm{Nu} = 0.0102\ \mathrm{Re}^{9/10}\ \mathrm{Pr}^{1/3} \tag{3.176}$$

Equation (3.176) is plotted in Figure 34, and is in excellent agreement with the experimental correlation of Harriott and Hamilton (1965) for $\mathrm{Pr} \gg 1$, i.e.,

$$\mathrm{Nu} = 0.0096\ \mathrm{Re}^{0.913}\ \mathrm{Pr}^{0.346} \tag{3.177}$$

4 RELATED STUDIES

Whatever approach taken, solution for the transfer rate, to or from a fluid at a phase boundary, requires to stipulate at least two assumptions; parameters, or experimental correlations regarding turbulent fluctuating flow characteristics near the transport surface. These "degrees of freedom" give birth to numerous studies which either introduce new assumptions, or attempt to validate or modify old ones. It is therefore impossible to encompass in one chapter all these studies, and the present work is limited to the exposition of various concepts—or basic assumptions—which lead to workable "models."

A short list of some theoretical and applied studies which were not included above but are closely related to the subject at hand is given here for future reference. Among these are works dealing with

1. non-Newtonian fluids: Skelland (1966), Hansen and Na (1967), Poreh and Paz (1968), Harris et al. (1971), Schwalter et al. (1972), Chmiel (1972), Edwards and Wilkinson (1972), Astarita and Nicodemo (1972), Ruckenstein (1972).
2. compressible fluids: Kutateladze and Leont'ev (1964), Dunbar et al. (1971), Price and Harris (1972); and fluids with variable properties: Cheng et al. (1972), Fand et al. (1972).
3. dispersions: Heyt, (1971); fluidized beds: Agrawal and Ziegler (1970), Chung et al. (1972); and packed beds: de Lingy (1970), Clark and Epstein (1970), Dewasch and Froment (1972).
4. turbulent jets: Davies and Ting (1967), Newman (1967), Spalding (1971), Coeuret and Ronco (1973).
5. turbulent boundary layer with transverse flow: Gill et al. (1966), Baker (1969), Tanscher et al. (1970), Maeda (1972), Simpson and Field (1972), Leslie (1973); localized: Libby (1966); and high mass fluxes: Szekely (1964), Nienow et al. (1969); hyperfiltration: Gill et al. (1971); phase change: Langford (1967), Sparrow and Yu (1970), Hakki and Lamb (1972); or ionic transfer through ion exchange membranes: Kitamoto and Takashima (1970).
6. high speed: Wilcox et al. (1972), Sivasegaram and Whitelaw (1971); accelerated flows: Kays and Moffat (1970), Brinich and Neumann (1971), Kearney et al. (1973); and fluctuating flow: Ghoshal (1970).
7. boundary layer at continuously moving surfaces: Tson et al. (1967), Roper et al. (1972); rotating discs: Owen (1971), Shimomura (1970), Holman (1971); compliant walls: Grosskreutz (1971); mobile interfaces: Stewart et al. (1970); and surface roughness: Townes et al. (1972), Dawson and Trass (1972), Davies and Warner (1969), Grass (1971).
8. entrance region: Masan and Jones (1971); and transient transport: Szekely (1964), Chung et al. (1971).

Also considered outside the limited scope of this review are

9. the extensive studies of transport rates with chemical reactions: These are partly summarized in the books of Danckwerts (1970), Astarita (1967), and others.

10. interfacial turbulence due to density (concentration) and/or surface tension gradients: Davies and Rideal (1963), Davies (1972); the effects of surfactants: Springer and Pigford (1970); or turbulence suppressors: Komasawa et al. (1972) on transport at free interfaces.

5. CLOSURE

The basic concepts, stipulated mechanisms, and workable models proposed in the literature for predicting and correlating turbulent heat and mass transfer rates at the transport boundary were presented above. This review does not pursue any one subject to completion, and the reader interested in a particular subject will have to consult the related literature. Nevertheless, it is hoped that the material collected will be stimulating to the reader, and successful in pointing out common problems and the wide variety of assumptions and stipulations made in order to arrive at rational solutions. It is also hoped that it will encourage those presently engaged in a particular line of work to examine a wider range of needs while formulating their solution and give prospective "students of the art" a feel for the subject and a proper perspective in viewing new assumptions and models.

NOTATION

A	dumping length, transfer area, dead time
A^+	dimensionless dumping length ($= Au^*/\nu$)
A_0	characteristic (constant) reference area
a	proportionality constant Eq. (3.42)
B^+	dimensionless dumping length Eq. (2.61) $B^+ = B^+(\mathrm{Pr})$
b	experimental constant (Table 4); proportionality constant Eq. (2.122); fraction of eddies actually reaching the interface
C_1, C_2, C_3, C_4	experimental constants (Table 4)
C_D	constant Eq. (2.31)
C_p	specific heat at constant pressure
c	concentration; c'–fluctuating and $\bar{c}$–time-averaged concentration
c_1, c_2	constants Eq. (2.36)
D	mass diffusivity
D_m	transform operator Eq. (3.66)
D_t	turbulent mass diffusivity
d	ratio dead time to residence time ($= A/\tau$)
E	King's eddy diffusivity, Eq. (3.41)
e	mean turbulence kinetic energy Eq. (2.26)
$F(\mathrm{Pr})$	function of the Prandtl number Eq. (2.120)
Fo	Fourier modulus, ($= \kappa t/l^2$)

Fo_τ	τ-averaged Fourier modulus ($= \kappa\tau/l^2$)
Fo'	modified (axial) Fourier modulus ($= \kappa\tau/L^2$)
$F(y)$	initial velocity profile
$F(\theta)$	residence time distribution function
$f(\theta)$	residence time frequency function
f	local coefficient of friction
$f(P)$	Laplace transform of $f(\theta)$ Eq. (3.64)
G	constant Eq. (2.44)
$G(S)$	transform function Eq. (3.66)
$g(P)$	inversed local flux Eq. (3.67)
H	average transfer coefficient, $= \Psi/(P_0 - P_b)$
H_i	instantaneous transfer coefficient, $= \Psi_i(\theta)/(P_0 - P_b)$
H_x	local transfer coefficient Eq. (3.158)
h_s	interfacial resistance
J_0, Y_0	Bessel functions
K	mass transfer coefficient
k	thermal conductivity of fluid
k_n	rate constant Eq. (3.2)
k_0	rate constant, $n = 0$
k_t	turbulent thermal conductivity Eq. (2.19)
L	thickness of mixing layer Eq. (2.77); L–length scale, Eq. (3.20)
L_0	molecular mean free path
l	thickness of fluid element or eddy Eq. (3.30); "approach" distance to wall
l^+	dimensionless distance from the interface ($= lu^*/\nu$)
l_c	characteristic turbulence length-scale
l_H	characteristic thermal mixing length
l_m	momentum mixing length Eq. (2.21)
m	system parameter
N	total mass flux ($= \Psi$)
Nu	Nusselt for Sherwood number; $\mathrm{Nu}(\theta)$–instantaneous, $\mathrm{Nu}(\tau)$–τ-averaged; Nu_x–local value
N_i	stirring rate (i = phase 1 or 2) (rpm)
N_t	turbulent component of the total mass flux Eq. (2.19)
n	exponent; experimental constant (Table 4)
P	concentration (or temperature)
P_b	bulk concentration (or temperature)
P_i	concentration of eddy approaching transport surface
P_0	interfacial concentration (or temperature)
Pr	Prandtl (or Schmidt) number (ν/α or ν/D or ν/κ)
Pr_t	turbulent Prandtl (or Schmidt) number ($= \nu_t/\alpha_t$ or ν_t/κ_t)
$\mathrm{Pr}_{t,e}$	turbulent Prandtl (or Schmidt) number for kinetic energy transfer

p	pressure
Q	volumetric flow rate
q	total heat flux
q_i	heat flux to or from fluid entity
q_t	turbulent component of the total heat flux Eq. (2.19)
R	tube radius
R_H	hydraulic radius
Re	Reynolds number
r_e	mean effective fluid entity radius
S	mean frequency of mixing of layer; average renewal rate of elements at the wall region ($= 1/\tau$)
S_a	specific renewal frequency Eq. (3.25)
Sc	Schmidt number
Sc_t	turbulent Schmidt number ($= \nu_t/D_t$)
St	Stanton number ($=$ Nu/Re $\cdot$ Pr)
S_Z	secondary source term Eq. (2.36)
T	temperature; T'–temperature fluctuating; $\bar{T}$–time-averaged fluctuating
T_b	bulk temperature
T_f	mixing temperature
T_i	initial temperature
T_0	surface temperature ($= T_w$)
$\bar{T}^+$	dimensionless time constant Eq. (3.141)
t	time
t_0	characteristic (constant) time
t_F	formation time
t^+	dimensionless time, ($= u^*\sqrt{t/\nu}$)
U	overall heat transfer coefficient
u	x-directed velocity component; u'–fluctuating velocity; $\bar{u}$–time-averaged
u_b	x-directed bulk velocity
u_i	initial mean velocity of fluid element
u_0	interfacial velocity
u_0'	pressure independent fluctuating velocity Eq. (2.46)
$u(\theta)$	unit step function
$\bar{u}_1$	average velocity in wall-layer
u^*	friction velocity ($= \sqrt{\sigma_w/\rho}$)
u^+	dimensionless universal velocity ($= \bar{u}/u^*$)
V	volume of mixing vessel
V_m	molecular velocity
V_t	characteristic random fluctuating velocity Eq. (2.21)
v	y-directed velocity component; v'–fluctuating; $\bar{v}$–time-averaged

v_0'	pressure independent fluctuating velocity Eq. (2.47)
W	work, required to replace unit interfacial area
X	dimensionless parameter ($= \mathrm{Fo}_\tau{}^{-0.5}$)
X_0	distance of eddy motion along x in wall region
X_1	distance of eddy motion along x in wall-layer
x	axial coordinate; dimensionless variable (Figure 24)
x_T	transitional distance ($= 0.4\, X_0 \mathrm{Pr}^{-1/3}$) Eq. (3.162)
x^+	dimensionless coordinate ($= xu^*/\nu$); Von Mises transform variable Eq (2.136)
y	normal coordinate, from interface
y_b	normal distance where bulk or core properties control
y^+	dimensionless normal coordinate ($= yu^*/\nu$)
y_1^+	distance from wall beyond which molecular effects are negligible Eq. (2.107)
Z	general functional variable Eq. (2.36); complimentary normal distance ($= 1 - y/R$)
z	orthogonal coordinate

Greek Symbols

α	thermal diffusivity
α_e	effective thermal diffusivity
α_t	turbulent thermal diffusivity
β	proportionality constant, Eq. (3.98)
β_n	eigenvalues, Eq. (3.77)
γ	normalized surface transfer coefficient, ($= h_s/\kappa$)
δ	boundary layer thickness; half jet width
δ_D	film thickness
δ_M	thickness of momentum boundary layer
$\delta_{\max}$	maximum thickness of viscous sublayer thickness (at time τ)
δ_T	thickness of thermal boundary layer
δ_k	thickness of concentration (thermal) boundary layer
δ_1	thickness of thin wall-layer
δ_1^+	dimensionless thickness of wall layer ($= \delta_1 u^*/\nu$)
ϵ	energy dissipation rate per unit fluid mass
$\bar{\epsilon}$	characteristic dimensionless amplitude factor
$\bar{\epsilon}_0$	characteristic factor Eq. (3.100)
ϵ^+	dimensionless effective viscosity ($= 1 + \nu_t/\nu$)
η	similarity parameter, dimensionless tube radius ($= y/R$)
θ	age of element; time of contact
κ	proportionality constant Eq. (2.24); molecular diffusivity
κ_C	diffusivities of the solute continuous phase
κ_D	diffusivities of the solute dispersed phase

λ	distance travel by fluid entity; proportionality coefficient Eq. (3.26)
λ_1	measure of the rate of equilibration approach at interface
μ	dynamic viscosity
μ_t	turbulent dynamic viscosity
ν	kinematic viscosity
ν_b	turbulent or eddy viscosity
ν_e, α_e, D_e	effective transport coefficients Eq. (2.20)
ζ	transformation variable Eq. (2.137)
ρ	fluid density
σ_e	turbulent Schmidt & Prandtl number for the transfer of turbulent kinetic
σ_0	interfacial surface stress
σ_t	turbulent component of the total shear stress
σ_w	wall shear stress ($= \sigma_0$)
σ_z	turbulent Schmidt or Prandtl number for the transfer of the function Z.
τ	mean contact or residence time ($= 1/S$)
τ_1	mean contact period of thin wall-layer Eq. (3.145)
τ^*	dimensionless time Eq. (3.47)
τ^+	mean dimensionless periodic time scale ($= \tau u^{*2}/\nu$)
τ_1^+	mean dimensionless periodic time scale of thin wall layer
$\phi(\theta)$	age distribution function
χ	function relating P_i to P_b Eq. (3.148)
Ψ	average transfer rate per unit area (= flux)
Ψ^*	shape or distortion factors of penetrating fluid entity; dimensionless transfer coefficient Eq. (3.47)
$\Psi_i(\theta)$	instantaneous flux
ω	oscillation frequency

REFERENCES

Acrivos, A., "Mass Transfer in Laminar-Boundary-Layer Flows with Finite Interfacial Velocities," *A.I.Ch.E.J.*, **6**, 410 (1960).

Agrawal, S. S., and Ziegler, E. N., "Surface to Fluidized-Bed Transport Processes," *Chem. Eng. Progr. Symp. Ser.*, **66**, 105 (1970).

Akiyama, T., "The Equivalence of the Penetration and Boundary Layer Theories," *Chem. Eng. Sci.*, **27**, 161 (1972).

Allen, R. W., and Eckert, R. G., "Friction and Heat-Transfer Measurements to Turbulent Pipe Flow of Water (Pr = 7 and 8) at Uniform Wall Heat Flux," *Trans. Am. Soc. Mech. Engrs.*, **86**, 301 (1964).

Angelo, J. B., Lightfoot, E. N., and Howard, D. W., "Generalization of the Penetration Theory for Surface Stretch: Application to Forming and Oscillating Drops," *A.I.Ch.E.J.*, **12**, 751 (4) (1966).

Armistead, R. A., and Keyes, J. J., "A Study of Wall-Turbulence Phenomena Using Hot-Film Sensors," *Trans. ASME J. Heat Transfer,* **13,** 90 (1968).

Astarita, G., *Mass Transfer with Chemical Reaction,* Amsterdam, Elsevier, (1967).

Astarita, G., and Nicodemo, L., "Transport Phenomena in Turbulent Flow of Rheologically Complex Fluids," *Prog. in Heat Mass Transfer,* **5,** 37 (1972).

Azer, N. Z., and Chao, B. T., "A Mechanism of Turbulent Heat Transfer in Liquid Metals," *Int. J. Heat Mass Transfer,* **1,** 121 (1960).

Baird, M. H. I., "Mass Transfer to Growing Drops," *Chem. Eng. Sci.,* **9,** 267 (1959).

Bakewell, H. P., and Lumley, J. L., "Viscous Sublayer and Adjacent Wall Region in Turbulent Pipe Flow," *Phys. Fluids,* **10,** 1880 (1967).

Berbente, C. P., and Ruckenstein, E., "Hydrodynamics of Wave Flow," *A.I.Ch.E.J.,* **14,** (5) 772 (1968).

Bénard, H., "Les Tourbillons Cellulaires dans une Nappe Liquide," *Rev. Gen. des Sci. Pures Appl.,* **11,** 1261 (1900).

Bird, R. B., Steward, W. E., and Lightfoot, E. N., *Transport Phenomena,* New York, John Wiley, (1958).

Black, T. J., in C. S. Wells, (Ed.), *Viscous Drag Reduction Examined in the Light of a New Model of Wall Turbulence,* New York, Plenum Press, p. 383, 1969.

Black, T. J., "Some Practical Applications of a New Theory of Wall Turbulence," *Proc. Heat Trans. & Fluid Mech. Inst.,* (1966).

Black, T. J., "A New Model of Turbulent Boundary layer/shock Interaction with Particular Reference to the Associated Wall Pressure Fluctuation," *A.I.A.A. 6th Aero. Sci. Meeting, A.I.A.A.,* No. 68–42 (1968).

Blom, J., "An Experimental Determination of the Turbulent Prandtl Number in a Developing Temperature Boundary Layer, presented at *The Fourth International Heat Transfer Conf.,* Paris, Versailles (1970).

Blom, J., and deVries, D. A., "On the Values of the Turbulent Prandtl Number," *Third All-Union Heat and Mass Transfer Conference,* Paper No. 1.8, Minsk (1968).

Boussinesq, J., "Theorie De L'Ecoulement Tourbillant," *Mem. Pres. Acad. Sci.,* XXIII. 46, Paris (1877).

Bradshaw, P., Ferris, D. H., and Atwell, N. P., "Calculation of Boundary Layer Development Using the Turbulent Energy Equation," *J. Fluid Mech.,* **28,** 593 (1967).

Braginski, L. N., and Pavlushensko, I. S., "Relation Between the Surface Removal Factor and Mixing Conditions in a Lewis Cell," *Z. Prikl. Khim.,* **38,** 1290 (1965).

Bremhorst, K., and Bullock, K. J., "Spectral Measurements of Turbulent Heat and Momentum Transfer in Fully Developed Pipe Flow," *Int. J. Heat Mass Transfer,* **16,** 2141 (1973).

Brinich, B. F., and Neumann, H. E., "Some Characteristics of Turbulent Boundary Layers in Rapidly Accelerated Flows," NASA-TN-D-6587 (1971).

Brinkworth, B. T., and Smith, P. C., "Velocity Distribution in the Core of Turbulent Pipe Flow," *Chem. Eng. Sci.,* **24,** 787 (1969).

Brinkworth, B. J., and Smith, P. C., "Radial Distribution of Diffusivity in the Core of Turbulent Flow," *Chem. Eng. Sci.,* **28,** 1847 (1973).

Brodkey, R. S., "Turbulent Motion, Mixing, and Kinetics," *ASME Meeting of the Fluids Eng. Div. & the Appl. Mech. Div.,* Atlanta, Georgia (1973).

Brodkey, R. S., Hershey, H. C., and Corino, E. R., "An Experimental Facility for the Visual Study of Turbulent Flows," *The Ohio State University preprint.* (1973).

Brusset, H., Depeyre, D., and Le Quang, T., "An Estimation of the Film-Penetration Model Parameters," *Chem. Eng. Sci.,* **28,** 31 (1973).

Buyco, E. H., "Heat and Momentum Transfer in Liquid Metals," *Ph.D. Thesis,* Purdue University (1961).

Cairns, R. C., Roper, G. H., "Heat and Mass Transfer at High Humidities in a Wetted Wall Column," *Chem. Eng. Sci.*, **3,** 97 (1954).

Calderbank, P. H., and Moo-Young, M. B., "The Continuous Phase Heat and Mass Transfer Properties of Dispersions," *Chem. Eng. Sci.*, **16,** 39 (1961).

Calderbank, P. H., and Patra, R. P., "Mass Transfer in the Liquid Phase During the Formation of Bubbles," *Chem. Eng. Sci.*, **21,** 719 (1966).

Carr, A. D., and Balzhiser, R. E., "Temperature Profiles and Eddy Diffusivities in Liquid Metals," *Brit. Chem. Eng.* **12,** 53 (1967).

Carslaw, H. S., and Jaeger, J. C., *Conduction of Heat in Solids,* 2nd ed., London, Oxford University Press, (1959).

Cebeci, T., "A Model for Eddy Conductivity and Turbulent Prandtl Number," *ASME J. Heat Transfer,* **95,** 227 (1973).

Chan, W. C., "Transfer Across Flowing Interfaces: Stagnation Flow Models," Ph.D. Thesis, University of Minnesota (1964).

Chan, W. C., and Scriven, L. E., "Absorption into Irrotational Stagnation Flow," *I&EC Fundamentals,* **9,** (1) (1970).

Chao, B. T., and Cheema, L. S., "Unsteady Heat Transfer in Laminar Boundary Layer over a Flat Plate," *Int. J. Heat Mass Transfer,* **11,** 1311 (1968).

Cheng, S. C., Birta, L. G., and Su, Y. L., "Application of a Variational Method to Flow over a Flat Plate in the Entrance Region with Variable Physical Properties," *Int. J. Heat Mass Transfer,* **15,** 1933 (1972).

Chi, S. W., and Spalding, D. B., "Influence of Temperature Ratio on Heat Transfer to a Flat Plate Through a Turbulent Boundary in Air," *Proc. of 3rd Intern. Heat Transfer Conf.*, Chicago, (AICHE) **11,** 41 (1966).

Chilton, T. H., and Colburn, A. P., "Mass Transfer (Absorption) Coefficients," *Ind. Eng. Chem.*, **26,** 1183 (1934).

Chmiel, H., Rautenbach, R., and Schmummer, P., "Theory of Heat Transfer in the Turbulent Flow of Viscoelastic Liquids in Pipes (in German)," *Chemie Ing. Tech.*, **44,** 543 (1972).

Chung, B. T. F., and Thomas, L. C., "Unsteady Heat Transfer for Turbulent Flow," *The Canadian Journal of Chem. Eng.*, **49,** 877 (1971).

Chung, B. T. F., Fan, L. T., and Hwang, C. L., "A Model of Heat Transfer in Fluidized Beds," *Journal of Heat Transfer,* **94,** 105 (1972).

Chung, B. T. F., Fan, L. T., and Hwang, C. L., "General Mathematical Models of Transport Processes with and without Chemical Reactions," *The Canadian Journal of Chem. Eng.*, **49,** 340 (1971).

Chung, B. T. F., Fan, L. T., and Hwang, C. L., "Surface Renewal and Penetration Models in the Transient State," *A.I.Ch.E.J.*, **17,** 154 (1971).

Ciborowski, J., and Rychlicki, R., "Momentum and Mass Transfer Analogy in Horizontal Cocurrent Gas and Liquid Wavy Flow," *Intern. Chem. Eng.*, **13,** (2), 287 (1973).

Clark, C., and Epstein, E., "End Effect on Interphase Heat and Mass Transfer in Cubically Packed Bed of Spheres," *Can. J. Chem. Eng.*, **48,** 538 (1970).

Clegg, G. T., and Mann, R., "A Penetration Model for Gas Absorption with First Order Chemical Reaction Accompanied by Large Heat Effects," *Chem. Eng. Sci.*, **24,** 321 (1969).

Coeuret, F., and Ronco, J. J., "Absorption Physique D'un Gaz Dans Un Jet Liquide Turbulent," *Chem. Eng. Sci.*, **28,** 315 (1973).

Coles, D., *The Law of the Wall in Turbulent Shear Flow,* 50 Jahre Grenzschichtforschung (Edited by H. Gortler and W. Tollmien), 153 (1955).

Corino, E. R., and Brodkey, R. S., "A Visual Investigation of the Wall Region in Turbulent Flow," *J. Fluid Mech.*, **37,** (1-30) (1969).

Coughlin, R. W., "Surface Resistance in Transport from Vapor to Liquid," *Chem. Eng. Sci.*, **22**, 1503 (1967).

Daly, B. J., and Harlow, F. H., "Transport Equations in Turbulence," *Physics of Fluids*, **13**, 2634 (1970).

Danckwerts, P. V., "Significance of Liquid-Film Coefficients in Gas Absorption," *Ind. & Eng. Chem.*, **43**, 1460 (1951).

Danckwerts, P. V., *Gas-liquid Reactions*, New York, McGraw-Hill (1970).

Davidov, B. I., "On the Statistical Dynamics of an Incompressible Fluid," *Dokl. AN SSSR*, **136**, 47 (1961).

Davies, J. T., "The Effects of Surface Films in Damping Eddies at a Free Surface of a Turbulent Liquid," *Proc. Roy. Soc. A.*, **290**, 515 (1966).

Davies, J. T., *Turbulence Phenomena: An Introduction to the Eddy Transfer of Momentum, Mass, and Heat, Particularly at Interfaces*, New York, Academic Press (1972).

Davies, J. T., "Turbulence Phenomena at Free Surfaces," *A.I.Ch.E.J.*, **18**, (1), 169 (1972).

Davies, J. T., and Rideal, E. K., *Interfacial Phenomena*, 2nd ed. New York, Academic Press (1963).

Davies, J. T., and Ting, S. T., "Mass Transfer into Turbulent Jets," *Chem. Eng. Sci.*, **22**, 1539 (1967).

Davies, J. T., and Warner, K. V., "The Effect of Large-Scale Roughness in Promoting Gas Absorption," *Chem. Eng. Sci.*, **24**, 231 (1969).

Dawson, D. A., and Trass, O., "Mass Transfer at Rough Surfaces," *Int. J. Heat Mass Transfer*, **15**, (7), 1317 (1972).

Deissler, R. G., "Analysis of Fully Developed Turbulent Heat Transfer at Low Reynolds Numbers in Smooth Tubes with Application to Liquid Metals," *N.A.C.A. R&M*, E52F05 (1952).

Deissler, R. G., *National Advisory Comm. Aeronaut. Tech.*, Note 3145, (1954).

Deissler, R. G., "Analysis of Turbulent Heat Transfer, Mass Transfer, and Friction in Smooth Tubes at High Prandtl and Schmidt Numbers," *NACA Report* 1210 (1955).

Deissler, R. G., "Turbulent Heat Transfer and Temperature Fluctuations in a Field with Uniform Velocity and Temperature Gradients," *Int. J. Heat Mass Transfer*, **6**, 257 (1963).

Denny, V. E., and Mills, A. F., "Nonsimilar Solutions for Laminar Film Condensation on a Vertical Surface," *Int. J. Heat Mass Transfer*, **12**, 965 (1969).

Denny, V. E., and Landis, R. B., "An Improved Transformation of the Patankar-Spalding Type for Numerical Solution of Two-Dimensional Boundary Layer Flows," *Int. J. Heat Mass Transfer*, **14**, 1859 (1971).

DeWasch, A. P., and Froment, G. F., "Heat Transfer in Packed Beds," *Chem. Eng. Sci.*, **27**, 567 (1972).

Dilman, V. V., "The Theory of Heat and Mass Transfer in Turbulent Flow," *Int. Chem. Eng.*, **8**, 156 (1968).

Dindorf, J. A., and Lightfoot, E. N., "Mass Transfer Across a Deformable Interface with Instantaneous Reversible Reaction," *Chem. Eng. Sci.*, **26**, 1833 (1971).

Dobbins, W. E., "Bod and Oxygen Relationships in Streams," *Journal of the Sanitary Engineering Division Proceedings of the American Society of Civil Engineers*, p. 53 (1964).

Dobbins, W. E., "The Nature of the Oxygen Transfer Coefficient in Aeration Systems," *Aeration-Theory and Design*, **2-1**, 141 (1958).

Doshi, M. R., and Gill, W. N., "A Note on the Mixing Length Theory of Turbulent Flow," *A.I.Ch.E.J.*, **16**, 885 (1970).

Doshi, M. R., and Gill, W. N., "An Improved Mixing Length Theory of Turbulent and Mass Transfer," *Int. J. Heat Mass Transfer*, **14**, 1355 (1971).

Downing, A. L., and Truesdale, G. A., "Some Factors Affecting the Rate of Solution of Oxygen in Water," *J. Appl. Chem.*, **5,** 570 (1955).

Dukler, A. E., and Bullin, J. A., "Random Eddy Models for Surface Renewal: Formulation as a Stochastic Process," *Chem. Eng. Sci.*, **27,** 439 (1972).

Dunbar, D. I. A., and Squire, L. C., "Correlations of Concentration, Temperature and Velocity Profiles in Compressible Turbulent Boundary Layers with Foreign Gas Injection," *Int. J. Heat Mass Transfer,* **14,** (1), 27 (1971).

Dwyer, O. E., "Eddy Transport in Liquid Metal Heat Transfer," *A.I.Ch.E.J.*, **9,** 261 (1963).

Eckelman, L. D., and Hanratty, T. J., "Interpretation of Measured Variations of the Eddy Conductivity," *Int. J. Heat Mass Transfer,* **15,** 2231 (1972).

Edge, R. M., "Surface Renewal Models in Mass Transfer," *Chem. Eng. Soc. J.*, **2,** 45 (1966).

Edwards, M. E., and Wilkinson, W. L., "Heat Transfer in Agitated Vessels. Part I: Newtonian Fluids, Part II: Non-Newtonian Fluids," *The Chem. Engr.*, **264,** 310, **265,** 328 (1972).

Einstein, H. A., and Li, H., "The Viscous Sublayer Along a Smooth Boundary," *Proc. Am. Soc. Civil Engrs., J. Eng. Mech. Div.*, **82,** 293 (1956).

Ellis, S. R. M., and Biddulph, M., "Interfacial Turbulence Measurements," *Chem. Eng. Sci.*, **21,** 1107 (1966).

Elrod, H. G., "Note on the Turbulent Shear Stress Near a Wall," *J. Aeronaut. Sci.*, **24,** 468 (1957). "Erratum–to Elrod [1957]," *J. Aeronaut. Sci.*, **27,** 145 (1960).

Escudier, M. P., "The Distribution of Mixing-Length in Turbulent Flows Near Walls," *Heat Transfer Section Rep.*, TWF/TN1. Imperial College, (1966).

Evans, G. V., "A Study of Diffusion in Turbulent Pipe Flow," *J. Bas. Engng.*, **89,** 624 (1967).

Fage, H., and Towend, H. C. H., "An Examination on Turbulent Flow with an Ultramicroscope," *Proc. Roy. Soc.*, **A135,** 656, (1932).

Fand, R. M., and Keswani, K. K., "The Influence of Property Variation on Forced Convection Heat Transfer to Liquids," *Int. J. Heat Transfer,* **15,** 1515 (1972).

Fortescue, G. E., and Pearson, J. R. A., "On Gas Absorption into a Turbulent Liquid," *Chem. Eng. Sci.*, **22,** 1163 (1967).

Fortuna, G., and Sideman, S., "Direct Contact Heat Transfer Between Immiscible Liquid Layers with Simultaneous Boiling and Stirring," *Chem. Eng. Sci.*, **23,** 1105 (1968).

Frantisak, F., Palade De Iribarne, A., Smith, J. W., and Hummel, R. L., "Nondisturbing Tracer Technique for Quantitative Measurements in Turbulent Flow," *I&EC Fundamentals,* **8,** (1), 160 (1969).

Friend, W. L., and Metzner, A. B., "Turbulent Heat Transfer Inside Tubes and the Analogy Among Heat, Mass, and Momentum Transfer," *A.I.Ch.E.J.*, **4,** 393 (1958).

Fulford, G. D., and Pei, D. C. T., "A Unified Approach to the Study of Transfer Processes," *Ind. & Eng. Chem.*, **61,** (5), 47 (1969).

Ghoshal, S., and Ghoshal, A., "Thermal Boundary-Layer Theory Near the Stagnation Point in Three-Dimensional Fluctuating Flow," *J. Fluid Mech.*, **43,** (3), 465 (1970).

Gill, W. N., Del Casal, E., and Zeh, D. W., "Free and Forced Convection in Conduits with Asymmetric Mass Transfer," *A.I.Ch.E.J.*, **12,** (2), 266 (1966).

Gill, W. N., Derzansky, L. J., and Doshi, M. R., "Convective Diffusion in Laminar and Turbulent Hyperfiltration (Reverse Osmosis) Systems," *Surface and Colloid Science,* **IV,** 263 (1971).

Gill, W. N., and Scher, M., "A Modification of the Momentum Transport Hypothesis," *A.I.Ch.E.J.*, **7,** 61 (1961).

Gilliland, E. R., and Sherwood, T. K., "Diffusion of Vapors into Air Streams," *Ind. & Eng. Chem.*, **26,** (5), 516 (1934).

Goldman, I. B., and Marchello, J. M., "Turbulent Schmidt Numbers," *Chem. Eng. Sci.*, **12,** 797 (1969).

Gosman, A. D., Pun, W. M., Runchal, A. K., Spalding, D. B., and Wolfshtein, M., *Heat and Mass Transfer in Recirculating Flows,* London, Academic Press (1969).

Gowen, R. A., and Smith, J. W., "Turbulent Heat Transfer from Smooth and Rough Surfaces," *Int. J. Heat Mass Transfer,* **11,** 1657 (1968).

Grant, H. L., "The Large Eddies of Turbulent Motion," *J. Fluid Mech.,* **4,** 149 (1958).

Grass, A. J., "Structural Features of Turbulent Flow over Smooth and Rough Boundaries," *J. Fluid Mech.,* **50,** (2), 233 (1971).

Groothuis, M., and Kramers, H., "Gas Adsorption by Single Drops During Formation," *Chem. Eng. Sci.,* **4,** 17 (1955).

Gross, R., and Thomas, L. C., "Significance of the Pressure Gradient on Fully Developed Turbulent Flow in a Pipe," *J. Heat Transfer, ASME,* **94,** 494 (1972).

Grosskreutz, R., "Interaction Between Turbulent Boundary Layers and Compliant Walls," (in German). *Max-Planck-Institut fur Stromungsforchung,* Report 53, (1971).

Hakki, T. A., and Lamb, J. A., "Simultaneous Heat and Mass Transfer in a Wetted-Wall Column," *Trans. Inst. Chem. Engrs.,* **50,** 1905 (1972).

Hanjalic, K., "Two-Dimensional Asymmetric Turbulent Flow in Ducts," *Ph.D. Thesis,* University of London (1970).

Hanjalic, K., and Launder, B. E., "A Reynolds Stress Model of Turbulence and Its Application to Asymmetric Shear Flows," *J. Fluid Mech.,* **52,** 609 (1972).

Hanjalic, K., and Launder, B. E., "Fully Developed Asymmetric Flow in a Plane Channel," *J. Fluid Mech.,* **51,** 301 (1972).

Hanna, O. T., and Sandall, O. C., "Developed Turbulent Transport in Ducts for Large Prandtl or Schmidt Numbers," *A.I.Ch.E.J.,* **18,** 527 (1972).

Hanratty, T. J., "Turbulent Exchange of Mass and Momentum with a Boundary," *A.I.Ch.E.J.,* **2,** (3), 359 (1956).

Hansen, A. G., "Possible Similarity Solutions of the Laminar, Incompressible, Boundary-Layer Equations," *Transactions of the ASME,* 1553 (1958).

Hansen, A. G., and Na, T. Y., "Similarity Solutions of Laminar, Incompressible Boundary Layer Equations of Non-Newtonian Fluids," *Transactions of the ASME,* Paper No. 67-WA/FE-2 (1967).

Harlow, F. H., and Nakayama, P. I., "Transport of Turbulence Energy Decay Rate," Los Alamos Sci. Lab., Report LA-3854, University of California (1968).

Harriott, P., "A Random Eddy Modification of the Penetration Theory," *Chem. Eng. Sci.,* **17,** 149 (1962).

Harriott, P., "A Review of Mass Transfer to Interfaces," *Can. J. Chem. Eng.,* **60,** (1962).

Harriott, P., "Mass Transfer to Particles: Part 1, Suspended in Agitated Tanks," *A.I.Ch.E.J.,* **8,** (1), 93 (1962).

Harriott, P., and Hamilton, R. M., "Solid-Liquid Mass Transfer in Turbulent Pipe Flow," *Chem. Eng. Sci.,* **20,** 1073 (1965).

Harris, J., and Wilkinson, W. L., "Momentum, Heat, and Mass Transfer in Non-Newtonian Turbulent Flow in Pipes," *Chem. Eng. Sci.,* **26,** 313 (1971).

Hatton, A. P., "Heat Transfer Through the Turbulent Incompressible Boundary Layer on a Flat Plate," *Int. J. Heat Mass Transfer,* **7,** 875 (1964).

Heertjes, P. M., Holve, W. A., and Talsma, H., "Mass Transfer between Isobutanol and Water in a Spray-Column," *Chem. Eng. Sci.,* **3,** 122 (1954).

Hettler, J. P., Muntzer, P., and Scrivener, O., *Compt. Rend. Mecanique des Fluids,* **258,** 4201 (1964).

Heyt, J. W., "Heat Transfer to Binary Mist Flow," *Int. J. Heat Mass Transfer,* **14,** (9), 1395 (1971).

Higbie, R., "Rate of Absorption of a Gas into a Still Liquid," *Trans. Am. Inst. Chem. Engrs.,* **31,** 365 (1935).

Hinze, J. O., *Turbulence–An Introduction to its Mechanism and Theory,* New York, McGraw-Hill (1959).

Hochreiter, L. E., and Sosenske, A., "Turbulent Structure of Isothermal and Non-Isothermal Liquid Metal Pipe Flow," *Int. J. Heat Mass Transfer,* **17,** 113 (1974).

Holman, K. L., and Ashar, S. T., "Mass Transfer in Concentric Rotating Cylinders with Surface Chemical Reaction in the Presence of Taylor Vortexes," *Chem. Eng. Sci.,* **26,** 1817 (1971).

Howard, D. W., and Lightfoot, E. N., "Mass Transfer to Falling Films: Part 1. Application of the Surface-Stretch Model to Uniform Wave Motion," *A.I.Ch.E.J.,* **14,** (3), 458 (1968).

Huang, C. J., and Kuo, C. H., "General Mathematical Model for Mass Transfer Accompanied by Chemical Reaction," *A.I.Ch.E.J.,* **9,** 161 (1963).

Hubbard, D. W., "Correlation of Mass Transfer Data: Comments on an Article by Son and Hanratty," *A.I.Ch.E.J.,* **14,** 354 (1968).

Hubbard, D. W., and Lightfoot, E. N., "Correlation of Heat and Mass Transfer Data for High Schmidt and Reynolds Numbers," *Ind. Eng. Chem. Fund.,* **5,** 370 (1966).

Huffman, G. D., Jones, C. D., and Brodkey, R. S., "Optimization of Turbulence Models by Means of a Logical Search Algorithm," *App. Sci. Res.,* **27,** 321 (1973).

Hughmark, G. A., "Eddy Diffusivity Close to a Wall," *A.I.Ch.E.J.,* **14,** 325 (1968).

Hughmark, G. A., "Momentum, Heat, and Mass Transfer Analogy for Turbulent Flow in Circular Pipes," *Ind. Eng. Chem. Fund.,* **8,** 31 (1969).

Hughmark, G. A., "Heat and Mass Transfer in the Wall Region of Turbulent Pipe Flow," *A.I.Ch.E.J.,* **17,** 51 (1971).

Hughmark, G. A., "Heat and Mass Transfer for Turbulent Pipe Flow," *A.I.Ch.E.J.,* **17,** (4), 902 (1971).

Hughmark, G. A., "Notes on Transfer in Turbulent Pipe Flow," *A.I.Ch.E.J.,* **18,** (5), 1072 (1972).

Hughmark, G. A., "Additional Notes on Transfer in Turbulent Pipe Flow," *A.I.Ch.E.J.,* **19,** (5), 1054 (1973).

Hutchinson, P., Hewitt, G. F., and Dukler, A. E., "Deposition of Liquid or Solid Dispersions from Turbulent Gas Streams: A Stochastic Model," *Chem. Eng. Sci.,* **26,** 419 (1971).

Ilkovic, D., *Colln. Czech. Chem. Engrs.,* **6,** 4908 (1937).

Isakoff, S. E., and Drew, T. B., "Heat and Momentum Transfer in Turbulent Flow of Mercury, General Discussion on Heat Transfer," *Inst. Mech. Engrs & ASME,* 405, (1951).

Jackson, M. L., and Ceaglske, H. N., "Distillation Vaporization and Gas Absorption in a Wetted Wall Column," *Ind. Chem. Eng.,* **42,** 1188 (1950).

Jayatilleke, C. L. V., "The Influence of Prandtl Number and Surface Roughness on the Resistance of the Laminar Sub-Layer to Momentum and Heat Transfer," *Progress in Heat and Mass Transfer,* **1,** 193 (1969).

Jenkins, R., "Variation of the Eddy Conductivity with Prandtl Modulus and its Use in the Prediction of Turbulent Heat Transfer Coefficients," *Proc. Heat Transfer & Fluid Mechanics Instit.,* Stanford University Press, p. 147 (1951).

Johnk, R. E., and Hanratty, T. J., "Temperature Profiles for Turbulent Flow of Air in a Pipe, Part I and II," *Chem. Eng. Sci.,* **17,** 867 (1962).

Johnson, A. I., and Huang, C. J., "Mass Transfer Studies in an Agitated Vessel," *A.I.Ch.E.J.,* **2,** (3), 412 (1956).

Jones, W. P., and Launder, B. E., "The Prediction of Laminarization with a 2-Equation Model of Turbulence," *Int. J. Heat Mass Transfer,* **15,** 301 (1972).

Kader, B. A., and Yaglom, A. M., "Heat and Mass Transfer Laws for Fully Turbulent Wall Flows," *Int. J. Heat Mass Transfer,* **15,** 2329 (1972).

Katsibas, P., and Gordon, R. J., "Momentum and Energy Transfer in Turbulent Pipe Flow: The Penetration Model Revisited," *A.I.Ch.E.J.,* **20,** (1), 191 (1974).

Kays, W. M., *Convective Heat and Mass Transfer,* New York, McGraw-Hill (1966).

Kays, W. M., Moffat, R. J., and Thielbahr, W. H., "Heat Transfer to the Highly Accelerated Turbulent Boundary Layer with and without Mass Additions," *J. Heat Transfer,* **92,** C(3), 499 (1970).

Kearney, D. W., Kays, W. M., and Moffat, R. J., "Heat Transfer to a Strongly Accelerated Turbulent Boundary Layer: Some Experimental Results, Including Transpiration," *Int. J. Heat Mass Transfer,* **16,** 1289 (1973).

Kestin, J., and Richardson, P. D., "Heat Transfer Across Turbulent, Incompressible Boundary Layers," *Int. J. Heat Mass Transfer,* **6,** 147 (1963).

Kim, H. T., Kline, S. J., and Reynolds, W. C., "The Production of Turbulence near a Smooth Wall in a Turbulent Boundary Layer," *J. Fluid Mech.,* **50,** 133 (1971).

King, J. C., "Mass Transfer During Short Surface Exposures in Countercurrent Flow," *I&EC Fundamentals,* **4,** (2), 125 (1965).

Kishinevskii, M. K., "Two Approaches to the Theoretical Analysis of Absorption Processes," *J. Appl. Chem.,* (USSR), **28,** 881 (1955).

Kishinevskii, M. K., "Some Results of Modern Theoretical Work in the Field of Absorption Complicated by Chemical Reactions," *Teoreticheski Osnovy Khimicheskoi Technologii,* **1,** (6), 759 (1967).

Kishinevskii, M. K., and Pamfilov, A. V., *Zh. Prikland. Khim,* **22,** 1173 (1949).

Kishinevsky, M. K., and Serebryansky, V. T., "The Mechanism of Mass Transfer at the Gas-Liquid Interface with Vigorous Stirring," *J. Appl. Chem.,* (USSR), **28,** 881 (1955).

Kitamoto, A., and Takashima, Y., "Ionic Mass Transfer in Turbulent Flow by Electrodialysis with Ion Exchange Membranes," *Journal of Chem. Eng. of Japan,* **3,** (2), 182 (1970).

Kline, S. L., and Runstadler, P. W., "Some Preliminary Results of Visual Studies of the Flow Model of the Wall Layers of the Turbulent Boundary Layer," *Trans. ASME,* 166 (1959).

Kline, S. L., Runstadler, P. W., and Reynolds, W. C., *Thermosciences Div. Dept. Mech. Eng.,* Report MD-8, Stanford University (1963).

Kline, S. J., Reynolds, W. C., Schraub, F. A., and Runstadler, P. W., "The Structure of Turbulent Boundary Layers," *J. Fluid Mech.,* **30,** 741 (1967).

Knudsen, J. G., and Katz, D. L., *Fluid Dynamics and Heat Transfer,* (1958).

Knuth, E. L., and Dershin, H., "Use of Reference States in Predicing Transport Rates in High Speed Turbulent Flows with Mass Transfers," *Int. J. Heat Mass Transfer,* **6,** 999 (1963).

Kokorev, L. S., and Ryaposov, V. N., "Turbulent Heat Transfer During the Flow of a Heating Medium of Small Prandtl Number Along a Tube," *Int. Chem. Eng.,* **2,** (4), 511 (1962).

Kolav, V., "Heat and Mass Transport in a Turbulent Medium," *Collection Czech. Chem. Communs.,* **26,** (1), 335 (1961).

Kolavandin, B. A., and Vatutin, I. A., "On the Statistical Theory of Non-Uniform Turbulence," *Int. Seminar on Heat & Mass Transfer,* Herceg Novi, Yugoslavia (1969).

Kolmogorov, A. N., "Equations of Turbulent Motion of an Incompressible Turbulent Fluid," *Izv. Akad. Nauk SSSR Ser. Phys.,* **VI,** (1-2), 56 (1942).

Komasaw, I., Saito, T., and Otake, T., "Mass Transfer across the Liquid-Liquid Interface. Interfacial Turbulence and its Elimination," *Int. Chem. Eng.,* **12,** (2), 34 (1972).

Koppel, L. B., Patel, R. D., and Holmes, J. T., "Statistical Models for Surface Renewal in Heat and Mass Transfer: Part 1. Dependence of Average Transport Coefficients on Age-Distribution," *A.I.Ch.E.J.,* **12,** (5), 941 (1966).

Kovasy, K., "Different Types of Distribution Functions to Describe a Random Eddy Surface Renewal Model," *Chem. Eng. Sci.,* **23,** 90 (1968).

Kovasznay, L. S. G., *Turbulence,* p. 194, Hinze, McGraw-Hill (1959).

Kozinski, A. A., and King, C. J., "The Influence of Diffusivity on Liquid Phase Mass Transfer to the Free Interface in a Stirred Vessel," *A.I.Ch.E.J.,* **12,** (1), 109 (1966).

Kropholler, H. W., and Carr, A. D., "The Prediction of Heat and Mass Transfer Coefficients for Turbulent Flow in Pipes at all Values of the Prandtl and Schmidt Numbers," *Int. J. Heat Mass Transfer,* **5,** 1191 (1962).

Kutateladze, S. S., and Leontev, A. I., *Turbulent Boundary Layers in Compressible Gases,* London, Edward Arnold Publ. Ltd., (1964).

Lamont, J. C., and Scott, D. S., "An Eddy Cell Model of Mass Transfer into the Surface of a Turbulent Liquid," *A.I.Ch.E.J.,* **16,** (4), 513 (1970).

Lamourelle, A. P., and Sandall, O. C., "Gas Absorption into a Turbulent Liquid," *Chem. Eng. Sci.,* **27,** 1035 (1972).

Langford, D., "Pseudo-Similarity Solutions of the One-Dimensional Diffusion Equation with Applications to the Phase Change Problem," *Quarterly of Applied Math.,* **25,** (1), 45 (1967).

Larson, R. I., and Yerazunis, S., "Mass Transfer in Turbulent Flow," *Int. J. Heat Mass Transfer,* **16,** 121 (1973).

Laufer, J., *NACA,* Report No. 32-119 (1961).

Laufer, J., and Badri Narayanan, M. A., "Mean Period of the Turbulent Production Mechanism in a Boundary Layer," *Phys. Fluids,* **14,** 182 (1971).

Launder, B. E., and Priddin, C. H., "A Comparison of some Proposals for the Mixing Length Near a Wall," *Int. J. Heat Mass Transfer,* **16,** 700 (1973).

Launder, B. E., and Spalding, D. B., *Lectures in Mathematical Models of Turbulence,* London, Academic Press (1972).

Leslie, D. C., "Analytical Treatment of Turbulent Pipe Flow with Wall Suction," *J. Brit. Nucl. Energy Soc.,* **12,** 323 (1973).

Levich, W. G., *Physicochemical Hydrodynamics,* New Jersey, Prentice Hall, p. 689, (1962).

Levins, D. M., and Glastonbury, J. R., "Application of Kolmogoroff's Theory to Particle-Liquid Mass Transfer in Agitated Vessels," *Chem. Eng. Sci.,* **27,** 537 (1972).

Lewis, J. B., "The Mechanism of Mass Transfer of Solutes across Liquid-Liquid Interfaces," *Chem. Eng. Sci.,* **3,** 248 (1954).

Lewis, W. K., and Whitman, W. G., "Principles of Gas Absorption," *Ind. Eng. Chem.,* **16,** 1215 (1924).

Libby, P. A., "The Laminar Boundary Layer with Arbitrarity Distributed Mass Transfer," *Int. J. Heat Mass Transfer,* **9,** 1109 (1966).

Licht, W. F., and Pansing, W. F., "Solute Transfer from Single Drops in Liquid-Liquid Extraction," *Ind. Eng. Chem.,* **45,** 1885 (1953).

de Ligny, C. L., "Coupling Between Diffusion and Convection in Radical Dispersion of Matter by Fluid Flow through Packed Beds," *Chem. Eng. Sci.,* **25,** 1177 (1970).

Lin, C. S., Denton, E. B., Gaskill, H. S., and Putnam, G. L., "Diffusion-Controlled Electrode Reactions," *Ind. Eng. Chem.,* **43,** 2136 (1951).

Lin, C. S., Moulton, R. W., and Putnam, G. L., "Mass Transfer between Solid Wall and Fluid Streams," *Ind. Eng. Chem.,* **45,** 636 (1953).

Linton, W. H., Jr., and Sherwood, T. K., "Mass Transfer from Solid Shapes to Water in Stream Line and Turbulent Flow," *Chem. Eng. Progr.,* **46,** 258 (1950).

Ludwieg, H., "Bestimmung Des Verhaltnisses Der Austauschkoeffizienten Fur Warme Und Impuls Bei Turbulenten Grenzschichten," *Z. Flugwissenschaften,* **4,** 73 (1956).

Lykoudis, P. S., and Toulukian, Y. S., "Heat Transfer in Liquid Metals," *Trans. Am. Soc. Mech. Engrs.,* **80,** 853 (1958).

Lynn, S., Straatemeier, J. R., and Kramers, H., "Absorption Studies in the Light of the Penetration Theory. III. Absorption by Wetted Spheres Singly and in Columns," *Chem. Eng. Sci.*, **4**, 63 (1955).

Maeda, T., "Fundamental Studies of Turbulent Boundary Layers with Injection or Suction through Porous Wall," *Bull. JSME,* **15,** 721 (1972).

Mahendra, R. D., and Gill, W. N., "A Note on the Mixing Length Theory of Turbulent Flow," *A.I.Ch.E.J.,* **16,** 885 (1970).

Malina, J. A., and Sparrow, E. M., "Variable-Property, Constant-Property, and Entrance-Region Heat Transfer Results for Turbulent Flow of Water and Oil in a Circular Tube," *Chem. Eng. Sci.,* **19,** 953 (1964).

Malkus, W. V. R., "Outline of a Theory of Turbulent Shear Flow," *J. Fluid Mech.,* **1,** 521 (1956).

Marchello, J. M., and Toor, H. L., "A Mixing Model for Transfer Near a Boundary," *I&EC Fundamentals,* **2,** (1), 8 (1963).

Martinelli, R. C., "Heat Transfer to Molten Metals," *Trans. Am. Soc. Mech. Engrs.,* **69,** 947 (1947).

Masan, D. T., Jones, W. O., and Von Behren, G. L., "Entrance Region Mass Transfer in Turbulent Pipe Flow," *A.I.Ch.E.J.,* **17,** 300 (1971).

McAdams, W. H., *Heat Transmission,* 3rd Ed., New York, McGraw-Hill (1954).

McCarter, R. J., and Stutzman, L. F., "Transfer Resistance and Fluid Mechanics," *A.I.Ch.E.J.,* **5,** 502 (1959).

McManamey, W. J., Davies, J. T., Woollen, J. M., and Coe, J. R., "The Influence of Molecular Diffusion on Mass Transfer Between Turbulent Liquids," *Chem. Eng. Sci.,* **28,** 1061 (1973).

Meek, R. L., "Mean Period of Fluctuations Near the Wall in Turbulent Flows," *A.I.Ch.E.J.,* **18,** 854 (1972).

Meek, R. L., and Baer, A. D., "The Periodic Viscous Sublayer Model and Heat Transfer to Drag Reducing Solutions," *A.I.Ch.E.J.,* **16,** 1100 (1970).

Meek, R. L., and Baer, A. D., "The Periodic Viscous Sublayer in Turbulent Flow," *A.I.Ch.E.J.,* **16,** 841 (1970).

Meek, R. L., and Baer, A. D., "Turbulent Heat Transfer and the Periodic Viscous Sublayer," *Int. J. Heat Mass Transfer,* **16,** 1385 (1973).

Mellor, G. L., and Herring, H. J., "A Survey of the Mean Turbulent Field Closure Models," *A.I.A.J.,* **11,** 590 (1973).

Merchuk, J. C., and Lupin, H. M., "Application of the Film-Penetration Model to Absorption with First Order Reaction is a Finite Volume," *Lat. Am. M. Chem. Eng. & Appl. Chem.,* 33 (1971).

Meter, D. M., "Tube Flow of Non-Newtonian Polymer Solutions: Part II. Turbulent Flow," *A.I.Ch.E.J.,* **10,** (6), 881 (1964).

Meyerink, E. S. C., and Friedlander, S. K., "Diffusion and Diffusion Controlled Reactions in Fully Developed Turbulent Pipe Flow," *Chem. Eng. Sci.,* **17,** 121 (1962).

Michels, H. M., *Mass Transfer During Drop Formation,* Ph.D. Thesis, University of Delaware (1960).

Miyauchi, T., and Kataoka, H., "Liquid Film Coefficient of Mass Transfer on Free Liquid Surface," *Journal of Chem. Eng. Japan,* **3,** (2), 257 (1970).

Mizushina, T., Ito, R., and Ogino, F., "Eddy Diffusivity Distribution Near the Wall," *Fourth Intern. Heat Transfer Conference,* Paris-Versailles (1970).

Mizushina, T., Ito, R. Ogino, R., and Muramoto, H., "Eddy Diffusivities for Heat in the Region Near the Wall," *Mem. Fac. Eng.,* Kyoto University, **31,** 169 (1969).

Mizushina, T., and Ogino, R., "Eddy Viscosity and Universal Velocity Profile in Turbulent Flow in a Straight Pipe," *J. Chem. Eng. Japan,* **3,** 166 (1970).

Mizushina, T., Ogino, F., Oka, Y., and Fukuda, H., "Turbulent Heat and Mass Transfer between Wall and Fluid Streams of Large Prandtl and Schmidt Numbers," *Int. J. Heat Mass Transfer,* **14,** 1705 (1971).

Mizushina, T., and Sasano, T., "The Ratio of the Eddy Diffusivities for Heat and Momentum and its Effect on Liquid Metal Heat Transfer Coefficients," *Proc. Intern. Heat Transfer Conf.,* Boulder, Colorado, USA, ASME Paper 78, (1962).

Moffat, R. J., and Kays, W. M., "The Turbulent Boundary Layer on a Porous Plate: Experimental Heat Transfer with Uniform Blowing and Suction," *Int. J. Heat Mass Transfer,* **11,** 1547 (1968).

Murphree, E. V., "Relation between Heat Transfer and Fluid Friction," *Ind. Eng. Chem.,* **24,** 726 (1932).

Na, T. Y., and Habib, I. S., "Heat Transfer in Turbulent Pipe Flow Based on a New Mixing Length Model," *Appl. Sci. Res.,* **28,** 302 (1973).

Nedderman, R. M., "The Measurement of Velocities in the Wall Region of Turbulent Liquid Pipe Flow," *Chem. Eng. Sci.,* **16,** 120 (1961).

Nedderman, R. M., "The Use of Stereoscopic Photography for the Measurement of Velocities in Liquids," *Chem. Eng. Sci.,* **16,** 113 (1961).

Nee, V. W., and Kovasznay, L. S. G., "The Calculation of the Incompressible Turbulent Boundary Layer by a Simple Theory," *Proc. of AFOSR/IFP Conf. on Computation of Turbulent Boundary Layers,* Vol. 1, Stanford University; also *Physics of Fluids,* **12,** 473 (1968).

Newman, B. G., "Turbulent Jets and Wakes in a Pressure Gradient," *Fluid Mechanics of Internal Flow,* (1967).

Ng, K. H., and Spalding, D. B., "Some Applications of a Model of Turbulence to Boundary Layers Near Walls," *Physics of Fluids,* **15,** (1972).

Nienow, A. W., Unahabhokha, R., and Mullin, J. W., "The Mass Transfer Driving Force for High Mass Flux," *Chem. Eng. Sci.,* **24,** 1655 (1969).

Nikuradse, J., "Gesetzmassigkeit Der Turbulenten Stromung in Glatten Rohren," *VDI–Forschungsheft,* 356. (1932).

Notter, R. H., and Sleicher, C. A., "The Limiting Form of the Eddy Diffusivity Close to a Wall," *A.I.Ch.E.J.,* **15,** 936 (1969).

Notter, R. H., and Sleicher, C. A., "The Eddy Diffusivity in the Turbulent Boundary Layer Near a Wall," *Chem. Eng. Sci.,* **26,** 161 (1971).

Notter, R. H., and Sleicher, C. A., "A Solution to the Turbulent Graetz Problem–III Fully Developed and Entry Region Heat Transfer Rates," *Chem. Eng. Sci.,* **27,** 2073 (1972).

Nusselt, W., "Das Grundgesetz Des Warmeuberganes," *Ges. Ing.,* **38,** 447 (1915).

Ohji, M., "Statistical Theory of Wall Turbulence," *Phys. Fluids,* **10,** 5153 (1967).

Onda, T., and Sade, S., *Chem. Proc. Plant,* Japan (1966).

Owen, J. M., "The Reynolds Analogy Applied to Flow Between a Rotating and a Stationary Disc," *Int. J. Heat Mass Transfer,* **14,** (3), 45 (1971).

Page, A., and Townend, H. C. H., *Proc. Roy. Soc.,* London **135A,** 656 (1932).

Pao, Y. H., and Daugherty, R. J., *IUTAM Int. Symp. On High-Speed Computing in Fluid Dynamics,* Monterey, California (1968).

Patankar, S. V., and Spalding, D. B., "A Finite-Difference Procedure for Solving the Equations of the Two-Dimensional Boundary Life," *Int. J. Heat Mass Transfer,* **10,** 1389 (1967).

Patankar, S. V., and Spalding, D. B., *Heat and Mass Transfer in Boundary Layers,* 2nd Ed., London, Intertext Books, (1970).

Penny, W. R., and Bell, K. J., "Close-Clearance Agitators. Part II: Heat Transfer Coefficients," *I&EC Fundamentals,* **59,** 47 (1967).

Perlmutter, D. D., "Surface-Renewal Models in Mass Transfer," *Chem. Eng. Sci.,* **16,** 287 (1961).

Persen, L. N., *Boundary Layer Theory,* Tapir, Forlay (1972).

Pinczewski, Z., and Sideman, S., "A Model for Mass (Heat) Transfer in Turbulent Tube Flow. Moderate and High Schmidt (Prandtl Numbers)," *Chem. Eng. Sci.,* (1974) in press.

Pohlhausen, E., angew. *Math. u. Mech.,* 1:115 (1921).

Popovich, A. T., "Statistical Analysis of Fluid Flow Fluctuations in the Viscous Layer Near a Solid Wall," *I&EC Fundamentals,* **8,** 609 (1969).

Popovich, A. T., and Hummel, R. L., "Experimental Study of the Viscous Sublayer in Turbulent Pipe Flow," *A.I.Ch.E.J.,* **13,** 854 (1967).

Popovich, A. T., and Lavallee, H. C., "Flow Conditions near Roughness Elements Obtained by Photolysis Method," *Chem. Eng. Sci.,* 1109 (1970).

Poreh, M., and Paz, U., "Turbulent Heat Transfer to Dilute Polymer Solutions," *Int. J. Heat Mass Transfer,* **11,** 805 (1968).

Porter, K. E., and Roberts, D., "Similarities Between the Effect of Different Flow Patterns on Diffusion with Chemical Reaction near an Interface," *Chem. Eng. Sci.,* **24,** 695 (1969).

Potter, C. E., "Mass Transfer Between Co-current Fluid Streams and Boundary Layer Solutions," *Chem. Eng. Sci.,* **6,** 170 (1957).

Prandtl, L., "Bericht Uber Untersuchungen Zur Ausgebildete Turbulenz," *angew. Math. u. Mech.,* **5,** 136 (1925).

Prandtl, L., "Uber Ein Neues Formelsystem Fur Die Ausgebildete Turbulenz," *Nachrichten von der Akad. der Wissenschaft in Gottingen.* (1945).

Prasher, B. D., "Gas Absorption into a Turbulent Liquid," *Chem. Eng. Sci.,* **28,** 1230 (1973).

Prasher, B. D., and Wills, G. B., "Mass Transfer in an Agitated Vessel," *Ind. Eng. Chem. Process Des. Develop.,* **12,** (3), 351 (1973).

Price, J. M., and Harris, J. F., "Computer Program for Solving Compressible Non-Similar Boundary-Layer Equations for Laminar, Transitional, or Turbulent Flows of a Perfect Gas," NASA-TM-X-2458; L-8037, Langley Research Center, NASA Langley Station, Virginia (1972).

Quarmby, A., and Anand, R. K., "Axisymmetric Turbulent Mass Transfer in a Circular Tube," *J. Fluid Mech.,* **38,** (3), 433 (1969).

Quarmby, A., and Quirk, R. K., "Measurements of the Radial and Tangential Eddy Diffusivities of Heat and Mass in Turbulent Flow in a Plain Tube," *Int. J. Heat Mass Transfer,* **15,** 2309 (1972).

Quarmby, A., and Quirk, R. K., "Aximmetric and Non-Aximmetric Turbulent in a Plain Circular Tube at High Schmidt Numbers," *Int. J. Heat Mass Transfer,* **17,** 143 (1974).

Rannie, W. D., "Heat Transfer in Turbulent Shear Flow," *J. Aero Sci.,* **23,** 485 (1956).

Ranz, W. E., and Marshall, W. R., "Evaporation from Drops," *Chem. Eng. Prog.,* **48,** 141 (1952).

Rao, M. A., and Brodkey, R. S., "Continuous Flow Stirred Tank Turbulence Parameters in the Impeller Stream," *Chem. Eng. Sci.,* **27,** 137 (1972).

Rasmussen, M., and Karamcheti, K., "On the Viscous Sublayer of an Incompressible Turbulent Boundary Layer," Dept. of Aero. & Astro., Stanford University (1965).

Reichardt, H., "Die Grundlagen Des Turbulenten Warmeuberganes," *Arch. Ges. Warmetech,* **2,** 129 (1951).

Reynolds, O., "On the Extent and Action of the Heating Surface for Stream Boilers," *Proc. Lit. Phil. Soc.,* Manchester, **14,** 7 (1874).

Richardson, S., "On the No-Slip Boundary Condition," *J. Fluid Mech.*, **59**, 707 (1973).

Rodi, W., and Spalding, D. B., "A Two Parameter Model of Turbulence and Its Application to Free Jets," *Warme und Stoffubertragung*, **3**, 85 (1970).

Roper, A. T., and Gentry, G. L., Jr., *Analysis of a Turbulent Boundary Layer Over a Moving Ground Plane*, NASA-TN-D-6788; L-7872, Langley Research Center, NASA, Langley Station, Virginia (1972).

Roper, A. T., and Gentry, G. L., Jr., *Turbulent-Boundary-Layer Development on a Moving Ground Belt of Rough Texture*, NASA-TM-X-2515; L-7845, Langley Research Center, NASA, Langley Station, Virginia (1972).

Rotta, J., "Statistische Theorie Nichthomogener Turbulenz," *Zeitsch. fur Physik*, **129**, 547, and **131**, 51 (1951).

Rotta, J., "Uber Eine Methode Zur Berechnung Turbulenter Scherstromungen," *Aerodynamische Versuchanstalt Gottingen*, Report 69 A 14. (1968).

Rotta, J., "Recent Attempts to Develop a Generally Applicable Calculation Method for Turbulent Shear Flow Layers," *Proceedings of AGARD Conference on Turbulent Shear Flows*, London (1971).

Rozenshtok, Y. L., in B. S. Petukhov, I. P. Ginsburg, and A. S. Kasperovich, (Eds.), *Heat & Mass Transfer*, **1**, Minsk (1965). English translation by Israel Program for Scientific Translations Ltd., Jerusalem (1967).

Ruckenstein, E., "A Note Concerning Turbulent Exchange of Heat or Mass with a Boundary," *Chem. Eng. Sci.*, **7**, 265 (1958).

Ruckenstein, E., "Some Remarks on Renewal Models," *Chem. Eng. Sci.*, **18**, 233 (1963).

Ruckenstein, E., "Physical Models for Mass or Heat Transfer Processes," *Intern. Chem. Eng.*, **7**, (3), 490 (1967).

Ruckenstein, E., *Chem. Eng. Sci.*, **22**, 474 (1967).

Ruckenstein, E., "A Generalized Penetration Theory for Unsteady Convective Mass Transfer," *Chem. Eng. Sci.*, **23**, 363 (1968).

Ruckenstein, E., "Mass or Heat Transfer to Turbulent Separated Flows for Large Schmidt or Prandtl Numbers," *Chem. Eng. Sci.*, **25**, 819 (1970).

Ruckenstein, E., "Effect of Interfacial Turbulence on the Rate of Absorption," *A.I.Ch.E.J.*, **16**, (6), 1098 (1970).

Ruckenstein, E., "Effect of Unsteady Convection on Mass Transfer Accompanied by a Chemical Reaction," *The Chem. Eng. Journal*, **2**, (1971).

Ruckenstein, E., "On the Mechanism of Drag Reduction in Turbulent Flow of Viscoelastic Liquids," *Chem. Eng. Sci.*, **26**, 1075 (1971).

Ruckenstein, E., "Transport Equations in Transfer Coordinates," *Chem. Eng. Sci.*, **26**, 1795 (1971).

Ruckenstein, E., "Momentum and Heat Transfer in Turbulent Pipe Flow of Solutions of Macromolecules," *The Chem. Eng. Journal*, **4**, 180 (1972).

Ruckenstein, E., "Turbulent Heat Transfer to Non-Newtonian Fluids," *Chem. Eng. Sci.*, **27**, 947 (1972).

Ruckenstein, E., and Berbente, C., "Mass Transfer in Wave Flow," *Chem. Eng. Sci.*, **20**, 795 (1965).

Ruckenstein, E., and Berbente, C., "Mass Transfer to Falling Liquid Films at Low Reynolds Numbers," *Int. J. Heat Mass Transfer*, **11**, 743 (1968).

Ruckenstein, E., and Berbente, C., "The Effect of Roll-Cells on Mass Transfer," *Chem. Eng. Sci.*, **25**, 475 (1970).

Ruckenstein, E., and Suciu, D. G., "Simulation of Physical Models for Turbulent Mass Transfer," *Chem. Eng. Sci.*, **24**, 1395 (1969).

Sage, B. H., and Venezian, E., "Temperature Gradients in Turbulent Gas Streams: Effect of Viscous Dissipation on the Evaluation of Total Conductivity," *AIChE J.*, **7**, 688 (1961).

Sani, R. L., "On Finite Amplitude Roll Cell Disturbances in a Fluid Layer Subject to Heat and Mass Transfer," *A.I.Ch.E.J.,* **11,** (6), 971 (1965).

Sawistowski, H., and Goltz, G. E., "The Effect of Interface Phenomena on Mass-Transfer Rates in Liquid-Liquid Extraction," *Trans. Instn. Chem. Engrs.,* **41,** 173 (1963).

Schlichting, H., *Boundary Layer Theory,* translated by J. Kestin (4th Ed.), New York, McGraw-Hill (1960).

Schubauer, G. B., and Tchen, C. M., "Turbulent Flow" in LINC. C., *Turbulent Flows and Heat Transfer,* London, Oxford University Press (1959).

Schwalter, W. R., et al., "Heat and Mass Transfer in Rheologically Complex Fluids," *Prog. In Heat Mass Transfer,* **5,** (1972).

Scriven, L. E., "Dynamics of a Fluid Interface," *Chem. Eng. Sci.,* **12,** 98 (1960).

Scriven, L. E., "Flow and Transfer at Fluid Interfaces," *Chem. Eng. Education,* **2,** 10 (1968); **3,** 26 (1969).

Scriven, L. E., and Pigford, R. L., "Fluid Dynamics and Diffusion Calculations for Laminar Liquid Jets," *A.I.Ch.E.J.,* **5,** 397 (1959).

Seban, R. A., and Doughty, D. L., "Heat Transfer to Turbulent Boundary Layers with Variable Free Stream Velocity," *Trans. ASME,* **78,** 217 (1956).

Senechal, M., *Contribution to Convection Theory for Liquid Metals,* Paris (1968).

Sesonski, A., et al., "Eddy Diffusivity Ratios for Mercury Flowing in a Tube," *Chem. Eng. Prog. Symp. Series,* **61,** (57), 101 (1965).

Sherwood, T. K., "Mass, Heat, and Momentum Transfer Between Phases," *Chem. Eng. Progr. Symp. Ser.,* **71,** 55 (25) (1959).

Sherwood, T. K., and Pigford, R. L., *Absorption and Extraction,* 2nd. Ed., New York, McGraw-Hill (1952).

Sherwood, T. K., Smith, K. A., and Fowles, P. E., "The Velocity and Eddy Viscosity Distribution in the Wall Region of Turbulent Pipe Flow," *Chem. Eng. Sci.,* **23,** 1225 (1968).

Shimomura, R., "Heat Transfer from a Rotating Disk in Parallel with Another Disk," *Bull. J.S.M.E.,* **13,** (66), 1449 (1970).

Shlanchyauskas, A. A., Vaitekunas, P. P., and Zhukauskas, A. A., "Calculation of Heat Transfer in the Turbulent Boundary Layer of Viscous Liquids and an Evaluation of the Effect of Pr_t," *Intern. Chem. Eng.,* **14,** (1), 33 (1974).

Sideman, S., "The Equivalence of the Penetration and Potential Flow Theories," *Indus. & Eng. Chem.,* 54 (1966).

Sideman, S., and Uliss, I., "Direct Contact Heat Transfer Between Immiscible Liquids in Stratified Flow with Simultaneous Interfacial Boiling," *Chem. Eng. Sci.,* **27,** 1245 (1972).

Sideman, S., and Barsky, Z., "Turbulence Effect on Direct-Contact Heat Transfer with Change of Phase," *A.I.Ch.E.J.,* **11,** (3), 539 (1965).

Simpson, R. L., and Field, R. L., "A Note on the Turbulent Schmidt and Lewis Numbers in a Boundary Layer," *Int. J. Heat Mass Transfer,* **15,** 180 (1972).

Simpson, R. L., Whitten, D. G., and Moffat, R. J., "An Experimental Study of the Turbulent Prandtl Number of Air with Injection and Suction," *Int. J. Heat Mass Transfer,* **13,** 125 (1970).

Sirkar, K. K., and Hanratty, T. J., "Limiting Behavior of the Transverse Turbulent Velocity Fluctuations Close to a Wall," *Ind. Eng. Chem. Fundamentals,* **8,** 189 (1969).

Sivasegarem, S., and Whitelaw, J. H., "The Prediction of Turbulent, Supersonic, Two-Dimensional Boundary-Layer Flows," *Aero. Quarterly,* **22,** 274 (1971).

Skelland, A. H. P., "Momentum, Heat, and Mass Transfer in Turbulent Non-Newtonian Boundary Layers," *A.I.Ch.E.J.,* **12,** (1), 69 (1966).

Sleicher, C. A., "Experimental Velocity and Temperature Profiles for Air in Turbulent Pipe Flow," *Trans. A.S.M.E.*, **80,** 693 (1958).

Sleicher, C. A., Awad, A. S., and Notter, R. H., "Temperature and Eddy Diffusivity Profiles in Nak," *Int. J. Heat Mass Transfer,* **16,** 1565 (1973).

Smith, J. W., and Hummel, R. L., "Studies of Fluid Flow by Photography Using a Non-Disturbing Light-Sensitive Indicator," *Journal of the SMPTE,* **82,** (4), 278 (1973).

Smith, A. G., and Shah, V. L., "The Calculation of Wall and Fluid Temperatures for the Incompressible Turbulent Boundary Layer with Arbitrary Distribution of Wall Heat Flux," *Int. J. Heat Mass Transfer,* **5,** 1179 (1962).

Son, J. S., and Hanratty, T. J., "Limiting Relation for Eddy Diffusivity Close to a Wall," *A.I.Ch.E.J.,* **13,** 689 (1967).

Spalding, D. B., "Heat Transfer to a Turbulent Steam From a Surface with Stepwise Discontinuity in Wall Temperature," *Intern. Developments in Heat Transfer, ASME,* **11,** 439 (1961).

Spalding, D. B., "Contribution to the Theory of Heat Transfer Across A Turbulent Boundary Layer," *Int. J. Heat Mass Transfer,* **7,** 743 (1964).

Spalding, D. B., "Heat Transfer from Turbulent Separated Flows," *J. Fluid Mech.,* **27,** 97 (1967).

Spalding, D. B., *The Prediction of Two-Dimensional, Steady Turbulent Flows,* Imperial College, Heat Transfer Sect., Rep. EF/TN/A16. (1969).

Spalding, D. B., "Concentration Fluctuations in a Round Turbulent Free Jet," *Chem. Eng. Sci.,* **26,** 95 (1971).

Sparrow, E. M., and Yu, H. S., "Heat and Mass Transfer in a Channel with Surface Mass Addition: Application to Phase Change Processes," *A.I.Ch.E.J.,* **16,** 588 (1970).

Springer, T. G., and Pigford, R. L., "Influence of Surface Turbulence and Surfactants on Gas Transport Through Liquid Interfaces," *Ind. Eng. Chem. Fundamentals,* **9,** 458 (1970).

Sternberg, J., "A Theory for the Viscous Sublayer of a Turbulent Flow," *Fluid Mech.,* 241 (1961).

Sternberg, J., "On the Interpretation of Space-Time Correlation Measurements in Shear Flow," *Phys. Fluids,* **10,** 146 (1967).

Stewart, W. E., "Mass Transfer in Laminar-Boundary-Layer Flows with Finite Interfacial Velocities," *A.I.Ch.E.J.,* **8,** (3), 421 (1962).

Stewart, W. E., "Forced Convection in Three-Dimensional Flows: 1. Asymptotic Solutions for Fixed Interfaces," *A.I.Ch.E.J.,* **9,** (4), 528 (1963).

Stewart, W. E., Angelo, J. B., and Lightfoot, E. N., "Forced Convection in Three-Dimensional Flows: II. Asymptotic Solutions for Mobile Interfaces," *A.I.Ch.E.J.,* **16,** 771 (1970).

Szekely, J., "Notes on Unsteady State Diffusion with a Large Flux of the Diffusing Substance," *Chem. Eng. Sci.,* **19,** 51 (1964).

Tauscher, W. A., Sparrow, E. M., and Lloyd, J. R., "Amplification of Heat Transfer by Local Injection of Fluid into a Turbulent Tube Flow," *Int. J. Heat Mass Transfer,* **13,** (4), 681 (1970).

Taylor, G. I., "Conditions at the Surface of a Hot Body Exposed to the Wind," *Brit. Adv. Com. Aero. R. and M.,* 272 (1916).

Thomas, L. C., "A Pseudo-Surface Rejuvenation Model for Turbulent Heat Transfer," *6th South-East Seminar Thermal Sci.,* (1970).

Thomas, L. C., "An Analysis for Low Turbulence Intensity Liquid Metal Heat Transfer," *Int. J. Heat and Mass Transfer,* **14,** 1747 (1971).

Thomas, L. C., "A Turbulent Transport Model with Emphasis on Heat Transfer to Liquid Metals," *The Can. Journal of Chem. Eng.*, **49,** 326 (1971).

Thomas, L. C., "Temperature Profiles for Turbulent Flow in Tubes and Basic Surface Renewal Models," *Chem. Eng. Sci.*, **26,** 1271 (1971).

Thomas, L. C., "A Preliminary Formulation for $^{\epsilon}H$ and $^{\epsilon}m$ Based on the Elementary Surface Renewal and Penetration Model," presented *AICHE Meeting,* Philadelphia (1973).

Thomas, L. C., and Fan, L. T., "The Surface Renewal Principle: A Consideration of Two Turbulent Heat Transfer Models," *Israel Journal of Tech.*, **8,** (6), 439 (1970).

Thomas, L. C., and Fan, L. T., "Heat and Momentum Transfer Analogy for Incompressible Turbulent Boundary Layer Flow," *Int. J. Heat Mass Transfer,* **14,** 715 (1971).

Thomas, L. C., and Fan, L. T., "A Model for Turbulent Transfer in the Wall Region," *Journal of the Eng. Mech. Div.*, **97,** 169 (1971).

Thomas, L. C., and Fan, L. T., "Adaptation of the Surface Rejuvenation Model to Turbulent Heat and Mass Transfer at a Solid-Fluid Interface," *Ind. Eng. Chem. Fundamentals,* **10,** (1), 135 (1971).

Thomas, L. C., and Greene, H. L., "An Experimental and Theoretical Study of the Viscous Sublayer for Turbulent Tube Flow," presented *Symp. on Turbulence in Liquids,* University of Missouri, Rolla (1973).

Thomas, L. C., and Rajagopal, R., "Adaptation of the Stochastic Formulation of the Surface Rejuvenation Model to Turbulent Convection Heat Transfer," *Chem. Eng. Sci.*, **29,** 1639 (1974).

Tien, C. L., "On the Eddy Diffusivities for Momentum and Heat," *Appl. Sci. Res.*, **A8,** 345 (1959).

Toor, H. L., and Marchello, J. M., "Film-Penetration Model for Mass and Heat Transfer," *A.I.Ch.E.J.*, **4,** (1), 97 (1958).

Townes, H. W., Gow, J. L., Powe, R. E., and Weber, N., "Turbulent Flow in Smooth and Rough Pipes," *J. Basic Eng.*, **94,** 353 (1972).

Townsend, A. A., "Equilibrium Layers and Wall Turbulence," *J. Fluid Mech.*, **11,** 97 (1961).

Tson, F. K., Sparrow, E. M., and Goldstein, R. J., "Flow and Heat Transfer in the Boundary Layer on a Continuous Moving Surface," *Int. J. Heat Mass Transfer,* **10,** 219 (1967).

Tyldesley, J. R., and Silver, R. S., "The Prediction of the Transport Properties of a Turbulent Fluid," *Int. J. Heat Mass Transfer,* **11,** 1325 (1968).

Van Driest, E. R., "On Turbulent Flow Near a Wall," *J. Aero. Sci.*, **23,** 1007 (1956).

Van Shaw, P., and Hanratty, T. J., "Fluctuations in the Local Rate of Turbulent Mass Transfer to a Pipe Wall," *A.I.Ch.E.J.*, **10,** 475 (1964).

Virk, P. S., Merrill, E. W., Mickley, H. S., and Smith, K. A., The Toms Phenomenon: Turbulent Pipe Flow of Dilute Polymer Solutions," *J. Fluid Mech.*, **30,** 305 (1967).

Vivian, J. E., and Schoenberg, T., "Mass Transfer Rates for Vaporization and Gas-Phase Controlled Desorption in a Short Wetted Wall Column," *A.I.Ch.E.J.*, **14,** 986 (1968).

Von Kármán, T., "Mechanische Ahnlichkeit Und Turbulenz," *Proc. 3rd. Int. Congress Appl. Mech.* Stockholm, **1,** 85 (1939).

Von Mises, R., "Bermekungen zur Hydrodynamik," *A. Angew. Math. Mech.*, **7,** 425 (1927).

Voskrenskii, U. P., *Proc. Int. Conf. Peaceful Uses of Atomic Energy,* U.N., New York (1956); S. S. Kutateladze, W. M. Borishanskii, I. I. Novikoc, and O. S. Fedynaskii, *Liquid Metal Heat Transfer Media,* Translation of Atomnaya Energy Suppl. 2 (1958).

Wallace, J. M., Eckelmann, H., and Brodkey, R. S., "The Wall Region in Turbulent Shear Flow," *J. Fluid Mech.*, **54,** 39 (1972).

Wasan, D. T., and Ahluwalia, M. S., "Consecutive Film and Surface Renwal Mechanism for Heat or Mass Transfer from a Wall," *Chem. Eng. Sci.*, **24,** 1535 (1969).

Wasan, D. T., Tien, C. L., and Wilke, C. R., "Theoretical Correlation of Velocity and Eddy Viscosity for Flow Close to a Pipe Wall," *A.I.Ch.E.J.*, **9**, 567 (1963).

Wasan, D. T., and Wilke, C. R., "Turbulent Exchange of Momentum, Mass, and Heat Between Fluid Streams and Pipe Wall," *Int. J. Heat Mass Transfer,* **7**, 87 (1964).

Wassel, A. T., and Cotton, I., "Calculation of Turbulent Boundary Layers over Flat Plates with Different Phenomenological Theories of Turbuelnce and Variable Turbulent Prandtl Number," *Int. J. Heat Mass Transfer,* **16**, 1547 (1973).

Webb, R. L., "A Critical Evaluation of Analytical Solutions and Reynolds Analogy Equations for Turbulent Heat and Mass Transfer in Smooth Tubes," *Warme-und Stoffubertragung Bd,* 4 S.197 (1971).

Whitman, W. G., "Two Film Theory of Gas Absorption," *Chem. & Met. Eng.*, **29**, 47 (1923).

Wieghardt in D. E. Coles and E. A. Hirst, Eds., *Proceedings–Computation of Turbulent Boundary Layers, 1968,* AFOSR–IFP, Stanford Conf., Vol. II, Compiled data, Stanford University, Palo Alto, California (1968).

Wilcox, D. C., and Alber, E., "Turbulence Model for High Speed Flows," *Proc. Heat Trans. & Fluid Mech. Inst.*, (1972).

Willmarth, W. N., and Tu, B. J., "Structure of Turbulence in the Boundary Layer Near the Wall," *Phys. Fluids,* **10**, S134 (1967).

Wygnanski, I. J., and Champagne, F. H., "On Transition in a Pipe. Part I. The Origin of Puffs and Slugs and the Flow in a Turbulent Slug," *J. Fluid Mech.*, **59**, 281 (1973).

Zaric, Z., "Wall Turbulence Statistical Analysis," presented at *The IV All-Union Heat & Mass Transfer Conf.*, Minsk, SSSR (1972).

Zweitering, T. N., "The Degree of Mixing on Continuous Flow Systems," *Chem. Eng. Sci.*, **11**, 1 (1959).

ACKNOWLEDGEMENT

This work was interrupted by the Yom-Kippur War. We owe more than just words to our students and their comrades in arms who enabled us to finish this study.

The study was partially sponsored by the Israel National Council for Research and Development and partly financed by a Grant from the Vice President for Research, Technion, Israel Institute of Technology.

Finally, thanks are due to Rosalie Greenspan, whose patience and devotion turned pages of scribbles into a readable manuscript.

PART TWO
Biotransport

3
Studies of Bio-Heat Transfer in Mammals

Avraham Shitzer
Faculty of Mechanical Engineering
Technion—Israel Institute of Technology
Haifa, Israel

CONTENTS

1 INTRODUCTION

The thermal system of mammals, and that of the human body in particular, has long been of interest to many investigators. The ability to maintain a fairly constant "deep body" temperature and to dissipate or preserve body thermal energy under widely varying environmental conditions and activity levels are only examples to the fascinating aspects of this system.

With the advent of scientific knowledge and methods, emphasis has shifted from mere observations, however elaborate, of this system and its functions to more systematic studies. Questions like—which part of the body is charged with the task of controlling temperatures and heat fluxes in the body and how is this task achieved—arose and awaited acceptable answers. In other words, efforts have been directed to provide more understanding and insight into the nature and thermal behavior of the thermoregulatory system.

In common with many branches of life sciences, the study of the thermoregulatory system has approached a stage in which work of descriptive nature will yield progressively less in the way of improved insight. Consequently, other approaches, e.g., analytical, should be employed. Analytical modeling of various aspects of the thermoregulatory system may be expected to provide improved understanding of the inherent mechanisms of thermoregulation and heat transfer. From these models new experimental directions may emerge.

Many problems in thermal physiology and bioengineering lend themselves to mathematical formulation and modeling. Among these the following demonstrate the diversity of topics already covered by scientists in these areas:

1. The study of the behavior or the mammalian thermoregulatory system as a control system.
2. The study of the physiological meaning of thermal comfort under varying environmental conditions.
3. The study of the interaction between the human body and an external cooling device, e.g., liquid cooled garments used in space exploration.
4. The study of skin burns.
5. Heat transfer associated with medical treatment, e.g., dental.
6. The measurement of various thermal and physiological properties of live tissue.
7. The study of thermoregulation in hibernation.

Methods of approach to modeling and solving the problem under consideration are diverse, too. They depend, in most cases, on the background of the investigator and the degree of approximation desired. Generally, analytical, numerical, and analog methods have been employed. The use of computers, digital and analog, is quite extensive, in common with other areas of science.

The objective of this work is to collate many of the bio-heat transfer studies in mammals. These works are reviewed and discussed. It was also felt that for completeness a short description of the thermoregulatory system in mammals should be included in the present work. Therefore, a section was devoted to the presentation of commonly accepted concepts and theories relating to mammalian thermoregulation in various situations. In another section bio-heat transfer mechanisms are discussed and the equation of thermal energy balance in a living tissue is derived. Also included is a list of dimensionless parameters associated with the transfer of thermal energy in living systems and a list of some *in vivo* and *in vitro* thermophysical properties of mammalian tissues.

A major area of bio-heat transfer, namely, mathematical models of the entire mammalian thermoregulatory system, is not included in the present work. Although this branch of thermal physiology is probably required for a more complete understanding of the mechanisms involved in bio-heat transfer, its ommission was necessitated by page limitation. For interested readers a separate list of references of works dealing with mathematical models of the mammalian thermoregulatory system is included at the end of this work. Also included in this bibliography are bio-heat transfer studies which were not discussed in the various sections of this work.

2 THE THERMOREGULATORY SYSTEM IN MAMMALS

2.1 Thermoregulation in homeotherms

Living creatures may be classified into two categories based on their blood temperature: poikilothermic and homeothermic. Poikilothermic, or "cold blooded" animals, are those whose body temperatures fluctuate with the temperature of the environment. Homeothermic, or "warm blooded" animals, constitute a group which exhibits a tendency to stability in the normal body states (fixed internal environment). There also exist certain animals that may be homeothermic or poikilothermic at times. These animals are referred to as hibernators.

All mammals are either homeothermic or hibernators. Consequently, the survival of mammals depends on their ability to control and maintain body temperatures within the range essential to life. This regulation has to be achieved under widely varying external, i.e., environment, and/or internal, i.e., exercise or disease, conditions. It is well known that the temperature of each of their organs and the entire organism should remain within the vital range of 0–42°C, with most of the internal temperatures controlled within the range of 35-39°C. Variability in endurance times to the extremes of this wider range among different organs is also great. Some organs, e.g., brain, may suffer irreversible damage should their temperature be allowed to exceed a much closer limit. The functioning of other organs, e.g., heart, may be slowed down or even impaired should their temperature be kept too low within that range. Typical body temperatures in relation to body weight of some mammals at rest are shown in Table 1.

Thermal regulation is essentially the continuous maintenance of body energy balance. This principle of conservation of energy may be formulated mathematically as follows (first law of thermodynamics for a control volume):[1]

$$\frac{d}{dt}\int_V \rho e dV = \int_S \rho h v_i n_i \, dA + \frac{\delta w}{\delta t} - \frac{\delta Q}{\delta t} \tag{2.1}$$

where the terms, from left to right, represent the rate of change in internal energy, net rate of flow of energy out of the body, the rate of work done by the body and the rate of energy exchange with the environment. In Eq. (2.1) the usual sign notation is adopted. Due to the complexity of the mammalian body, a much simpler expression for the enrgy equation is often used (modified from ref. [2]):

$$\dot{M} + \dot{Q}_{\text{gain}} = \dot{Q}_{\text{loss}} + \dot{Q}_{\text{store}} + \dot{W} \tag{2.2}$$

Table 1 Typical Body Weights and Normal Deep Body Temperatures at Rest of Some Mammals* Listed Alphabetically

Species	Body weight	Deep body temperature, °C
African elephant	–	36.4
Black bellied hamster	400 gr	39.5
Camel	281 kg	38.1
Deer mouse	100 gr	37.8
Dolphin	95 kg	38.2
Domestic rabbit	2.4 kg	38.5
Eland	200 kg	38.7
Goat	–	39.0
Ground squirrel	100 gr	36.5
Guanaco	58 kg	38.4–39.0
Hippopotamus	–	36.5
Man	–	36.6–37.0
Mole-rat	90 gr	35.0
Muskrat	1.10 kg	39.5
Ox	499 kg	38.5
Pig	–	39.7
Reindeer	83 kg	38.6–38.9
Sheep	–	38.5
Shrew	4 gr	38.8
Waterbuck	95–105 kg	38.7
White rhinoceros	3000 kg	35.2
Wood rat	190 gr	36.6

*Reproduced in part from "Comparative Physiology of Thermoregulation," Vol. 2–Mammals, G. C. Whittow, Editor, Academic Press, New York, 1971.

The subsystem charged with the control of body energy balance and temperatures is the thermoregulatory system. This highly complex system is composed of three interconnected elements: sensory, control, and regulatory.

Hypothalamic and cutaneous thermal receptors form the major components of the sensory system.[3,7] In addition, there are indications that temperature-sensitive structures exist in the spinal cord,[4] in the voluntary muscles,[3] in the hypothalamus,[5] and in the respiratory tract,[6] which probably play a subsidiary role in thermoregulation. These temperature-sensitive structures, which subdivide into cold and warm receptors,[12] connect to the center of thermal control through the afferent neural pathways. Changes in skin temperatures, and probably also in heat flux at the skin,[5,8] trigger neural signals which are communicated to the hypothalamus.

The hypothalamus is a structure of gray matter which lies at the base of the brain forming the floor and part of the lateral wall of the third ventricle.[9–11]

It is generally accepted that neural control of body temperatures is one of the functions of the hypothalamus. In addition, the hypothalamus plays an important role in water, fat, and carbohydrate metabolism.[2] There appear to be two centers in the hypothalamus which are concerned with temperature control; the posterior one concerned with protection against cold and the anterior one concerned with protection against heat.[5] These two centers, depicted schematically in Figure 1, are mutually inhibitory.

Regulatory mechanisms in the body make up the third element of the thermoregulatory system. These mechanisms are activated in response to need and, in turn, by the control through the nervous system. The maintenance of appropriate control is achieved, to some extent, by control of heat production but for the most part is affected by regulation of heat loss.[5] It should be noted, however, that at best, physiological control of body temperatures accounts only for a small part of mammalian ability to inhabit regions of extreme climatic stresses. Most of the ground has been won by behavioral (or in some cases also cultural) adaptation. The development of proper diets, sleeping habits, selection (or erection) of suitable habitations and, particularly applicable to humans, the design and use of garments and the utilization of artificial heating/cooling are representative of this aspect of thermoregulation. As an example, a naked native sitting in the shade of a tree and dabbling his feet in a brook has made an almost perfect adaptation to heat. Further details on adaptations of various peoples to climate are beyond the scope of this article and may be found in a paper on the subject by Wulsin.[13]

Our attention here is turned back to the no less fascinating physiological aspects of thermoregulation. Commonly, a distinction may be made between long-term and short-term thermoregulatory responses. Long-term adaptations to environmental changes are affected through the endocrine glands. It appears that

Figure 1 Schematic representation of mammalian temperature control center and its functions.[2]

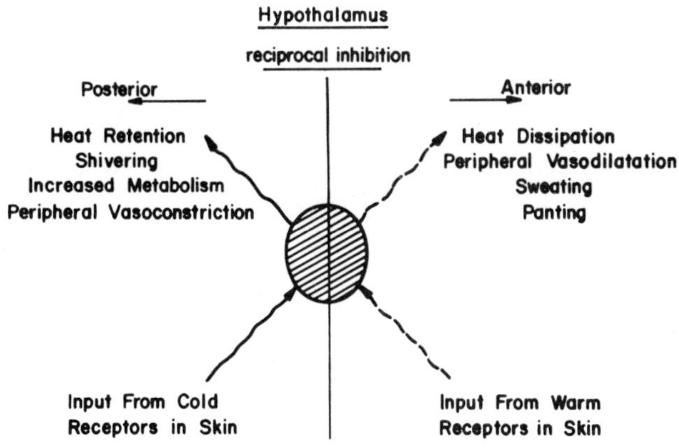

the pituitary gland, situated in the cranium, is the master of control of the endocrine system. It exerts control through chemical mediators which cause secretion of metabolites or hormones.[5] As an example, in response to long exposures to cold climates, adrenaline is secreted by the thyroid, medulla of the suprarenal gland and possibly by the cortex of the suprarenal gland. The function of the circulating adrenalin is to cause modifications in the cardiac system which favor blood supply to active muscles, liberation of glucose from liver glycogen, increased capacity of the tissue to assimilate glucose and a considerable increase in heat production.[5] Also, blood volume is increased in warmth and decreased in cold.

Short-term responses to varying conditions, internal and/or external, are produced by three basic types of mechanisms: metabolic, vasomotor, and sudomotor. The natural sequences of adjustment to both hot and cold stresses are summarized in Table 2.

Table 2 Natural Sequences of Adjustment to Cold and Hot Exposures

Type of response	Sequential step no.	Cold exposure	Hot exposure
Short-term	1	Superficial vasoconstriction to diminish the amount of heat transported by the blood stream to the skin. This constriction is balanced by splanchnic dilatation, i.e., dilation of blood vessels in the internal cavities of the body.	Superficial vasodilation to enhance blood flow and, consequently, the amount of heat transported by the blood stream to the skin.
	2	Pilomotor activity ("goose flesh") to reduce air movement at the skin surface and thereby diminish heat convection.	Sudomotor activity (sweating) to enhance the removal of heat by evaporation of water vapor from the surface of the skin.
	3	Increased heat production that can take either or both of two forms: mechannical and glandular. The mechanical part is achieved by high frequency muscle contractions and is manifest as shivering: the glandular part involves secretion of adrenalin (see text).	Increased respiration to promote heat removal by the expired air stream. In mammals other than humans, e.g., dogs, this mechanism, known as panting, is of great importance.
Long-term	4	Decreased blood volume and accumulation of fat in the outer layers to increase thermal insulation.	Increased blood volume and decreased basal metabolism due to reduced activity of the thyroid.

2.2 Thermoregulation during hibernation[14]

As mentioned above, some mammals, e.g., ground squirrel, possess the capability of lowering their body temperatures to levels which are lethal to other non-hibernating mammals. Hibernation (from the Latin word *hibernatus,* pp. wintered) is one of two most common biological strategies to meet the challenge of winter shortage of food; the other complex chain of events is migration.

The state of the hibernating animal can best be described as follows:[14]

> The temperature of the animal's body hovers only a few degrees above the freezing point (above 5°C, more than 30°C below the normal level). Curled up, immobile, the squirrel appears to have withdrawn from nature. In comparison with its active state, nearly every system in its body is radically altered. The composition of the blood, the functioning of the digestive tract, the state of fat reserves, the output of hormones–these and much else are changed so that a new physiological state is achieved.

In spite of the seemingly passive state of the hibernating animal, it still regulates and maintains body temperatures within closely defined limits. It appears that the temperature of the brain remains constant and consistently a few degrees above the ambient level in spite of fluctuations in skin temperatures. The range of controlled temperatures is evidently quite different than that of summer months but the animal's "thermostat" has not been turned off; it has merely been turned down.[14]

Entry into hibernation follows quite a remarkable natural sequence. As winter approaches the animal's food consumption soars far above normal levels. This phenomenon, known as obesity, "is a kind of physiological hoarding that protects the animal when food supplies run out."[14] This desire for food is selective, however, and whereas available food will be consumed voraciously, the animal will be reluctant to expand energy in a food quest that may go unrewarded. Concurrently, diurnal fluctuations in body temperatures become progressively more pronounced until, at the end of the fall, the coolness and torpor of the night hours continue through the following day. The hibernator then remains in this state, except for occasional arousals, for the rest of the winter.

Coming out of hibernation, or self-rewarming, "will take place in spite of such handicaps as cold room environment only a few degrees above freezing . . . . Whether arousal is spontaneous (another way of saying that the cause is unknown, although it is presumably internal) or is in response to external disturbances, the main occurence is a rapid rise in body temperature."[14] This considerable achievement in the rate of heat production, e.g., 30°C or more over a period of 3 hours in the hamster, appears to be made possible largely by a tissue called brown fat.[15]

It is characteristic of all arousing hibernators that the anterior portion warms before the posterior. This phenomenon is probably due to a differential vasoconstriction.[16] Energy for rewarming is generated in large measure from muscular movement with gross shivering as the main mechanism. During this period "both heart rate and oxygen consumption can increase more than 100 fold."[16]

Evidence presented in the literature apparently implicates the anterior regions of the hypothalamus with the control of hibernation.[14] Some investigators view hibernation as a state of exaggerated swing in the controlled range of temperatures. What precisely triggers entry into hibernation is, however, yet to be answered satisfactorily. Two probable explanations are commonly accepted. The first, the more traditional one, views hibernation as a response to winter's shortened daylight hours and lower temperatures. This explanation was shown to be a simple minded one since animals in controlled laboratory environments were found to hibernate in rooms illuminated 24 hours a day and held at a constant temperature. They even hibernated at normal room temperature (21°C). It was suggested, therefore, that "some internal biological clock, with a periodicity of about a year, is responsible for the animal's hibernation."[14]

The second plausible explanation to hibernation is due to Morovsky. He suggested that the animal's fat deposits, rather than any biological clock, are responsible for the triggering of hibernation. According to him there exists a certain individual critical weight margin which triggers either entry into or arousal from hibernation.

It is thus seen that hibernation is another possible natural state in which some mammals may exist through some astonishing physiological modifications. Thermoregulation is nonetheless still maintained although temperature levels are drastically altered.

2.3 Thermoregulation in pathological states[17]

The opposite of hibernation in mammals are those cases in which body temperatures are regulated at levels higher than normal. The range of elevated temperatures caused by exercise, hyperthermia or febrile diseases, are depicted in Figure 2. According to Minard and Copman "efficient mechanisms of thermal homeostasis permit the healthy human subject to live and work in hot environments with his body temperature regulated at levels only a few degrees below the thermal death point of the tissue cells."[17] It also appears that in most cases of abnormally high body temperatures the thermoregulatory system functions well at the new "thermostatic" level set by the particular stage of the disease or heat exposure. The critical limit, encoded in some unknown way, is probably around 41.1°C where an "emergency regulatory mechanism in fever protects the body from extreme hyperthermia with great efficiency."[17]

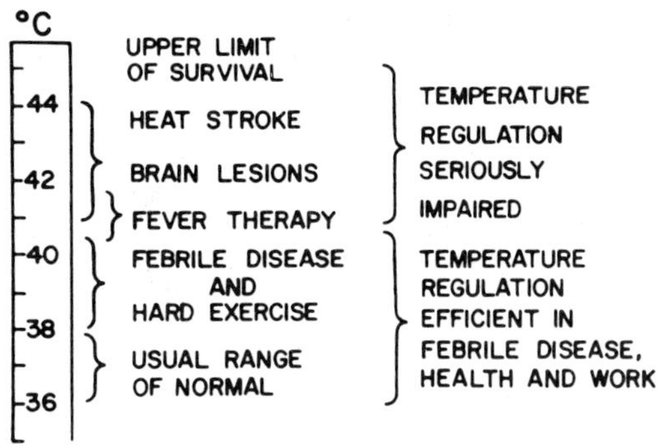

Figure 2 Ranges of elevated body temperature showing the zones of temperature regulation and thermoregulatory failure.[17]

Heat illnesses may be classified into four categories:[18]

1. Skin disorders including sunburn and prickly heat.
2. Heat syncope (fainting) resulting from peripheral vasoldilation.
3. Disorders of water and electrolyte metabolism which include heat exhaustion and heat cramps.
4. Failures of heat regulation, i.e., anhidrotic heat exhaustion and hyperpyrexia (heat stroke).

Of these illnesses only the last two are major disorders and only in the last of the four is body temperature significantly elevated.

Sustained exposures to heat stresses, which may lend to failures of the thermoregulatory system, are probably affected by the degree of acclimatization of the individual to heat exposure and by some inherent susceptibility found in a small fraction of the population. Other contributing factors are obesity, age, alcoholism, degenerative diseases of the heart and blood vessels, and water restriction.

The biochemical process that causes either permanent disablement or even death may be described as follows:

> At normal body temperature the rate of destruction of cell proteins occurs so slowly that the normal concentrations of specific species of native proteins, e.g., an oxidative enzyme, is readily maintained by synthesis. As the temperature rises, both the synthetic repair process and the destruction of the protein increases, but at widely different rates until a temperature level is reached at which the rate of destruction equals the rate of repair. According to Ratin this is the temperature of greatest life activity, or what he designates as the optimum temperature. . . . Any further increment in temperature, however, leads to the rate of protein destruction rapidly exceeding that for repair because of the great disparity in the temperature coefficients of the opposing processes.[17]

Denaturation is a rate process involving a biochemical reaction between the protein molecules and water. This process is extremely temperature dependent. Thermal injury, is therefore, time dependent; the longer the temperature of a specific body organ or tissue is allowed to stay at the supercritical level the greater the injury. Hardy has compared pain and tissue damage in thermal injury to a speedometer, in which pain corresponds to the position of the speedometer needle while tissue injury corresponds to the mileage accumulated.[19]

Obviously, slowing down and stopping the destruction of proteins depends also on the rate of temperature reduction. This is normally attempted by the thermoregulatory system through the activation of profuse sweating. When this effective mechanism of heat dissipation is insufficient in bringing about the desired change, sweating ceases and natural thermoregulation fails. The disorder of thermoregulation in heat stroke seems to be primarily a central failure. Fatigue of the sweat glands occurring peripherally may be only a precipitating factor. At times the heat stroke patient seems to behave in a poikilothermic manner with his body temperatures sinking to hypothermic levels after excessive cooling without eliciting a suitable metabolic reaction. Other commonly found reasons for heat death are those attributed to circulatory collapse and hypoxia associated with a state of shock due to loss of fluid from the blood and also due to accumulation of toxic factors found primarily in cases of tissue injury.[20]

In summary "physiological and clinical manifestation of heat effects on man depend on inherent and also acquired patterns of thermoregulatory response to heat stress. The normal subject, whether acclimatized or not, tolerates hyperthermia poorly. He maintains thermal homeostasis under heat stress at the cost of circulatory strain."[17] His thermoregulatory system functions well at an elevated "thermostatic" level except in heat stroke cases when it fails.

2.4 Theories on thermoregulation

2.4.1 The "set-point" theory J. D. Hardy in his celebrated paper of 1953[21] formulated and discussed the "set-point" concept in physiological temperature regulation. According to this concept there exist temperature "set-points" for different tissues and thermal exposures.* For man, Stolwijk and Hardy propose these encoded temperatures at 36.6°C for the head core, 34.1°C for mean skin and 35.9°C for the resting muscle.[22] Other investigators propose slightly different temperatures.

Widely accepted as describing thermoregulatory responses in mammals the "set-point" theory assumes corrective thermoregulatory measures to be taken when a deviation from "normal" occurs. As previously mentioned, the "normal" levels are not fixed, however, and depend on internal, i.e., disease, as well as

*The term thermal exposure is used here to describe the variety of heat stresses emanating from either exercise, febrile disease or hyperthermia.

external, i.e., environment, factors. The simultaneous existence of two or more of the factors influencing the level of the "set-point" is probably cumulative as suggested by Hardy.[23] He reported on an experiment performed by MacPharson who was observing body temperature rise during exercise in a group of men. One of the group contracted some illness and developed a mild fever but volunteered to continue with the experiments. The results showed that this individual's body temperature followed the same course of change as the others but was shifted about 1.5°C upward as a result of the two combined heat stresses.

A block diagram of the automatic regulator is depicted in Figure 3. In this figure a distinction is made between the passive controlled system and the active controlling one. The controlling system

> receives signals from some reference elements which may contain set-point and command information. The actuating signal S_e, contains in some form the error signal $(S_1 - S_2)$ and the information upon which the controlling elements act in the conservation of energy into controller action, A_c, the effect of which on the passive system, G, is to oppose the effects of the environmental disturbance and to restore V_c to its desired level.[22]

Another block diagram for regulation of hypothalamic temperature was proposed by Hammel.[24] This investigator suggests that "the temperature of the hypothalamus is the regulated internal body temperature."[24] According to Hammel "the set temperature (threshold) decreases at the onset of sleep and increases again upon awakening." It also "appears to shift downward immediately with the onset of exercise and to shift upward again when exercise stops."[24] Hammel's model is shown in Figure 4 in a block diagram form. In this figure the body of the homeotherm is divided again into two systems: the controlling and the controlled. The set-point information, influenced by inputs from the skin, exercise rate, etc., is contained in the reference input adjusting elements. The reference signal, T_{set_R}, is then combined with the temperature of the hypothalamus to generate the actuating signal which is fed to the controller.

These two theories on thermoregulation are essentially similar in that they both assume the existence of temperature "set-points" by which thermoregulation is maintained.

2.4.2 The combined set-point and rate of change in skin temperature theory The important role which the skin of the homeotherm plays in thermoregulation has long been recognized. The investigations which led to the formulation of the role of skin were based primarily on steady-state situations. "Since all thermoregulatory responses are activated under unsteady-state, dynamic conditions, it is hazardous to attempt to postulate an adequate control mechanism on the basis of steady state observations."[25]

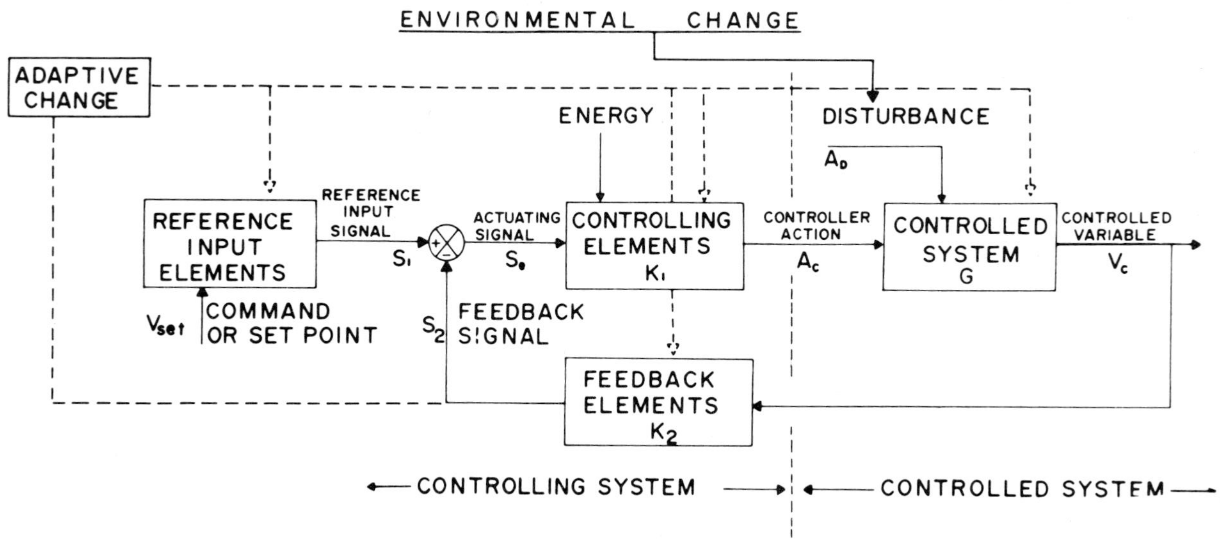

Figure 3 Block diagram of regulatory system.[22]

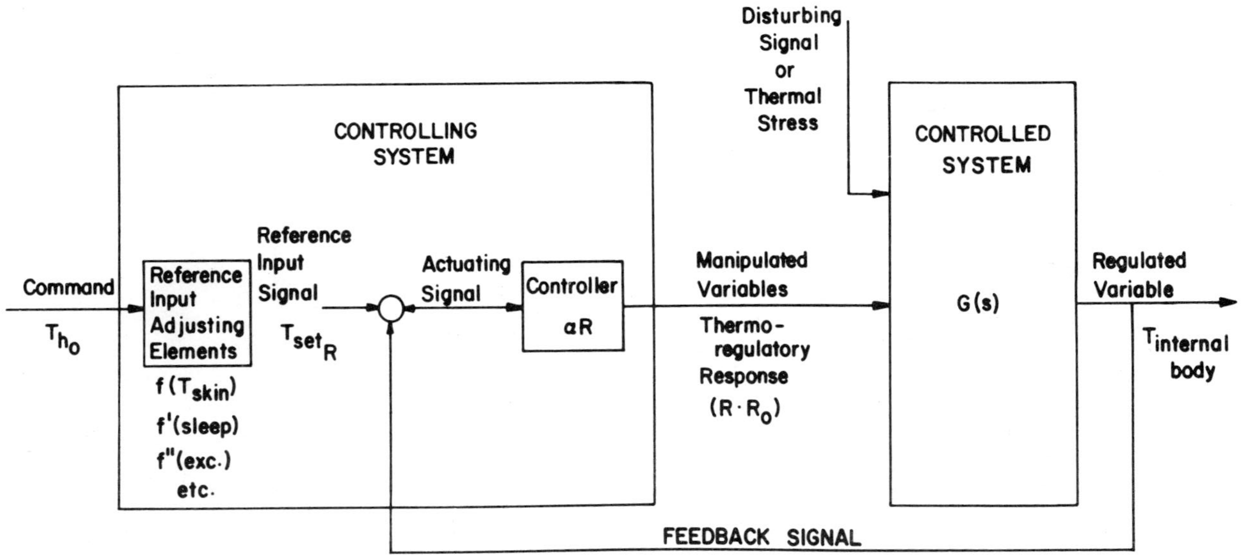

Figure 4 Block diagram for regulation of hypothalamic temperature.[24]

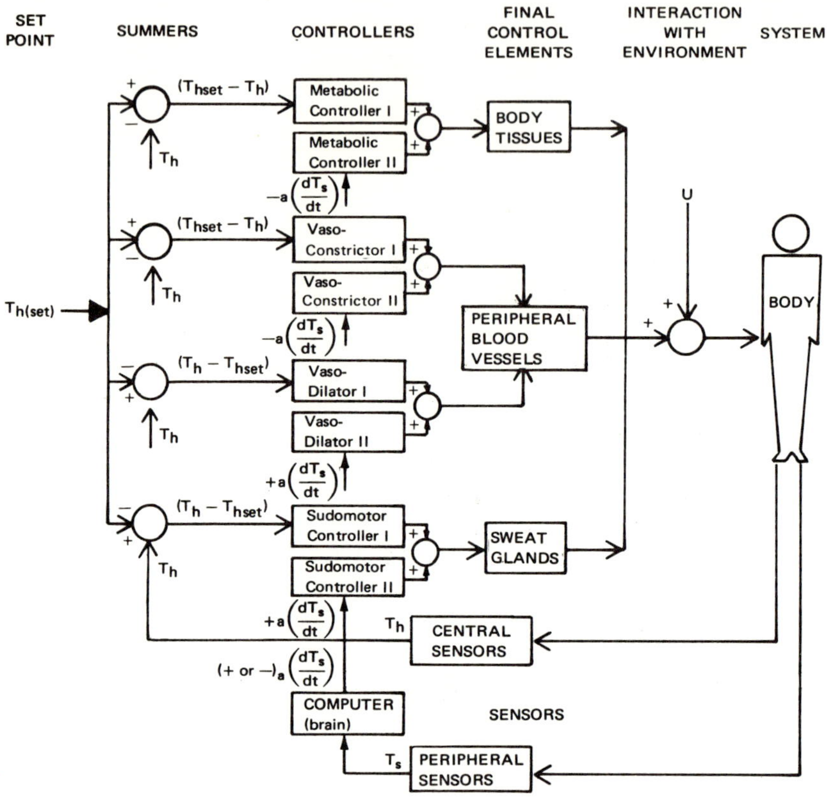

Figure 5 Human thermoregulation, normal short-term control.[29]

Evidence in the literature[26–28] suggests that the rate of change of skin temperature, dT_s/dt, exerts an effect in activating some responses. Huckaba and co-workers[29] presented a normal, short-term temperature control model for humans. Shown in Figure 5, the proposed control mechanism assigns a "logical anticipatory role for skin receptors under changing environmental conditions. Furthermore, this diagram utilizes dT_s/dt as the motivating potential for anticipatory responses, which is in agreement with the distinction between cold and warm receptors described by Hensel."[25]

These investigators state the generally accepted short-term human thermoregulation experimental facts as follows:[29]

1. The preoptic hypothalamic tissue is like no other part of the nervous system in being essential for regulation of internal body temperatures. If deafferented, this tissue can reduce changes in internal body temperature caused by external thermal stress.

2. The mean hypothalamic temperature and its variability do not depend upon the environmental temperature whether the latter be low with accompanying shivering or high with sweating.
3. Normal temperature control usually involves cycling of 0.1°C to 0.2°C around a mean value.
4. Vasomotor action provides effective narrow range control; shivering and sweating come into play at fairly definite threshold points beyond this narrow range.
5. Thermoregulatory responses are affected by both central and skin sensors.

In addition these investigators present a list of questionable concepts prevalent among workers in the area. These are:

1. Is there a single regulated temperature with a corresponding localized set-point?
2. Is there any physiological significance to average skin temperature?
3. Cold receptors vs. warm receptors, are they actually different?
4. Do the reported cooling experiments involving the hypothalamus, spinal cord, etc., provide a valid means of determining the specific locations of sensors?
5. Can quasi-steady state experimental data be used in elucidating dynamic control mechanisms?

Based on the above information Huckaba et al. reason that

> the localized hypothalamic temperature . . . (is) the best candidate to serve this function (set-point) in the light of presently available evidence.

They continue to suggest that this hypothesis is based on the fact that

> the idea of a mean skin temperature was invented by investigators as an artifice for conveniently comparing different thermal states; certainly the body does not compute such a quantity and exhibits it as a discrete point for use in thermoregulation such as is believed to be the case for hypothalamic temperature. Furthermore, since all parts of the body are in continuous thermal communication with one another, it is difficult to conceive of more than one set point which is subject to independent specification. Once the set point for T_h, the hypothalamic temperature, is specified, all other temperatures are dependent.[29]

The model shown in Figure 5 is based on the observation that extremely precise control of internal temperatures is achieved even in the face of major and rapid variations in thermal exposures. From engineering experience these investigators propose that

> it is difficult to imagine that such smooth and orderly compensations for drastic changes in the environment could be achieved solely by means of a feedback control mechanism. An inherent weakness of feedback control is that the system must already by in trouble before any corrective response can be initiated.[29]

And, since physiological evidence points out to the existence of some anticipatory mechanism in thermoregulation, Huckaba et al. suggest that "the central sensors control the action of the feed-back loops, whereas the peripheral sensors, by virtue of their direct contact with the environment, serve an anticipatory (early warning) function."[29] This feedforward function is assigned to the time derivative of skin temperature. Skin temperature itself is assigned only the role of influencing the threshold activation (dead-band effect of the sudomotor and shivering responses). The action of each of the individual controllers is assumed to be basically proportional with possible modifying effects both integral or derivative in origin. For the feedback controllers

$$R = K_{pi}(T_{h(\mathrm{set})} - T_h) \tag{2.3}$$

and for the feedforward controllers

$$R = K_{pj}\left(\pm \frac{dT_s}{dt}\right) \tag{2.4}$$

or alternatively,

$$R = K_p\left(\pm \frac{dT_s}{dt}\right) + T_D\left(\pm \frac{d^2T_s}{dt^2}\right) \tag{2.5}$$

which also includes the derivative action.

This feedback-feedforward theory on thermoregulation which includes both a single internal temperature set-point and peripheral modifying effects based on the temporal course of skin temperature appears to be feasible from both physiological and engineering view points. However, its validity and refining details are yet to be decided pending further experimental work currently underway in various laboratories.[29]

3 THERMAL BEHAVIOR OF BIOLOGICAL MEDIA

3.1 Heat transfer mechanisms

A problem of prime importance in thermal physiology is the mathematical description of the thermal state of the organism. Living tissues are highly complex in composition, non-homogeneous and non-isotropic. Heat is generated in the tissues (metabolic reactions) and they also include transporting streams i.e., blood, which, among other things, transport thermal energy. Dramatic support to the complexity of living tissue may be found in any text-book on physiology and particularly in the illustration by F. Netter.[30] Consequently, a number of simplifying assumptions may be necessary for the purpose of obtaining a basic modeling equation which describes the *in vivo* thermal state of the organism.

Among the factors to be considered when attempting such a model are the following:

1. Geometry of the organism.
2. Heat capacity of the organism (thermal inertia).
3. Conduction of heat due to thermal gradients.
4. Metabolic heat production.
5. The role of blood flow in heat transfer, i.e., transport of thermal energy by circulating blood and countercurrent heat exchange between large blood vessels.
6. Thermoregulatory mechanisms in the organism and their functions, i.e., vasomotor activity, sweating, shivering, increased metabolism due to glandular activity, pilo-motor activity (goose flesh), and panting.
7. Thermophysical and physiological properties of various organs and tissues, e.g., thermal conductivity, specific heat, density, blood perfusion rates, etc., and their dependence upon temperature and location.
8. The interaction with the environment and its condition, e.g., dry and wet bulb temperatures, pressure, and air motion relative to the body.

Consider a representative control volume of tissue, Figure 6. This basic element is in continuous thermal communication with surrounding elements. To

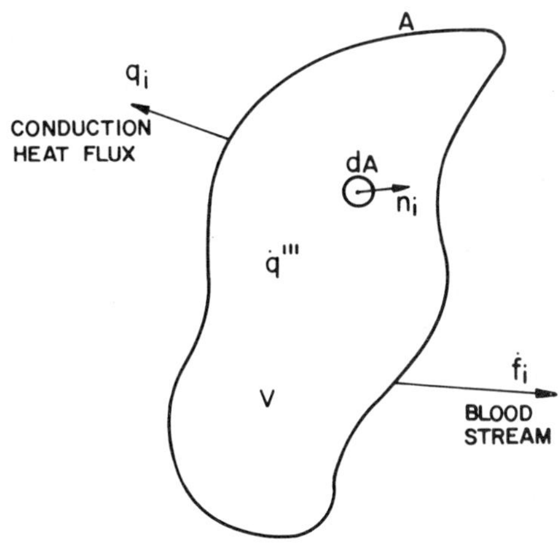

Figure 6 Control volume of tissue element.

maintain equilibrium, thermal energy which is generated inside the element is conducted to adjacent elements, transported (convected) by the blood stream or stored inside the element. These mechanisms will now be considered separately.

3.1.1 Tissue conduction The conductive heat flux vector is assumed to obey Fourier's law of conduction. Accordingly,

$$\dot{q}_i = -k_{ij}\frac{\partial T}{\partial x_j} \tag{3.1}$$

where k_{ij} is the proportionality constant referred to as thermal conductivity. Due to the general non-homogeneity and anisotropy of mammalian tissue, k_{ij} will generally be a second order tensor also dependent upon tissue temperature and location. The total conductive heat through the envelope A is given by

$$\dot{q}_c = -\int_A k_{ij}\frac{\partial T}{\partial x_j} n_i \, dA \tag{3.2}$$

In Eq. (3.2) it should be noted that heat is being exchanged by conduction both with surrounding tissues and the environment. The distinction between these two entities may be made through the appropriate specification of boundary conditions. Another approach is possible by setting the thermal conductivity, k_{ij}, identically equal to zero at the external surfaces of the organism where the interaction with the environment takes place.[31] This method requires the inclusion of a specific term in the energy balance equation which describes heat exchange with the environment and is described in a subsequent paragraph.

3.1.2 Metabolic heat generation The fact that living organisms generate heat has long been recognized. This thermal energy is liberated by either aerobic or anaerobic metabolic processes (the sum of biochemical processes by which food is broken down into simpler compounds with the exchange of energy) and the conversion of kinetic and other forms of energy into heat. Thus the total amount of heat generated (thermogenesis) in the control volume is given by:

$$\dot{q}_g = \int_V \dot{q}_m''' \, dV \tag{3.3}$$

where $\dot{q}_m'''$ is the specific (volumetric) rate of heat production which may generally be a function of tissue temperature and location.

3.1.3 Storage of thermal energy (thermal inertia) Under transient conditions part of the thermal energy generated or transferred to the control volume may go to alter the amount stored inside it. "This alteration will be reflected in a temperature change, since presumably no change of physical state is possible within the physiological range of temperatures"[31] found in mammals. Thus the rate of change in storage of thermal energy is given by:

$$\dot{q}_s = \frac{\partial}{\partial t} \int_V cT\, dV \tag{3.4}$$

where c is the specific heat of the various tissues comprising the organism and is also generally a function of tissue temperature and location.

3.1.4 Transport (convection) by the circulatory system The transport of thermal energy throughout the organism is an important concomitant of the circulatory system. As previously mentioned, the thermoregulatory system employs the cardiac system as a vital mechanism for thermoregulation, i.e., vasoconstriction and vasodilation. From the anatomy of mammals it is known that the circulatory system is constructed from vessels of different dimensions. A list of these vessels and their typical sizes for humans is given in Table 3.[32] As may be construed from Table 3 the capillary bed forms the major site for the *in vivo* exchange of mass and energy between the blood stream and surrounding tissue. This exchange is dependent upon the temperature and perfusion rate of the blood stream and the thermophysical properties of the tissues through which it occurs. Stated mathematically, the net amount of thermal energy carried away from the control volume by the circulatory system is given by:

$$\dot{q}_B = \int_A c_b T_b \dot{f}_i n_i\, dA \tag{3.5}$$

Table 3 Typical Sizes of Human Blood Vessels[32]

Blood vessel	Luman diameter	Wall thickness	Approximate total cross section area in the body cm^2
Aorta	2.5 cm	2 mm	4.5
Artery	0.4 cm	1 mm	20.
Arteriole	30 μ	20 μ	400.
Capillary	6 μ	1 μ	4500.
Venule	20 μ	2 μ	4000.
Vein	0.5 cm	0.5 mm	48.
Vena Cava	3 cm	1.5 mm	18.

where $\dot{f}_i$ is the blood perfusion rate vector. It should be noted that $\dot{f}_i$ is identically zero except at those locations where a blood vessel is traversed on the surface of the control volume.

An alternate form for the term representing the amount of thermal energy transport by the circulatory system is based on Fick's principle. This principle when applied to biological systems, can be stated as "the amount of substance taken up by an organ (or by the whole body) per unit time is equal to the arterial level of the substance minus the venous level times the (rate of) blood flow."[32] First suggested in this connection by Pennes in 1948,[33] the amount of thermal energy convected by the blood stream from the control volume may be written as follows (modified from Pennes):[33]

$$\dot{q}_b''' = \dot{w}_b c_b (T_{\text{art}} - T_{\text{ven}}) \tag{3.6}$$

where $\dot{w}_b$ is the tissue average volumetric blood perfusion rate. T_{ven}, the temperature of the venous blood leaving the tissue, is undoubtedly a function of the temperature of the tissue at the point of exit. This interdependency of temperatures is not known explicitly. However, it appears reasonable to assume that in the capillary bed the existing condition of creeping flow favors almost complete thermal equilibrium between the blood stream and the surrounding tissue. Since our attention is actually focused on this major site of exchange of heat in the organism, a further assumption of equality in temperatures may be deemed justifiable. Thus, setting $T_{\text{ven}} = T$, Eq. (3.6) may be rewritten

$$\dot{q}_b''' = \dot{w}_b c_b (T_{\text{art}} - T) \tag{3.7}$$

The validity of Eq. (3.7) is not general. It may provide a good approximation for the amount of heat transported by the blood perfusing the tissue when most of the flow is through the capillary vessels (or maybe even through the venules). When this is not the case Eq. (3.7) may be rendered invalid.

The physical interpretation of Eq. (3.7) is as follows. Blood is supplied to tissues from the arterial side from a major vessel which gradually branches off into the smaller vessels leading to the capillaries. The arterial blood, at some presumably known temperature T_{art}, flows through the capillary bed where complete thermal equilibrium between blood and tissue is attained. Consequently, the blood stream either warms up or cools down during its passage through the tissue as a function of the temperature difference prevalent at the specific tissue under consideration. Upon draining into the venous side, the temperature of the blood, at each point in the tissue where the drainage occurs, is equal to the temperature of the tissue. Further exchange of heat between the major venous vessels and the tissue surrounding them is still possible. Since this exchange, as well as the exchange between major arteries and enveloping tissues, is presumably not significant compared to the exchange in the capillary bed, its effect is not included in the equation of thermal energy balance. However, this effect should somehow be accounted for in the gross balance of the entire organism.

When integrated over the entire control volume the total amount of thermal energy transported by the blood stream when represented by the modified Fick principle becomes:

$$\dot{q}_b = \int_V \dot{w}_b c_b (T_{\text{art}} - T)\, dV \tag{3.8}$$

3.1.5 Heat exchange with the environment A mammal exchanges heat with the environment through a combination of the following mechanisms: convection, radiation, evaporation, and conduction (of particular significance for clothed humans). If all these mechanisms are lumped together to yield a single parameter, $\dot{h}_i$, the total amount of heat exchanged between the control volume and the environment is given by[31]

$$\dot{q}_e = \int_A \dot{h}_i n_i \, dA \tag{3.9}$$

In performing energy balances on mammals the determination of h_i, presents major experimental difficulties. These difficulties are due to the existence of multiple parameters, e.g., temperature, sweat rates, etc., which keep varying in time. For example, a complete steady-state would not be achieved in the human body in less than 3 hours.[34] Furthermore, since the time constant is so large, the thermoregulatory system may never reach a true steady-state and the study and analysis of transient conditions become of prime importance. Nevertheless, numerous studies have been performed to evaluate the functional relationships pertinent to the various heat exchange mechanisms. A very brief summary of useful results is given below.

(a) Convection The overall equation describing convection exchange between a mammal and his environment is given by

$$\dot{q}_{\text{conv}} = h_c A_D f_c (\bar{T}_s - T_a) \tag{3.10}$$

where h_c is the convective heat transfer coefficient which is dependent on the type of convection, i.e., natural (or free) and forced, and also on geometrical factors, i.e., size, shape, orientation, etc. Typical values of h_c for humans are shown in Table 4.[35] A_D is the total surface area of the body. For humans A_D is given by the famous DuBois equation[36]

$$A_D\,[\text{cm}^2] = 71.84\, W[\text{kg}]^{0.425} \cdot h[\text{cm}]^{0.725} \tag{3.11}$$

Table 4 Typical Values of the Convective Heat Transfer Coefficients for Humans[35]

Situation	h_c, [kcal/m^2 hr°C]
Free convection, nude, standing	2.3
Free convection, nude, seated	2.0
Forced convection, standing	$7.5v^{0.67}$; v[m/s]
Forced convection, seated	$6.4v^{0.67}$; v[m/s]

f_c is the percentage of total surface area through which convective exchange takes place. This factor is also affected by clothing. $\bar{T}_s$ is mean skin surface temperature typically given by the following equation:

$$\bar{T}_s = \sum_{i=1}^{n} a_i T_i \tag{3.12}$$

where n is the number of weighted regions, a_i is the weight assigned to the various areas satisfying $\sum_{i=1}^{n} a_i = 1$ and T_i is a representative local skin temperature of a given region. Commonly used are a number of different distributions of the parameter a_i. One of the more often used distributions for humans is given in Table 5.[35]

(b) Radiation Energy exchange through radiation is given by the following relationship:

$$\dot{q}_{\text{rad}} = \sigma \epsilon A_D f_c (\bar{T}_s^4 - T_a^4) \tag{3.13a}$$

where σ is the Stefan-Boltzman constant, ϵ the emittance of the outer surface of the body typically taken at 0.99[34] or 0.97.[37] f_r is the effective radiation area factor determined at 0.696 for sedentary body posture and 0.725 for standing posture for humans by Fanger.[37]

An alternative method for treating radiation heat exchange is also commonly used. Accordingly, a relationship similar to the one given in Eq. (3.10) is written[34]

$$\dot{q}_{\text{rad}} = h_r A_D (\bar{T}_s - T_a) \tag{3.13b}$$

Table 5 Typical Weights for Computing Mean Skin Temperature[35]

Body region	Head	Arms	Hands	Feet	Legs	Thighs	Trunk
a_i	0.07	0.14	0.05	0.07	0.13	0.19	0.35

where h_r is a "radiative" exchange coefficient determined experimentally and typically given by [34]

$$h_r = 3.0[1 + 0.0055(\bar{T}_s + T_a)] \tag{3.14}$$

For the range of environmental temperatures $0°C < T_a < 50°C$ h_r varies between 3.3 and 4.4 kcal/m^2 hr °C.

(c) Evaporation Removal of body heat by evaporation occurs both at the skin surface and through the respiratory tract. A distinction is usually made between heat loss by skin diffusion, a process which is not subject to thermoregulatory control, and heat loss by evaporation of sweat secretion.

Water vapor diffusion through the skin is given by[37]

$$\dot{q}_{\text{diff}} = \lambda m A_D(p_s^* - p_a) \tag{3.15}$$

where λ is heat of vaporization of water, 575 kcal/kg at 35°C, m is the permeance coefficient of skin typically taken at $6.1{*}10^{-4}$ kg/hr m^2 mmHg. $(p_s^* - p_a)$ is the difference between saturated water vapor pressure at skin temperature and the ambient water vapor pressure in mmHg.

The amount of heat removed by evaporation of sweat is given by the following equation[37]

$$\dot{q}_{sw} = \alpha A_D\left(\frac{\dot{Q}_{\text{met}}}{A_D} - \dot{q}_0\right) \tag{3.16}$$

According to Eq. (3.16) sweating will not occur for specific metabolic rates less than $\dot{q}_0$ (a typical value for humans is 50 kcal/m^2 hr). Above this value, which corresponds to average sedentary conditions in humans, sweating is an increasing function of thermal exposure. α is an experimental coefficient relating the amount of heat exchange by sweating to physiological parameters. A typical value of α for humans is 0.42.[37] In other mammals where panting replaces sweating as the prime defense mechanism against hyperthermia values of the parameters α and $\dot{q}_0$ will be different according to the species under consideration.

Equation (3.16) is often rewritten in a way similar to Eq. (3.10)

$$\dot{q}_{sw} = h_e A_D f_w(\bar{T}_s - T_a) \tag{3.17}$$

where h_e is the evaporative heat loss coefficient and f_w is the wettedness coefficient defined[38]

$$f_w = \frac{p_s - p_a}{p_s^* - p_a} \tag{3.18}$$

Respiratory heat exchange is again commonly divided between sensible (or dry) and latent (or wet) exchange mechanisms. Heat and water vapor are exchanged between the inspired air stream and the mucosal lining of the respiratory tract. "On reaching the alveoli the air is (assumed) at deep body temperature and saturated with water vapor. As the air moves outward through the respiratory tract some heat is transferred back to the body and water is condensed but the expired air emerging from the nose still contains more heat and water than the inspired air in comfortable environments."[37] Accordingly, the following two expressions were suggested. For latent respiration heat loss[37]

$$\dot{q}_{\text{res}}^{\text{lat}} = \lambda \dot{m}(w_{ex} - w_a) \tag{3.19}$$

where $\dot{m}$ is mass rate of pulmonary ventilation and w is the humidity ratio (or specific humidity) of the appropriate air streams. Substitution of approximate expressions in Eq. (3.19) yields the following relationship for humans[37]

$$\dot{q}_{\text{res}}^{\text{lat}} = 0.0023\, \dot{Q}_{\text{met}}(44 - p_a) \tag{3.20}$$

with p_a in mmHg. Sensible respiration heat loss is given by[37]

$$\dot{q}_{\text{res}}^{\text{sen}} = \dot{m}c_a(T_{ex} - T_a) \tag{3.21}$$

$$\dot{q}_{\text{res}}^{\text{sen}} = 0.0014\, \dot{Q}_{\text{met}}(34 - T_a) \tag{3.22}$$

with T_a in °C.

(d) Conduction Direct conduction heat exchange between mammals and their environment is usually very small if not negligible. However, conduction heat transfer may be more significant in cases of direct contact between a mammal and a solid surface, cf. Ref. [39], or when the mammal wears clothes which are in direct contact with his body. The conductive heat transfer is then calculated from (modified from Ref. [37]):

$$\dot{q}_{\text{cond}} = h_{cl} A_D f_{cl}(\bar{T}_s - T_{cl}) \tag{3.23}$$

where h_{cl} is the overall heat exchange coefficient of a given clothing ensemble. In the literature h_{cl} is replaced by I_{cl} which is the total thermal resistance offered by the clothing ensemble. This term was suggested by Gagge et al.[40] and is defined

$$I_{cl} \equiv \frac{5.56}{h_{cl}} \tag{3.24}$$

The experimental determination of I_{cl} is rather difficult. As Fanger puts it "it is not sufficient to measure the conductive resistance of each textile layer, analogous to the measuring of conductive resistances for building materials."[37] The most commonly used method for measuring I_{cl} is through the aid of full size electronically instrumented manikins. These manikins are expensive to manufacture and as a result only a limited number of I_{cl} values for clothing have been published. I_{cl} values are ordinarily given in clo units. This unit is defined as "a unit of insulation and is the amount of insulation necessary to maintain comfort and a mean skin temperature of 33.3°C (92°F) in a room at 21.1°C (70°F) with air movement not over 0.05 m/s (10 ft/min), humidity not over 50%, with a metabolism of 50 kcal/cm^2 hr."[40] Physically, if one assumes 76% of total metabolic heat to be lost through the clothing the clo unit may be redefined as:[40]

$$1\,[\text{clo}] = 0.18\,°\text{C/kcal m}^2\,\text{hr}$$

Typical values of I_{cl} and f_{cl} for different clothing ensembles are shown in Table 6.[37]

3.2 Derivation of the "bio-heat" equation

The principle of conservation of energy requires that the rate of heat loss, given by the sum of Eqs. (3.2), (3.5), and (3.9), be equal to the rate of internal energy production plus storage, Eqs. (3.3) and (3.4).* Written mathematically

$$\int_A k_{ij} \frac{\partial T}{\partial x_j} n_i \, dA + \int_A c_b T_b \dot{f}_i n_i \, dA = \int_V \dot{q}_m''' \, dV - \frac{\partial}{\partial t} \int_V cT \, dV \qquad (3.25)$$

Equation (3.25) is the fundamental equation of *in vivo* conservation of thermal energy of an organism. Applying the divergence theorem and observing that the above statement must hold for any arbitrary volume element within the organism, Eq. (3.25) may be rewritten in a differential form:

$$\frac{\partial}{\partial t}(cT) = \frac{\partial}{\partial x_i}\left(k_{ij} \frac{\partial T}{\partial x_j}\right) - \frac{\partial}{\partial x_i}(c_b T_b \dot{f}_i) + \dot{q}''' \qquad (3.26)$$

or in vector form (modified from Brown[31])

*In the above statement work performed by the control volume is neglected. This assumption may be deemed valid due to the very low thermal efficiency of biological organisms.

$$\frac{\partial}{\partial t}(cT) = \nabla \cdot (\bar{k}\nabla T) - \nabla \cdot (c_b T_b \bar{f}) + \dot{q}''' \tag{3.27}$$

In addition the energy balance between the organism and its surroundings must also be specified. A typical mathematical statement for this balance might be

Table 6 Data for Different Clothing Ensembles[37]

Clothing ensemble	I_{cl} [clo]	f_{cl}
Nude	0	1.0
Shorts	0.1	1.0
Typical tropical clothing ensemble: Shorts, open-neck shirt with short sleeves, light socks and sandals	0.3–0.4	1.05
Apollo constant wear garment (astronauts): Light cotton undergarment with short sleeves and ankle length legs, cotton socks	0.35	1.05
Light summer clothing: Long light-weight trousers, open neck shirt with short sleeves	0.5	1.1
Light working ensemble: Athletic shorts, woollen socks, cotton work shirt (open-neck), and work trousers, shirt tail out	0.6	1.1
U.S. Army "fatigues," man's: Light-weight underwear, cotton shirt and trousers, cushion sole socks and combat boots	0.7	1.1
Combat tropical uniform: Same general components as U.S. Army fatigues but with shirt and trousers of cloth, wind resistant, poplin	0.8	1.1
Typical business suit	1.0	1.15
Typical business suit + cotton coat	1.5	1.15
Light outdoor sportswear: Cotton shirt, trousers, T-shirt, shorts, socks, shoes and single ply poplin (cotton and dacron) jacket	0.9	1.15
Heavy traditional European business suit: Cotton underwear with long legs and sleeves, shirt, woollen socks, shoes, suit including trousers, jacket and vest	1.5	1.15–1.2
U.S. Army standard cold-wet uniform: Cotton-wool undershirt and drawers, wool and nylon flannel shirt, wind resistant, water repellant trousers and field coat, cloth mohair and wool coat liner and wool socks	1.5–2.0	1.3–1.4
Heavy wool pile ensemble: (Polar weather suit)	3–4	1.3–1.5

$$\int_s h_i n_i \, ds = \int_s k_{ij} \frac{\partial T}{\partial x_j} n_i \, ds \tag{3.28}$$

where the integration is carried out on the external surface of the entire organism, s.

The general solution of Eq. (3.27) may be very difficult to obtain due to the geometrical and physiological parameters involved. Lack of detailed and accurate data contribute to the complexity of the problem. Particularly, the blood flow vector $\dot{f}_i$, may not be known even approximately although volumetric average blood perfusion rates for various organs and tissues are better known. Consequently, an alternate form of Eq. (3.28) is commonly used. In this equation the term representing transport of heat by the blood stream is represented by Eq. (3.8) to yield:

$$\frac{\partial}{\partial t}(cT) = \frac{\partial}{\partial x_i}\left(k_{ij} \frac{\partial T}{\partial x_j}\right) - \dot{w}_b c_b (T_{\text{art}} - T) + \dot{q}''' \tag{3.29}$$

Equation (3.29) is referred to as the "bio-heat" equation and was first suggested in this form by Pennes.[33] When studying temperature distributions in tissues which are perfused primarily by capillary vessels the result obtained from the solution of Eq. (3.29) may agree well with experimental data. In any case, the validity of this equation is dependent upon the nature of equilibrium between the blood and surrounding tissues. The "bio-heat" equation has received extensive attention in the literature and some solutions obtained therefrom are presented in later chapters.

4 ANALYTICAL BIO-HEAT TRANSFER STUDIES

Several reviews on heat transfer studies in mammals have been published.[2,41–43] In the following chapters a number of these studies are classified and discussed.

4.1 Mathematical models for single elements of the organism

The study of single elements of the organism, e.g., forearm, probably ranks among the earliest attempts to obtain an analytical description of the temperature distribution in biological tissues. Pennes in 1948[33] studied the human forearm depicting it as a cylinder. Other assumptions were:

a. The tissues of the forearm contain two distributed heat sources: heat produced by tissue metabolism and heat transferred from blood to tissue at each point in the forearm.

b. As far as the rate of metabolic production, bone is treated mathematically the same as soft tissue. This assumption is primarily due to lack of contradicting experimental evidence, but is known today as being inaccurate.
c. Blood flow rate is assumed uniform throughout the forearm.
d. Tissue specific thermal conductivity is also assumed uniform throughout the forearm.
e. Axial invariance and angular symmetry are assumed.
f. The analysis is performed for a steady-state.

Mathematically the problem is stated as follows:

$$-k\left(\frac{d^2T}{dr^2}+\frac{1}{r}\frac{dT}{dr}\right)=\dot{q}_m+\dot{q}_b \tag{4.1}$$

subjected to the boundary conditions

$$\left.\frac{dT}{dr}\right|_{@\ r=0}=0 \tag{4.2}$$

$$-k\left.\frac{dT}{dr}\right|_{@\ r=R}=h(T|_{@\ r=R}-T_a) \tag{4.3}$$

Pennes further assumed the term representing transport of heat by the circulatory system to be given by:

$$\dot{q}_b=\dot{w}_bc_b(\beta-1)(T-T_{\text{art}}) \tag{4.4}$$

where $0<\beta<1$ is a so-called equilibration constant between blood and surrounding tissue. He then substituted Eq. (4.4) into (4.1) to obtain

$$\frac{d^2T}{dr^2}+\frac{1}{r}\frac{dT}{dr}+aT=b \tag{4.5}$$

where $$a=\frac{\dot{w}_bc_b(\beta-1)}{k} \tag{4.6}$$

$$b=\frac{\dot{w}_bc_b(\beta-1)T_{\text{art}}-\dot{q}_m}{k} \tag{4.7}$$

Equation (4.5) is probably the first version of the steady-state "bio-heat" equation found in the literature.

The solution for Eqs. (4.5) and (4.2) as obtained by Pennes is given by

$$T = \frac{\left(T_{\text{skin}} - \frac{b}{a}\right) J_0(i\sqrt{a}\, r)}{J_0(i\sqrt{a}\, R)} + \frac{b}{a} \tag{4.8}$$

It should be noted that the temperature of the skin at the surface of the forearm, T_{skin}, appears as a parameter in Eq. (4.8). This temperature is not known but can be calculated by employing the remaining boundary condition, Eq. (4.3), to obtain

$$T_{\text{skin}} = \frac{(bk/\sqrt{a})[-iJ_1(i\sqrt{a}\, R)] + 1.21\, hJ_0(i\sqrt{a}\, R) T_a}{k[-iJ_1(i\sqrt{a}\, R)] + 1.21\, hJ_0(i\sqrt{a}\, R)} \tag{4.9}$$

Figure 7 shows the correlation between analytical and experimental results as obtained by Pennes.[33] The lower curve in this figure represents the condition where no blood flows through the tissue, $\dot{q}_b = 0$ in Eq. (4.1). The solution of

Figure 7 Comparison of experimental and theoretical curves of temperature distribution in the human forearm.[33]

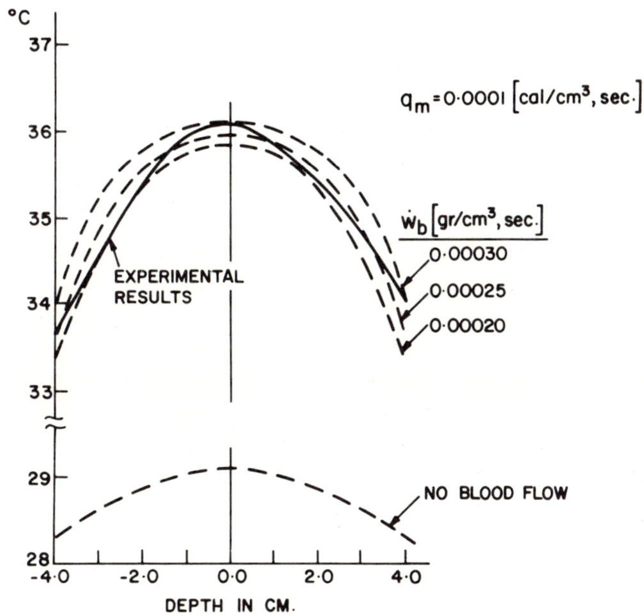

this equation is a second order polynomial (parabola) and for the same rate of metabolic heat production, i.e., $\dot{q}_m = 0.0001$ cal/cm^3 sec, yields temperatures which are considerably lower than those obtained experimentally. When transport of heat by the circulating blood is taken into account, a better approximation of experimental data was possible as shown by the dotted curves in Figure 7. Full equilibration between blood and surrounding tissue was assumed ($\beta = 0$) and volumetric blood flow rates were chosen to conform to values found in the literature.

In 1957 Vendrik and Vos proposed a model of heat transfer in the human skin for the purpose of measuring its thermal conductivity.[44] Heat flow in the presumed homogeneous skin was assumed to be governed by the heat equation (no heat generation and no allowance for blood flow). When the skin is brought into contact with another medium, e.g., glass, which is also assumed homogeneous, an expression for the temperature of the contact plane, T_c, is given by

$$\frac{T_2 - T_c}{T_c - T_1} = \left(\frac{k_1 \rho_1 c_1}{k_2 \rho_2 c_2}\right)^{1/2} \tag{4.10}$$

where T_c remains constant with respect to time if the two media are each semi-infinite and at uniform, but different initial temperatures, T_1, and T_2.[45]

Three basic assumptions were made in the analysis, namely:

a. the materials are homogeneous;
b. the initial temperatures in each are uniform but different;
c. the two media are infinitely large.

Justification of the last assumption is dependent on the duration of the measurement. If the time of measurement is short, even a relatively thin object may be assumed infinitely large for any practical purposes, e.g., 1 cm thick glass for measuring time of up to 30 sec. The first two assumptions are obviously not valid for any biological tissue. To obtain an approximation for the deviation caused by assuming the tissue homogeneous and uniform in temperature, Vendrik and Vos assumed both quantities to vary linearly with depth. By doing so, the temperature of the contact plane is not constant any more but varies with time. Measurement of T_c then enables the calculation of the assumed linear gradients of thermal conductivity and temperature of the tissue. These authors suggest that this method is "rather laborious and does not give very accurate results."[44] Another shortcoming of this work is the fact that blood flow effects were not explicitly included in the analysis. Nevertheless, in a number of experiments they were able to obtain a value for thermal conductivity of human skin (inner forearm) at $1.5 \pm 0.3 * 10^{-3}$ cal/cm sec °C. This value conforms well with other results reported in the literature.

Richardson and Whitelaw described a somewhat similar method for measuring the thermal conductivity of human skin.[46]

An analysis of the thermal response of the human leg during cooling was performed by Mitchell et al.[47] Refrigeration anesthesia for leg amputation or treatment of leg injury is being employed in various medical institutions. Thus, information on the thermal response of various limbs is important for medical treatment. The practicing surgeon should be able to predict "the time required for cooling the leg, the temperature levels in the leg, and the heat flow from the leg."[47] Specifically the effects of leg size and those of metabolism and blood flow rate cannot be predicted from experiments alone.

These investigators proposed an analytical model of the leg divided into two sections: the thigh and the calf. These two sections are depicted as cylinders, as shown in Figure 8. The thigh was assumed to be represented by three concentric cylindrical shells: the inner depicting the femur (bone), the outer the skin and fat, and the intermediate the muscle. The calf was described

Figure 8 Anatomical relations for human legs.[47]

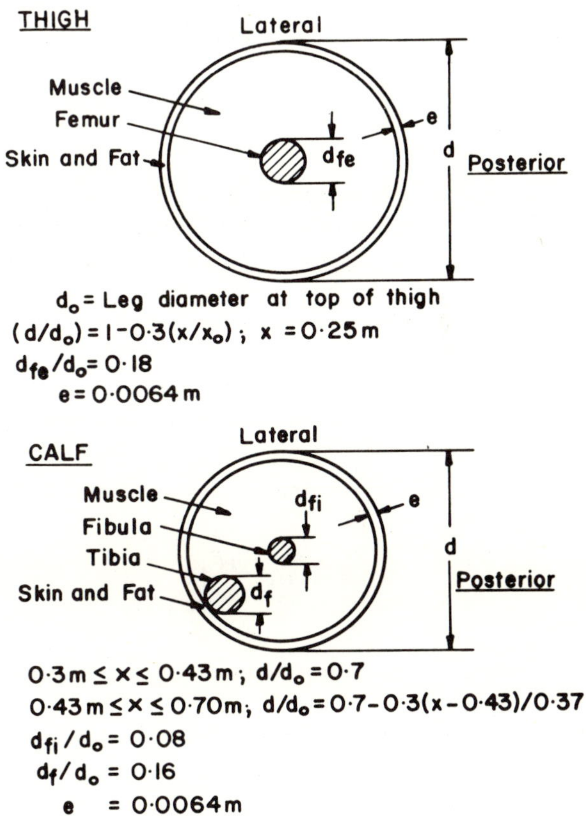

Table 7 Thermal Properties of Various Tissues in the Human Leg[47]

	Thermal conductivity k[W/(m°C)]	Density ρ[kg/m³]	Specific heat c[J/(kg°C]	Thermal diffusivity α[m²/S]
Muscle	0.50	1.08×10^3	3.8×10^3	12.2×10^{-8}
Fat	0.21	0.85×10^3	3.8×10^3	6.5×10^{-8}
Skin	0.35	1.00×10^3	3.8×10^3	9.0×10^{-8}
Bone	2.10	1.92×10^3	2.2×10^3	50.0×10^{-8}
Blood	–	1.05×10^3	3.8×10^3	–

in a similar fashion except that an additional cylinder, representing the tibia, was assumed to be present at some location tangent to the inner radius of the skin and fat layer. Thermal properties values for the various tissues are given in Table 7.

Mitchell et al. assumed the heat flow in the leg to be governed by the "bio-heat" equation. Using cylindrical coordinates and assuming both axial and radial heat flow they obtained (modified from Ref. [47]):

$$\frac{1}{\alpha}\frac{\partial T}{\partial t} = \frac{\partial^2 T}{\partial z^2} + \frac{1}{r}\frac{\partial}{\partial r}\left(r\frac{\partial T}{\partial r}\right) + \left[\frac{\dot{w}_b c_b}{k}(T_{re} - T) - \frac{\dot{q}_m}{k}\right] \tag{4.11}$$

where T_{re} is the temperature of the core (rectal). In bone tissue, these investigators suggest that "the energy contributions of blood flow and metabolism are negligible,"[47] thus setting $\dot{w}_b$ and $\dot{q}_m$ to zero in Eq. (4.11). In their analysis of the heat flow in the leg, Eq. (4.11) was solved numerically subject to the following conditions:

a. At the interfaces of the various regions inside the leg matching conditions of both temperature and heat flux are satisfied

$$T|_{@\, r \to R_i^-} = T|_{@\, r \to R_i^+} \tag{4.12}$$

$$k_i \left.\frac{\partial T}{\partial r}\right|_{@\, r \to R_i^-} = -\, k_{i+1} \left.\frac{\partial T}{\partial r}\right|_{@\, r \to R_i^+} \tag{4.13}$$

where i is an integer representing a given region and $i + 1$ represents the adjacent region.

b. The outer surface of the skin of the leg is assumed to be exposed to a coolant at temperature, T_{cool}. Thus

$$-k \left. \frac{\partial T}{\partial r} \right|_{@\, r=R_0} = h(T|_{@\, r=R_0} - T_{\text{cool}}) \tag{4.14}$$

c. Initially the leg is assumed at some temperature which may be a function of location

$$T|_{@\, t=0} = T_i(r, z) \tag{4.15}$$

The anatomy of the leg was used as a basis for the calculation. For simplicity it was assumed that both the human leg and the bones inside it are very close to circular and can be treated as such. "All dimensions are referenced to the leg diameter measured at the top of the thigh. This one measurement, combined with the (geometrical) relations of Figure 8, appears to be sufficient to characterize any given leg." It was further assumed that the "typical leg tapers from thigh to knee, is of relatively constant diameter from the knee to about one-third of the way down the calf, and then tapers to the ankle. The bone diameters are essentially constant with length."[47] Mean leg length was taken at 0.81 m for adult man and 0.74 m for adult female.

Variations in metabolic rates were also included in the analysis. It was assumed that the metabolic heat generation rate of resting skin and muscle is $\dot{q}_{m_0} = 0.17$ w/kg. The effect of temperature of metabolism was assumed to be given by:

$$\dot{q}_m = \dot{q}_{m_0} Q_{10}^{-(T-T_0)/10} \tag{4.16}$$

where $Q_{10} = 2$, T is local temperature in °C, and $T_0 = 37$°C.

The model described in the previous paragraphs was employed to generate temperature-time and heat flow-time relations for typical leg cooling situations. The objectives were to determine the parameters that govern the response of the leg to external thermal changes and to shed some insight into the nature of this response. Numerical solutions via a finite-difference technique were obtained by assuming the entire leg to be initially at the temperature of the core, $T_i(r,z) = T_{re}$, and the temperature of the skin to be stepped to the temperature of the coolant at $t = 0$. Representative results are shown in Figures 9 and 10. In Figure 9 the non-dimensional temperature at the center of the leg is given as a function of cooling time and axial length. For this calculation "the leg is taken to be uniform in diameter with no bone. The distribution of blood flow is representative of an extreme leg amputation case in which the upper leg is perfused with a blood flow value equal to that for a healthy leg while the lower leg has no blood flow."[47] It is seen that axial heat flow is less important than radial flow due to the poor thermal conductivity of the tissue. It is also obvious that the effects of the circulating blood on temperature distribution inside the

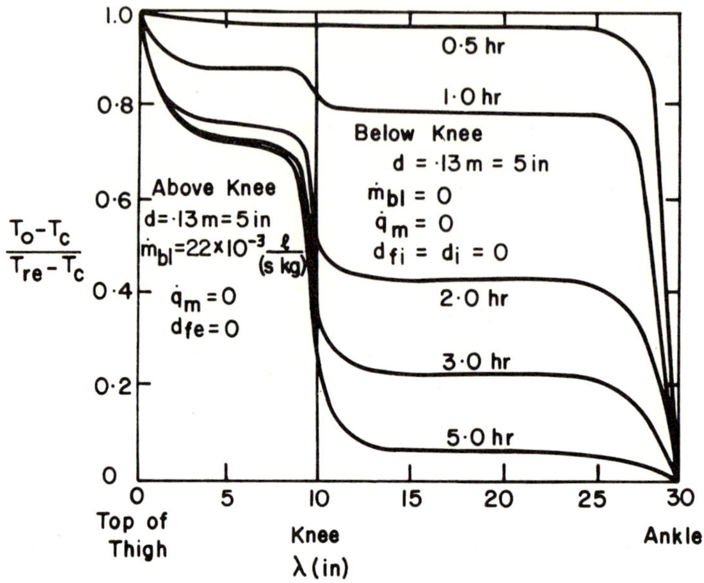

Figure 9 Model results for axial distribution of center human leg temperature as a function of time.[47]

leg are of prime significance. It was concluded that steady-state would not be reached in the unperfused section in less than 5 hours and that leg diameter and blood flow rate bear strong effects on the time constant of the human leg. Furthermore, effects of metabolic rates on leg temperatures were estimated to be of the order of less than 1% at most, thus rendering accurate information on metabolism not essential.

Model predictions were compared with experimental data and are shown for one of the cases studied in Figure 10. In this figure the time histories of the temperatures in three locations are compared to the analytical results for which blood flow rate was estimated from the measurements, and metabolic rates were neglected. The agreement obtained between the experimental and analytical results is quite good indicating the validity of the analytical method employed in this work.

A measure of the time needed to reach a steady-state condition in the cooled leg was also developed. This criterion was found to be given by

$$\frac{\alpha t_{ss}}{R_0^2} = 0.5 \tag{4.17}$$

or, in terms of leg diameter,

$$t_{ss}\,[\text{hrs}] = 310\, D^2\, [\text{m}^2] \tag{4.18}$$

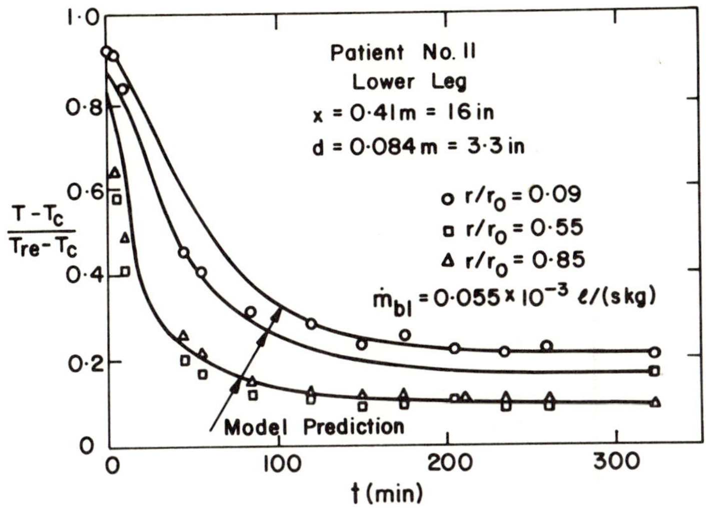

Figure 10 Leg temperatures as a function of time for patient 11.[47]

where the thermal diffusivity was taken at $\alpha = 11.4*10^{-8}\,\mathrm{m}^2/\mathrm{s}$ which is close to the value given for muscle tissue in Table 7. It should be noted that the dimensionless group given in Eq. (4.17) is the Fourier modulus for the human leg. From either of the above two equations it may be calculated that a 0.076 m diameter leg requires about 2 hr, 0.13 m leg about 5 hr, and 0.18 m leg about 10 hr to cool. Blood flow and metabolism were not included in these equations and are assumed to have negligible effects on the time required to cool the human leg.

The following conclusions were drawn from this study:

a. The axial conduction of heat in a human leg during cooling is not significant relative to other avenues of heat flow.
b. The interior temperatures in the leg are strongly influenced by both blood flow and leg diameter.
c. Metabolism does not significantly affect the temperature during cooling. Thus knowledge of exact bone size is not essential.
d. The heat flux from the leg is strongly influenced only by blood flow; leg diameter, metabolism, and the presence of bone have minor effects.
e. The relatively good agreement between the analytical and experimental results suggests that once cooling is initiated, the leg cools passively without active process.

Effects of cold wind on the temperature of fingers were considered by Molnar.[48] He assumed the rate of cooling of the human digit to be given by

$$\ln\left[\frac{T_{\text{skin}} - T_a}{(T_{\text{skin}} - T_a)|_{@\, t=0}}\right] = -\,ht \tag{4.19}$$

The fingers were cooled by inserting them into a wind tunnel in which the air temperature varied between −10°C and −15°C and wind velocity was 10 m/s. Further details are given in Ref. [48]. A study of the oscillations of the cutaneous temperatures in the human digit via an analog method was presented by Gautherie.[49]

4.2 Countercurrent heat exchange in blood vessels

The thermal energy balance of any limb in general, and the extremity of a homeotherm in particular, is dependent, among other mechanisms, upon the environmental conditions and the energy transported by the blood stream. If the animal is exposed to an environment whose temperature is lower than the average temperature of the skin, heat will be lost to the environment. To combat such loss, the thermoregulatory system may employ vasoconstriction, thus practically "abandoning" the extremities to save the more vital central parts. In opposite situations warm blood is shunted from the warmer parts to the superficial regions to enhance heat loss.

Examination of the anatomy of various organs of homeothermic animals reveals the natural close proximity of arteries and veins, e.g., leg of a bird. Thus countercurrent heat exchange between blood vessels is likely to occur. This mechanism of countercurrent arterial-venous heat exchange has often been postulated as an important factor in decreasing heat transfer to the environment. An analytical model was formulated by Mitchell and Myers[50] to study the transfer parameters that govern this mechanism. The basic configuration on which the analysis was performed is shown in Figure 11. In general, consideration is given to thermal energy exchange between arteries and veins and between these two streams and their environment. Other assumptions are:[50]

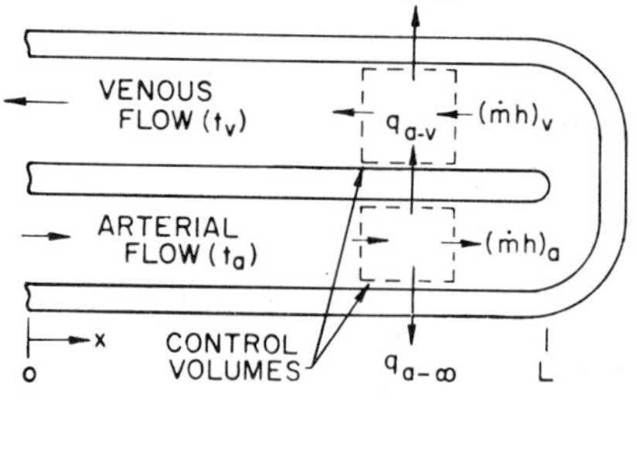

Figure 11 Analytical model for counter-current heat exchange in blood vessels.[50]

a. The temperature of the arterial and venous flows vary with distance in the flow direction only. The analysis will apply to a multiple vessel system (e.g., a rete) if the heat transfer behavior for all flow passages is similar.
b. The thermal conductances between artery and vein, artery and environment, and vein and environment are independent of distance along the limb.
c. The mass flow rates of the arterial and venous flows are equal and constant with distance. This will be valid for situations such as a rete, a bird's leg, a man's forearm, etc., where there is no significant recurrent branching of the vessels.
d. Thermal energy generated due to metabolism is small relative to heat transfer terms. This is reasonably valid for a resting, non-shivering animal in which the major portion of the metabolic heat generated does not occur in the skeletal muscles.
e. The limb is in a steady state; temperatures and flow rates are not changing with time.
f. The thermal properties of the blood and tissue are constant. That is, they are independent of temperature, distance, and time.

Application of the principle of conservation of energy to the two control volumes indicated in Figure 11 yielded the following governing equations:

For arterial flow:

$$\dot{m}_b c_b \frac{dT_{\text{art}}}{dx} + (UA')_i(T_{\text{art}} - T_{\text{ven}}) + (UA')_{\text{art}}(T_{\text{art}} - T_a) = 0 \qquad (4.20)$$

For venous flow:

$$\dot{m}_b c_b \frac{dT_{\text{ven}}}{dx} + (UA')_i(T_{\text{art}} - T_{\text{ven}}) - (UA')_{\text{ven}}(T_{\text{ven}} - T_a) = 0 \qquad (4.21)$$

where A' is the heat transfer area per unit length.

These equations were subject to the boundary conditions:

$$T_{\text{art}}|_{@\,x=0} = T_0 \qquad (4.22)$$

$$T_{\text{art}}|_{@\,x=L} = T_{\text{ven}}|_{@\,x=L} \qquad (4.23)$$

Equations (4.20) and (4.21) are two coupled, first order, ordinary differential equations. Their solution, which satisfies the assumed boundary conditions (4.22) and (4.23), was obtained by elementary analytical techniques to yield:

$$u = \frac{T_{\text{art}} - T_a}{T_0 - T_a} = e^{(N_v - N_a)\xi/2}\left\{\frac{B\cosh[A(1-\xi)] + \sinh[A(1-\xi)]}{B\cosh A + \sinh A}\right\} \qquad (4.24)$$

and

$$\nu = \frac{T_{\text{ven}} - T_a}{T_0 - T_a}$$

$$= e^{(N_v - N_a)\xi/2} \left\{ \frac{B \cosh [A(1-\xi)] - \sinh[A(1-\xi)]}{B \cosh A + \sinh A} \right\} \tag{4.25}$$

where $\xi \equiv \dfrac{x}{L}$ (4.26)

$$N_k = \frac{(UA')_k L}{\dot{m}_b c_b} \tag{4.27}$$

where the index k may be exchanged by art., ven., or the index i. Also:

$$A = 0.5\,[(N_{\text{art}} + N_{\text{ven}})(N_{\text{art}} + N_{\text{ven}} + 4N_i)]^{1/2} \tag{4.28}$$

$$B = \left(\frac{N_{\text{art}} + N_{\text{ven}} + 4N_i}{N_{\text{art}} + N_{\text{ven}}} \right)^{1/2} \tag{4.29}$$

Due to the appearance of three parameters in the general solution, i.e., N_{art}, N_{ven}, and N_i, a graphical presentation of the results appeared impractical. Therefore, and to gain some insight into the nature of the mechanism of countercurrent heat exchange, two representative cases were studied. It is claimed by the authors that those two cases are representative of the majority of the situations found in nature.[50]

Case I: $N_{\text{art}} = N_{\text{ven}}$ This case represents the situation found in the arm of man, leg of a bird, and a rete of a sloth leg. The schematic diagram for this situation is given in Figure 12. Arteries and veins, either one of each, or a multiplicity of them, are assumed to occupy the central portion of the extremity. Thus it appears reasonable to assume the conductances of these vessels to the environment to be equal, $N_{\text{art}} = N_{\text{ven}} = N_0$. Temperature variations for a range of values of this parameter, $0 < N_0 < 1$, and ratios of the conductance parameter between the vessels, N_i, and N_0, $0 < N_i/N_0 < 10$, are shown in Figure 12. According to the authors

> it is seen that for low values of N_0, (e.g., $N_0 = 0.1$), there is little cooling of either the arterial or venous flow, and the counter-current heat exchange effect is small. For large N_0 the temperature drop of the arterial flow is large, and the venous flow at the end of the limb is relatively cold. There is, then, a large temperature difference between the arterial flow and the

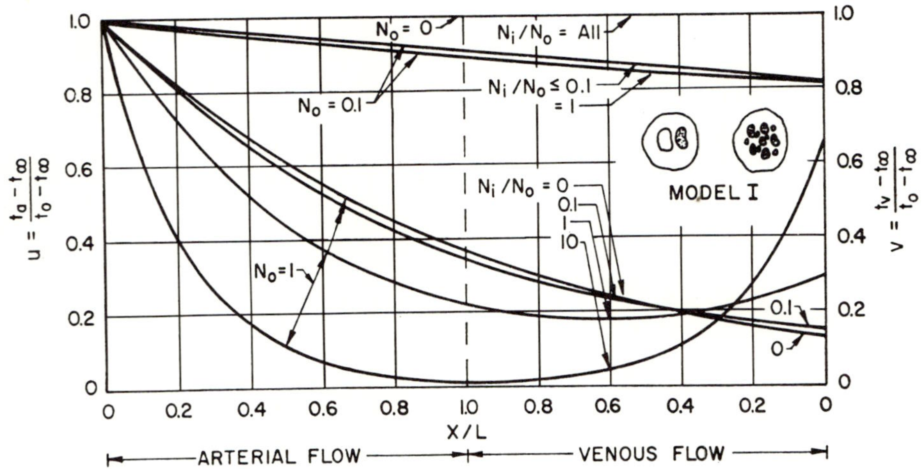

Figure 12 Normalized temperature distributions along major blood vessels in a limb; model I.[50]

returning venous flow and thus a significant potential exists for rewarming the venous flow through counter-current heat exchange. As indicated in Figure 12, the conductance between the artery and vein must be about the same size or greater than that between the vessels and the surroundings in order to obtain heating of the venous flow. It should also be noted that even for low values of N_i/N_0, there will be heat transfer from artery to vein even though the venous temperature continues to decrease. For example, for $N_0 = 1$ and $N_i/N_0 = 0.1$, the venous return temperature is slightly higher than that for $N_i/N_0 = 0$.[50]

Case II: $N_{art} = 0$ This case represents the situation found in the fins of whales and porpoises, and possibly other species.[50] A schematic diagram for this case is shown inside Figure 13. Arteries, arranged in a circle, are assumed to completely surround a single central vein. For this case, Mitchell and Myers suggest that heat transfer from the artery to the environment is negligible as compared to that between the arteries and the vein.[50] It is felt that a more careful analysis to gain further support for this assumption was necessary. Nevertheless, a graphical evaluation of the model for situations where this assumption may be shown to be applicable is shown in Figure 13. Parameter values in this figure are $0 < N_0 < 1$ and $0 < N_i/N_0 < 50$. The plots indicate that

for low values of N_0, the results are very similar to those for Case I; the temperature change of either the arterial or the venous flow is small. For large N_0 and N_i, the counter-current effect is significant, and a considerable rewarming of the venous flow may be obtained. As in Case I, the effect may exist and not necessarily warm the venous flow, but still serve to reduce heat transfer to the environment. For example, for $N_0 = 1$ and $N_i/N_0 = 1$, the venous temperature continually decreases in the flow direction, but the exit temperature is 10% higher than if there were no counter-current heat transfer ($N_i/N_0 = 0$).[50]

The authors further present analytical results for the amount of heat lost from the limb, the origin of which is the blood stream. They also present a method based on heat transfer principles, by which various parameters employed in the study may be evaluated. They conclude their study with the following observations:[50]

(a) The data currently available are insufficient for an adequate comparison with the theory developed in this paper. Additional experimentation, in which the quantities incorporated in this theory are measured, is needed in order to gain further insight into the counter-current mechanism.

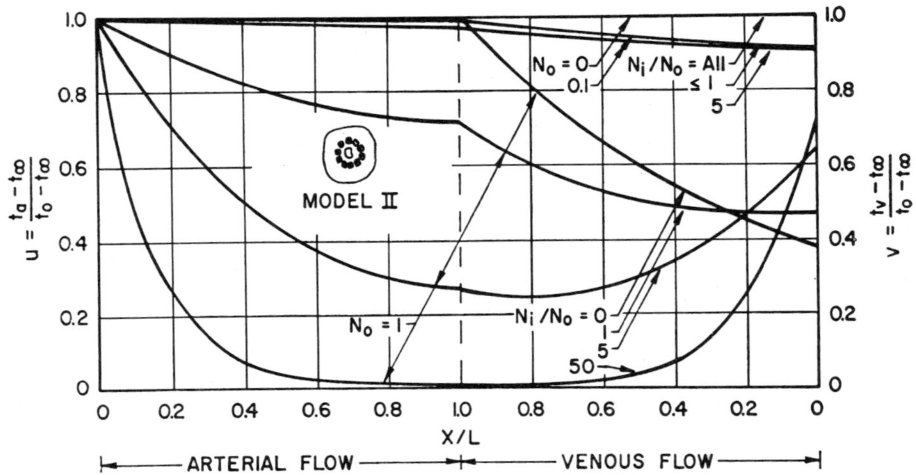

Figure 13 Normalized temperature distributions along major blood vessels in a limb; model II.[50]

(b) The counter-current effect becomes more significant as the blood flow rate decreases. In terms of the parameters of this analysis, reduced blood flow increases the non-dimensional conductance between the veins or arteries and the environment (N_0), the ratio of conductances (N_i/N_0) remaining the same. This effect has been noted by Scholander and Krog in their experiments with the sloth. Thus the vasoconstriction response to cold serves to protect the animal in two ways. The blood flow rate in the limb and, correspondingly, the total amount of thermal energy carried by the blood, is reduced. In addition, the venous return temperature is raised more than normally through counter-current heat exchange, further reducing the heat loss.

(c) The use of a non-dimensional temperature parameter allows generalizing the results for any ambient temperature. However, the effect of ambient temperature on the thermal control system of the animal may produce changes in blood flow rate, blood vessel diameter, and blood flow distribution and must be accounted for.

(d) The results of this analysis indicate that there is not a significant counter-current effect in the arm of a man, or the fluke of a porpoise. The effect is shown to be significant in the sloth rete, however. Further experiments in which all of the important parameters are measured are needed to confirm or disprove these conclusions.[50]

Keller and Seiler present a steady-state analysis on peripheral heat transfer in human tissue, coupling countercurrent heat exchange occurring in blood vessels with conduction, convection, and metabolic heat production.[51] They divided the skin into two regions: an isothermal core and a peripheral region where the temperature varies from that of the core to that of the region. In the peripheral region it was assumed that "tissue composition varies in a more-or-less continuous manner from the relatively well vascularized connective tissue and fascia to the poorly vascularized fat."[51] The analysis was restricted to regions in which the temperature varies only in a direction normal to the surface, i.e., one-dimensional.

The steady-state energy balance on the tissue, artery and vein, yielded the following equations, respectively:

$$k\frac{d^2T}{dx^2} + (h_{\text{art}}A_{\text{art}} + \dot{w}_b c_b)(T_{\text{art}} - T) + h_{\text{ven}}A_{\text{ven}}(T_{\text{ven}} - T) + \dot{q}_m = 0 \tag{4.30}$$

$$\frac{d}{dx}(\dot{m}_{\text{art}}c_bT_{\text{art}}) + \dot{w}_b c_b T_{\text{art}} + h_{\text{art}}A_{\text{art}}(T_{\text{art}} - T) = 0 \tag{4.31}$$

$$-\frac{d}{dx}(\dot{m}_{\text{ven}} c_b T_{\text{ven}}) + \dot{w}_b c_b T + h_{\text{ven}} A_{\text{ven}}(T - T_{\text{ven}}) = 0 \tag{4.32}$$

with the boundary conditions:

$$T|_{@\, x=0} = T_{\text{art}} = T_b \tag{4.33}$$

$$T|_{@\, x=\delta} = T_{\text{ven}} = T_s \tag{4.34}$$

where T_b and T_s are the temperatures of the inner core and the skin surface, respectively, and δ is the depth of the tissue.

To facilitate a solution to the above set of equations the following assumptions were made:

a. metabolic heat generation is negligible, i.e., $\dot{q}_m = 0$;
b. blood flow rate is constant, i.e., $\dot{w}_b = \text{const.}$;
c. conductances of both arteries and veins are equal, i.e., $h_{\text{art}} A_{\text{art}} = h_{\text{ven}} A_{\text{ven}} = hA$.

Rewriting, in nondimensional form, gives:

$$(1 + 2\mu)\frac{d^2\theta}{d\xi^2} = (1 + 2\mu)\theta - (1 + \mu)\theta_{\text{art}} - \mu\theta_{\text{ven}} \tag{4.35}$$

$$\xi \frac{d\theta_{\text{art}}}{d\xi} = \mu(\theta_{\text{art}} - \theta) \tag{4.36}$$

$$\xi \frac{d\theta_{\text{ven}}}{d\xi} = (1 + \mu)(\theta - \theta_{\text{ven}}) \tag{4.37}$$

$$\text{where } \theta = \frac{T - T_s}{T_b - T_s} \tag{4.38}$$

$$\mu = \frac{hA}{c_b \dot{w}_b} \tag{4.39}$$

$$\xi = (\lambda\delta)(1 + 2\mu)^{1/2}\left(1 - \frac{x}{\delta}\right) \qquad \xi_0 = \xi|_{@\, x=0} \tag{4.40}$$

$$\lambda = \left(\frac{\dot{w}_b c_b}{k}\right)^{1/2} \tag{4.41}$$

These equations were transformed into a single fourth order equation by eliminating θ_{art} and θ_{ven}:

$$\theta^{IV} + \frac{2}{\xi}\theta''' - \left[1 + \frac{\mu(1+\mu)}{\xi^2}\right]\theta'' - \frac{2}{\xi}\theta' = 0 \tag{4.42}$$

Two additional* boundary conditions are required on θ. The first, self suggesting one, was that no extremely large heat fluxes are to be allowed:

$$\theta'|_{@\,\xi=\xi_0} \neq \infty \tag{4.43}$$

The other one was obtained by performing an overall heat balance:

$$\theta''|_{@\,\xi=\xi_0} = \frac{\mu}{\xi_0}\,[\theta'|_{@\,\theta=\theta_0} - \theta'|_{@\,\xi=\xi_0}] \tag{4.44}$$

Under these conditions the general solutions are given by:

$$\theta = \frac{f(\xi,\mu)}{f(\xi_0,\mu)} \tag{4.45}$$

where
$$f(\xi,\mu) = \int_0^{\xi}\Big\{ z^{1/2}\Big[I_\nu(z)\int_0^z z'^{-3/2} I_{-\nu}(z')\,dz' - I_{-\nu}(z)\int_0^z z'^{-3/2} I_\nu(z')\,dz + C_0 I_\nu(z)\Big]\Big\}\,dz \tag{4.46}$$

where C_0 is a constant of integration, and $\nu = \frac{1}{2} + \mu$. For the purpose of obtaining a solution amenable for analysis Keller and Seiler further investigated a limiting case of no vascular interaction, i.e., no heat exchange between arterial and venous blood, $hA = 0$. For this case

$$\theta_{art} = 1 \qquad \text{for } x < \delta \tag{4.47}$$

and

$$\theta = \frac{1}{2\sinh(\lambda\delta)}\,\{[1 - \phi(e^{-\lambda\delta} - 1)]e^{\lambda x} - [1 - \phi(e^{\lambda\delta} - 1)e^{-\lambda x}\} - \phi \tag{4.48}$$

*The original boundary conditions, Eqs. (4.33) and (4.34) are transformed into $\theta|_{@\,\xi=\xi_0} = 1$ and $\theta|_{@\,\xi=0} = 0$, respectively.

$$\phi = \frac{\dot{q}_m}{(T_b - T_s)\dot{w}_b c_b} \tag{4.49}$$

From this solution the so-called effective conductivity of the tissue may be calculated. This conductivity is defined by:

$$k_{\text{eff}} \equiv \frac{Q\delta}{T_b - T_s} \tag{4.50}$$

where Q is the heat flux from the body to the skin. Thus, for the case of no vascular interaction

$$\left.\frac{k_{\text{eff}}}{k}\right|_{@\ hA=0} = \frac{\lambda\delta}{\tanh(\lambda\delta)} + \frac{\dot{q}_m \delta^2 \tanh(\lambda\delta/2)}{k\lambda\delta(T_b - T_s)} \tag{4.51}$$

Equation (4.51) was plotted for various values of the parameters present in it. Results are shown in Figure 14. It is seen that the effective conductivity of the tissue increases both with metabolic heat production rate and capillary blood perfusion rate. It should also be noted that the effect of metabolic heat production on k_{eff} attains a maximum at $\lambda\delta \to 0$

> and for extreme cases may give rise to a value of k_{eff} 1.65 times as great as that in the absence of such heat generation. However, as $\lambda\delta$ increases, the effect of heat generation markedly diminishes; for the upper curve, the deviation is a factor of 1.4 at $\lambda\delta = 1$ and only a factor of 1.1 (or 10%) at $\lambda\delta = 3$. In resting muscle, the maximum deviation is only 5% and it is negligible for values of $\lambda\delta$ greater than 1.[51]

Another limiting case investigated was that describing the condition of a resting muscle in which both metabolic production rate and vascular interaction are neglected. Thus, for $hA = 0$ and $\dot{q}_m = 0$,

$$\theta = \frac{\sinh(\lambda x)}{\sinh(\lambda\delta)} \tag{4.52}$$

and

$$\frac{k_{\text{eff}}}{k} = \frac{\lambda\delta}{\tanh(\lambda\delta)} \tag{4.53}$$

Equation (4.52) is plotted in Figure 15 for various values of λ. It should be noted

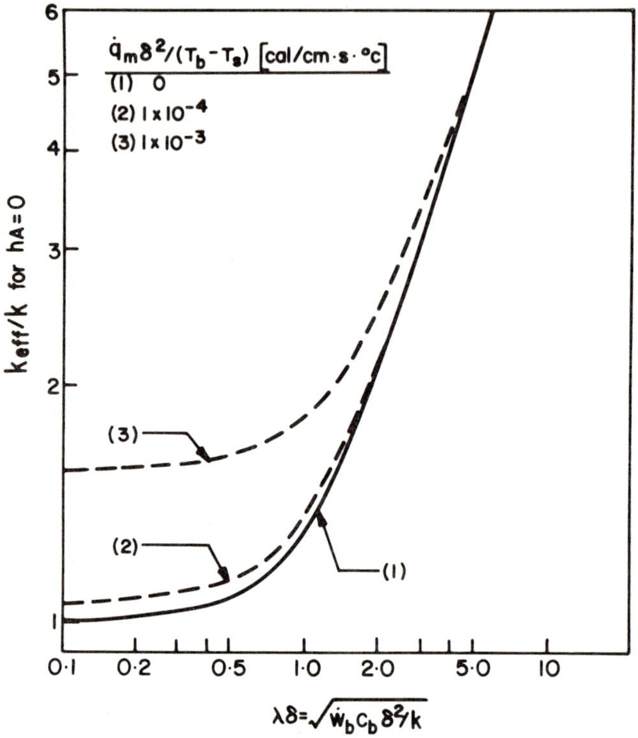

Figure 14 Illustration of dependence of effective thermal conductivity on dimensionless quantity, $\lambda\delta$, in the limiting case where arterial-venous heat exchange is negligible. Curves are for different rates of metabolic heat generation.[51]

that when $\lambda \ll 1/\delta$ the temperature profile is essentially linear as in a simple conductive process and, indeed, according to Eq. (4.53), $k_{\text{eff}} = k$, the tissue conductivity. Under these circumstances, capillary perfusion is unimportant in heat transfer. As λ increases, the deviation from linearity becomes pronounced and for $\lambda \ll 1/\delta$ the variation in temperature from that of the interface to that of the core occurs within a distance of the order of $1/\lambda$ from the interface. For such large values of λ, Eq. (4.53) indicates that

$$k_{\text{eff}} = k\lambda\delta = \sqrt{k\dot{w}_b c_b}\,\delta$$

and since the heat flux is proportional to k_{eff}/δ (which is the thermal conductance K) Q is independent of the choice of δ. Thus the product $\lambda\delta$, which may be interpreted as the ratio of perfusion-induced transfer to conductive heat transfer (see Appendix A), provides a measure of the

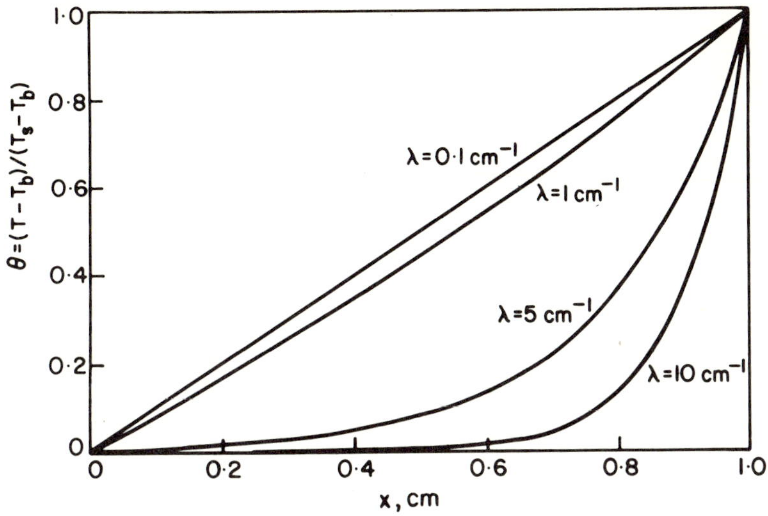

Figure 15 Temperature profiles for $\delta = 1$ cm, $hA = 0$, and various values λ, ratio of capillary-induced-perfusion heat transfer to conductive heat transfer.[51]

relative importance of capillary perfusion on heat transfer and, in certain situations, of the thickness of the nonisothermal layer. Note in Figure 15 how perfusion begins to have a significant effect when $\lambda\delta$ reaches a value of the order of 1.[51]

Analytical results of this study were compared to data found in the literature. Agreement found, although limited to the few cases for which data could be found, is quite good.

Both these studies demonstrate the important role that counter-current heat exchange between major blood vessels plays in the energy balance within a biological tissue. More detailed analyses of this phenomenon seem warranted.

4.3 Localized heat sources (sinks)

The study of temperature fields in live tissues emanating from an artificial local heat source (sink) is important for the evaluation of the local thermophysical properties of the tissue. Information thus gained may also aid in determining the tissue characteristics relevant for medical treatment, e.g., freezing times. In 1968 Chato developed a method for the measurement of the thermal properties of biological materials.[52] This method employs a single, small thermistor bead (about 1.5 mm in diameter) which is completely imbedded in the tissue. The thermistor bead acts both as the heat source and the temperature measuring device. A specially controlled power supply, the circuit diagram for which is given in Ref. [52], operates the thermistor bead in one of the following two modes:

1. Sudden step-up of bead temperature above that of its surrounding tissue and maintenance of this temperature by varying the power unit.
2. A sudden application of a fixed heat input and monitoring the resulting changes in bead temperature.

Chato assumed the model for the temperature field in the tissue to be depicted as a spherical heat source of radius R located in an infinite medium. The bio-heat equation was assumed to be the governing equation. Other assumptions were:

a. the tissue is isotropic and homogeneous;
b. spherical symmetry.

Stated mathematically, the problem then becomes:

$$\frac{\partial(T-T_0)}{\partial t}=\alpha\nabla^2(T-T_0)-\frac{\dot{w}_b c_b}{\rho c_p}(T-T_0)+\frac{Q'}{\rho c_p} \tag{4.54}$$

where $Q' = \dot{q}_m + \dot{w}_b c_b(T_{\text{art}} - T_0)$ (4.55)

with the boundary conditions:

$$T|_{@\ t=0 \text{ and } r>R} = T_0 \tag{4.56}$$

$$T|_{@\ t>0 \text{ and } r=R} = T_R \tag{4.57}$$

$$T|_{@\ t>0 \text{ and } r\to\infty} = T_0 \tag{4.58}$$

Substituting the following equation

$$T^* = \left(T - T_0 - \frac{Q'}{\dot{w}_b c_b}\right) r \exp\left[\left(\frac{\dot{w}_b c_b}{\rho c_p}\right) t\right] \tag{4.59}$$

into Eq. (4.54) reduces it to

$$\frac{\partial T^*}{\partial t} = \alpha \frac{\partial^2 T^*}{\partial r^2} \tag{4.60}$$

with the boundary conditions:

$$T^*|_{@\ t=0 \text{ and } r>R} = -\left(\frac{Q'}{\dot{w}_b c_b}\right) r \tag{4.61}$$

$$T^*|_{@\, t>0 \text{ and } r>R} = \left(T_R - T_0 - \frac{Q'}{\dot{w}_b c_b}\right) R \exp\left[\left(\frac{\dot{w}_b c_b}{\rho c_p}\right) t\right] \tag{4.62}$$

$$T^*|_{@\, t>0 \text{ and } r\to\infty} = -\left(\frac{Q'}{\dot{w}_b c_b}\right) r \tag{4.63}$$

The solution to the above set is given by

$$\begin{aligned}
&\frac{T - T_0 - Q'/wc_b}{T_R - T_0 - Q'/wc_b} \\
&= \frac{R}{2r} \left\{ \exp\left[(r-R)\left(\frac{wc_b}{k}\right)^{1/2}\right] \operatorname{erfc}\left[\frac{r-R}{2\sqrt{\alpha t}} - \left(\frac{wc_b}{\rho c_p} t\right)^{1/2}\right] \right. \\
&\quad \left. + \exp\left[(r-R)\left(\frac{wc_b}{k}\right)^{1/2}\right] \operatorname{erfc}\left[\frac{r-R}{2\sqrt{\alpha t}} + \left(\frac{wc_b}{\rho c_p} t\right)^{1/2}\right] \right\} \\
&\quad - \frac{Q'/wc_b}{T_R - T_0 - Q'/wc_b} \exp\left(-\frac{wc_b}{\rho c_p} t\right)
\end{aligned} \tag{4.64}$$

From Eq. (4.64) the heat flow from the bead may be calculated to yield

$$\begin{aligned}
q = \left(T_R - T_0 - \frac{Q'}{wc_b}\right) &\left[4\pi R k + \frac{4\sqrt{\pi} R^2 k}{\sqrt{\alpha t}} \exp\left(-\frac{wc_b}{\rho c_p} t\right)\right. \\
&\left. + 4\pi R^2 (kwc_b)^{1/2} \operatorname{erf}\left(\frac{wc_b}{\rho c_p} t\right)^{1/2}\right]
\end{aligned} \tag{4.65}$$

as $t \to \infty$ a steady state will be reached and the above equation will be reduced to

$$q|_{@\, t\to\infty} = \left(T_R - T_0 - \frac{Q'}{wc_b}\right) \left\{ 4\pi R k \left[1 + R\left(\frac{wc_b}{k}\right)^{1/2}\right] \right\} \tag{4.66}$$

From this equation it is seen that

> the apparent thermal conductivity increases by a factor of $[1 + R(wc_b/k)^{1/2}]$ as a result of the uniform flow of blood. It is important to note that this factor is a function of the geometry, R, as well as the blood flow rate, heat capacity and tissue conductivity. It is not to be expected

that one set of measurements performed on a given tissue with blood flowing will yield both thermal property values as well as the blood flow rate. However, if the thermal properties are known, then the blood flow rate can be determined by the application of the above equations to the data obtained.[52]

The method developed in the preceding paragraphs was tested on a variety of cases. Details of the experimental set-up are given by Chato.[52] The probe, which consisted of thermistor bead 0.7 mm in radius and soldered into the end of two stainless steel tubes, was first calibrated in a material whose thermal properties are known, e.g., Agar solution. Typical test results of the power dissipated from the bead versus $t^{-1/2}$ are shown in Figure 16. The bead was then inserted into the material to be tested with the aid of a hypodermic needle which protected the sensitive bead. When the needle was withdrawn, the tested material closed around the bead.

Experimental results, similar to those shown in Figure 16, were obtained on a number of materials. By assuming no blood flow and no heat generation (or the use of biological materials *in vitro*) the expression for the heat flow was recalculated. Accordingly, setting $\dot{w}_b$ and $\dot{q}_m$ to zero in Eq. (4.65) yields

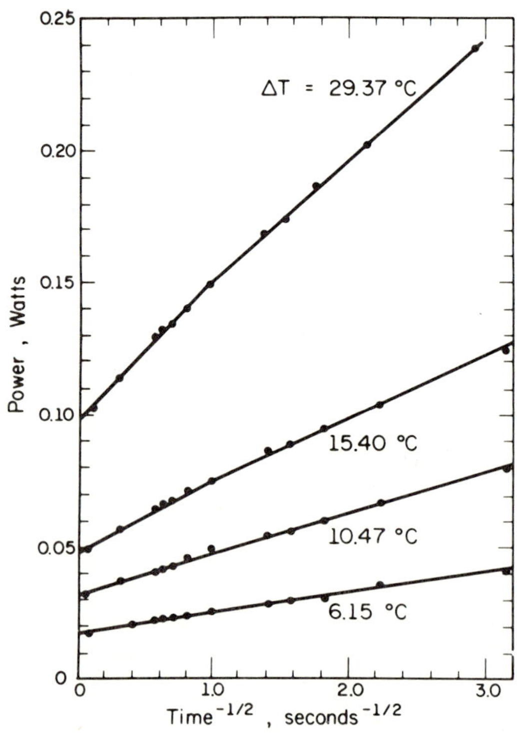

Figure 16 Thermistor power dissipation curves in a 1.75% Agar solution.[52]

$$q = 4\pi R \Delta T \left[k + R \left(\frac{k\rho c_p}{\pi} \right)^{1/2} t^{-1/2} \right] \tag{4.67}$$

Thus a graph of q versus $t^{-1/2}$ is a straight line whose intercept as $t \to \infty$ is directly related to k and the slope to the thermal inertia, $k\rho c_p$. Due to difficulties in obtaining a geometrically accurate bead, only the thermal conductivity could be determined with a sufficient degree of confidence. Additional data, mostly in vivo, were obtained via this technique by Chato et al.[53]

The author concludes with the observation that the method presented has a great potential for measuring local blood flow rates and thermophysical properties of biological tissues *in vivo* with a single small insertion of the measuring probe. The duration of the measurement is relatively short, about 2 mins., which is another advantage of this method over steady-state methods.

Trezek and coworkers[54–56] reported on the development of an analytically-based method for determining thermal diffusivity data for various biological tissues. Similar to the arguments presented above

> knowledge of the thermal properties of brain tissue are essential to the further development of (a) Cryo or radio frequency surgical brain probes where a prime consideration is the rate of the temperature field development, and (b) other neurophysiological research tools such as reversible blocking where localized cooling will stop the transmission of an action potential in a nerve or axion.[54]

A schematic diagram of the probe developed for this method is shown in Figure 17.

> In essence, this probe is a needle-like copper cylinder which acts as a thermocouple when a junction with a constantan wire is created at its center. With the probe initially at a temperature different from that of the tissue, it can be easily and rapidly inserted into the tissue and the thermal properties of the medium can be deduced from its temperature-time response. The distinguishing characteristic of the needle probe is that it has a response time which is quite large compared to the response time of a typical thermocouple. It is this length response time which forms the basis of the thermal property determination.[56]

To facilitate an analytical solution for the temperature field resulting from the insertion of this probe, the following assumptions were made:[56]

a. The "bio-heat" equation describes the temperature variations in the tissue for this case.

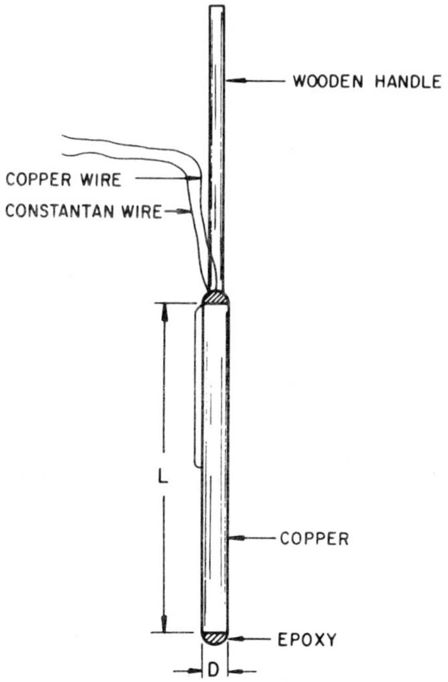

Figure 17 Needle probe assembly.[56]

b. Initially, the probe is at a uniform temperature different than that of the tissue in which it is inserted.
c. The probe is infinite in extent in the axial direction.
d. The tissue is homogeneous and infinite in extent.
e. The temperature of the tissue far from the probe remains unchanged at its initial value.
f. There is no resistance to heat flow between the probe and the surrounding tissue.
g. The probe is a perfect conductor of heat.
h. All thermophysical properties are constant.

Using cylindrical coordinates the problem becomes:

$$\rho c_p \frac{\partial T}{\partial t} = \frac{k}{r}\frac{\partial}{\partial r}\left(r\frac{\partial T}{\partial r}\right) + \dot{w}_b c_b (T_{\text{art}} - T) + \dot{q}_m \tag{4.68}$$

with the initial and boundary conditions:

$$T_p|_{@\ t=0 \text{ and } 0<r<r_0} = T_{po} \tag{4.69}$$

$$T|_{@\ t=0 \text{ and } r>r_0} = T_0 \tag{4.70}$$

$$T_p|_{@\ r=r_0 \text{ and } t>0} = T \tag{4.71}$$

$$T|_{@\ r\to\infty \text{ and } t>0} = T_0 \tag{4.72}$$

$$\rho' c' V \frac{\partial T}{\partial t} = kA \left.\frac{\partial T}{\partial r}\right|_{@\ r=r_0 \text{ and } t>0} \tag{4.73}$$

where p' and c' are the properties of the probe, V and A are the probe volume and surface area, respectively.

The above set of equations was solved by employing a standard Laplace transformation technique to yield

$$\theta_p = \frac{T_p - T_0}{T_{po} - T_0} = \frac{4\alpha \exp(-\beta t)}{\pi^2} \int_0^\infty \frac{\exp(-\tau u^2)}{u\Delta(u)}\, du \tag{4.74}$$

$$\text{where } \alpha = \frac{2\rho c_p}{\rho' c'} \tag{4.75}$$

$$\beta = \frac{\dot{w}_b c_b r_0^2}{k} \tag{4.76}$$

$$\tau = \frac{kt}{r_0^2} \tag{4.77}$$

$$\Delta(u) = \left[\left(u + \frac{\beta}{u}\right) J_0(u) - \alpha J_1(u)\right]^2 + \left[\left(u + \frac{\beta}{u}\right) Y_0(u) - \alpha Y_1(u)\right]^2 \tag{4.78}$$

A numerical method using Simpson's rule was used to integrate Eq. (4.74). Results are given in Ref. [56] in tabular form. A number of probes, with different L/D ratios, were constructed and first tested in a 1% Agar solution. The purpose of this series of experiments was to determine the ratio that will yield diffusivity values closest to those of the calibration material. With experimental compromises primarily necessitated by the characteristic dimensions of various organs, the selected probe was employed both *in vitro* and *in vivo*. Among the organs tested *in vitro* were human heart, kidney, liver, spleen, and brain; canine liver was tested *in vivo*.

The 1.5 mm diameter probe with L/D ratio of 15 was brought to a certain temperature, i.e., 23°C, and then quickly interred into a 100 gr sample which was precooled to 5°C after removal from the cadaver. Thermal diffusivity values were obtained by plotting the temperature-time response data of the experimental probe on a family of analytical curves, with the diffusivity as a

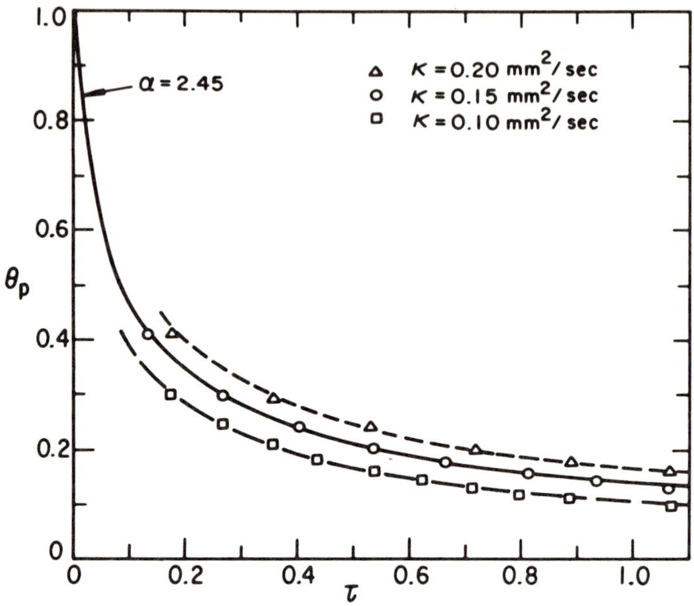

Figure 18 Probe temperature response versus nondimensional times for a 1.5 mm diameter probe in a 1% Agar solution.[56]

parameter. The curve that yields best agreement between analytical and experimental results determines the desired thermal diffusivity. A representative plot, obtained in a 1% Agar solution, is shown in Figure 18. Results obtained via this method, both *in vivo* and *in vitro,* are in good agreement with values found in the literature indicating the basic validity of this method. Similar to the method suggested by Chato, one insertion is sufficient for the determination of the thermal diffusivity of the tissue, and the time required for the measurement is relatively short, i.e., about 90 sec. However, the probe has to be maintained at a uniform temperature prior to insertion and, due to its inherent length, may yield average values rather than local values of minute volumes of tissue as obtained by Chato's method. On the other hand, the technique developed by Trezek et al., pertains directly to cylindrical cryosurgical probes in use for the destruction of tissue and thus has a greater potential for this application.

In their earlier studies, Trezek and coworkers measured the radial temperatures around a cylindrical probe imbedded in the tissue by means of thermocouples at regular intervals from the probe. Details are given in Trezek and Cooper[54] and in Trezek, Jewett, and Cooper.[55] Priebe and Betz[57] also discussed temperature fields in a biological tissue emanating from a small spherical heat source imbedded in the tissue.

4.4 Some applications of heat transfer in medicine

Various medical treatments involve generation or dissipation of thermal energy. In other cases, the cause of damage to the tissue may be a heat source. In these cases, knowledge of the thermal behavior of the tissue *in vivo* is very important, as was already demonstrated in the preceding paragraphs. Other applications of the analytical study of heat transfer in medicine follow.

4.4.1 The study of skin burns Recent wars and the increasing use of flame-causing warfare have dramatically increased the need for better understanding of the processes involved in thermal injury. Specific information regarding tolerance times of living tissue and the rate of propagation of irreversible tissue damage is required. To meet these ends Henriques and Moritz initiated a series of investigations into the nature of thermal injury.[58]

These investigators first sought an approximation of the time-temperature relationship existing in the layers of the epidermis. Assuming heat balance in the tissue to be described by the regular one-dimensional heat equation,

$$\frac{\partial T}{\partial t} = \alpha \frac{\partial^2 T}{\partial x^2} \tag{4.79}$$

they studied, theoretically and experimentally, the following two cases:

a. Initially the skin surface is stepped up and maitained at a predetermined temperature level until the threshold of irreversible epidermal injury is reached.
b. The entire animal is placed in a test chamber whose ambient temperature and radiant properties are known.

Mathematically these two cases were combined to form a single boundary condition:

$$k \left.\frac{\partial T}{\partial x}\right|_{@\ t>0 \text{ and } x=0} = h(T_{HS} - T|_{@\ t>0 \text{ and } x=0}) \tag{4.80}$$

where T_{HS} is the temperature of the heat source and h is infinity for case (a) above and assumes some finite value for case (b).

In addition the skin surface was assumed constant initially

$$T|_{@\ t=0 \text{ and } x=0} = T_0 \tag{4.81}$$

after assuming the tissue to be infinitely large in extent, the solution to the above set was obtained[58]

$$\frac{T_{HS} - T}{T_{HS} - T_0} = \operatorname{erf}\left(\frac{\gamma}{\sqrt{t}}\right)\left\{\exp\left[\frac{hL}{k}\left(1 + \frac{hL}{4\gamma^2 k}t\right)\right]\right\}$$

$$\cdot\left\{1 - \operatorname{erf}\left[\frac{\gamma}{\sqrt{t}}\left(1 + \frac{hL}{2\gamma^2 k}\right)\right]\right\} \tag{4.82}$$

$$\text{where } \gamma = \frac{L}{(k/\rho c_p)^{1/2}} \tag{4.83}$$

Henriques and Mortiz continued their basically experimental studies on the various aspects of thermal injury[59,60] and thus laid the foundation to this branch in bio-heat transfer studies. In 1951 Buettner undertook a rather elaborate study of the effects of extreme heat on man. He analyzed the temperature changes caused by different kinds of heat application to the skin.[61] In this connection, Buettner defines extreme heat as "an influx of heat into man with such an intensity that temperatures of the skin never reach a steady-state until all safety limits are exceeded."[61] He considered both applications of high temperature sources, e.g., flames, glowing materials, and atom bombs, and low temperature sources, e.g., wind at 30 kph and −40°C hitting the bare skin.

Similar to Henriques and Mortiz, Buettner assumed the energy balance of the skin to be that given by the general heat equation, (Eq. [4.79] above). He also assumed one-dimensional flow of heat, constant properties, and an infinite solid to be representative of the skin. A brief presentation of the various cases studied by Buettner follows.

Case A Constant non-penetrating heat flux absorbed by the upper surface of the skin, e.g., a hot radiator.

Initital and boundary conditions for this case are given by

$$T|_{@\ t=0} = T_0 + Kx \tag{4.84}$$

where K is the assumed linear gradient of the initial temperature in the skin

$$\left.\frac{\partial T}{\partial x}\right|_{@\ t>0 \text{ and } x=0} = -\frac{Q}{k} \tag{4.85}$$

where Q is the constant net influx of heat caused by the combined effect of radiation, convection, and evaporation.

The solution to the above set is given by:

$$T = \left(\frac{Q}{k} + K\right)\left\{\frac{2\sqrt{\alpha t}}{\pi}\exp\left(\frac{x^2}{4\alpha t}\right) - x\left[1 - \operatorname{erf}\left(\frac{x}{2\sqrt{\alpha t}}\right)\right]\right\} + T_0 + Kx \tag{4.86}$$

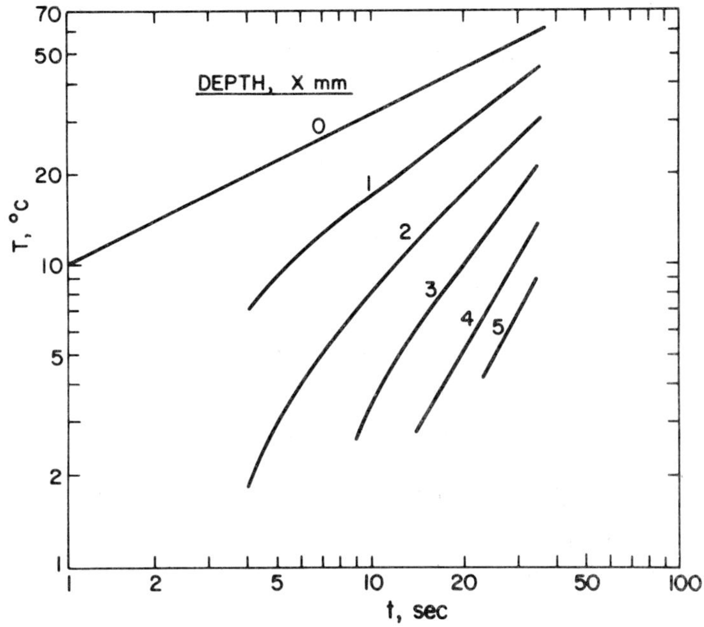

Figure 19 Calculated temperature-time increase of different skin depths when skin receives nonpenetrating heat of 0.044 cal/cm² sec. $k = 0.0025$ kcal/m sec °C, $\rho c = 1$.[61]

Temperature changes at different skin depths as calculated from Eq. (4.85) are shown in Figure 19.

Case B Constant penetrating radiant heat flux absorbed by the surface and also by the deeper layers, e.g., solar radiation.

Assuming the radiation flux to be given by I and the absorption coefficient by r Eq. (4.79) becomes:

$$\frac{\partial T}{\partial t} = \alpha \frac{\partial^2 T}{\partial x^2} + \frac{Ire^{-rx}}{\rho c_p} \tag{4.87}$$

with the initial and boundary conditions:

$$T|_{@\ t=0} = T_0 \qquad \text{(constant initial temperature)} \tag{4.88}$$

$$\left.\frac{\partial T}{\partial x}\right|_{@\ t>0 \text{ and } x=0} = -\frac{R}{k} \tag{4.89}$$

The additional term in Eq. (4.87) indicates the contribution of the penetrating energy to the energy balance of the tissue which decays exponentially with depth. In addition there is a nonpenetrating heat flux at the skin, Eq. (4.89). The solution to the above set is given by:

$$T = \frac{I}{2kr}\left\{\exp(\alpha r^2 t - rx)\left[1 - \operatorname{erf}\left(r\sqrt{\alpha t} - \frac{x}{2\sqrt{\alpha t}}\right)\right]\right.$$
$$\left. + \exp(\alpha r^2 t + rx)\left[1 - \operatorname{erf}\left(r\sqrt{\alpha t} + \frac{x}{2\sqrt{\alpha t}}\right)\right]\right\}$$
$$- \frac{I\exp(-rx)}{kr} + \frac{2(I + R)}{k\sqrt{\pi}}\sqrt{\alpha t}\exp\left(\frac{-x^2}{4\alpha t}\right)$$
$$- \frac{x(I + R)[1 - \operatorname{erf}(x/2\sqrt{\alpha t})]}{k + T_0} \tag{4.90}$$

Equation (4.90) is plotted for various values of the parameters in Figure 20. Of the curves shown in this figure, the one for the absorption coefficient $r =$ 10, is the one corresponding to solar energy.[61]

Figure 20 Increase of skin temperature with time when a penetrating radiation is applied. r is the absorption coefficient of the radiation.[61]

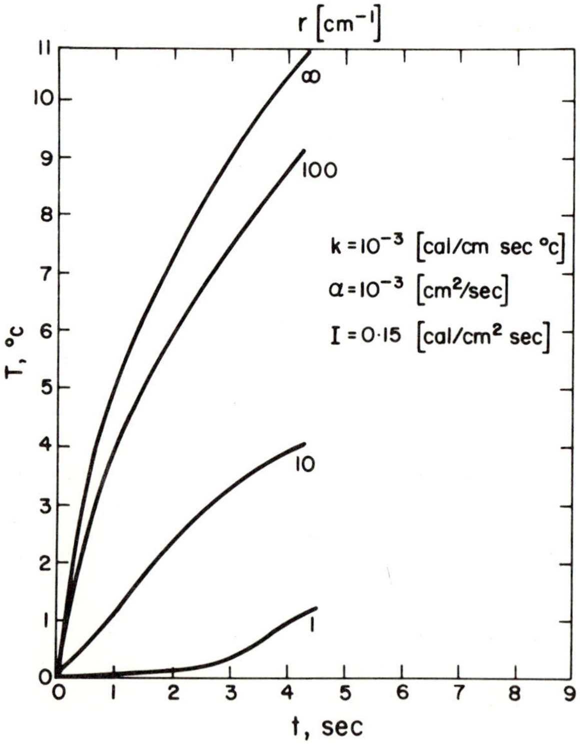

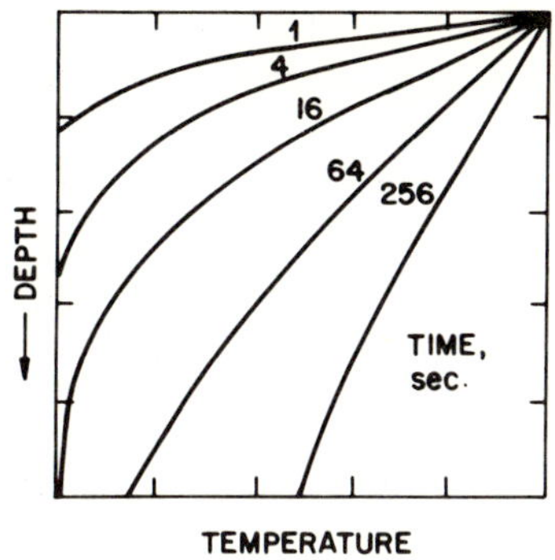

Figure 21 Temperature rises at different skin levels after application of a warm contacting body to the surface. The surface temperature then stays constant.[61]

Case C Convective skin heating (or cooling) proportional to the temperature difference between the environment and the skin, e.g., warm air.

This case is identical to the one studied by Henriques and Mortiz.[58] The solution is given by Eq. (4.82) above by replacing L with x. Figure 21 shows qualitatively the temperature rises at different skin levels after application of a warm contacting body to the surface ($h \to \infty$ in Eq. [4.79]).

Case D Skin heating through a protective layer, e.g., cloth.

This case, typical to many applications of extreme heat in reality, is treated by assuming that upon entering a fire, the temperature of which is T_f, the temperature of one's protective suit is raised immediately to this temperature. It is further assumed that the skin is initially at a constant temperature. Placing the origin of the coordinate system at the interface between the surface of the skin and the protective layer, the initial and boundary conditions may be stated:

$$T|_{@\ t=0 \text{ and } x<0} = T_{so} \tag{4.91}$$

$$k \left.\frac{\partial T}{\partial x}\right|_{@\ x=0} = \frac{k(T_f - T_{so})}{d} G(z^2) \tag{4.92}$$

where d is the thickness of the protective layer, and

$$G(z^2) = 1 - 2 \sum_{n=1}^{\infty} (-1)^{n-1} \exp(-z^2 n^2) \tag{4.93a}$$

$$\text{where } z^2 = \frac{\alpha_{cl} \pi t}{d} \tag{4.93b}$$

Buettner distinguishes among three cases for z:

1. Initially, for $0 < z^2 < 0.2$, G remains zero, i.e., no heat penetrates the skin.
2. For $0.2 < z^2 < 1.0$, G and, consequently the heat flux, increase rapidly.

This increase is approximately given by:

$$k\left.\frac{\partial T}{\partial x}\right|_{@\, x=0} = \frac{1.2k_{cl}(T_f - T_{so})}{d} \sin\,[m(z^2 - 0.2)] \tag{4.94}$$

where $m = 0.872\ \text{cm}^{-1}$.

3. For $z^2 > 1.0$, G approaches a steady value and eventually

$$k\left.\frac{\partial T}{\partial x}\right|_{@\, x=0} = \frac{1.2k_{cl}(T_f - T_{so})}{d} \tag{4.95}$$

With Eq. (4.94) replacing the more general boundary conditions, Eq. (4.92), Buettner obtained an analytical solution to the temperature of the tissue. This solution, which serves as a first estimate of tissue time-temperature history, is given by:

$$T_s = \frac{1.2k_{cl}(T_f - T_{so})(k_{cl}\rho_{cl}c_{cl}m)^{1/2}}{d}\left\{\sin\left[m(z^2 - 0.2) - \frac{\pi}{4}\right] + \frac{I}{\sqrt{2}}\right\} + T_{so} \tag{4.96}$$

Equation (4.96) is plotted in Figure 22 for a selected set of parameters.

After removal of the heat source, the temperature of the various layers of skin may remain at levels beyond those compatible with reversible damage. Stoll

Figure 22 Increase of heat flow, G, and skin temperature under a protective layer of $d = 0.99$ cm thickness. Circles indicate $c\sin[m(z^2 - 0.2)]$; cross marks indicate $T_0 - T_1 = 1000°\text{C}$.[61]

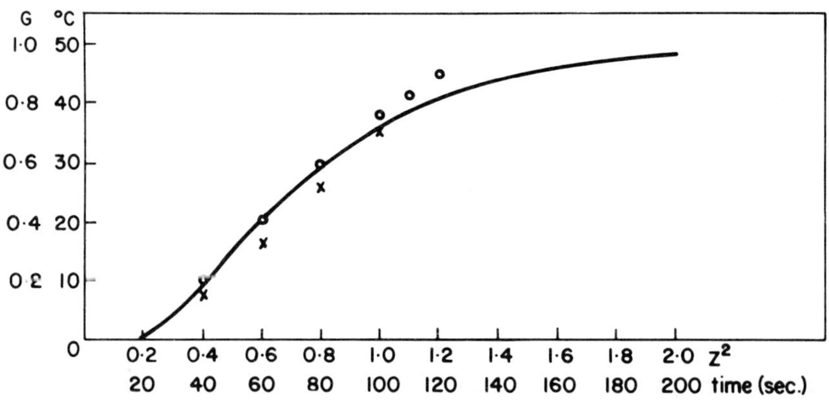

Table 8 Thermal Sensations and Associated Effects Throughout Range of Temperatures Compatible with Tissue Life[62]

Sensation	Skin color	Tissue temperature °C	Tissue temperature °F	Process	Injury
Numbness	White	72	162	Protein Coagulation	Irreversible
	Mottled red & white	68 64		Thermal inactivation of tissue constituents	Possibly reversible
Maximum pain Severe pain Threshold pain	Bright red Light red	60 56 52 48	140		Reversible
Hot Warm	Flushed	44 40	111	Normal metabolism	None
Neutral	Flesh	36	93		
Cool Cold	Blanched red	32 28 24 20	82		
Threshold pain Severe pain	Bluish red Reddish purple	16 12	64	Physico-chemical inactivation of tissue constituents	Reversible
Numbness	Bright pink	8 4	50		Possibly reversible
	White	0 −4	32 25	Protein coagulation	Irreversible

and Chianta indicate these temperatures to be 0°C and 72°C or, more likely, 10°C and 60°C.[62] Details of sensations, skin colors, bio-chemical processes, and tissue injury, as functions of tissue temperature are given in Table 8.

It is apparent from this table that damage to the tissue and a definite sensation of pain will continue as long as the range of 10°C to 60°C is exceeded. Thus the study of time history of the temperature of the tissue after the removal of a heat source is rendered of prime importance. Buettner also considered a number of cases of the decrease of skin temperature after heat application.[61] However, it appears that the general expression, giving in one equation the time-temperature history of the tissue, was first developed by Weaver and Stoll.[63] In their analysis, based on Buettner's work,[61] they assumed a constant heat flow and an initial isothermal condition, Eqs. (4.85) and (4.84) with $K = 0$. The solution is then divided into two time domains:

1. $0 < t < \tau$, the heating phase, $Q > 0$ in Eq. (4.85).
2. $t > \tau$, the cooling phase, $Q = 0$ in Eq. (4.84).

Accordingly, the real part of the following equation is the solution for the heating-cooling process of a biological tissue[63]

$$T - T_0 = \frac{Q}{k}\left\{\frac{2\sqrt{\alpha t}\exp(-x^2/4\alpha t)}{\sqrt{\pi}} - x\left[1 - \operatorname{erf}\left(\frac{x}{2\sqrt{\alpha t}}\right)\right]\right\}$$
$$- \frac{Q}{k}\left\{\frac{2\sqrt{\alpha(t-\tau)}\exp[-x^2/4\alpha(t-\tau)]}{\sqrt{\pi}}\right.$$
$$\left. - x\left[1 - \operatorname{erf}\left(\frac{x}{2\sqrt{\alpha(t-\tau)}}\right)\right]\right\} \tag{4.97}$$

Equation (4.97) was evaluated at the skin surface, $x = 0$, and at a depth of 80μ. Results for an assumed irradiance of 100 mc/cm^2 sec, are shown in Figure 23. For this particular case the approximate lower limit of injurious temperatures is reached after about 12 sec from the onset of the radiation. Radiation continues for another 22 sec and then ceases. Injurious temperatures prevail for another 24 sec bringing the total to about 46 sec while the duration of the exposure is only 34 sec. In another case[64] skin was subjected to an irradiance four fold larger. For a total exposure time of about 6 sec (about $\frac{1}{6}$ of the previous duration) injurious temperatures were found to exist in the tissue for about 10 sec along with a much higher peak temperature, i.e., 60°C versus 53°C in the previous case.

This method of calculating temperature distributions in the skin does not allow the inclusion of metabolic heat production and blood flow effects. Moreover, it assumes uniform temperatures initially and constant properties along with infinite dimensions of the tissue. For better understanding of the mechanisms involved these effects should be explicitly included and a more realistic geometrical representation of the tissue employed. Nevertheless, due to the relatively short times involved in radiation cases, experimental results conform well to the theoretical predictions as indicated by Weaver and Stoll.[63]

As suggested by Buettner "two kinds of heat traumata may be considered, pain and burns."[61] As may be construed from Table 8, the threshold of pain is reached when the skin is brought to 19°C or 44°C. Both these temperatures, with some individual variations, may be termed "critical" or "thresholds of thermally induced pain" temperatures. On the other hand, the occurrence of tissue damage is harder to predict since both time and temperatures play a role.[61] Henriques attempted to define mathematically a damage rate function in the following way.[60] Assume that the biochemical process involved in the eventual death of the tissue (epithelium) conforms to that of most chemical and

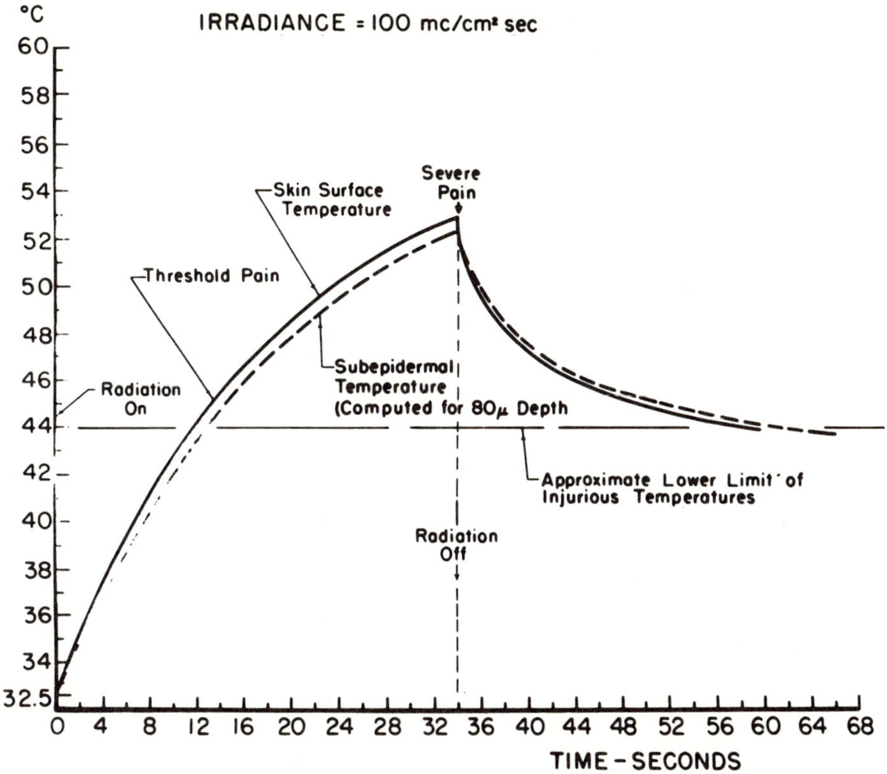

Figure 23 The heating and cooling pattern of skin exposed to a square-wave pulse of radiant energy.[63]

physical rate processes. Thus, if Ω denotes an arbitrary function of epidermal injury as determined by histologic examination, the rate at which this injury is produced is given by:[60]

$$\frac{d\Omega}{dt} = A \exp\left[-\frac{\Delta E(T + 273)}{R_0}\right] \tag{4.98}$$

where A is an experimental constant, ΔE energy of activation, and R_0 the universal gas constant. Henriques suggested the value for the energy of activation to be given by $\Delta E = 150{,}000$ cal/mol with $R_0 = 2$ cal/mol deg K, the numerical constant in the exponent of Eq. (4.98) becomes quite large, i.e., 75,000 deg K^{-1}. Thus the rate of tissue damage becomes a strong function of the absolute temperature of the tissue. Henriques further assumed the value of $\Omega = 1.0$ to represent complete transepidermal necrosis and obtained $A = 3.1*10^{98}$ sec^{-1}.

It follows that in order to obtain the damage done to a tissue, one should integrate Eq. (4.98) to obtain:

$$\Omega = A \int_0^t \exp\left[-\frac{\Delta E(T + 273)}{R_0}\right] dt \tag{4.99}$$

with the substitution of the numerical constant derived by Henriques, the result of the integral indicates one of the three possibilities:[61]

1. For $\Omega < 0.53$ the time-temperature relationship is such that can be tolerated without the occurrence of irreversible epidermal damage.
2. For $\Omega > 1.0$ complete epidermal necrosis is produced.
3. $0.53 < \Omega < 1.0$ epidermal injury is irreversible but the degree of the damage is not complete.

Figure 24 Tissue damage rate ($d\Omega/dt$) change with temperature increase.[64]

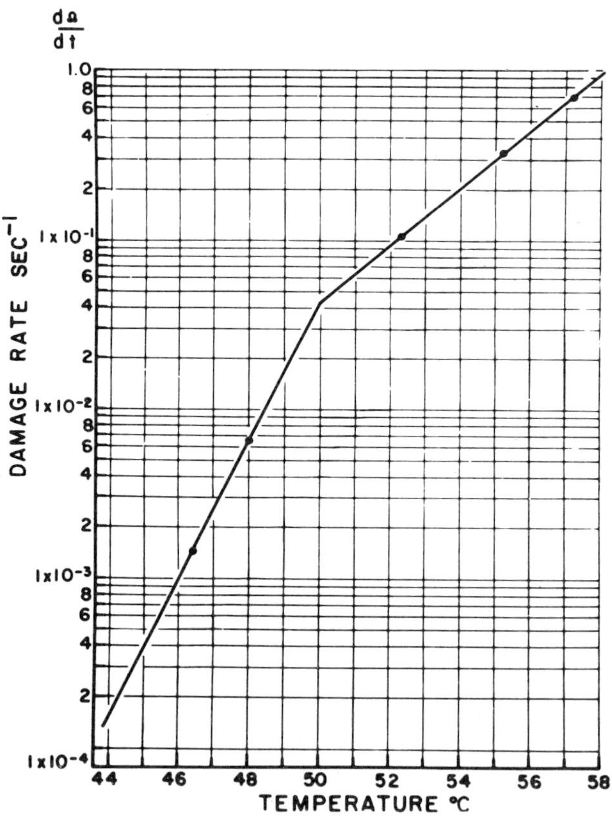

Henriques compared his analytical results with experimental data and reports on "excellent" agreement.[61] Stoll and Greene[65] further discussed tissue damage rates as a function of its temperature. A typical plot obtained by them is shown in Figure 24.

Stoll[66] and Weaver[67] also reported on the development of computer solutions for the estimation of thermal tissue damage integrals from experimental data. Dussan and Weiner studied the burn hazard in human tissue and its implications to consumer product design.[68] More recently, Wu discussed material properties criteria for thermal safety.[69]

4.4.2 Heat transfer in teeth (dentistry) The thermal environment of teeth of mammals varies over a wide range of temperatures. Consumption of food at various temperatures, open mouth breathing and dental treatment are only examples to the sources which may cause extreme changes in teeth temperature over a short period of time. These changes may lead to initiation of caries or may cause pain. Thermally induced fatigue may also initiate fracture in the enamel coating of the tooth. Also, high speed dental instruments generate heat while cutting the tooth and may cause undesired stresses without adequate cooling.

To obtain some information on the mechanism of heat transfer in teeth, Jacobs et al. analyzed this phenomenon via a numerical method.[70] They depicted a typical maxillary right central incisor as shown in Figure 25. The tooth was assumed to be composed of the two most basic components, i.e., enamel and dentin. Both these materials are assumed isotropic and homogeneous and their physical properties are given in Table 9. Other assumptions are:

a. Energy balance in the tooth is governed by the heat equation.
b. Initially the tooth is at nominal body temperature, e.g., 37°C.
c. All the tooth below the enamel crown is constantly maintained at the normal body temperature. This assumption is "due to the ample supply of blood to the supportive tissue and the good insulative properties of the gingiva."[70]
d. The problem may be treated as a two dimensional problem due to the good insulative properties of the tooth materials and the relatively short times of exposure to environmental temperatures.

Mathematically the problem may be stated:

$$\nabla^2 T_d = \frac{1}{\alpha_d} \frac{\partial T_d}{\partial t} \tag{4.100}$$

$$\nabla^2 T_e = \frac{1}{\alpha_e} \frac{\partial T_e}{\partial t} \tag{4.101}$$

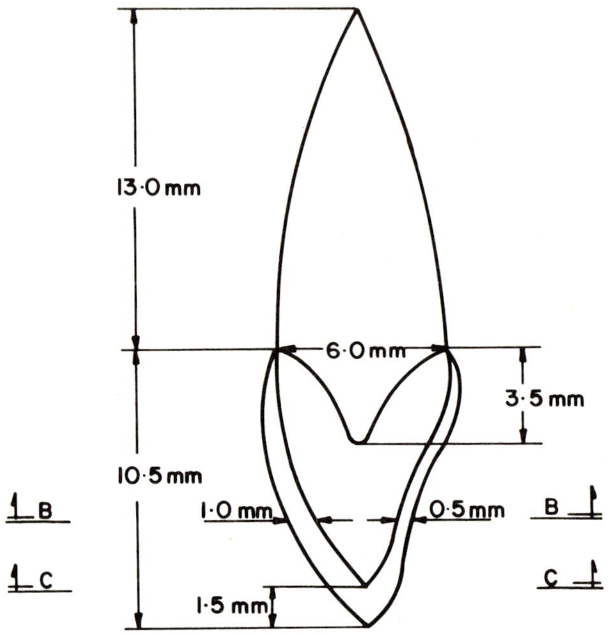

Figure 25 Maxillary right central incisor.[70]

where subscripts *d* and *e* refer to dentin and enamel, respectively.

The initial condition

$$T_d = T_e|_{@\ t=0} = T_0 \tag{4.102}$$

and the boundary and matching conditions:

$$-k_e \frac{\partial T_e}{\partial n} = h(T_e - T) \quad @ \quad \text{external surface} \tag{4.103}$$

Table 9 Physical Properties of Teeth[70]

	ρ gm/cm³	c_p cal/gm°C	k, cal/ sec cm °C	α cm²/sec
Enamel	2.8	0.17 ± 0.01	2.23×10^{-3}*	2.23×10^{-3}
Dentin	1.96 ± 0.01	0.38 ± 0.01	1.39×10^{-3}** 1.36×10^{-3}***	$1.87 \times 10^{-}$

*Longitudinal.
**Perpendicular to tubules.
***Parallel to tubules.

$$T_e = T_d \quad @ \quad \text{interface} \tag{4.104}$$

$$k_e \frac{\partial T_e}{\partial n} = k_d \frac{\partial T_d}{\partial n} \quad @ \quad \text{interface} \tag{4.105}$$

where n is the outward normal vector.

The above set of equations was solved by a finite element method. Experiments were also conducted on freshly extracted teeth to correlate the analytical and experimental data. The tooth was prepared for the experiment by installing two chromel-alumel thermocouples and imbedding the root in an epoxy putty. Locations of the thermocouples are shown in Figure 26. A stream of hot water at about 65°C (150°F) was then directed at the tooth from above, completely covering it and offering symmetric flow. Comparison of the experimental and analytical results is shown in Figure 27. Agreement obtained at both locations is within 5%.

These investigators further considered the problem of cracking of enamel. This cracking may be caused by tensile stresses that exceed the yield stress of the material. Such stresses are encountered by sudden cooling of the tooth and, since enamel is not self restorative, it is very important to have a method of estimating safe exposure times. It was found, by assuming the tooth to be

Figure 26 Locations of thermocouples of maxillary right central incisor.[70]

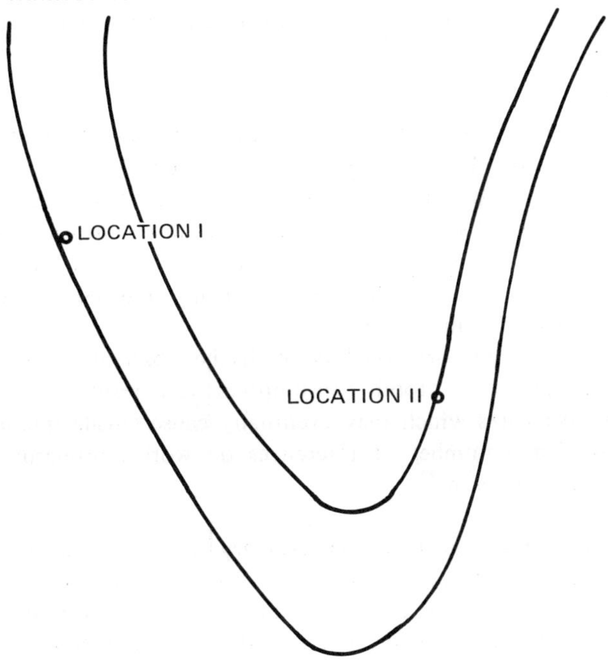

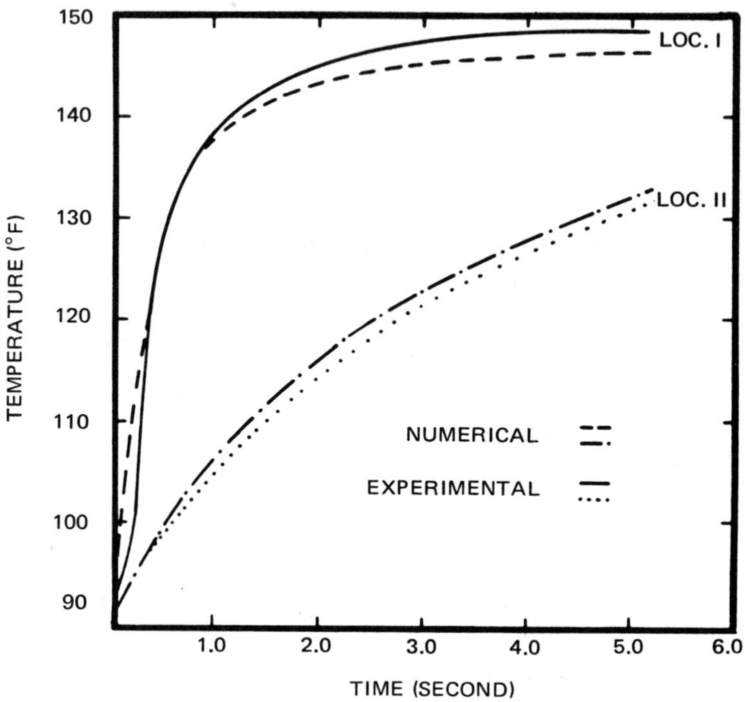

Figure 27 Numerical and experimental results of the temperature distributions in incisor.[70]

suddenly subjected to an icewater environment at about 1°C, that tensile stresses in excess of 1500 psi may result in less than $\frac{1}{2}$ sec. This is $\frac{2}{3}$ of the ultimate stress in enamel and may cause fracturing with a few thousand repetitive loadings.[70] Typical results of the calculated stress distributions for the axis of symmetry of one of the cross-sections of a tooth are shown in Figure 28. It is seen that in a very short time, i.e., 0.4 sec, thermally induced tensile stresses in the enamel nearly double in magnitude.

The authors conclude with the observation that thermal environments encountered in food consumption can result in sufficiently high or low temperatures which may eventually cause tensile fracture of the enamel. They also list a number of references on works, primarily experimental, on heat transfer in teeth.[70]

4.4.3 Heat transfer in cryosurgery Applications of refrigeration in medical treatment are numerous, e.g., hypothermic anesthesia, conservation of tissues and blood, and surgery. This paragraph is intended to discuss briefly a number of applications of refrigeration in medicine, particularly cryosurgery.

The prefix cryo comes from Greek and means "icy cold." In the physical sciences this prefix usually indicates very low temperatures, −150°C and below. In medicine, however, the implication of this expression is a temperature capable of freezing the tissue, i.e., −1°C and below. Thus cryosurgery indicates the use of freeze-causing instruments in surgery. Among the applications of cryosurgery are treatment of Parkinson's disease, removal of tumors (particularly cancerous brain tumors), gynecological diseases, cataract removal, ear disease, etc. As discussed by Chato, the cooling effect in the temperature range of 0°-27°C may be used for effectively eliminating the normal nerve functions.[71] With the removal of the cooling source, the temperature of the nerve returns to normal thus restoring also complete nerve function provided that the temperature of the tissue was not allowed to drop below 0°C. These reversible lesions enable the surgeon to accurately fix the desired location of a surgical probe before destroying the tissue. In many cases cryosurgery has distinct advantages over other methods.

One of these advantages stems from the previously described nerve deadening effect of cold: the cooling provides its own anesthesia. Another advantage is that since the only physical cutting occurs when the cryoprobe is introduced into the tissue, whereas the volume of the tissue to be destroyed is much larger than that of the probe, usually very few

Figure 28 Stresses distribution at the axis of symmetry section B-B.[70]

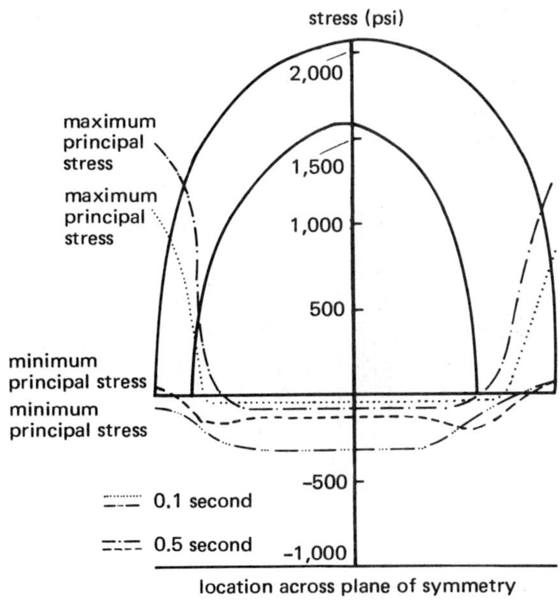

blood vessels are ruptured and little bleeding occurs. In special applications other advantages may occur. For example, if a cataract is to be removed from the eye, a cryoprobe cannot only freeze the tissue but will also freeze to it, thus providing two desirable effects simultaneously. First, the hard frozen tissue is much easier to cut than the normal, jelly-like substance; and second, the probe acts as a holder of the tissue and tremendously facilitates its handling and final removal.[71]

It is apparent that precise information on both thermophysical properties and the local effects of blood perfusion and metabolic heat generation are of utmost importance for cryosurgical applications. The extent of freezing and the rate at which the freezing front propagates are mandatory before any attempt to freeze the tissue is made. While the extent of freezing is bascially a physiological problem, the rate of propagation is a problem that is amenable to engineering methods.

Cryoprobes generally assume either a spherical or a cylindrical shape, as depicted in Figures 29 and 30.[72] A cryoprobe which is inserted into the tissue will produce a lesion, the front of which will propagate until a steady state is reached. These, and related problems, were considered analytically by Cooper and Trezek.[73] They formulated and solved the steady-state temperature fields which are established around spherical and cylindrical probes. Cooper and Trezek further assumed constant properties and a homogeneous and isotropic tissue.

Dividing the tissue into two phases, the unfrozen and the frozen, the steady state problem, for a spherical geometry, may be stated as follows:[73]

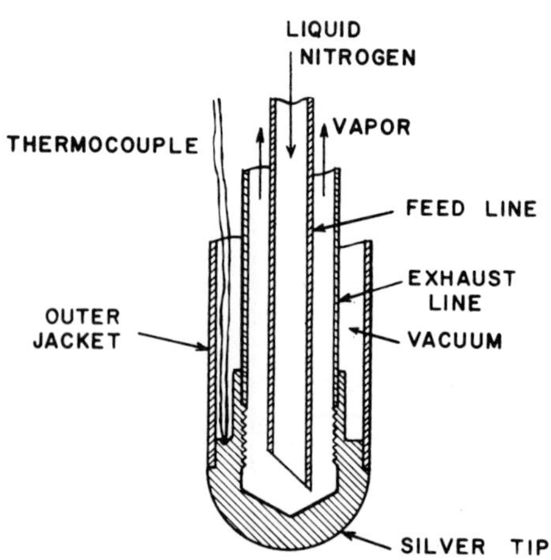

Figure 29 Section view of a typical cryosurgical probe tip.[72]

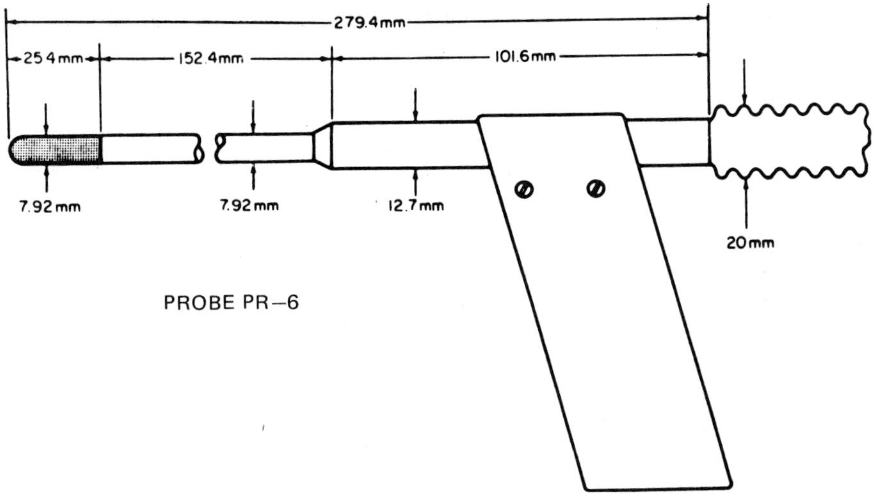

Figure 30 A cylindrical cryoprobe used with the Linde cryosurgery system. Shaded area denotes the low-temperature surface.[72]

frozen phase:

$$\frac{d}{dr}\left(r^2 \frac{dT_{fr}}{dr}\right) = 0 \tag{4.106}$$

unfrozen phase:

$$\frac{k}{r^2}\frac{d}{dr}\left(r^2 \frac{dT}{dr}\right) + \dot{w}_b c_b (T_{\text{art}} - T) + \dot{q}_m = 0 \tag{4.107}$$

with the boundary conditions:

$$T_{fr}|_{@\ r=r_0} = T_{pr} \tag{4.108}$$

where r_0 is the probe radius. This condition implies no contact resistance to heat transfer between the probe and immediately surrounding tissue.

$$T|_{@\ r\to\infty} = T_0 = T_{\text{art}} + \frac{\dot{q}_m}{\dot{w}_b c_b} \tag{4.109}$$

and matching conditions at the interface of the two phases:

$$T_{fr}|_{@\ r=\bar{R}} = T_{pc} = T \tag{4.110}$$

$$k_{fr} \left.\frac{dT_{fr}}{dr}\right|_{@\ r=\bar{R}} = k \left.\frac{dT}{dr}\right|_{@\ r=\bar{R}} \tag{4.111}$$

The solutions to this set of equations was obtained in non-dimensional form:

frozen phase:

$$\theta_{fr} = 1 + (\theta_{pc} - 1)\frac{r^*}{1 - r^*} \cdot \frac{1 - R}{R} \tag{4.112}$$

unfrozen phase:

$$\theta = \frac{\theta_{pc} r^*}{R} \exp\left[-\beta^{1/2}(R - r^*)\right] \tag{4.113}$$

$$\text{where } \theta = \frac{T - T_0}{T_{pr} - T_0} \tag{4.114}$$

$$r^* = \frac{\bar{R}}{r_0} \tag{4.115}$$

$$R = \frac{r}{r_0} \tag{4.116}$$

$$\beta = \frac{\dot{w}_b c_b r_0^2}{k} \tag{4.117}$$

with the location of the frozen front given by (for $\beta > 0$):

$$r^* = \frac{\beta^{1/2} - 1}{2\beta^{1/2}} + \left[\frac{(1 - \beta^{1/2})^2}{4\beta} + \frac{1 + \phi}{\beta^{1/2}}\right]^{1/2} \tag{4.118}$$

$$\text{where } \phi = -\frac{k_{fr}(T_{pc} - T_{pr})}{k(T_{pc} - T_0)} \tag{4.119}$$

For a cylindrical geometry the steady-state bio-heat equation is modified accordingly, and with the same boundary and matching conditions, the solutions become:

frozen phase:

$$\theta_{fr} = \frac{(\theta_{pc} - 1)\ln R}{\ln r^*} + 1 \tag{4.120}$$

unfrozen phase:

$$\theta = \frac{\theta_{pc} K_0(\beta^{1/2} R)}{K_0(\beta^{1/2} r^*)} \tag{4.121}$$

and the location of the frozen front may be calculated from the following transcendental equation:

$$\phi = \frac{\beta^{1/2} r^* \ln r^* K_1(\beta^{1/2} r^*)}{K_0(\beta^{1/2} r^*)} \tag{4.122}$$

Steady-state temperature distributions in the tissue as calculated from Eqs. (4.112), (4.113), (4.120), and (4.121) are shown in Figure 31 (spherical

Figure 31 Steady-state spherical probe temperature fields in the frozen and unfrozen phases for three ice front positions ($r^* = 2, 4, 10$) and blood parameters ($\beta = 0.01, 0.10, 1.0$).[73]

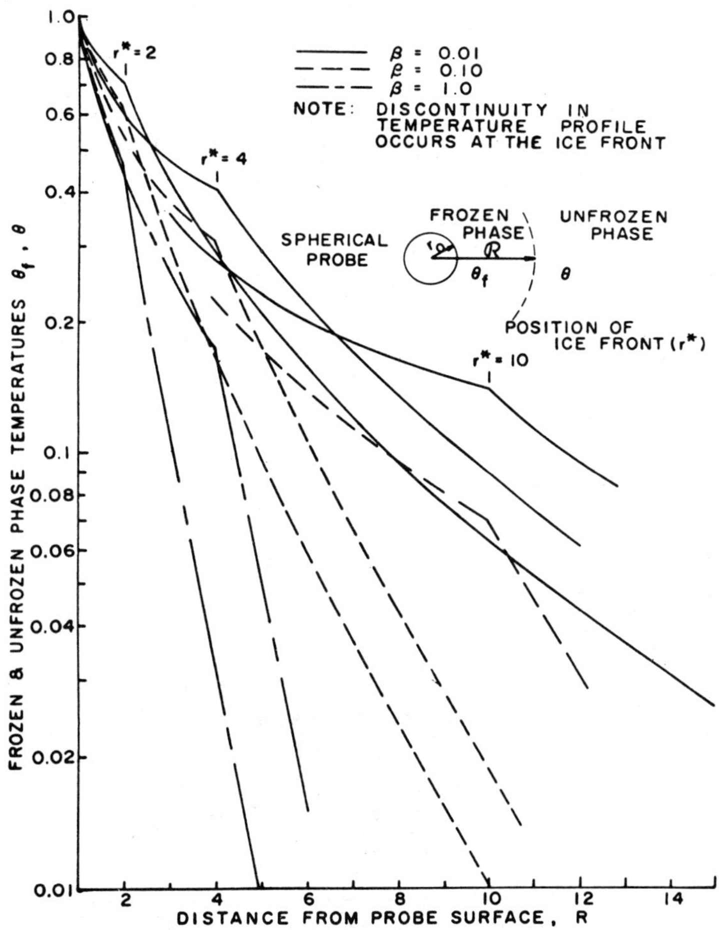

geometry) and Figure 32 (cylindrical geometry). It is seen that blood circulating in the tissue has the effect of impeding the spread of the ice front.

Rate of lesion growth in tissue was obtained by a numerical solution of the transient bio-heat equation. In the numerical analysis blood flow rate, metabolic heat generation and the frozen phase heat capacity were assumed functions of local tissue temperature. Also, the latent heat of fusion was included, in the boundary conditions. Typical results for the spherical probe are shown in Figure 33. The discontinuity in the temperature profile indicates the lesion size at a particular time after initiation of cooling.

Results of this study were compared to experimental data found in the literature in another article by the same two authors.[74] They report good

Figure 32 Steady-state cylindrical probe temperature fields in the frozen and unfrozen phases for three ice front positions ($r^* = 2, 4, 10$) and blood flow parameters ($\beta = 0.01, 0.10, 1.0$).[73]

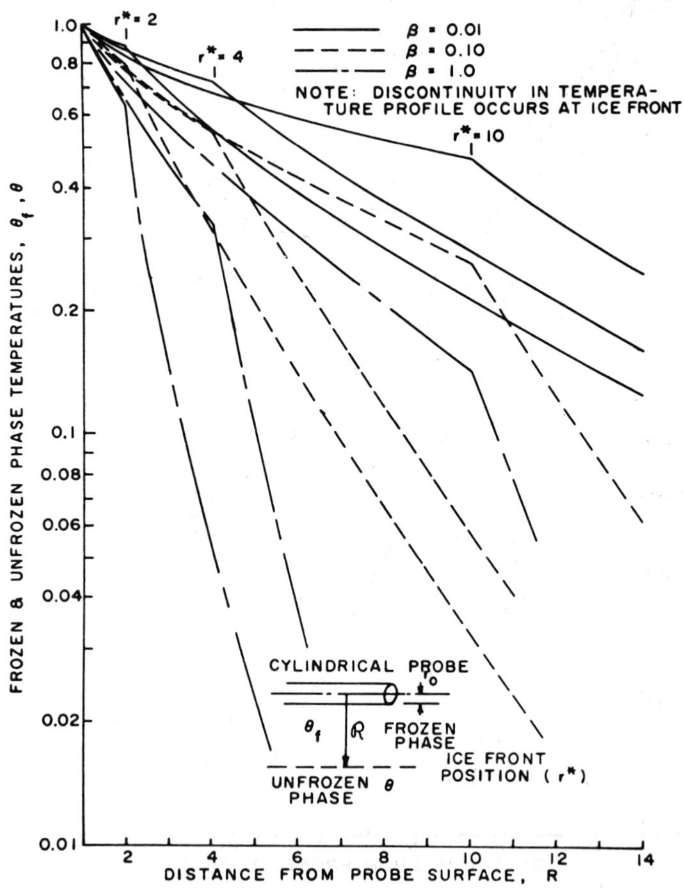

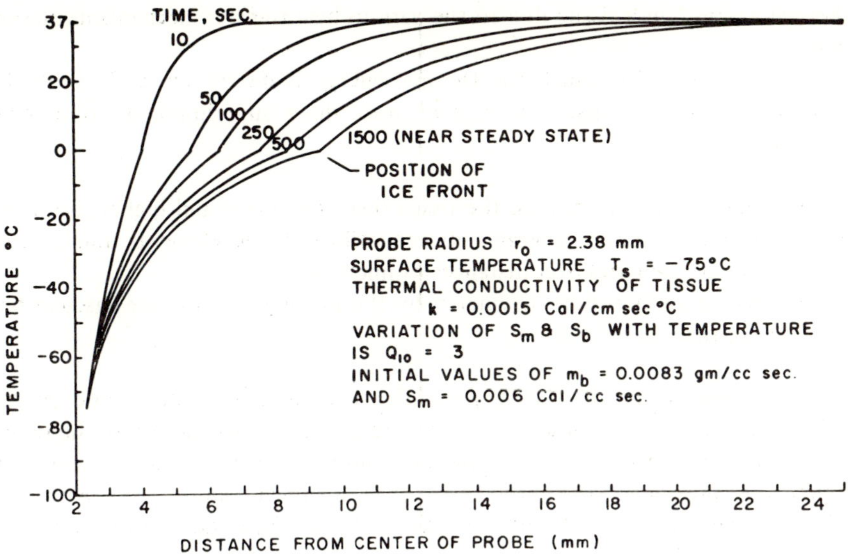

Figure 33 Typical transient temperature profiles for a spherical probe having a surface temperature of −75°C in tissue.[73]

agreement between theoretical and experimental data which probably indicates the validity of this analytical approach to heat transfer problems in cryosurgery. Others works pertinent to the subject are the study by Chato on temperature fields emanating from small spherical bead mentioned above;[52] and other works by Trezek and Jewett[75,76] and Trezek.[77]

Surgical applications of lesion producing probes are not limited to low temperatures; high temperatures, i.e., temperatures in excess of about 55°C, produced by resistively heated probes, are capable of achieving the same result. Detailed discussion of the application of heated surgical probes is not presented here. Interested readers are referred to a recent paper by Cooper and Groff[78] who analyzed this problem by solving the bio-heat equation in a tissue heated by a surgical probe.

4.5 Determination of blood perfusion

One of the applications of heat transfer theory in biological systems has been the *in vivo* determination of blood perfusion rates. The objective is to use invasive or non-invasive measurements and obtain the desired information by analysis of the collected data. One of the earliest attempts was made by Perl, who reported his studies in a series of articles.[79–82] Perl considered the case of dissipation of thermal energy* in a tissue due to the various heat transfer mechanisms present in it.

The process by which the thermal energy is distributed and eventually cleared out of the tissue is divided into three superimposed component processes:[79]

a. Diffusion of energy through the tissue over millimeter and higher distances.
b. Exchange-diffusion of energy between cellular tissue elements and moving blood in capillaries over distances of microns.
c. Convection, or transport, of energy by the moving blood in the vascular bed of the tissue.

By application of Fick's principle, Perl obtained the "*basic partial differential equation*"[79] describing *in vivo* temperature distribution in the tissue. This equation is essentially the so-called bio-heat equation the derivation of which was previously presented. Perl then assumed, similar to Chato,[52] a spherical heat source to be situated inside the tissue. Other assumptions were:

a. The tissue is uniform in properties.
b. The tissue is large in extent (mathematically infinite).
c. The problem is spherically symmetric and thus is only a function of the distance from the heat source.
d. The temperature of the spherical heat source may change in steps from T_b to T_0 (or reversed) by either switching on or off the power supply.
e. The temperature of the tissue is uniform and constant, T_0, when the probe is switched off.

Perl first obtained a steady-state solution to the problem of temperature distribution in the tissue with the heat source turned on, subject to the following boundary conditions:

$$T|_{@\ r=b} = T_b \text{ (no contact resistance at tissue-heat source interface)} \tag{4.123}$$

$$T|_{@\ r\to\infty} = T_0 \tag{4.124}$$

further denoting

$$\omega = \frac{\rho_b c_b \dot{w}_b}{\rho c_p} \tag{4.125}$$

*Perl also considered diffusion of matter in the tissue.[79] This subject is not treated in the present work.

the solution to the steady-state bio-heat equation subject to Eqs. (4.123) and (4.124) becomes

$$u(r) = u_b\left(\frac{b}{r}\right)\exp[-\eta(r-b)] \tag{4.126}$$

where $u = T - T_0$ (4.127)

$$\eta = \left(\frac{\omega}{\alpha}\right)^{1/2} = \left(\frac{\dot{w}_b}{\alpha_b}\right)^{1/2} \tag{4.128}$$

$$\alpha_b = \frac{k}{\rho_b c_b} \tag{4.129}$$

The power dissipated from the heat source is given by

$$\gamma_b = -b\left.\frac{du}{dr}\right|_{@\, r=b} = \frac{I^2R}{4\pi bk} \tag{4.130}$$

Where I is the electrical current through and R is the resistance of the heat source. Steady-state non-dimensional temperature distribution in the tissue as a function of the imbedded heat source for different blood perfusion rates are depicted in Figure 34. The apparent effect of increased blood perfusion rate, for fixed heat source radius, b = const., is to decrease temperature levels in the tissue. In other words, the equivalent, or apparent, thermal conductivity of the tissue is increased due to the effect of blood flow.

Two types of transient solutions to the problem were obtained by Perl.[79] These were:

1. Step changes at $t = 0$ assumed for u_{bi} and η_i, i.e., $u_{bi} \to u_b$ and $\eta_i \to \eta = (\phi/\alpha_b)^{1/2}$ with γ_b remaining constant.

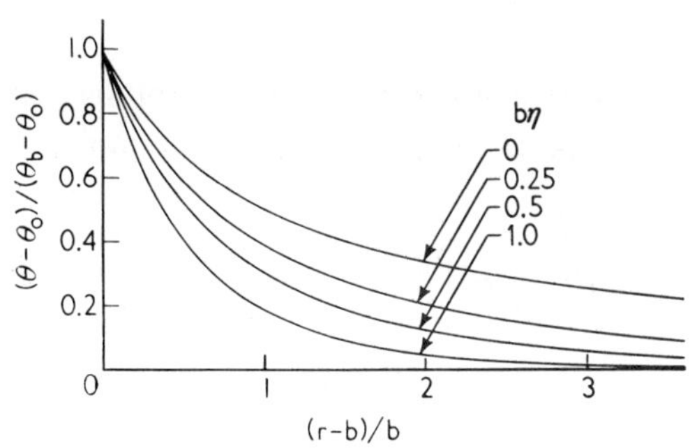

Figure 34 Theoretical steady-state solutions for temperature distribution in tissue due to heat source, for several values of non-dimensional perfusion parameter.[79]

2. Step changes at $t = 0$ assumed for γ_{bi} and η_i, i.e., $\gamma_{bi} \rightarrow \gamma_b$ and $\eta_i \rightarrow \eta$ with u_b remaining constant.

In both of these cases initial temperature distribution in the tissue was assumed to be the solution to the steady-state equation given by Eq. (4.126) above. This is one of the aspects in which Perl's analysis differs from Chato's,[52] making it more general. The solutions to both cases were obtained by employing the Laplace transform.

Case I The expression for the transformed $\bar{u}(r, t)$ is given by

$$\bar{u}(r,p) = \frac{b}{r}\left\{\frac{u_b}{p}\exp[\zeta(b-r)] + \frac{u_{bi}}{(p+\omega-\omega_i)}\exp[\eta_i(b-r)] - \exp[\zeta(b-r)]\right\} \tag{4.131}$$

in which

$$\zeta(p) = \left(\frac{p+\omega}{\alpha}\right)^{1/2} \tag{4.132}$$

The inverse of Eq. (4.131) is not given by Perl who comments that it might be obtained by employing a standard Laplace transforms table. However, he develops the inversed expression for the power dissipated by the heat source which is given by:[79]

$$\gamma_b(t) = u_b(1+b\eta) + \left\{(u_b - u_{bi})\left(\frac{\pi\alpha t}{b^2}\right)^{-1/2} + \omega_{bi}b\eta_i \exp(\omega_i t)\,\mathrm{erfc}[(\omega_i t)^{1/2}] - u_b b\eta\,\mathrm{erfc}[(\omega t)^{1/2}]\right\}\exp(-\omega t) \tag{4.133}$$

Perl considers a number of limiting cases for Eq. (4.133).[79]

Case II Similar to Case I, Perl only gives the transformed solution,

$$\bar{u}(r,p) = \frac{\gamma_b \exp[\zeta(b-r)]}{rp(b^{-1}+\zeta)} + \frac{\gamma_{bi}}{r(p+\omega-\omega_i)}\left\{\frac{\exp[\eta_i(b-r)]}{(b^{-1}+\eta_i)} - \frac{\exp[\zeta(b-r)]}{(b^{-1}+\zeta)}\right\} \tag{4.134}$$

For this case the temporal temperature variations of the heat source is required. This expression is given by:

$$u_b(t) = \frac{\gamma_b}{1 + b\eta} - \left[\operatorname{erfc}\left(\frac{\alpha t}{b^2}\right) - b\eta \operatorname{erfc}(\omega t)\right]\frac{\gamma_b \exp(-\omega t)}{1 - b^2\eta^2}$$
$$+ \left[\operatorname{erfc}\left(\frac{\alpha t}{b^2}\right) - b\eta_i \operatorname{erfc}(\omega_i t)\right]\frac{\gamma_{bi} \exp(-\omega t)}{1 - b^2\eta_i^2} \tag{4.135}$$

This equation is plotted, for a constant power input, $\gamma_b = \gamma_{bi}$, in Figure 35. Perl concludes his study with the observation that the analysis presented above is compatible with measurements made by the Gibbs heater-thermocouple method. Other details, particularly on experiments performed by Perl and coworkers, are reported in subsequent papers.[80–82]

Stow, in a paper in 1965,[83] describes a method of measuring tissue blood flow. In this method a heat source, along with a temperature measuring device,

Figure 35 Theoretical transient responses of heat clearance probe at constant power input. Ordinate: ratio of probe temperature increment at time t to probe temperature increment at zero tissue perfusion. Abscissa: time, t. Occlusion transient is labeled 3.0 denoting $w_i = (\rho_b c_{pb}/\rho c_p)w_{bi} = 3$ and $w = (\rho_b c_{pb}/\rho c_p)w = 0$, (upper time scale). Initial reactive hyperemia transient is labeled 0, w for various values of w, (lower time scale). Final reactive hyperemia transient is labeled 30, 3 denoting $w_i = 30$, $w = 3$, (lower time scale). Asymptotic values of each curve labeled correspondingly at right. Initial slopes of 2.0 and 30.3 transients indicated by straight lines.[79]

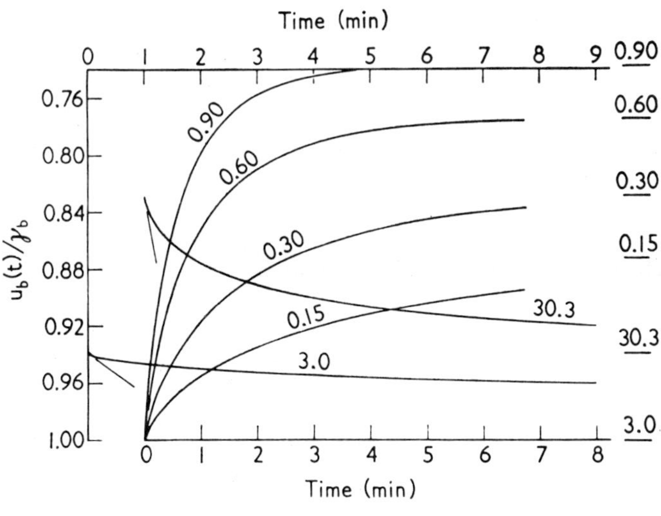

are applied externally to a given skin region. Another temperature measuring device is attached at a known distance and the transient response of both thermometers is continuously monitored. This response is correlated with results of heat transfer theory to obtain the approximate regional blood flow rate.

Eberhart, Jackson, and Trezek[84] applied a heat transfer model to determine regional blood flow rate. This method was developed for the purpose of obtaining information on the recovery of patients who have undergone open-heart surgery. According to Eberhart et al., body heat is conserved during the first post-operative hours by vasoconstriction in response to low hypothalamic and skin temperatures. Most of the body metabolic heat production and the natural decrease of heat loss in the extremities enhance elevation of core temperature. This increase continues until some temperature set-point is reached. At that instance the vessels in the extremities dilate and warm blood flows to these regions of the body followed by a significant temperature rise (3–10°C). This normal post-operative response was found, statistically, to be necessary for full recovery. The present work was undertaken to "*explore the basis*"[84] for this phenomenon. A secondary objective was to develop a "detailed heat transfer model to the indirect computation of regional blood flow, in the context of the general post-operative course of cardiac surgery patients."[84]

The bio-heat equation was assumed to model the heat transfer process in the great toe. This toe was chosen as the site for measuring and modeling blood flow effects in an extremity. Heat transfer in the toe is assumed to be in the radial direction only, an assumption which is partially supported by temperature measurements along the toe. Boundary conditions are as follows:

$$\left.\frac{\partial T}{\partial r}\right|_{@\, r=0} = 0 \qquad \text{(centerline symmetry)} \tag{4.136}$$

$$-k\left.\frac{\partial T}{\partial r}\right|_{@\, r=r_0} = h(T|_{@\, r=r_0} - T_i) \tag{4.137}$$

As an initial condition a constant temperature in the toe was specified

$$T|_{@\, t=0} = T_b \tag{4.138}$$

The solution to this set of equations is

$$\theta = \left(\theta_b + \frac{\dot{q}_m}{\dot{w}_b \dot{c}_b}\right)\left\{1 - \frac{hr_0 I_0(\beta^{1/2}R)}{k\beta^{1/2}I_1(\beta^{1/2}) + hr_0 I_0(\beta^{1/2})} - \frac{2\beta hr_0}{k}\sum_{m=1}^{\infty}\frac{\exp[-\alpha/r_0^2(\lambda_m^2+\beta)t]J_0(\lambda_m R)}{(\lambda_m^2+\beta)[(hr_0/k)^2+\lambda_m^2]J_0(\lambda_m)}\right\} \tag{4.139}$$

where $\theta = T - T_a$ (4.140)

$$R = \frac{r}{r_0} \tag{4.141}$$

β is given by Eq. (4.76), r_0 is the radius of the cylinder depicting the toe, and λ_m is an eigenvalue given by

$$\lambda_m J_1(\lambda_m) - \frac{hr_0}{k} J_0(\lambda_m) = 0 \tag{4.142}$$

Equation (4.139) may be evaluated for any values of the parameters present in it, e.g., β. Agreement between experimental and analytical data, for known values of T_b and h, may be expected if the proper value for the blood perfusion rate, $\dot{w}_b$, were chosen. However, it was found that treating $\dot{w}_b$ as a step function with time does not yield surface temperature profiles which are in agreement with experimental data. This observation is a little surprising in view of the physiological description of the sudden dilation of blood vessels leading to the toe, around which the analysis is presented. It might be argued, in favor of Eberhart, Jackson, and Trezek's work, that their experiments were conducted on post-operative subjects who probably did not exhibit the same phenomena as might be expected in healthy subjects whose blood flow to the toe was temporarily occluded.

To circumvent this discrepancy the authors introduce an iterative procedure which enables the extension of Eq. (4.139) for variable blood flow rate.[84] This is done by holding blood flow rate constant over a time interval, typically 500 sec, and then increasing (or decreasing) it to the new level where the process is repeated. At each time interval a new approximate initial condition for the temperature of the toe is assumed, namely a parabola,

$$\theta|_{@\, r=0} = T_c - ar^2 \tag{4.143}$$

with this modification Eq. (4.139) becomes

$$\theta = \left(\theta_b + \frac{\dot{q}_m}{\dot{w}_b c_b}\right)\left\{1 - \frac{hr_0 I_0(\beta^{1/2} R)}{k\beta^{1/2} I_1(\beta^{1/2}) + hr_0 I_0(\beta^{1/2})}\right.$$
$$\left. + \sum_{m=1}^{\infty} A_m J_0(\lambda_m R) \exp\left[-\frac{\alpha}{r_0^2}(\lambda_m + \beta)t\right]\right\} \tag{4.144}$$

$$\text{where } A_m = \frac{2hr_0\,[T_c(\lambda_m^2 + \beta) - (\theta_b + \dot{q}_m/\dot{w}_b c_b)(1 - 4\lambda_m^2)(\lambda_m^2 + \beta)]}{k(\lambda_m^2 + \beta)[(hr_0/k)^2 + \lambda_m^2]J_0(\lambda_m)} - \frac{4ak(\lambda_m^2 + \beta)}{k(\lambda_m^2 + \beta)[(hr_0/k)^2 + \lambda_m^2]J_0(\lambda_m)} \tag{4.145}$$

This infinite series was found to have a rapid rate of convergence so that only 5–6 terms are required to yield a value of temperature within 10^{-7}.

The model developed was tested on three cases. A typical result is shown in Figure 36. In this figure toe surface temperatures are shown in a post-operative subject exhibiting the typical step rise in toe temperature. Toe blood flow, as measured by a venous occlusion plethysmography, is compared with two analytical curves computed by Eq. (4.144). These two curves, which yield reasonably good agreement, are calculated by changing the heat transfer coefficient as indicated in the figure. Other cases studied gave worse agreements between theoretical and experimental data.

Figure 36 Comparison of analytical and experimental blood flow rates for case I.[84]

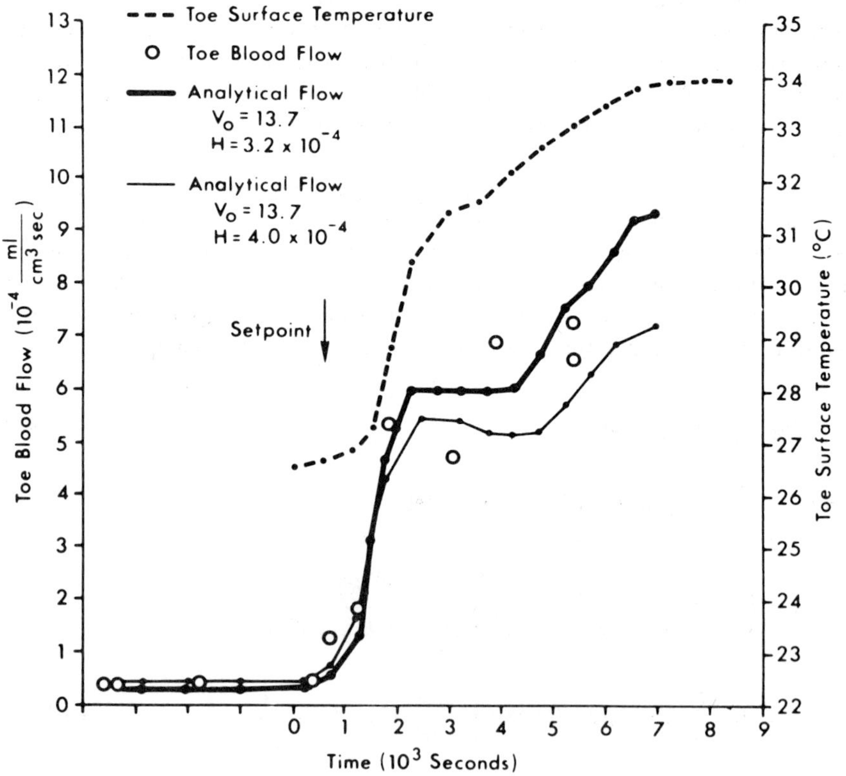

The authors conclude this study by stating that their model predicts reasonable values of time-varying blood flow for patients with normal post-operative rewarming patterns.

4.6 Thermographic and infrared studies

> The use of the scanning infrared camera in the detection of breast cancer places emphasis on the observation of detailed thermal patterns on the female thorax. These patterns are associated with the basic tissue temperature and the vascular patterns caused by veins which course near the surface of the skin. The vein temperatures are related to the average tissue temperature in the region drained by the vein. Thus both a prediction of skin temperature and average tissue temperature are of interest.[85]

This paragraph, excerpted from a paper presented by Love, Haberman, and Francis, introduces the basic need for an analytical tool by which the diagnostic physician might be able to estimate physiological variables.

Indeed, thermographic studies have been used to measure, non-invasively, surface temperatures.[86] In the paper by Love et al.,[85] a model of heat transfer is developed for the purpose of relating physiological parameters to convective and radiative heat exchange with the environment. The modeling equation is assumed to be the one dimensional, steady-state, bio-heat equation with the following boundary conditions:

$$T|_{@\, y=0} = T_{\text{art}} \tag{4.146}$$

where the origin of the coordinate system is chosen at a depth $y = L$ below the surface of the skin such that the local temperature is equal to core, or arterial temperature.

$$-k \left.\frac{dT}{dy}\right|_{@\, y=L} = \phi h(T|_{@\, y=L} - T_{\infty}) \tag{4.147}$$

where ϕ is a " 'view factor' which varies from 0 to 1 to account for reduced heat transfer from protected regions such as under the neck."[85]

Since L is not a known quantity, the authors elect to specify a condition of no heat flux at the core, namely

$$\left.\frac{dT}{dy}\right|_{@\, y=0} = 0 \tag{4.148}$$

which will enable the elimination of L. This assumption does not seem accurate, however, since there is exchange of heat by conduction from the core to overlying tissue due to temperature difference which normally exists.

The solution to the above set was obtained:

$$\theta = Q \cosh(m\xi) + \frac{B[1 - Q\cosh(m) + Q - Qm\sinh(m)/B]}{m\cosh(m) + B\sinh(m)} \sinh(m\xi) - Q \quad (4.149)$$

$$\text{where } \theta = \frac{T_{\text{art}} - T}{T_{\text{art}} - T_\infty} \quad (4.150)$$

$$Q = \frac{\dot{q}_m}{\dot{w}_b c_b (T_{\text{art}} - T_\infty)} \quad (4.151)$$

$$\xi = \frac{y}{L} \quad (4.152)$$

$$m = \left(\frac{\dot{w}_b c_b L^2}{k}\right)^{1/2} \quad (4.153)$$

$$B = \frac{\phi h L}{k} \quad (4.154)$$

The additional relationship which eliminates L is given by:

$$m\sinh(m) + \cosh(m) = 1 + \frac{1}{Q} \quad (4.155)$$

From Eqs. (4.149) and (4.155) the skin surface temperature and the average skin temperature, both at $\xi = 1$, may be calculated. A typical plot of Eq. (4.155) is shown in Figure 37. This figure establishes the depth below the skin surface at which the local temperature equals that of the core. A typical plot of Eq. (4.149), for a fixed value of Q, is shown in Figure 38. Other plots, including a number for average skin temperature are given in the article by Love et al.[85]

These investigators employed a thermovision scanning infrared camera to obtain experimental data on skin temperatures of a female thorax. Typical pictures for a normal breast are shown in Figure 39. In these thermograms bright areas represent isothermal regions. The bright bar on the lower scale represents the camera isotherm setting.

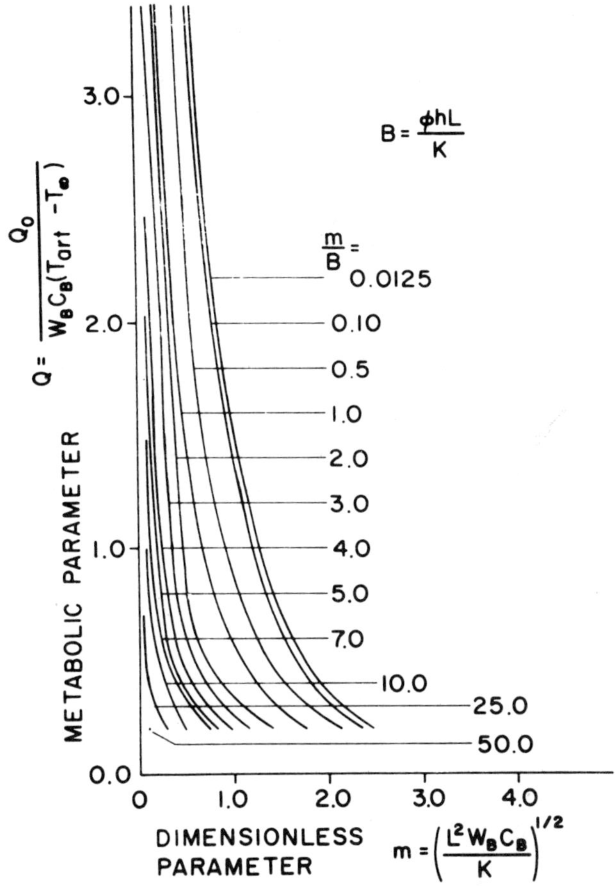

Figure 37 Metabolic parameter as a function of m and m/B.[85]

Note the cooler temperatures represented by the fatty tissues of the breast with increased temperatures in the inframammary fold and the axillary. The medial and superior abdomen also represent slightly higher temperatures since the tissues in these parts are primarily muscle and bone, which have higher thermal conductivities. The regions where veins course near the surface are at elevated temperatures typical of the average tissue temperature.[85]

The surface temperatures shown in these thermograms were compared to the analytical model. Comparison was made by first selecting thermal property values for the tissue to conform to common values. Next, the various parameters, i.e., Q, m, B, and ϕ were estimated. By varying the viewfactor ϕ, agreement between analytically predicted and experimentally determined data could be found. This is undoubtedly one of the shortcomings of this method since a wrong selection of ϕ may throw the result off by tens of percents. Extension of the model to two, or maybe even three, dimensions may also improve its validity in helping the diagnostic physician design and interpret thermograms for detection of abnormal skin temperature patterns.

An investigation by infrared thermometry of the changes in human skin temperature as a function of ambient temperature is presented by Gros and Gautherie.[87] These investigators employed a one-dimensional, time dependent heat transfer model with no allowance for blood convection or metabolic heat generation. Assuming constant thermal properties and an isotropic and homogeneous tissue, they specified a constant uniform temperature in the skin initially. Boundary conditions were a constant ambient temperature and convective heat exchange with the environment. An analytical solution to the problem, in terms of error functions and exponents, was obtained and compared to results obtained by an infrared bolometer. Agreement between model predictions and experimental data is quite good in spite of the relative simplicity of the analytical approach. It may be expected that a number of modifications in the modeling equation, e.g., inclusion of blood perfusion effects, and the other assumptions may improve the applicability of such a model to predict skin temperatures.

Figure 38 Temperature distribution as a function of the parameter m and Biot number for $Q = 0.05$.[85]

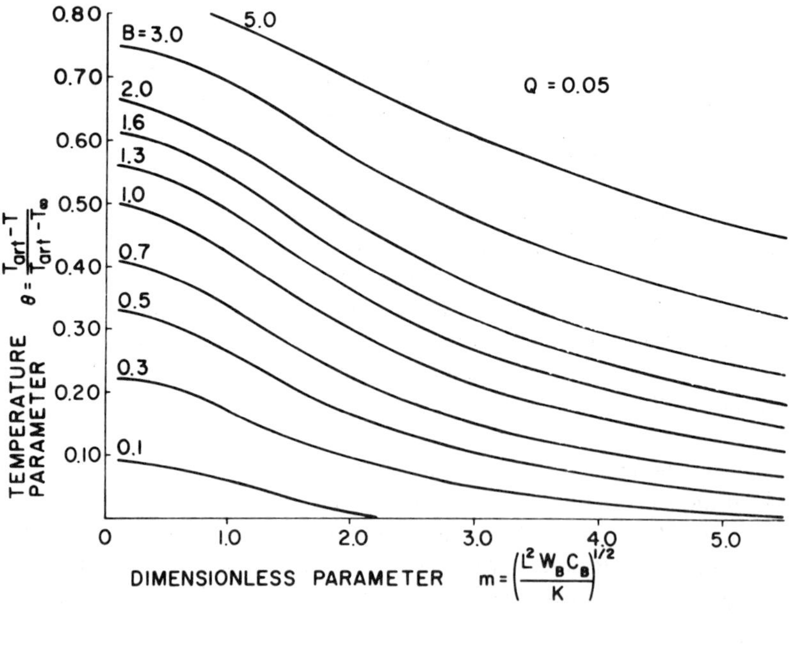

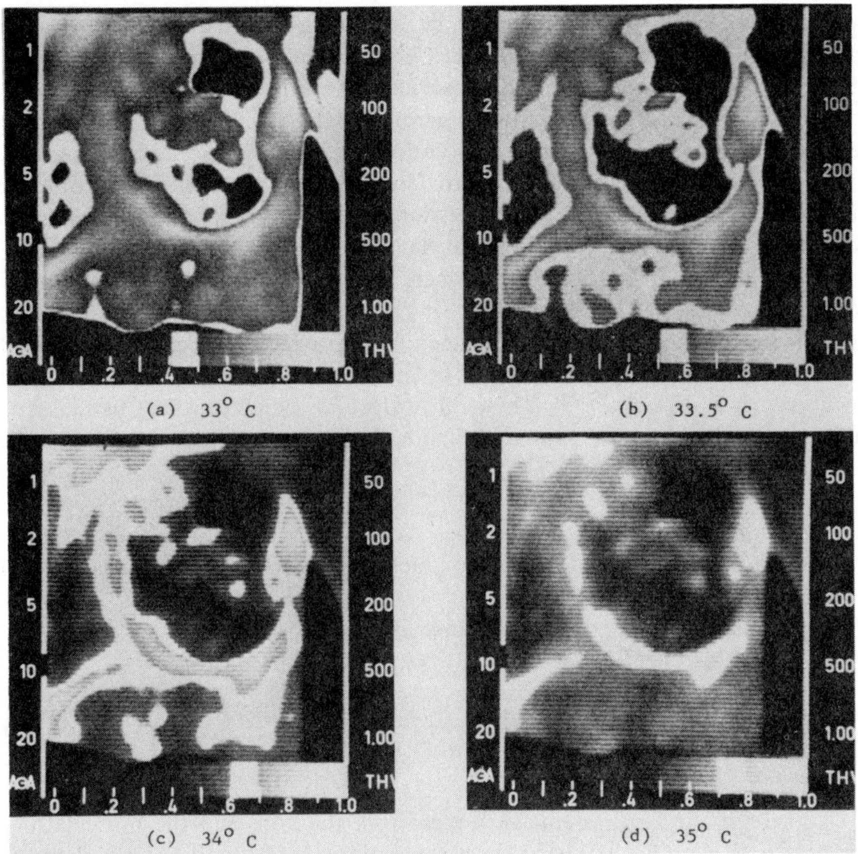

Figure 39 Thermogram pictures of selected isotherms for an avascular breast. Room temperature 19.8°C.[85]

4.7 Thermal interaction with external devices

As indicated above, in all bio-heat transfer studies, consideration must be given to energy exchange with the environment in which the mammal lives. One specific area of the broad subject, particularly applicable to humans, is the thermal interaction with artificial apparel. Under this category, regular clothing and fluid cooled garments may be listed. Clothing is basically a passive element, inasmuch as heat transfer is concerned, whereas, fluid cooled garments may be classified active for this purpose. Therefore, the thermal interaction with these two distinct external devices is considered separately.

4.7.1 Heat exchange with fluid cooled garments In recent years man has been venturing into thermally hostile environments, e.g., space. These environments pose extreme conditions that cannot be handled by the natural thermoregulatory system. Thus, artificial means must be employed to augment the capability of the natural system such that these environments become habitable. First applications of fluid, i.e., air, cooled garments which achieve this goal were in aeronautics. Subsequently, space exploration required the development of more sophisticated garments, capable of providing the astronaut with the complete microclimate he needs to fulfill his missions. A rather comprehensive review on liquid cooled garments is given by Nunneley.[88]

Commonly, a liquid cooled garment is constructed of a parellel, flexible network of tubes, in direct contact with the skin. Conditioned water is pumped through the tubes and serves as the medium with which exchange of body heat takes place. It appears that the first to model this heat transfer process were Buchberg and Harrah.[89,90] These investigators depicted the tissue as a multilayered slab model as shown in Figure 40. The tissue layers are, from inside outward:

a. A constant temperature central zone.
b. Musculature zone, the thermal conductivity of which varies linearly with metabolism.
c. Functional periphery, the conductance of which is given by

$$k_{fp} = \left[\frac{q_c + q_m}{A} + \frac{Q_{fp} L_{fp}}{2}\right] \frac{L_{fp}}{T_R - T_f} \tag{4.156}$$

where q_c and q_m are heat fluxes from the core and the musculature, respectively, Q_{fp} is internal heat generation of the functional periphery, L

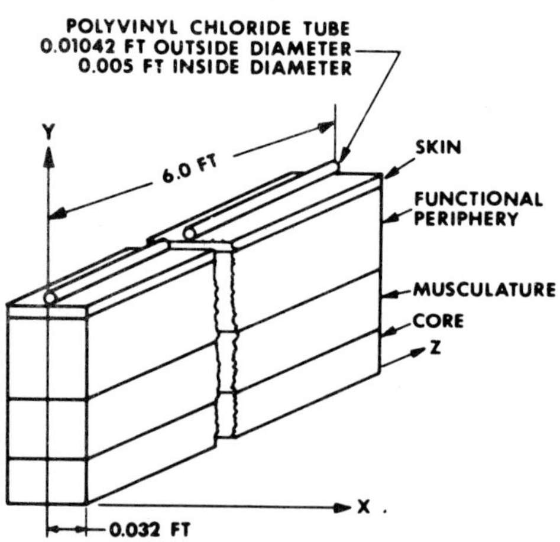

Figure 40 Multilayered slab model.[89]

Table 10 Physical Properties of the Biothermal Slab Model[89]

Zone	Area (ft^2)	Volume (ft^3)	Thickness (ft)	Mass (lb)	Density (lb/ft^3)
Skin	15.4	0.125	0.0081	8.25	66.1
Functional periphery	15.4	1.000	0.0650	64.4	64.3
Musculature	15.4	0.721	0.0468	47.4	66.1
Core	15.4	0.514	0.0334	34.6	66.1

is length, T_R and T_f are temperatures of the boundaries between musculature and functional periphery, and functional periphery and skin zone, respectively.

d. Skin zone with constant properties and no heat generation. Physical properties of the slab model are shown in Table 10.

Additional assumptions are as follows:

1. Heat exchange with the environment occurs only along the area of contact with the cooling tubes.
2. Heat transfer parallel to the axis of the tubes is negligible as compared to the directions perpendicular to it. Thus the problem becomes two dimensional.
3. Due to the repetitive pattern of the cooling tubes and their regular spacing only a solution to a representative section of the slab model is required.
4. The interface between the musculature and the functional periphery zones is maintained at a constant temperature, T_r, which is a function of core temperature and metabolic rate. This assumption cannot be justified *a priori,* however, since at elevated metabolic rates most of the heat is generated in the muscle zone.

Assuming the x axis to coincide with the interface between the musculature and the functional periphery and the y axis to pass through the center of a cooling tube, the steady-state problem may be stated as follows:

skin zone:

$$\frac{\partial^2 T}{\partial x^2} + \frac{\partial^2 T}{\partial y^2} = 0 \tag{4.157}$$

functional periphery zone:

$$\frac{\partial^2 T}{\partial x^2} + \frac{\partial^2 T}{\partial y^2} = -\frac{Q_{fp}}{k_{fp}} \tag{4.158}$$

with the boundary conditions:

$$\left.\frac{\partial T}{\partial x}\right|_{@\ x=0} = 0 \tag{4.159}$$

$$\left.\frac{\partial T}{\partial x}\right|_{@\ x=a} = 0 \tag{4.160}$$

$$T|_{y=0} = T_R \tag{4.161}$$

$$\left.\frac{\partial T}{\partial y}\right|_{@\ y=b} = \begin{cases} \text{const.}, 0 < x < \beta a \\ 0 \quad , \beta a < x < a \end{cases} \tag{4.162}$$

The authors employed a numerical finite difference method of successive over-relaxation to solve the problem. Typical results of the temperature distribution in the representative cross-section of the slab model are shown in Figure 41. This figure shows the isotherms for a high metabolic rate (760 w) and an area factor $f_A = 0.061$. This factor is defined as the ratio of the contact area to the total outside tube area. It is seen that for this high metabolic rate, skin surface temperatures are too low (10°C) to be comfortable. The authors further calculated the mean contact surface temperature as a function of metabolic rate and contact surface area. They also estimated inlet water temperatures.

Buchberg and Harrah's work seems to lack in a number of aspects. First of these is the fact that no allowance was made in their work for convection by the blood stream. This mechanism has already been demonstrated to be very important in the *in vivo* heat transfer processes and even a crude first approximation which includes this mechanism explicitly is required. Second, as was already mentioned above, the assumption of the musculature zone as a zone of constant temperature does not seem valid. At elevated metabolic rates most of the heat is generated in this region entailing high temperature gradients. Third, it appears that in the original version of this work the mathematical problem was over-specified; six boundary conditions were specified to a second order, two dimensional problem.[89] It is not clear how this condition was satisfied by the numerical scheme and although in a subsequent paper the authors amended this condition,[90] the fact remains that their results are presented for this mathematically invalid problem. Fourth, Buchberg and Harrah present their analysis for a steady-state, whereas in reality, transient states more closely describe the behavior of an active living system.

Shitzer and Chato addressed themselves to these and other problems.[91,93–100] Utilizing essentially the same geometry as in Ref. [89], they first considered the effects of blood flow on the temperature distributions in the living tissue.[91] They assumed two regions to represent the tissue; skin and

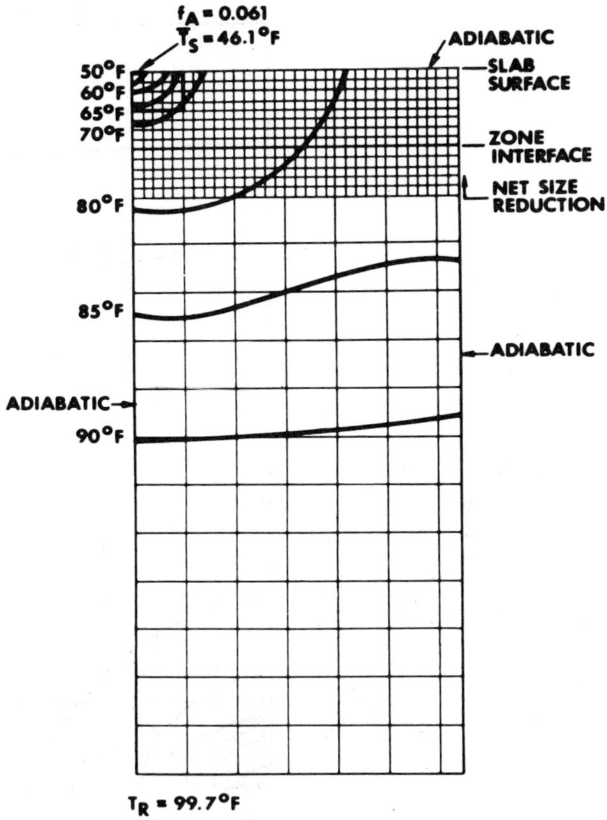

Figure 41 Diagram of the finite difference grid showing the selected net spacing and the temperature field for $f_A = 0.061$ and a metabolic rate of 2600 Btu/hr.[89]

skeletal muscle. This region is a combination of Buchberg and Harrah's functional periphery and musculature. Also the two regions were assumed homogeneous and isotropic and possessing uniform and constant but different properties. Analytical solutions were obtained for two cases:

1. A constant temperature at the interface with the inner core.
2. A constant heat flux at the interface with the inner core.

Results were compared to the previous work where blood flow effects were not included and are shown in Figure 42. It is seen that surface temperatures are significantly higher for this case (15°C vs. 8°C). These temperatures are also closer to values found in the literature for the assumed metabolic rate and are more compatible with the thermal comfort of the wearer

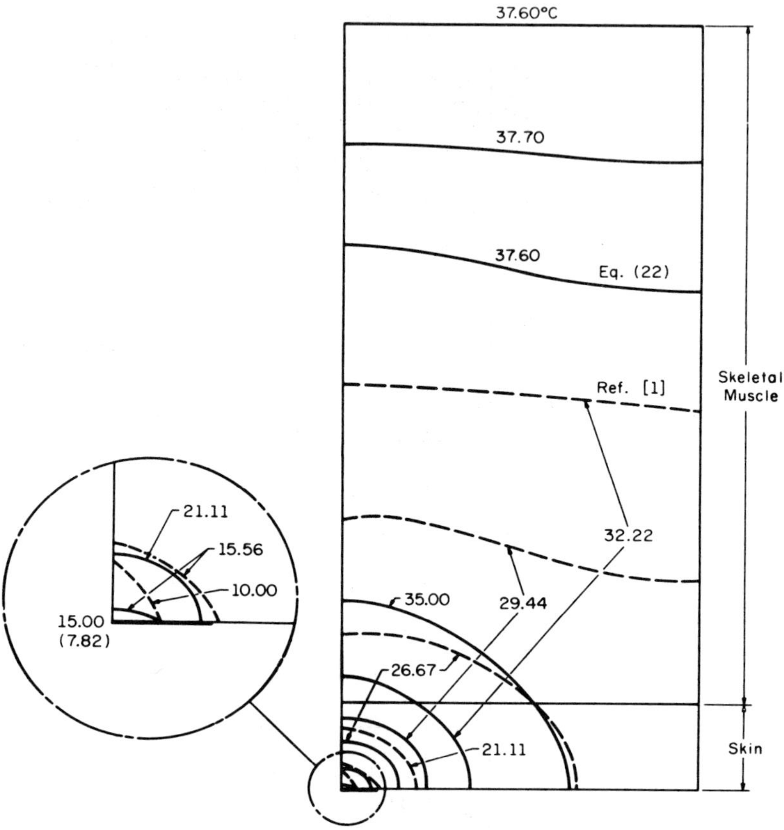

Figure 42 Temperature distribution in the separated tissue for the rectangular model. Dashed curves indicate temperature distribution obtained by Buchberg and Harrah[89] without blood flow.[91]

of the cooled garment. Also, a maximum temperature is found to occur inside the tissue when blood flow effects are included in the modeling equation. The existence of this maximum temperature inside the tissue was observed experimentally in a working muscle by Saltin et al.[92] Were it not for convective transport by the circulating system, the appearance of a maximum temperature in the tissue would mean heat flow into the inner core with subsequent accumulation of heat and inevitable collapse. Indeed, when blood perfusion rate is diminished, i.e., $w_b \rightarrow 0$ in the steady-state bio-heat equation, the maximum temperature is found on the boundaries resulting in sharper temperature gradients and lower temperature levels in the tissue in order to sustain removal of the same amount of heat.[93] Location of this maximum temperature, as a function of the various thermophysical parameters, is discussed in Ref. [94].

Steady-state temperature distribution in living tissue modeled as cylindrical shells was considered by Shitzer and Chato.[95] The basic configuration for which the analysis was performed is shown in Figure 43. Assuming angular symmetry and other assumptions as were made before,[91] an analytical solution to the problem was obtained.[95] Results obtained were essentially similar to those of the rectangular slab due to the inherently identical geometry of the representative cross-sections. Comparison of results for these two geometries is shown in Figure 44. It was shown that for some "critical" internal radius of the cylindrical model, $R_1 \rightarrow 7$ cm, results obtained from the easier for calculation rectangular model, are essentially identical to those predicted by the mathematically more accurate cylindrical model. Another case of cylindrical geometry was solved analytically by assuming the tubes to run on the skin parallel to the axis of the cylinder.[95] Results are shown in Figure 45.

Leo et al. attempted to partially validate the analytical predictions of both the rectangular and cylindrical models.[96] They constructed a number of cooling pads with parellel cooling tubes and tested them on the thigh of an exercising individual. Temperatures of the skin between adjacent cooling tubes were monitored and then compared to the analytical results. A typical result of this study is shown in Figure 46. Considering the uncertainty existing in the estimation of the various parameters, agreement between experimental and theoretical data is quite encouraging. Further experimental validation of the model, particularly in depth measurement, have to be delayed for the development of non-interfering measuring techniques.

Shitzer and Chato also considered the more important problem of transient heat transfer in living tissue.[97] This problem was solved analytically for both the rectangular and cylindrical geometries. Following is a presentation of

Figure 43 Representative section of cylindrical model.[99]

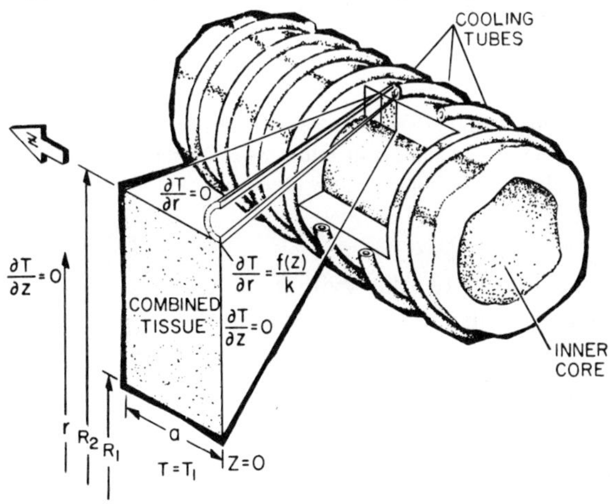

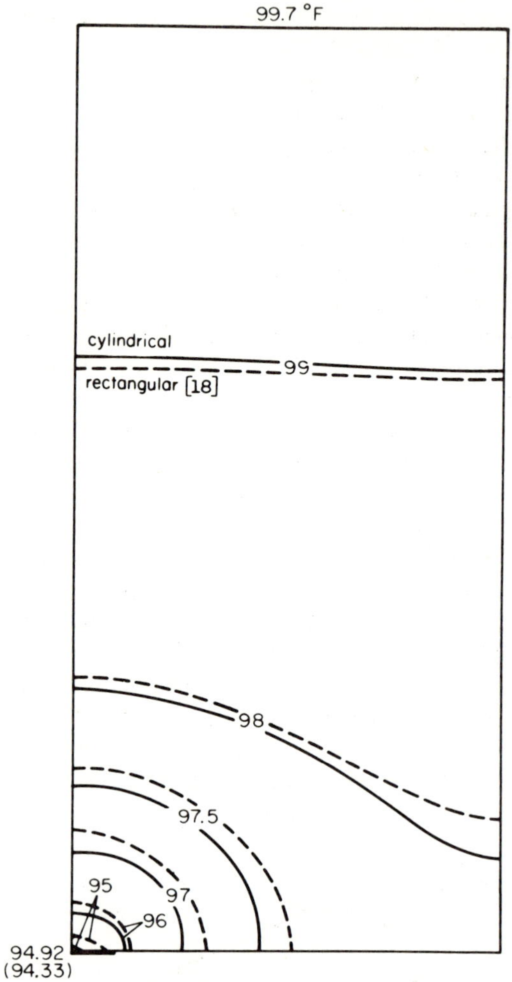

Figure 44 Comparison between the steady-state, two-dimensional analytical solutions for rectangular and cylindrical models for low metabolic rate, 85 watt.[95]

the solution to the cylindrical geometry as depicted in Figure 43. For a homogeneous, constant, and uniform properties region, for which angular symmetry is assumed, the problem may be stated as follows:

modeling equation:

$$\frac{1}{\alpha}\frac{\partial \theta}{\partial t} = \nabla^2\theta - w_2{}^2\theta + Q_2 \tag{4.163}$$

where $\theta = T - T_{\text{core}}$ (4.164)

$$w_i^2 = \frac{\dot{w}_{b,i} c_b}{k} \tag{4.165}$$

$$Q_i = \frac{\dot{q}_{m,i}}{k} + \frac{\dot{w}_{b,i} c_b (T_{\text{art}} - T_{\text{core}})}{k} \tag{4.166}$$

boundary conditions:

$$\left.\frac{\partial \theta}{\partial r}\right|_{@\ r=R_2} = \begin{cases} -f_2(z)/k & @\ 0 < z < \beta a \\ 0 & @\ \beta a < z < a \end{cases} \tag{4.167}$$

$$\theta|_{@\ r=R_1} = 0 \tag{4.168}$$

Figure 45 Comparison of analysis (lines) with experimental results (circles) for cooling pad No. 2. Metabolic rate 528 watt.[96]

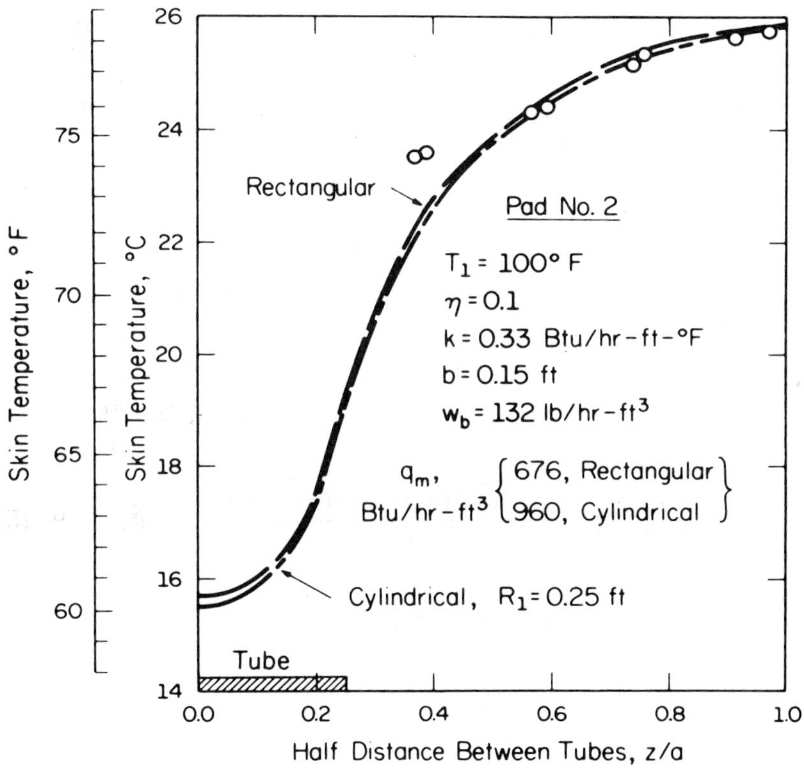

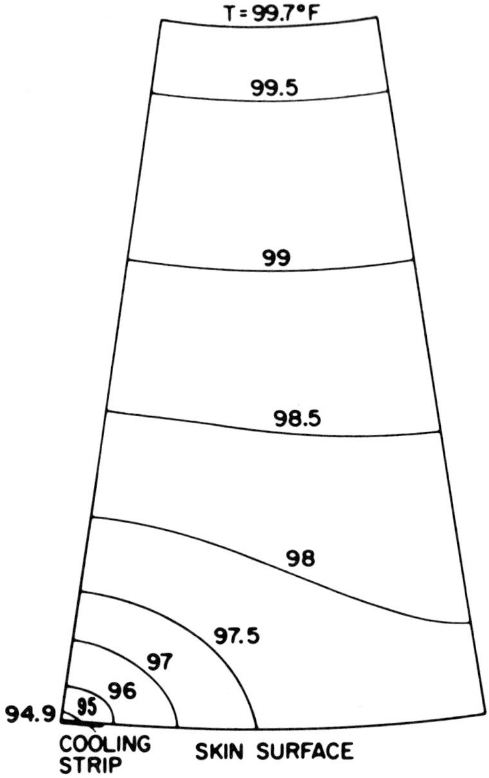

Figure 46 Steady-state temperature distribution in the combined tissue for cylindrical model with tubes on the skin running parallel to cylinder axis for low metabolic rate, 85 watt.[95]

$$\left.\frac{\partial\theta}{\partial z}\right|_{@\ z=o} = 0 \tag{4.169}$$

$$\left.\frac{\partial\theta}{\partial z}\right|_{@\ z=a} = 0 \tag{4.170}$$

initial condition (the steady-state solution to Eq. [4.163])[95]

$$\theta|_{@\ t<0} = Q_1\left[\frac{1-\psi_{01}{}^2(w_1 r)/\psi_{01}{}^2(w_1 R_1)}{w_1{}^2}\right] - \frac{\beta f_{a1}\psi_{00}{}^1(w_1 r)}{k w_1 \psi_{01}{}^2(w_1 R_1)} - \sum_{n=1}^{\infty}\left[\frac{\alpha_{n,1}\psi_{00}{}^1(\zeta_1 r)}{\zeta_1\psi_{01}{}^2(\zeta_1 R_1)}\right]\cos(\lambda_n z) \tag{4.171}$$

where $\psi_{km}{}^{j}(\zeta_i r) = I_k(\zeta_i r)K_m(\zeta_i R_j) - (-1)^{k+m} I_m(\zeta_i R_j)K_k(\zeta_i r)$ (4.172)

where β is the ratio of effective width of cooling tube to cooling tube spacing, and,

$$f_{a,i} = \frac{1}{\beta a}\int_0^{\beta a} f_i(\xi)d\xi \tag{4.173}$$

where a is half the distance between cooling tubes, $f_i(\xi)$ is the heat flux function specified at the skin-cooling tube interface, ξ is the dummy variable of integration, and

$$\alpha_{n,i} = \frac{2}{ak}\left[\int_0^{\beta a} f_i(\xi)\cos(\lambda_n \xi)d\xi\right] \tag{4.174}$$

where $\lambda_n = n\pi/a$, and

$$\zeta_i = \lambda_n^2 + w_i^2 \tag{4.175}$$

The solution to the above set is given by:

$$\begin{aligned} T(r,z,t) = {} & T_{\text{core}} + \frac{Q_2}{w_2^2}\left[1 - \frac{\psi_{01}{}^2(w_2 r)}{\psi_{01}{}^2(w_2 R_1)}\right] - \frac{\beta f_{a,2}}{kw_2}\,\frac{\psi_{00}{}^1(w_2 r)}{\psi_{01}{}^2(w_2 R_1)} \\ & + \sum_{n=1}^{\infty} \frac{\alpha_{n,2}\,\psi_{00}{}^1(\zeta_2 r)}{\zeta_2\,\psi_{01}{}^2(\zeta_2 R_1)}\cos(\lambda_n z) \\ & + \frac{2\pi}{k}\sum_{n=1}^{\infty}\left\{\frac{2[(q_{m,2}/\epsilon_2^2) - (q_{m,1}/\epsilon_1^2)]}{\Omega_n^2(\mu_n R_2) - 4}\right. \\ & \left. - \frac{\mu_n\Omega_n(\mu_n R_2)[(\beta f_{a,2}/\epsilon_2^2) - (\beta f_{a,1}/\epsilon_1^2)]}{\Omega_n^2(\mu_n R_2) - 4}\right\} \\ & \cdot \chi_n(\mu_n r)\exp(-\alpha\epsilon_2^2 t) + 2\pi \sum_{m=1}^{\infty}\sum_{n=1}^{\infty}\left[\frac{\alpha_{m,2}}{\zeta_2^2 + \mu_n^2}\right. \\ & \left. - \frac{\alpha_{m,1}}{\zeta_1^2 + \mu_n^2}\right]\frac{\mu_n\Omega_n(\mu_n R_2)}{\Omega_n^2(\mu_n R_2) - 4} \\ & \cdot \chi_n(\mu_n r)\cos(\lambda_m z)\cdot \exp[-\alpha(\epsilon_2^2 + \lambda_m^2)t] \end{aligned} \tag{4.176}$$

where $$\chi_n(\mu_n r) \equiv J_0(\mu_n r) Y_0(\mu_n R_1) - J_0(\mu_n R_1) Y_0(\mu_n r) \tag{4.177}$$

$$\Omega_n(\mu_n R_2) \equiv \pi \mu_n R_2 \chi_n(\mu_n R_2) \tag{4.178}$$

and μ_n's are the roots of

$$J_0(\mu_n R_1) Y_1(\mu_n R_2) - J_1(\mu_n R_2) Y_0(\mu_n R_1) = 0 \tag{4.179a}$$

and

$$\epsilon_i^2 = \mu_n^2 + w_i^2 \tag{4.179b}$$

Equation (4.176) was evaluated for one- and two-dimensional cases. Results of the one-dimensional case, and for assumed steps in metabolic rate from high to low and reversed, are shown in Figure 47. The observed transient response of the temperatures in the tissue may be divided into two time domains:

1. A relatively short time, of about 5 min., where most of the change takes place.
2. A longer time, dependent on the direction of the change in activity, where very little changes in tissue temperatures occur.

It should be noted that the model presented here assumes step changes in all physiological parameters. This assumption may be deemed valid only as a first approximation. Extension of the model to include arbitrary variations in these

Figure 47 Transient temperature distributions in the one-dimensional, cylindrical model. Step changes are assumed from low (85 w) to high (760 w) metabolic rate for left figure and reversed for right figure.[97]

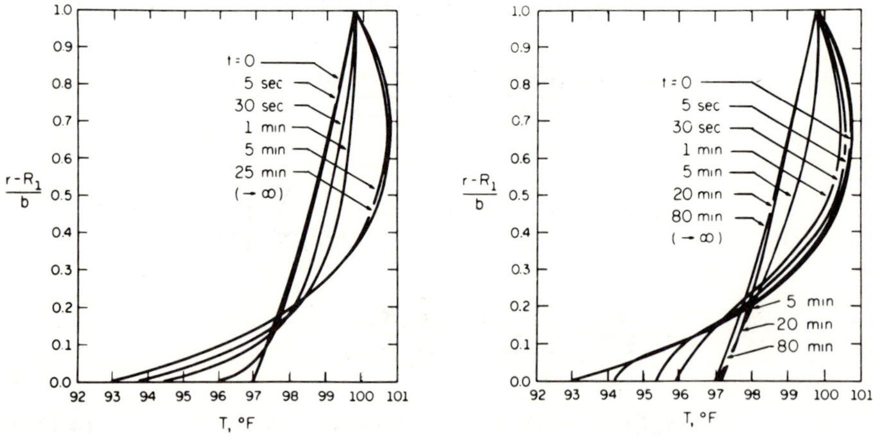

parameters is given in Ref. [97]. Nevertheless, model predictions conform well with scant observed data.

Extensions of this model to include variations in blood perfusion temperature were made by Shitzer et al.[98,99] As discussed in Chapter 3, the temperature of the blood perfusing the tissue is commonly assumed to be constant at arterial temperature. However, this assumption is incorrect, particularly at a distance away from an artery or arteriole. It seemed reasonable, therefore, to attempt to solve the bio-heat equation for a variable arbitrary temperature of blood perfusing the tissue. Thermal equilibrium between leaving venous blood and surrounding tissue was still assumed. Details of the analytical solutions obtained for cylindrical, as well as for rectangular geometries, are given in Ref. [99]. Only one representative result is presented here. In this example, an artery is assumed to bc located at a predetermined depth in the tissue below the cooling strip. Blood temperature is then assumed to vary exponentially in the direction parallel to the skin and parabolically in the direction perpendicular to it with the assumed artery as the center. Mathematically, blood temperature is given by:[99]

$$T_b = T_{\mathrm{art}} \left[1 - \left(\frac{y}{b} - \epsilon \right)^2 \right] \exp\left(- \frac{\gamma x}{a} \right) \tag{4.180}$$

where ξ and γ are numerical parameters, a is half the distance between the adjacent cooling tubes, and b is the depth of the tissue. A typical plot of the temperature distribution in the tissue under these assumptions is shown in Figure 48. Dashed lines in this figure depict the assumed variations in entering blood temperature. Solid lines show tissue temperature changes. Again, maximum temperature occurs inside the tissue. Due to the choice of blood temperature variations, this maximum temperature is at a point (mathematical) instead of at a line.

The heat transfer from a blood vessel near the skin cooled by a symmetrical liquid perfused strip was considered by Chato and Shitzer.[100] This problem is of significance in cases where localized, rather than overall, cooling is employed, e.g., head and neck cooling. Recent studies have indicated the significant effect on human thermal comfort resulting from cooling the skin of the neck about the carotid artery. Therefore, an analytical study estimating the amount of heat extracted from the artery by the cooling patch was undertaken.

Generally, assumptions made were similar to those in Ref. [93]. Also, based on a previous analysis of the cylindrical geometry,[95] this present problem was formulated and analyzed in rectangular coordinates. A typical steady-state temperature distribution in the tissue is shown in Figure 49. It was found that the optimum width of a cooling strip is about 3 times the depth of the centerline of the blood vessel. Also, the heat extracted from a blood vessel similar to the

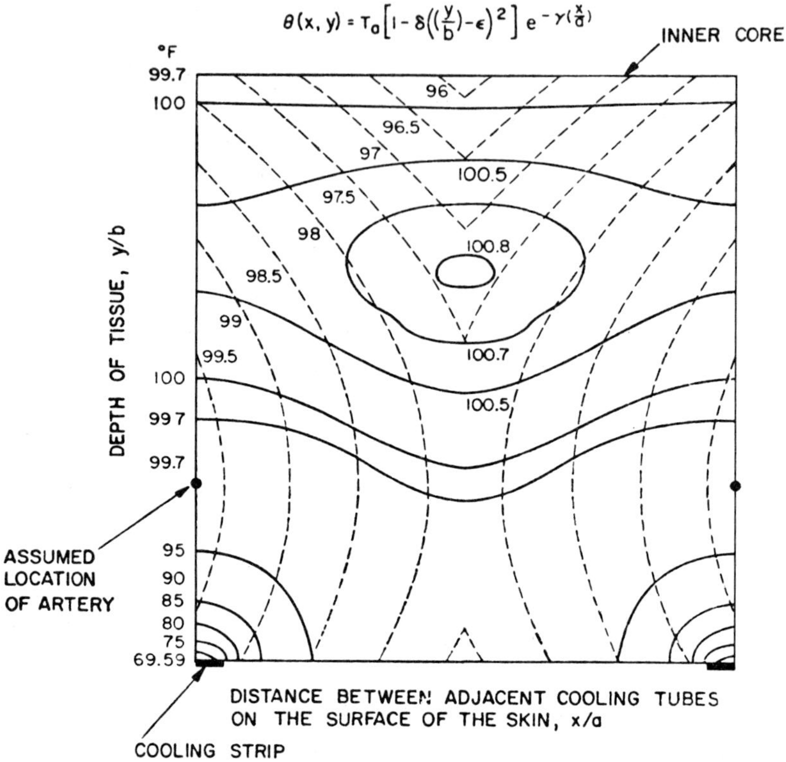

Figure 48 Temperature distribution in the tissue (rectangular model) for high metabolic rate (760 w) and specified blood temperature.[99]

Figure 49 Dimensionless temperature distribution in the vicinity of a cooled artery under assumed "normal" conditions.[100]

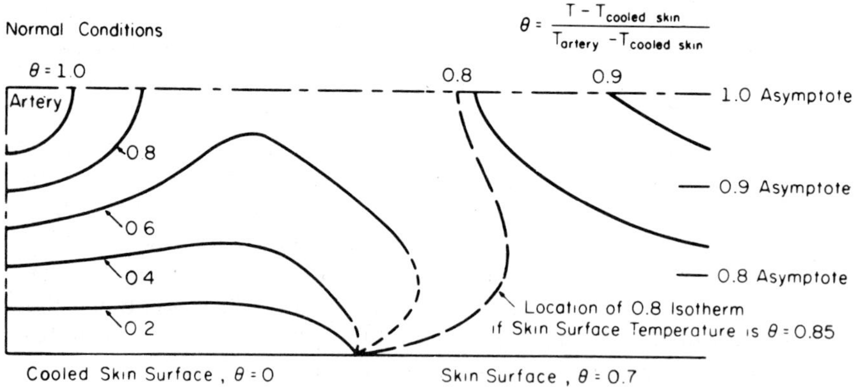

carotid artery cooled by such a strip was calculated to be too small to significantly affect the temperature of the blood flowing in it.

It appears that these models are good first approximations to the thermal behavior of living tissue interacting with an external liquid cooled garment.

Further improvements may be in allowing variable properties in the modeling equation. This improvement, however, must await the availability of realistic detailed, and accurate *in vivo* thermophysical data.

4.7.2 Heat exchange with clothing Humans wear clothes under most normal conditions. It is, therefore, obvious that the transfer of heat and moisture from the skin through the clothing to the environment is an important factor in thermal comfort. As discussed previously, removal of body heat is accomplished primarily by convection and radiation in low activity levels. Evaporation of perspiration is employed at intermediate and high metabolic rates and also in environments which pose high heat stresses.

Ma and coworkers have undertaken to investigate the complex problem of clothing-airspace-skin heat and mass transfer.[101] These investigators considered the steady-state problem in a one-dimensional model composed of five layers. This model is depicted in Figure 50. Other assumptions made are:

a. The body may be considered as an infinite slab. This assumption amounts to no more than a first approximation since conditions on the skin vary from one location to another.
b. The thermal conductivity of the skin is a function of average body temperature which is a function of time alone. This assumption probably stems from the fact that no blood flow effects are included in the model. Thus, as is sometimes customary in such cases, cf. Buchberg and Harrah,[89]

Figure 50 Model of fabric-airspace-skin system.[101]

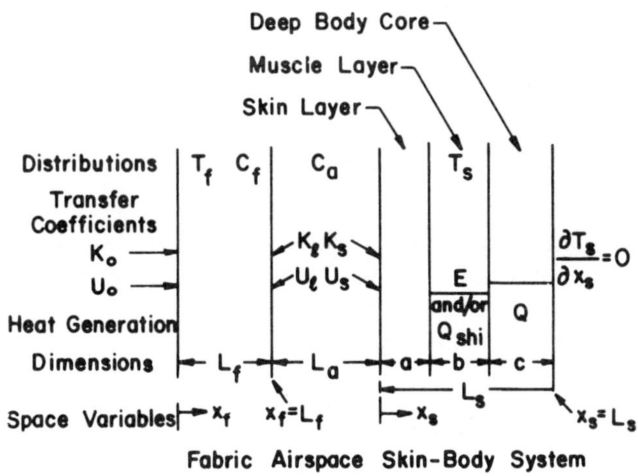

effects of the circulating blood are assumed to influence the "effective," thermal conductivity of the tissue. This assumption is inaccurate since blood flow effects should be explicitly included in the model.

c. The tissue is assumed to possess constant physical properties.
d. No energy can penetrate beyond the deep core. This assumption amounts to assuming two symmetrical slabs to represent the body with the interface inside the body.

For the various elements of the model the problem can be stated mathematically as follows:

The skin-body system

$$\rho c_p \frac{\partial T_s}{\partial t} = k_s \frac{\partial^2 T_s}{\partial x_s^2} + \dot{q}_m \tag{4.181}$$

initial condition:

$$T_s|_{@\ t=0} = T_s(x_s) \tag{4.182}$$

boundary conditions:

$$\left.\frac{\partial T_s}{\partial x_s}\right|_{x_s=L_s} = 0 \tag{4.183}$$

$$k_s \left.\frac{\partial T_s}{\partial x_s}\right|_{@\ x_s=0} = \dot{m}_s[c_{pv}(T_{so} - T_{\text{ref}}) + \lambda] + U_s(T_{so} - T_a) \tag{4.184}$$

where $\dot{m}_s$ is the mass flux from skin surface given by:

$$\dot{m}_s = \begin{cases} K_s(c_{so} - c_a) & \text{for} \quad M_{\text{avai}} > 0 \\ \text{minimum}\,[S, K_s(c_{so} - c_a)] & \text{for} \quad M_{\text{avai}} = 0 \end{cases} \tag{4.185}$$

where K_s is the mass transfer coefficient at the skin surface, c_{so} water vapor concentration at the same location, c_a is the water vapor concentration at the airspace, S is the sweat generation rate, and M_{avai} is the mass of liquid available for evaporation at the skin surface,

$$M_{\text{avai}} = M_{\text{avai}}|_{@\ t=0} + \int_0^t [S(\xi) - \dot{m}_s(\xi)]\, d\xi \tag{4.186}$$

Also, λ is heat of vaporization and U_s is the overall heat transfer coefficient at skin surface.

The authors utilize physiological data to establish empirical relations for the various thermophysical parameters. These are

$$k_s = k_{so}\left(1 + \alpha_1 \Delta T_B + \alpha \frac{dT_B}{dt}\right) \leqslant 1.511\, k_{so} \qquad \text{for } \Delta T_B > 0 \qquad (4.187)$$

$$k_s = k_{so}\left(1 + \alpha_2 \Delta T_B + \alpha \frac{dT_B}{dt}\right) \geqslant 0.675\, k_{so} \qquad \text{for } \Delta T_B \leqslant 0 \qquad (4.188)$$

$$\text{where } T_B = \frac{1}{L_s}\int_0^{L_s} T_s(\xi, t)d\xi \qquad (4.189)$$

$$\Delta T_B = T_B - 96.35\,[°\mathrm{F}] \qquad (4.190)$$

k_{so} is the reference skin conductivity, 7.995×10^{-5} Btu/ft sec °F, α is the geometric porosity of the fabric, and α_1 and α_2 are the proportionality constants, 0.405 and $0.181°\mathrm{F}^{-1}$, respectively.

Ma et al. assume metabolic heat generation rates to be distributed in the following manner:

1. No heat is generated in the skin layer, i.e., $\dot{q}_m = 0 \;@\; 0 \leqslant x_s \leqslant a$.
2. Heat due to exercising or shivering is produced in the muscle layer, i.e., $\dot{q}_m = E + \Delta Q_{SHI} \;@\; a < x_s < (a + b)$.
3. Basal metabolism is uniformly produced inside the inner core, i.e., $\dot{q}_m = 3.595 \times 10^{-2}$ Btu/ft^3 sec @ $(a + b) \leqslant x_s \leqslant L_s$.

Further details on variations of heat produced by shivering are given by Ma et al.[101]

Evaporation of perspiration is assumed to be given by:

$$S = S_0 + \alpha_s \Delta T_B + \lambda_s (\Delta T_B)^4 \leqslant 60S \qquad \text{for } \Delta T_B > 0 \qquad (4.191)$$

$$S = S_0 \qquad \text{for } \Delta T_B \leqslant 0 \qquad (4.192)$$

where α_s and λ_s are the first power proportionality constant for sweating, 7.946×10^{-7} lb/ft^2 sec °F^4, and fourth power proportionality constant for sweating, 5.251×10^{-7} lb/ft^2 sec °F^4, respectively. S_0 is the reference sweat generation rate, 5.729×10^{-7} lb/ft^2 sec.

The fabric The following assumptions are made:

(1) The surface diffusion of vapor or liquid water through the solid portions of the fabric is negligible in comparison to diffusion through the pores of the fabric.
(2) An effective diffusion coefficient can be used for the entire fabric slab.
(3) The diffusion of mass due to thermal effects is small and can be taken into account by a modification of the effective diffusion coefficient.
(4) The porosity of the fabric is independent of time, that is, the slab does not swell.
(5) There are no capillary effects.
(6) Quasi-equilibrium exists at all times; that is, the solid portions of the slab reach equilibrium with their immediate surroundings instantaneously.[101]

If M denotes the mass of water vapor absorbed on the surface of the fiber per unit area, T_f the temperature and C_f the concentration of the vapor in the fabric, the following equations describe concentration and temperature variations:

$$\frac{\partial C_f}{\partial t} = \frac{\partial}{\partial x_f}\left(D\frac{\partial C_f}{\partial x_f} + \frac{\dot{V}_1 C_f}{\alpha A_s}\right) - \frac{1-\alpha}{\alpha}\rho_{fi}\frac{\partial M}{\partial t} \tag{4.193}$$

$$\left[(\rho c_p)_{\mathrm{air}} + c_{pv}C_i + \frac{1-\alpha}{\alpha}\rho_{fi}(C_{fi} + C_1 M)\right]\frac{\partial T_f}{\partial t}$$
$$= c_{pv}\left(D\frac{\partial C_f}{\partial x_f} + \frac{\dot{V}_1 C_f}{\alpha A_s}\right)\frac{\partial T_f}{\partial x_f} + (\rho c_p)_{\mathrm{air}}\frac{\partial}{\partial x_f}\left[\frac{\dot{V}_1(T_f - T_{\mathrm{ref}})}{\alpha A_s}\right]$$
$$+ \frac{\partial}{\partial x_f}\left(\frac{k_f}{\alpha}\frac{\partial T_f}{\partial x_f}\right) + \frac{1-\alpha}{\alpha}\rho_{fi}[\lambda - (c_1 - c_{pv})(T_f - T_{\mathrm{ref}})] \tag{4.194}$$

where D is water vapor diffusion coefficient, $\dot{V}_1$ is volumetric flow rate through the fabric and other symbols are listed in the Notation at the end of the article.

Equations (4.193) and (4.194) are assumed to be subjected to the following initial and boundary conditions:

$$T_f|_{@\,t=0} = T_f(x_f) \tag{4.195}$$

$$C_f|_{@\,t=0} = C_f(x_f) \tag{4.196}$$

$$D\left.\frac{\partial C_f}{\partial x_f}\right|_{@\,x_f=0} = K_0(C_{fo} - C) \tag{4.197}$$

$$k_f \left. \frac{\partial T_f}{\partial x_f} \right|_{@\, x_f=0} = U_0(T_{fo} - T_\infty) \tag{4.198}$$

$$D \left. \frac{\partial C_f}{\partial x_f} \right|_{@\, x_f=L_f} = -K_1(C_{fi} - C_a) \tag{4.199}$$

$$k_f \left. \frac{\partial T_f}{\partial x_f} \right|_{@\, x_f=L_f} = -U_1(T_{fi} - T_a) \tag{4.200}$$

The rate of water vapor absorbed by the fabric, $\partial M/\partial t$, is a function of the temperature and relative humidity of the air surrounding it. Thus,

$$M = M(C_f, T_f) \tag{4.201}$$

and

$$\frac{\partial M}{\partial t} = \frac{\partial M}{\partial C_f}\frac{\partial C_f}{\partial t} + \frac{\partial M}{\partial T_f}\frac{\partial T_f}{\partial t} \tag{4.202}$$

Equation (4.202) was substituted into Eqs. (4.193) and (4.194) to eliminate $\partial M/\partial t$. Details may be found in Ref. [101].

The airspace The airspace couples the heat and mass processes transfer occurring in the skin and in the fabric. Two phenomena may be distinguished in the airspace: free and forced convection. In the latter case a strong mechanism for the removal of water vapor and heat is provided. Utilizing mass and energy balances, the following equations are obtained:

$$\frac{dC_a}{dt} = \frac{\dot{m}_s}{L_a} + K_1(C_{fi} - C_a)L_s + \frac{\dot{V}_1(C_c - C_{fi})}{A_s L_a} + \frac{\dot{V}_2(C_c - C_a)}{A_s L_a} \tag{4.203}$$

$$\begin{aligned}
[c_{pv}C_a + (\rho c_p)_{\text{air}}]\frac{dT_a}{dt} &= \frac{\dot{m}_s c_{pv}(T_{so} - T_a)}{L_a} \\
&+ \frac{K_1 c_{pv}(C_{fi} - C_a)(T_{fi} - T_a)}{L_a} \\
&+ \frac{U_s(T_{so} - T_a)}{L_a} + \frac{U_1(T_{fi} - T_a)}{L_a} \\
&+ \frac{\dot{V}_1 c_{pv}[C_c(T_c - T_a) - C_{fi}(T_{fi} - T_a)]}{A_s L_a} \\
&+ \frac{\dot{V}_2 c_{pv} C_c(T_c - T_a)}{A_s L_a} + \frac{\dot{V}_1(\rho c_p)_{\text{air}}(T_c - T_{fi})}{A_s L_a} \\
&+ \frac{\dot{V}_2(\rho c_p)_{\text{air}}(T_c - T_a)}{A_s L_a}
\end{aligned} \tag{4.204}$$

with the initial conditions:

$$T_a|_{@\ t=0} = T_{ai} \tag{4.205}$$

$$C_a|_{@\ t=0} = C_{ai} \tag{4.206}$$

The solution to the complete set of equations and boundary and initial conditions describing the skin-airspace-fabric system was obtained by a numerical method. The method employed is the explicit finite forward difference. Results were obtained for an assumed man clad in a poplin cloth working in a tropical environment. In the airspace only natural convection was assumed. Typical results are shown in Figures 51 and 52.

Figure 51 shows the effect of different activity levels on skin and rectal temperatures. Ambient temperature for this figure was about 29°C and relative

Figure 51 Computer solution for the effects of two intensities of work on man clothed in poplin tropical suits. The environment was a typical tropical condition (73.2% relative humidity and 85°F).[101]

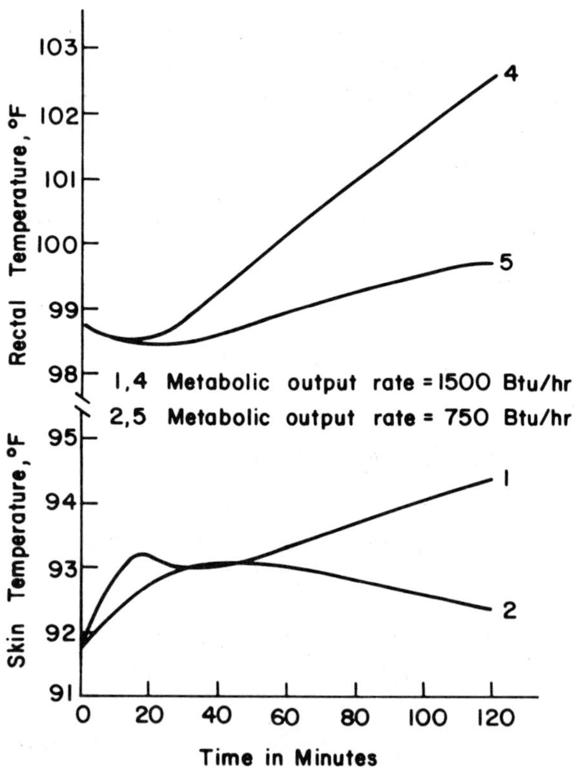

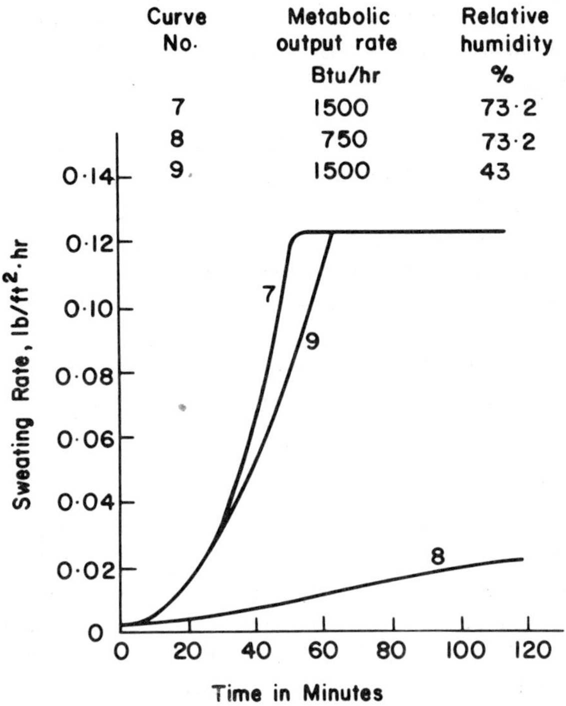

Figure 52 Computer solution for the transient sweating rate responses of a clothed subject exposed to a tropical environment under different conditions.[101]

humidity 73.2%. At the higher metabolic rate, which corresponds to an individual exercising rigorously, skin temperature rises for the first period of the exercise (about 20 min) because of insufficient perspiration. As time progresses, sweating becomes pronounced and sufficient cooling is provided resulting in a drop in skin temperature. Subsequently, however, skin temperature starts to rise again due to the high relative humidity of the environment which causes a decrease in the rate of evaporation. Unevaporated sweat starts to accumulate on the skin causing what is customarily known as a condition of over-saturated skin. Changes in rectal temperature are also shown in Figure 51. For the same high metabolic rate, rectal temperature drops initially and then rises without reaching a steady state for the entire duration of the experiment (2 hrs). This initial drop in rectal temperature is well documented and is due to the vasodilation occurring in the body when exposed to a high heat stress.

Calculated sweat rates for various tropical conditions are depicted in Figure 52. It is seen that sweating rate reaches its assumed maximum in a relatively short time (about 40 min) when the exercise rate is high with little influence of the relative humidity, curves 7 and 9 in Figure 52. When activity level is reduced by a factor of two, however, sweat rate increases gradually at about 50 gr/m^2 but is far below the assumed maximum rate after two hours of exercise (this exercise level of about 220 watts corresponds approximately to light bench work).

The work of Ma et al., described above, presents a significant development in the study of the simultaneous energy and mass transfer processes occurring in the skin-airspace-fabric system interacting with the environment. The model can probably be applied directly, or with little modification, to study the transient effects in human subject exposed to continuously changing environments and exercise levels. No information is given in the paper on the computer time required for a complete simulation run.

A much simpler model of the human body as an inconstant heat source and its relation to clothing insulation was considered by Iberall.[102]

4.8 Respiratory heat transfer

The mechanics of respiration in land-living mammals is such that inhaled air flows through the various nasal and lung passages where both heat and mass are exchanged in addition to the exchange of other gases. Mammals residing in environments where temperatures are lower than their average body temperature, lose heat to the inspired air. The amount of heat lost is dependent on the humidity of the inhaled air and the respiratory rate. In man, exhaled air was found to be about 80% saturated upon exiting the nostrils.[103] Of the total metabolic heat generated in the body, about 10% was found to be carried away by the exhaled stream.[5]

In other mammals that lack the mechanism of sweating, e.g., dog, respiration is employed as the prime defense mechanism against hyperthermia, i.e., panting. Yet other animals employ opposite mechanisms to save body moisture, which is primarily typical of mammals residing in arid areas.

Collins and coworkers developed a model of respiratory heat transfer in a small mammal, the kangaroo rat.[104] This animal inhabits the arid regions of Southwest America and employs a remarkable water economy which enables it to survive in such a hostile environment.

> One aspect of this water economy is a very low rate of respiratory water loss, which is attributable to remarkably low temperatures in the air passages rather than to any unusual ability to reduce respiratory ventilation rate by an increased extraction of oxygen from the pulmonary air. The nasal passageways have been described as a heat exchanger in which during inspiration heat and water are added to the inspired air, and during expiration part of the heat and water are redeposited on the now cooler walls.[104]

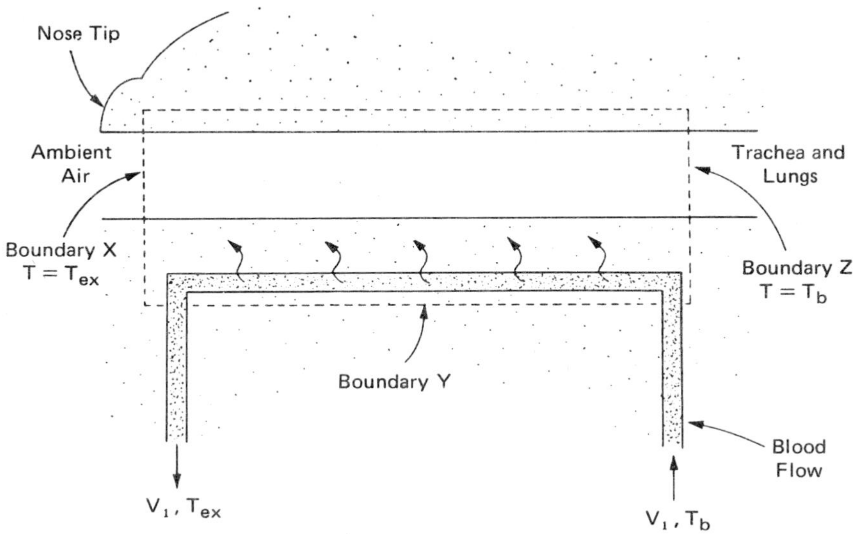

Figure 53 Diagram of nasal passage region of kangaroo rat. Dotted lines denote boundary of system for which energy balance is considered.[104]

In the model studied by Collins and coworkers, the steady-state heat and water vapor transfer in the upper respiratory tract (based on the kangaroo rat) was considered. This model, which is assumed applicable to other small mammals as well, is depicted in Figure 53. It was found, experimentally, that nasal temperatures in the kangaroo rat change primarily in response to environmental conditions and, to a lesser extent, changes in activity. Thus the model studied by Collins et al. related respiratory parameters to environmental conditions. First, the nasal passageways were modeled as two infinitely large parallel plates held at a constant temperature, T_1. At some reference time a stream of air, initially at a temperature $T_2 \neq T_1$, is suddenly introduced into the space between the two plates. The equations which describe energy and mass diffusion are assumed, respectively:

$$\nabla^2 T = \frac{1}{\alpha} \frac{\partial T}{\partial t} \tag{4.207}$$

$$\nabla^2 C = \frac{1}{D} \frac{\partial C}{\partial t} \tag{4.208}$$

Solutions to these two equations were taken from a standard heat and mass transfer textbook. From these solutions, by introducing proper numerical values, the times to reach equilibrium were estimated. Next, fluctuations in wall temperatures and mass concentrations were estimated to be very small, leading to the conclusion that a steady-state analysis for this case may be applicable. It should be noted that this conclusion may hold for animals which inhabit arid zones. These animals depend on their natural ability to conserve the essential moisture. Extension of the steady-state analysis to other mammals (e.g., man) who do not exercise such water vapor economy may cause significant errors.

A single flow model was then developed. This model includes the following mechanisms:

1. Energy transfer to expired vapor entering as liquid. It was assumed that inspired air entered the boundary x in Figure 53 at ambient absolute humidity and that expired air crosses the boundary saturated with water vapor. Both these streams are assumed at approximately the same temperature, T_{ex}. If all water lost during the air passage is assumed to enter as liquid at body temperature, the rate of energy transfer is given by:

$$E_v = \dot{m}_a(w_{ex} - w_a)(h_{v,ex} - h_{w,a}) \tag{4.209}$$

or by using common thermodynamic data:

$$E_v = \dot{m}_a(w_{ex} - w_a)(597 + 0.444T_{ex} - T_b) \tag{4.210}$$

with T_{ex} and T_b in deg. C.

2. Energy transfer to remainder of inspired air. It was estimated that this transfer due to the nominal exchange of oxygen and carbon dioxide is negligible.
3. Energy transfer through the nose tip. The nose tip is assumed to be depicted by a hemisphere with the inner surface maintained at temperature T_{ex}. Assuming free convection cooling of the outer surface, the equation for energy loss through the nose tip is obtained:

$$E_n = 5.05 \times 10^{-5}(T_{ex} - T_a) \tag{4.211}$$

4. Energy transfer from blood and from the body core. Treating blood as a stream which crosses into the control volume shown in Figure 53 at temperature T_b and leaving at T_{ex}, the energy transfer to this stream is given by:

$$E_b = \dot{m}_b c_b(T_b - T_{ex}) \tag{4.212}$$

Conduction of heat from the body is included in Eq. (4.212) by modification of the effective value of blood flow rate. The energy balance equation is the combination of the above mechanisms

$$\dot{m}_a(w_{ex} - w_a)(597 + 0.444T_{ex} - T_b) + 5.05 \times 10^{-5}(T_{ex} - T_b) = \dot{m}_b c_b(T_b - T_{ex}) \tag{4.213}$$

All parameters in Eq. (4.213), except $\dot{m}_b$, can be either measured directly or estimated from physiological measurements. Collins et al. performed the calculation of $\dot{m}_b$ as a function of ambient conditions.[104]

Model predictions were compared to experimental data. Results are shown in Figures 54 and 55. In these figures respiratory energy losses and respiratory water vapor losses are shown to be well within ±20% range of observed data. Thus the internal consistency of the model was demonstrated.

In view of its simplicity, the model developed by Collins et al. yields reasonably good agreement with observed data. This model, however, does not provide any local details on the various parameters associated with energy and mass transfer processes in respiration. Furthermore, the assumption on steady-state in the nasal passageways does not permit the direct extension of the model

Figure 54 Calculated net rate of respiratory energy loss vs. estimated values based on physiological measurements and assumptions, normalized to 100 g body weight. Line A, calculated loss = estimated loss; lines B, B′ represent ±10%; lines C, C′ represent ±20%.[104]

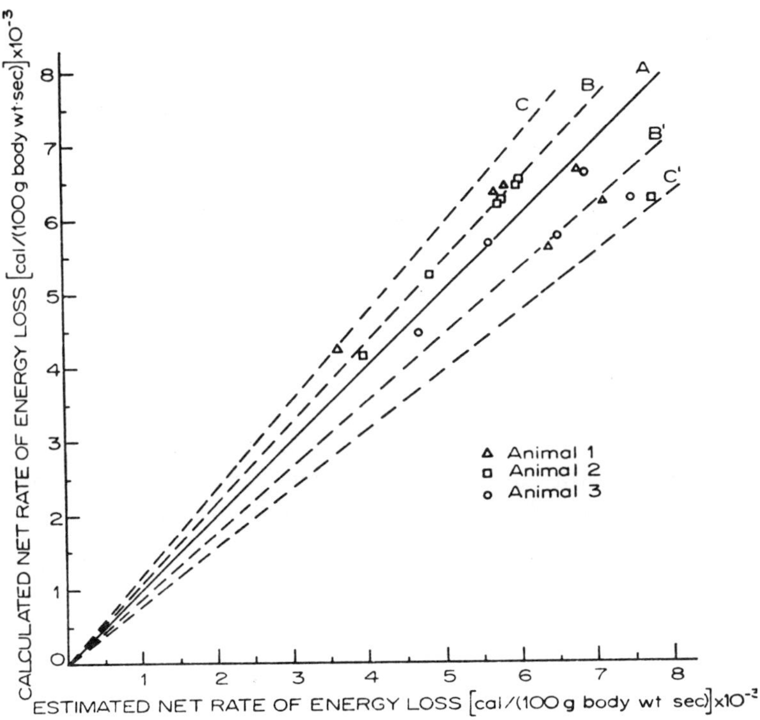

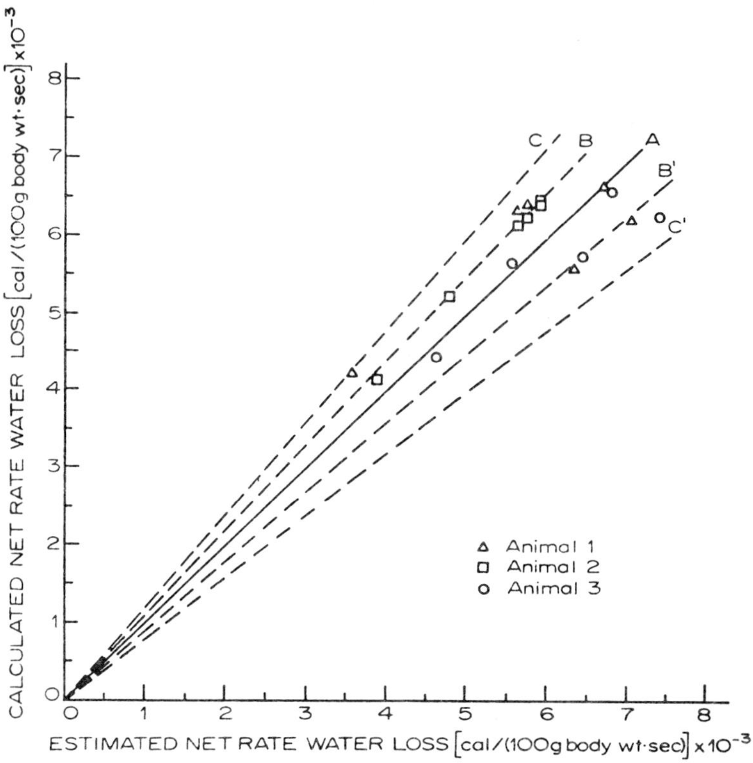

Figure 55 Calculated values of net rate of respiratory water loss vs. estimated values based on physiological measurements and assumptions, normalized to 100 g body weight. Line A, calculated loss = estimated loss; lines B, B′ represent ±10%; lines C, C′ represent ±20%.[104]

to other mammals in which the physiology of the respiratory tract dictates a transient state of the inspired-expired stream.

5 SUMMARY

In the preceding chapters bio-heat transfer studies are reviewed and discussed. From the diversity of the topics covered by investigators during the years, it becomes apparent that a separate branch of thermal physiology, namely bio-heat transfer, has emerged. At first, works in this area were relatively simple, both in nature of the problems studied and in the complexity of the methods employed. In recent years however, much ground has been won and new applications of bio-heat transfer have been introduced.

In spite of all the progress made, a major problem yet to be solved hampers future developments in this area. This problem is the lack of satisfactory non-invasive measuring techniques. It is obvious that no theoretical model, be it the most elaborate and physically sound, may be considered completely valid until verified by experiments. However, when dealing with live, active systems, one should realize that available measuring techniques cannot do; the introduction of a measuring probe into a live tissue will inevitably cause pooling of blood around it, inflammation, etc., resulting in a distortion of the measured quantity. Some promise lies in miniature transmitting elements which may be introduced into large organs, particularly the digestive tract, thereby monitoring body temperatures. Temperature of blood in major vessels may probably be evaluated from tracer data. Superficial temperature can be measured non-invasively by thermography. But these and other widely used measuring techniques do not provide the solution for measurements of local temperatures anywhere in the tissue with accurate geometrical fixation.

Another problem which requires much attention is the lack of accurate and detailed *in vivo* thermophysical data. This again delays the development of more exact models of bio-heat transfer.

In spite of all these handicaps, considerable fundamental work has been accomplished in bio-heat transfer. Vital information needed by physicians has been generated along with other studies, e.g., the thermal interaction with external cooling devices. Blood perfusion has, time and again, been demonstrated as playing an important role in the thermal energy balance of the organism, etc. It is believed that many more applications of bio-heat transfer will emerge in the future. This area also appears to be ripe for new approaches and the introduction of more modern techniques.

NOTATION

a	coefficient in Eq. (4.143)
a	half distance between cooling tubes
a	parameter, defined by Eq. (4.6)
a	skin layer thickness
a_i	nondimensional weight assigned to various skin areas for computing mean skin temperature
A	area
A	constant
A	parameter, defined by Eq. (4.28)
A_m	Coefficient in an infinite series, defined by Eq. (4.145)
A_D	total surface area of body, calculated by DuBois' equation (Eq. [3.11])
b	depth (thickness) of tissue
b	muscle layer thickness
b	parameter, defined by Eq. (4.7)
b	radius of spherical probe

B	defined by Eq. (4.154)
B	parameter, defined by Eq. (4.29)
c	deep body core thickness
c	specific heat
C	vapor concentration
C_0	constant
d	diameter
d	thickness of protective layer
D	vapor diffusion coefficient
e	specific internal energy
e	thickness of skin and fat layer
E	energy production rate due to exercise
E	thermal energy
ΔE	energy of activation
$f(x)$	function
$f(\xi,\eta)$	defined by Eq. (4.46)
$f_{a,i}$	defined by Eq. (4.173)
f_A	area factor, ratio of contact area to total outside cooling tube area
f_c	percentage of surface area through which convective exchange occurs
$\dot{f}_i, \overline{f}$	blood perfusion rate vector
f_r	effective radiation area factor
f_w	wettedness coefficient, defined by Eq. (3.18)
G	defined by Eq. (4.92)
h	heat transfer coefficient
h	height
h	specific enthalpy
h_{fg}	latent heat of vaporization
$\dot{h}_i, \overline{h}$	lumped heat transfer coefficient vector between organism and environment
i	integer
i	$\sqrt{-1}$
I	electrical current
I	radiation flux
Icl	total thermal resistance offered by clothing ensemble, defined by Eq. (3.24)
$I_\nu(x)$	modified Bessel function of the first kind of order ν
j	integer
$J_0(x), J_1(x)$	Bessel functions of the first kind of zero and first orders, respectively
k	thermal conductivity
k_{eff}	effective thermal conductivity, defined by Eq. (4.50)
$k_{ij}, \overline{\overline{k}}$	thermal conductivity tensor

K	linear gradient of initital temperature in the skin
K	mass transfer coefficient
$K_i(x)$	modified Bessel function of the second kind of order i
K_p	proportional control constant in general terms
K_{pi}	individual constant for a given feedback controller
K_{pj}	individual constant for a given feedforward controller
L	length
m	defined by Eq. (4.153)
m	integer
m	numerical constant
m	permeance coefficient of skin
$\dot{m}$	mass flow rate
M	dimensionless mass of absorbed water vapor
$\dot{M}$	metabolic rate
n	integer
n_i	outward unit vector normal to surface element
N_k	defined by Eq. (4.27)
p	water vapor pressure
p	Laplace transform parameter
$\dot{q}'''$	rate of heat generation per unit volume
$\dot{q}_c$	total conductive heat
$\dot{q}_e$	rate of heat exchange with the environment
$\dot{q}_i$	conductive heat flux vector
Q	defined by Eq. (4.55)
Q	total amount of heat
Q_i	defined by Eq. (4.166)
Q'	defined by Eq. (4.55)
r	coordinate in radial direction
r	radiation absorption coefficient
r_0	radius of probe
r^*	location of frozen fron defined by Eq. (4.115)
R	controller
R	defined by Eq. (4.116) or Eq. (4.141)
R	nonpenetrating radiation heat flux
R, R_i	radius
R_0	universal gas constant
s	surface area
S	sweat generation rate
t	time
T	temperature
T_D	derivative control constant
$\bar{T}_s$	mean skin temperature
T^*	transformed temperature, defined by Eq. (4.59)

u	defined by Eq. (4.24)
u	temperature, defined by Eq. (4.127)
U	overall heat transfer coefficient
v	defined by Eq. (4.25)
v	velocity
v_i	velocity vector
V	volume
$\dot{V}_i$	volumetric flow rate
w	specific humidity or humidity ratio
w	work
$\dot{w}_b$	volumetric blood perfusion rate
w_i	defined by Eq. (4.165)
W	weight
x	coordinate
y	coordinate
$Y_i(x)$	Bessel function of the second kind of order i
z	coordinate in axial direction
z^2	defined by Eq. (4.93)

Greek symbols

α	defined by Eq. (4.75)
α	experimental coefficient relating the amount of heat exchanged by sweating to physiological parameters (Eq. [3.16])
α	geometric porosity of fabric
α	thermal diffusivity
α_b	defined by Eq. (4.129)
$\alpha_{n,\,i}$	defined by Eq. (4.174)
α_s	first power proportional control coefficient for sweating
α_1, α_2	physiological conductivity constants
β	defined by Eq. (4.76) or (4.117)
β	equilibration constant between blood and surrounding tissue (cf. Eq. [4.4])
β	ratio of effective width of cooling tube to cooling tubes spacing
γ	defined by Eq. (4.83)
γ	numerical parameter $0, < \gamma < 1.0$
γ_b	defined by Eq. (4.130)
δ	depth of tissue, Eq. (4.34)
$\Delta(u)$	defined by Eq. (4.78)
ϵ	emittance
ϵ	numerical parameter, $0 < \epsilon < 1.0$
ϵ_i	defined by Eq. (4.179b)
$\zeta(x)$	defined by Eq. (4.132)

ζ_i	defined by Eq. (4.175)
η	defined by Eq. (4.128)
θ	dimensionless temperature
λ	heat of vaporization
λ	parameter, defined by Eq. (4.41)
λ_m	eigenvalue, given by Eq. (4.142)
λ_n	$n\pi/a$
μ	defined by Eq. (4.39)
μ_n	defined by Eq. (4.179a)
ν	defined under Eq. (4.46)
ξ	dimensionless length, defined by Eq. (4.26)
ξ	dimensionless length, defined by Eq. (4.152)
ξ	dummy variable of integration
ξ	parameter, defined by Eq. (4.40)
ρ	density y
σ	Stefan Boltzman constant
τ	defined by Eq. (4.77)
τ	time
ϕ	defined by Eq. (4.119)
ϕ	parameter, defined by Eq. (4.49)
ϕ	view factor, Eq. (4.147)
$\chi_n(\mu_n r)$	defined by Eq. (4.177)
ψ^j_{km}	defined by Eq. (4.172)
ω	defined by Eq. (4.125)
Ω	function of epidermal injury, defined by Eq. (4.99)
Ω_n	defined by Eq. (4.178)

Subscripts

a	airspace
a, art	artery
a	environment
avai	available
b, B	blood
c, conv	convection
cl	clothing
cond	conduction
cool	coolant
d	dentin
diff	diffusion
e	enamel
e	evaporation
ex	expired, exhaled

f	fabric
f	fire
f	tibia
fe	femur
fi	fibula
fp	functional periphery
fr	frozen
g	generation
h	hypothalamus
HS	heat source
l	liquid
m, met	metabolic
n	nasal
po	probe
pr	probe
r, rad	radiation
R, re	rectal
ref	reference
res	respiratory
S	skin
SHI	shivering
ss	steady-state
sw	sweat
v	vapor
v, ven	vein
o	reference

Superscripts

lat	latent
sen	sensible
*	saturation
–	mean
•	derivative with respect to time

APPENDIX A Dimensionless Parameters Associated with the Transport of Thermal Energy Inside a Biological Tissue, and Their Physical Interpretation[105]

No.	Parameter	Range used	Physical interpretation
1	$\dfrac{c_b \dot{w}_b b^2}{k}$ $\dfrac{c_b \dot{w}_b V}{kb}$	0–18	$\dfrac{\text{Energy transported by blood stream (capillary perfusion)}}{\text{Energy transferred by conduction through the tissue}}$
2	$\dfrac{hb}{k}$ *	1–14	$\dfrac{\text{Energy transferred by convection at the surface of the tissue}}{\text{Energy transferred by conduction through the tissue}}$
3	$\dfrac{UA}{c_b \dot{w}_b V}$	0–60	$\dfrac{\text{Energy exchanged between arteries and surrounding tissue or counter-current arterial-venous heat exchange}}{\text{Energy transported by blood stream (capillary perfusion)}}$
4	$\dfrac{kt}{\rho c b^2}$ **	0–0.5	$\dfrac{\text{Energy transferred by conduction through the tissue}}{\text{Heat capacity of the tissue (thermal inertia)}}$
5	$\dfrac{c_b \dot{w}_b t}{\rho c}$	>0	$\dfrac{\text{Energy transported by blood stream (capillary perfusion)}}{\text{Heat capacity of the tissue (thermal inertia)}}$
6	$\dfrac{\dot{q} t}{\rho c \overline{\overline{T}}^*}$	>0	$\dfrac{\text{Local metabolic heat generation}}{\text{Heat capacity of the tissue (thermal inertia)}}$
7	$\dfrac{\dot{q} V}{Q_m}$	.2–1	$\dfrac{\text{Local metabolic heat generation}}{\text{Total metabolic heat generation}}$
8	$\dfrac{\dot{q} V k^{1/2}}{Q_m (c_b \dot{w}_b)^{1/2} b}$	.2	$\dfrac{\text{(Local metabolic heat generation)} \times \text{(Energy transferred by conduction through the tissue)}}{\text{(Total metabolic heat generation)} \times \text{(Energy transported by blood stream)}}$

APPENDIX A Dimensionless Parameters Associated with the Transport of Thermal Energy Inside a Biological Tissue, and Their Physical Interpretation[105] (*Continued*)

No.	Parameter	Range used	Physical interpretation
9	$\frac{\dot{q}b^2}{k\overline{T}^*}$	.03–.5	Local metabolic heat generation / Energy transferred by conduction through the tissue
10	$\frac{m_e h_{fg} b}{k\overline{T}^*}$	.02–.27	Evaporative heat flow at the surface of the tissue / Energy transferred by conduction through the tissue
11	$\frac{\dot{q}}{c_b \dot{w}_b \overline{T}^*}$	0–250	Local metabolic heat generation / Energy transported by blood stream

*Nusselt modulus.
**Fourier modulus.

APPENDIX B Thermophysical Properties of Biological Media*

Material	Thermal conductivity, k w/(cm°C) × 10^3	Thermal diffusivity, α cm²/sec × 10^3	Thermal inertia $k\rho c_p$ w² sec/(cm⁴ °C²) × 10^3
Internal tissues, in vitro			
Animal muscle	4.06–4.18		
Calf muscle (veal)	5.45		
Dog muscle	21.8–24.3		
Human muscle	3.85		
Human muscle	4.40		
Dog liver	12.6–14.5		
Dog lung	2.19–2.64		
Dog brain	7.12–7.95		
Rat and rabbit livers	4.95		
Rabbit kidney	5.02		
Rabbit liver	11.7		
Beef kidney	5.25		
Beef brain	4.97 and 4.15		
Beef liver	4.88		
Bovine liver	5.06 and 5.53		
Beef lung	2.82		
Beef bone marrow	2.20		
Bone			8.75
Animal fat	1.32–1.56		
Beef fat	2.04		
Pork fat, fresh	3.59–3.71		
Subcutaneous fat	1.60		
Human and beef fat	2.04		
Human fat	2.00		4.55 ± 0.5
Cat brain, 1½ hours after expiration		0.63–0.64	
Cat brain, 2½ hours after expiration		0.44–0.50	
Skin tissue			
Skin	2.09		
Epidermis	2.09		
Dermis	3.70		
0–2 mm skin	3.76		
Cool living skin	5.45		
Very warm living skin	28.0		
Cold living hand	3.35		
Normal living hand	9.60		

*Reproduced in part from a paper by J. C. Chato.[71]

APPENDIX B Thermophysical Properties of Biological Media *(Continued)*

Material	Thermal conductivity, k w/(cm°C) × 10^3	Thermal diffusivity, α cm²/sec × 10^3	Thermal inertia $k\rho c_p$ w² sec/(cm⁴ °C²) × 10^3
Chicken skin	3.56		
Excised skin (dry)			9.62
Excised skin (moist)			13.1
Living skin (no blood flow)			15.8
Living skin (with blood flow)			15.8–70.0
Living forearm skin			22.8
Living average skin			17.5
Living fingertip skin			40.0
Living skin			21.9
Living skin			21.0
Living inner forearm skin			24.5 ± 3.5
Irradiated skin			13.1–31.7
Porcine living skin	4.14	0.82–0.86	
Internal tissues, in vivo			
Cat brain		1.12–1.24	
Living bone	22.8		
Bovine bones:			
Midshaft of tibia	8.88–30.8		
Rib	4.03–11.1		
Frontal bone	4.87–21.5		
Caprine bones:			
Midshaft of tibia	4.63–5.04		
Neck of tibia	3.30–2.84		
Rib	1.58–2.84		
Frontal bone	4.72–6.50		
Periosteum of the tibia:			
Bovine	19.0–27.6		
Caprine	8.10–22.5		
Cartilage of the scapula:			
Bovine	17.9–27.6		
Caprine	13.7–19.0		
Fluids			
Dog blood (anticoagulant EDTA)	6.70–7.54		
Dog plasma (anticoagulant EDTA)	6.28–7.12		
Blood (42 and 43% hematocrit)	5.09 and 5.30		
Human blood	$k \times 10^3 \cong 5.7 - (1.21 \times 10^{-2})$ (% hematocrit)		
Plasma	5.71		

APPENDIX B Thermophysical Properties of Biological Media *(Continued)*

Material	Thermal conductivity, k w/(cm°C) × 10^3	Thermal diffusivity, α cm^2/sec × 10^3	Thermal inertia $k\rho c_p$ w^2 sec/ (cm^4 $°C^2$) × 10^3
Dog blood (1% heparin)	$k \times 10^3 = 6.0 \times$ [1–0.2 volume fraction of red blood cells)] –0.0226 (37–T°C)		
Rat blood	5.27		
Human blood	5.06		
Human plasma	5.82		
Human blood corpuscles	4.81		
Cow's milk	5.31		
Skimmed milk	5.74		
Top of milk	4.23		
Cream (double Devon)	3.11		
Egg white	5.56		
Egg yolk	3.38		
Cod-liver oil	1.70		
Beef vitreous humor	5.94		
Beef aqueous humor	5.78		
Urine	5.61		
Gastric juice	4.45		
PVP additive	4.74		
Frozen and cold materials			
Eye of loin beef, U.S. good grade, aged 69.5% water, frozen at:			
0°C	2.94		
–3°C	9.90		
–10°C	10.4		
–17°C	10.7		
Beef at –130°C	15.5		
Frozen beef	22.5		
Dehydrated beef	0.35		
Lean neck beef –7°C	4.15–4.73		
Beef fat, –9°C	2.25–2.30		
Animal integuments			
Fur and feathers (considerable scatter	$k \times 10^3 = 0.252$ +0.0478 × thickness in cm)		

REFERENCES

1. Van Wylen, G. J. and Sonntag, R. E.: *Fundamentals of Classical Thermodynamics.* John Wiley and Sons Inc., New York, 1967.
2. Chato, J. C. and Hertig, B. A.: Thermal behavior of biological media. *AIAA Paper,* No. 70-813, 1970.
3. Robinson, S., Meyer, F. R., Newton, J. L., Ts'ao, C. H., and Nolgersen, L. O.: Relations between sweating, cutaneous blood flow and body temperature in work. *J. Appl. Physiol.,* Vol. 20, 1965, pp. 575-582.
4. Thauer, R.: Der Nervose Mechanismus der Chemichen Temperaturregulation des Warmbluters. *Naturwissenschaften,* Vol. 51, 1964, pp. 73-80.
5. Bazett, H. C.: The regulation of body temperatures. In *Physiology of Temperature Regulation and the Science of Clothing,* L. H. Newburgh, ed., Hafner Publishing Co., New York, 1968, pp. 109-192.
6. Bligh, J.: Localization of the thermal stimulus of polypnoea. *J. Physio.,* Vol. 135, 1957, pp. 48p-49p.
7. Horowitz, J. M. and Erskine, L. K.: Central regulation of temperature in cold environments. I. A dynamic model with two temperature inputs. *Computers and Biomedical Research,* Vol. 6, 1973, pp. 57-73.
8. Huckaba, C. E., Downey, J. A., and Darling, R. C.: A feedforward-feedback mechanism for human thermoregulation. *Chemical Engineering Symposium Series,* Vol. 67, 1971, pp. 1-7.
9. Hardy, J. D.: The 'set point' concept in physiological temperature regulation. In *Physiological Controls and Regulations,* W. S. Yamamoto and J. R. Brobeck, eds., Saunders, Philadelphia, 1965, pp. 88-116.
10. Nagayama, T. H., Eisenman, J. S., and Hardy, J. D.: Single unit activity of anterior hypothalamus during local heating. *Science,* Vol. 134, 1961, pp. 560-561.
11. Grollman, S.: *The Human Body–Its Structure and Physiology.* 2nd ed., The MacMillan Co., London, 1969.
12. Hensel, H.: Electrophysiology of thermosensitive nerve endings. In *Temperature–Its Measurement and Control in Science and Industry,* J. D. Hardy, ed., Vol. 3, Part 3, Reinhold, New York, 1963, pp. 191-198.
13. Wulsin, F. R.: Adaptations to climate among non-European peoples. In *Physiology of Heat Regulation and the Science of Clothing,* L. H. Newburgh, ed., Hafner Publishing Co., New York, 1968, pp. 3-69.
14. Morovsky, N.: The adjustable brain of hibernators. *Scientific American,* Vol. 218, pp. 110-118.
15. Dawkins, J. R. and Hull, D.: The production of heat by fat. *Scientific American,* Vol. 213, 1965, pp. 62-67.
16. Lyman, C. P.: Homeostasis in hibernation. In *Temperature–Its Measurement and Control In Science and Industry,* J. D. Hardy, ed., Vol. 3, Part 3, Reinhold, New York, 1963, pp. 453-457.
17. Minard, D. and Copman, L.: Elevation of body temperature in disease. In (same as 16 above), pp. 253-273.
18. Ladell, W. S. S.: Disorders due to heat. *Transac. Royal Society of Tropical Medicine and Hygiene,* Vol. 51, 1957, pp. 189.
19. Hardy, J. D.: Cutaneous pain and tissue damage by thermal radiation. *Proceedings, 1st National Biophysics Conference,* 1959.
20. Hart, J. S.: Rodents. In *Comparative Physiology of Thermoregulation.* G. C. Whittow, ed., Vol. 2–Mammals, Academic Press, New York, 1971, pp. 1-149.

21. Hardy, J. D.: Control of heat loss and heat production in physiologic temperature regulation. *Harvey Lectures,* Series XLIX, 1953, pp. 242–270.
22. Stolwijk, J. A. J. and Hardy, J. D.: Temperature regulation in man–a theoretical study. *Pflügers Archiv.,* Vol. 291, 1966, pp. 129–162.
23. Hardy, J. D.: (Same as 9 above).
24. Hammel, H. T.: The regulator of body temperature. *Brody Memorial Lecture VI,* University of Missouri, 1966.
25. Downey, J. A., Huckaba, C. E., Bonilla, L. E., and Luke, W. K.: The role of skin temperature in human thermoregulation. *Chemical Engineering Symposium Series,* Vol. 67, 1971, pp. 211–216.
26. Banerjee, M., Elizondo, R., and Bullard, R.: Reflex response of human sweat glands to different rates of skin cooling. *J. Appl. Physiol.,* Vol. 26, 1969, pp. 787–792.
27. Brenglemann, G. L.: Human temperature regulation: dynamic characteristics of metabolic response. Ph.D. Thesis, University of Washington, 1967.
28. Wurster, R. D. and McCook, R. D.: Influence of rate of change in skin temperature on sweating. *J. Appl. Physiol.,* Vol. 27, 1969, pp. 237–240.
29. Huckaba, C. E.: A feedforward-feedback mechanism for human thermoregulation. *Chemical Engineering Symposium Series,* Vol. 67, 1971, pp. 1–7.
30. Netter, F.: Cross section of the normal skin (illustration). *Clinical Symposia,* Vol. 19, 1967, p. 38.
31. Brown, A. C.: Equations of heat distribution within the body. *Bull. Of Math. Biophys.,* Vol. 27, 1965, pp. 67–78.
32. Ganong, W. F.: *Review of Medical Physiology.* 3rd ed., Lange Medical Publications, Los Altos, 1967.
33. Pennes, H. H.: Analysis of tissue and arterial blood temperatures in the resting human forearm. *J. Appl. Physiol.,* Vol. 1, 1948, pp. 93–122.
34. Iberall, A. S. and Schindler, A. M.: On the physical basis of a theory of human thermoregulation. *Trans. ASME, J. Dynamic Systems, Measurement and Control,* Vol. 95, 1973, pp. 68–75.
35. Seagrave, R. C.: *Biomedical Applications of Heat and Mass Transfer.* The Iowa State University Press, Ames, Iowa, 1971.
36. Dubois, D. and Dubois, E. F.: A formula to estimate the approximate surface area if height and weight be known. *Archiv. International of Medicine,* Vol. 17, 1961, pp. 863–871.
37. Fanger, P. O.: *Thermal Comfort–Analysis and Applications in Environmental Engineering.* McGraw Hill Book Co., New York, 1970.
38. Ibamoto, K.: The use of a model in the analysis of the relation between skin temperature and the rate of sweating. *Proceedings, Symposium International de Thermoregulation Comp.,* 1970, pp. 284–287.
39. Spillman, C. K. and Hinkle, C. N.: Conduction heat transfer from swine to controlled temperature floors. *Transac. ASME,* 1971, pp. 301–303.
40. Newburgh, L. H., ed.: *Physiology of Temperature Regulation and the Science of Clothing.* Hafner Publishing Co., New York, 1968.
41. Shitzer, A., Chato, J. C., and Hertig, B. A.: A study of the thermal behavior of living biological tissue with application to thermal control of protective suits. Technical report No. ME–TR-207, Department of Mechanical and Industrial Engineering, University of Illinois, Urbana, Illinois, January 1971.
42. Fan, L. T., Hsu, F. T. and Hwang, C. L.: A review on mathematical models of the human thermal system. *IEEE Trans on Bio-Medical Eng.*, Vol. BME-18, 1971, pp. 218–234.

43. Shitzer, A.: Mathematical models of thermoregulation and heat transfer in Mammals—a compendium of research. Nasa TMX-62,172, July 1972.
44. Vendrik, A. J. H., and Vos, J. J.: A method for the measurement of the thermal conductivity of human skin. *J. Appl. Physio.,* Vol. 11, 1957, pp. 211-215.
45. Carslaw, H. S. and Jaeger, J. C.: *Conduction of Heat in Solids.* Oxford University Press, London, 1959.
46. Richardson, P. D. and Whitelaw, J. H.: Transient heat transfer in human skin. *J. Franklin Inst.,* Vol. 286, 1968, pp. 169-181.
47. Mitchell, J. W., Galvez, T. L., Hengle, J., Myers, G. E., and Siebecker, K. L.: Thermal response of human legs during cooling. *J. Appl. Physio.,* Vol. 29, 1970, pp. 859-865.
48. Molnar, G. W.: Analysis of the rate of digital cooling. *Proc. Symp. Inter. de Thermoreg. Comp.,* 1971, pp. 350-352.
49. Gautherie, M.: Analog studies on the mechanisms and factors of the phenomenon of periodic oscillation of cutaneous temperatures of the fingers' skin. *Acta Electronica,* Vol. 12, 1969, pp. 313-338.
50. Mitchell, J. W. and Myers, G. E.: An analytical model of the counter-current heat exchange phenomena. *Biophysical Journal,* Vol. 8, 1968, pp. 897-911.
51. Keller, K. H. and Seiler, L. Jr.: An analysis of peripheral heat transfer in man. *J. Appl. Physio.,* Vol. 30, 1971, pp. 779-786.
52. Chato, J. C.: A method for the measurement of the thermal properties of biological materials. *Thermal Problems in Biotechnology,* ASME Symposium Series, 1968, pp. 16-25.
53. Chato, J. C., Shitzer, A., and Fry, F. J.: Measurement of thermal properties of living tissue. *Proceedings, 23rd Annual Conference on Engineering in Medicine and Biology,* 1970, p. 156.
54. Trezek, G. J. and Cooper, T. E.: Analytical determination of cylindrical source temperature fields and their relation to thermal diffusivity of brain tissue. *Thermal Problems in Biotechnology,* ASME Symposium Series, 1968, pp. 1-15.
55. Trezek, G. J., Jewett, D. L., and Cooper, T. E.: Measurements of *in vivo* thermal diffusivity of cat brain. *Preceedings, 7th Thermal Conductivity Conference,* 1967, pp. 749-754.
56. Cooper, T. E. and Trezek, G. J.: A probe technique for determining the thermal conductivity of tissue. *ASME Paper,* No. 70-WA/HT-18, 1970.
57. Priebe, L. and Betz, E.: Heat transport in homogeneously and isotropically perfused tissue. *Arctliche Forschung,* Vol. 1, 1969, pp. 18-30.
58. Henriques, F. C. and Mortiz, A. R.: Studies of thermal injury. I. The conduction of heat to and through skin and the temperatures attained therein. A theoretical and an experimental investigation. *American J. Pathology,* Vol. 23, 1947, pp. 531-549.
59. Mortiz, A. R. and Henriques, F. C.: Studies of thermal injury. II. The relative importance of time and surface temperature in the causation of cutaneous burns. *American J. Pathology,* Vol. 29, 1947, pp. 695-720.
60. Henriques, F. C.: Studies of thermal injury. V. The predictability and the significance of thermally induced rate processes leading to irreversible epidermal injury. *Archives of Pathology,* Vol. 43, 1947, pp. 489-502.
61. Buettner, K.: Effects of extreme heat and cold on human skin. I. Analysis of temperature changes caused by different kinds of heat application. *J. Appl. Physiol.,* Vol. 3, 1951, pp. 691-702.
62. Stoll, A. M. and Chianta, M. A.: Application of biophysical heat transfer studies in protection systems. U.S. Department of the Navy, NADC-MR-6722, 1967.
63. Weaver, J. A. and Stoll, A. M.: Mathematical model of skin exposed to thermal radiation. *Aerospace Medicine,* Vol. 40, 1969, pp. 24-30.

64. Stoll, A. M. and Chianta, M. A.: Heat transfer through fabrics as related to thermal injury. *Trans. New York Academy of Sciences,* Series II, Vol. 33, 1971, pp. 649–670.
65. Stoll, A. M. and Greene, L. C.: Relationship between pain and tissue damage due to thermal radiation. *J. Appl. Physiol.,* Vol. 14, 1959, pp. 373–382.
66. Stoll, A. M.: A computer solution for determination of thermal damage integrals from experimental data. *IRE Transactions on Medical Electronics,* Vol. ME-7, 1960, pp. 355–393.
67. Weaver, J. A.: Calculation of time-temperature histories and prediction of injury to skin exposed to thermal radiation. NADC-MR-6623, pp. 1–27.
68. Dussan, B. I. and Weiner, R. I.: Study of burn hazard in human tissue and its implication on consumer product design. *ASME Paper* 72-WA/HT-39, 1971.
69. Wu, Y. C.: Material properties criteria for thermal safety. *J. Materials,* Vol. 7, 1972, pp. 572–579.
70. Jacobs, H. R., Brown, W. S., and Ting, V. C.: Influence of heat transfer on teeth. *ASME Paper,* No. 71-WA/HT-40, 1971.
71. Chato, J. C.: Heat transfer in bioengineering. In *Advanced Heat Transfer,* B. T. Chao, ed., University of Illinois Press, 1969, pp. 395–414.
72. Barron, R. F.: Cryoinstrumentation. In *Cryogenics in Surgery,* H. von Leden and W. G. Cahan, eds., Medical Examination Publishing Co. Inc., New York, 1971, pp. 80–127.
73. Cooper, T. E. and Trezek, G. J.: Mathematical Predictions of Cryogenic Lesion. In *Cryogenics in Surgery,* H. Von Leden and W. G. Cahan, eds., Medical Examination Publishing Co., Inc., New York, 1971, pp. 128–149.
74. Cooper, T. E. and Trezek, G. J.: Analytical prediction of the temperature field emanating from cryosurgical cannula. *J. Cryobiology.*
75. Trezek, G. J. and Jewett, D. L.: Thermal modeling of cylindrical source transient temperature fields in brain. *Proc. 21st Conference on Engineering in Medicine and Biology,* 1968, pp. 53.4.
76. Trezek, G. T. and Jewett, D. L.: Nodal network simulation of transient temperature fields from cooling sources in anesthesized brain. *IEEE Trans. on Biomedical Engineering,* Vol. BME-17, 1970, pp. 281–286.
77. Trezek, G. J.: Thermal and electrical modeling for brain surgical probes. *ASME Paper,* No. 69-WA/HT-40, 1969.
78. Cooper, T. E. and Groff, J. P.: Thermal mapping via liquid crystals of the temperature field near a heated surgical probe. *Trans. ASME, J. Heat Transfer,* Vol. 95, 1973, pp. 250–256.
79. Perl, W.: Heat and matter distribution in body tissues and the determination of tissue blood flow by local clearance methods. *J. Theo. Biol.,* Vol. 2, 1962, pp. 201–235.
80. Perl, W.: An extension of the diffusion equation to include clearance by capillary blood flow. *Ann. of the New York Academy of Sciences,* Vol. 108, 1963, pp. 92–105.
81. Perl, W. and Cucinell, S. A.: Local blood flow in human leg muscle measured by a transient response thermoelectric method. *Biophys. J.,* Vol. 5, 1965, pp. 211–230.
82. Perl, W. and Hirsch, R. L.: Local blood flow in kidney tissue by heat clearance measurement. *J. Theoret. Biol.,* Vol. 10, 1966, pp. 251–280.
83. Stow, R. W.: Thermal measurement of tissue blood flow. *Trans. New York Academy of Sciences,* Vol. 27, 1963, pp. 748–758.
84. Eberhard, R. C., Jackson, M., and Trezek, G. J.: Application of bio-heat-transfer model to determine regional blood flow rate. *ASME Paper,* 71-WA/HT-38, 1971, also *Trans. ASME, J. Heat Transfer,* Vol. 95, 1973, pp. 113–199.
85. Love, T. J., Haberman, J. D., and Francis, J. E.: A comparison of predicted skin temperatures with thermographic measurements. *Temperature–Its Measurement and Control,* 1971.

86. Haberman, J. D.: The present status of mammary thermobraphy. *Ca–A Cancer Journal of Clinicians*, Vol. 18, 1968, pp. 314–321.
87. Gros, C. and Gautherie, M.: Investigation by infrared thermometry of the evolution in the course of time of human skin temperature as a function of ambient temperature. Idea of cutaneous thermal adaptance. *Rev. Franc. Etudes Cliniques et Biol.*, Vol. 13, 1968, pp. 697–703.
88. Nunneley, S.: Water cooled garments–a review. *Space Life Sciences*, Vol. 2, 1970, pp. 335–360.
89. Buchberg, H. and Harrah, C. B.: Conduction cooling of the human body–a biothermal analysis. *Thermal Problems in Biotechnology*, ASME Symposium Series, 1968, pp. 82–95.
90. Harrah, C. B. and Buchberg, H.: Prediction of the skin temperature distribution and gross body conductance for conduction cooling of the human body. *Proceedings of the Symposium on Individual Cooling*, 1969, pp. 92–100.
91. Shitzer, A. and Chato, J. C.: Analytical modeling of the thermal behavior of living human tissue. *Proceedings, 4th International Heat Transfer Conference*, 1970, pp. Cu 3.9–1–10.
92. Saltin, B., Gagge, A. P., and Stolwijk, J. A. J.: Muscle temperature during submaximal exercise in man. *J. Appl. Physiol.*, Vol. 25, 1968, pp. 679–688.
93. Chato, J. C. and Shitzer, A.: Thermal modeling of the human body–further solutions to the steady-state heat equation. *AIAA J.*, Vol. 9, 1971, pp. 365–369.
94. Chato, J. C. and Shitzer, A.: On the dimensionless parameters associated with heat transport within living tissue. *Aerospace Medicine*, Vol. 41, 1970, pp. 390–393.
95. Shitzer, A. and Chato, J. C.: Steady-state temperature distribution in living tissue modeled as cylindrical shells. *ASME Paper*, No. 71–WH/HT–34, 1971.
96. Leo, R. J., Shitzer, A., Chato, J. C., and Hertig, B. A.: Steady and transient temperature distributions on the skin of a human thigh covered by a water-cooled pod. *ASHRAE Trans.*, Vol. 79, Part I, 1973, pp. 62–74.
97. Shitzer, A. and Chato, J. C.: Analytical solutions to the problem of transient heat transfer in living tissue. *ASME Paper*, No. 71–WA/HT–36, 1971.
98. Shitzer, A. and Chato, J. C.: Effect of variable blood supply temperature on the temperature distribution in a biological tissue. *Proceedings, 24th Conference on Engineering in Medicine and Biology*, Vol. 13, 1971, p. 57.
99. Shitzer, A.: A model of heat transfer in a biological tissue perfused by blood of arbitrary temperature. *Israel J. Tech.*, Vol. 11, 1973, pp. 169–177.
100. Chato, J. C. and Shitzer, A.: Analytical prediction of the heat transfer from a blood vessel near the skin surface cooled by a symmetrical strip. *Trans. ASME, J. Fluids Engng.*, (to be published).
101. Ma, Y. H., Rust, L. W., Larson, R. E., and Spano, L. A.: Mathematical simulation of simultaneous energy and mass transfer process in a clothing-airspace-skin system. *Proceedings, Summer Computer Simulation Conference, N.Y. Assn. for Computing Mach.*, Vol. 2, 1970, pp. 859–867.
102. Iberall, A. S.: Human body as an inconstant heat source and its relations to clothes insulation. Part I–descriptive models of the heat source. *Trans. ASME, J. Basic Engng.*, Vol. 81, 1960, pp. 96–102.
103. Mitchell, J. W., Nadel, E. R., and Stolwijk, J. A. J.: Respiratory weight loss during exercise. *J. Appl. Physiol.*, Vol. 32, 1972, pp. 474–476.
104. Collins, J. C., Pilkington, T. C., and Schmidt-Nielsen, K.: A model of respiratory heat transfer in a small mammal. *Biophysical Journal*, Vol. 11, 1971, pp. 886–914.
105. Shitzer, A. and Chato, J. C.: Further studies on the dimensionless parameters associated with the *in vivo* transport of heat within biological media. *Aerospace Medicine*, Vol. 42, 1971, pp. 1279–1283.

BIBLIOGRAPHY

Atkins, A. R.: An electrical analogue of heat flow in the human body. *The South African Mechanical Engineer,* Vol. 13, 1963, pp. 40–46.

Atkins, A. R. and Wyndham, C. H.: A study of temperature in the human body with the aid of an analogue computer. *Pflügers Arch.,* Vol. 307, 1969, pp. 104–119.

Barker, R. S.: Status report biothermal man Fortran program. Douglas Missile and Space Systems Division, 1966.

Beckman, W. A., Mitchell, J. W., and Porter, W. P.: Thermal model for prediction of a Desert Iguana's daily and seasonal behavior. *ASME Paper,* 71-WA/HT-35, 1971.

Birkebak, R. C.: Heat transfer in biological systems. *International Review of General and Experimental Zoology,* Vol. II, 1966, pp. 269–344.

Birkebak, R. C., Creamers, C. J., and Lefeevre, E. A.: Thermal modeling applied to animal systems. *Trans. ASME, J. Heat Transfer,* Vol. 86, 1966, pp. 125–130.

Bligh, J. and Moore, R. E.: *Essays on Temperature Regulation,* North Holland Publishing Co., Amsterdam, 1972.

Brown, A. C.: Analog computer simulation of temperature regulation in man. Report No. AMRL-TDR-63-116, 1963.

Brown, A. C.: Further development of the biothermal analog computer. *Aero. Med. Res. Lab.,* University of Washington, 1966, AMRL-TR-66-197.

Burriss, W. L., Lin, S. H., and Berenson, P. J.: Study of the thermal process for man-in-space. NASA-CR-216, 1965.

Burton, A. C.: The application of the theory of heat flow to the study of energy metabolisms. *J. Nutrition,* Vol. 7, 1934, pp. 497–533.

Chambers, A. B., Blackaby, J. R., and Miles, J. B.: A study of the thermoregulatory characteristics of a liquid-cooled garment with automatic temperature control based on sweat rate: Experimental investigation and bio-thermal man-model development. Nasa TN D-7311, 1973.

Chan, A., Sigelmann, R. A., Guy, A. W., and Lehman, J. F.: Calculation of temperature distribution in layered tissues due to distributed thermal sources by finite difference method. *Proc. 22nd Conference on Engineering in Medicine and Biology*, 1969, p. 27.3.

Chan, A. K., Sigelmann, R. A., Guy, A. W., and Lehman, J. F.: Calculation by the methods of finite differences of the temperature distribution in layered tissues. *IEEE Transactions on Biomedical Engineering,* Vol. BME-20, 1973, pp. 86–90.

Crosbie, R. J., Hardy, J. D., and Fessenden, E.: Electrical analog simulation of temperature regulation in man. *Temperature, Its Measurement and Control in Science and Industry,* J. D. Hardy, ed., Vol. 3, 1963, pp. 627–636.

Herrington, L. P.: The biotechnical problem of the human body as a heat exchanger. *ASME Trans.,* Vol. 80, 1958, pp. 343–346.

Herrington, L. P.: Full scale human body-model thermal exchange compared with equational condensation of human calorimetric data. ASME, *J. Heat Transfer,* Vol. 81, 1959, pp. 187–194.

Houdas, Y. M., Sauvage, M. A., and Fessard, M. A.: Regulation of heat loss in man submitted to external heat. *C. R. Acad. Sc. Paris,* Vol. 273, 1971, pp. 2294–2297.

Hsu, F. T., Fan, L. T., and Hwang, C. L.: Unsteady-state simulation of the human thermal system. *Proceedings of the 23rd Conference on Engineering in Medicine and Biology,* Vol. 12, 1970, p. 80.

Hsu, F. T., Fan, L. T., and Hwang, C. L.: Steady-state simulations of the human thermal system. Institute for Systems Design and Optimization, Kansas State University, Report No. 23, 1970.

Kandror, I. S. and Novosel'tzev, V. N.: A possible approach to simulation of the human organism's thermoregulatory system. *Isdatel'stvo Nauka,* 1960, pp. 35–42.

Layne, R. S. and Barker, R. S.: Digital computer simulation of the biothermal man. Douglas Missile and Space Systems Division, Paper No. 3369, 1965.

Lee, W. K.: Distributed-parameter model for passive thermal behavior of a cooling biological system. University of Missouri, Columbia, Department of Chemical Engineering, 1970.

Lipkin, M. and Hardy, J. D.: Measurement of some thermal properties of human tissue. *J. Appl. Phys.*, Vol. 7, 1954, pp. 212–217.

Luecke, R. H., Gray, E. W., and South, F. E.: Simulation of passive thermal behavior of a cooling biological system: Entry into hibernation. *Pflügers Archiv.*, Vol. 327, pp. 37–52.

Miles, J. B., Chambers, A. B., and Blackaby, J. R.: Computer simulation of the physiological parameters utilized in the automatic temperature control of liquid cooled garments. *Proceedings of the Summer Computer Simulation Conference,* 1972, p. 1129.

Molnar, G. W., Hurley, M. J., Jr., and Ford, R.: Application of Newton's law of body cooling. *Pflügers Arch.,* Vol. 311, 1969, pp. 16–24.

Montgomery, L. D.: A simulation of heat transfer in man under immersed conditions. Ph.D. Dissertation, University of California, Los Angeles, 1972.

Morgan, L. W., Collett, G., and Cook, D. W., Jr.: Computer program documentation transient metabolic simulation program. Lockheed Electronics Company, Report No. LEC/672-22-08068, 1970.

Nevins, R. G. and Darwish, M. A.: Heat transfer through subcutaneous tissue as heat generation porous material. *Physiological and Behavioral Temperature Regulation,* J. D. Hardy et al., eds., 1971, pp. 281–301.

Osman, D. B.: An electrical analog of the biothermal system. M.Sc. Thesis, University of California at Los Angeles, 1962.

Piironen, P. and Takalo, K.: Assessment of regional heat losses for verification of mathematical analogues of the human thermal system. AD 720830, 1970.

Piironen, P. and Takalo, K.: The control of human thermoregulatory heat production. AD 720831, 1970.

Poppendick, H. F., Randall, R., Breeden, J. A., Chambers, J. E., and Murphy, J. R.: Thermal conductivity measurements and predictions for biological fluids and tissue. *Cryobiology,* Vol. 3, 1966, pp. 318–327.

Potters, W. P. and Gates, D. M.: Thermodynamic equilibria of animals with environment. *Ecological Monographs,* Vol. 39, 1969, pp. 227–244.

Priebe, L.: Heat transport and specific blood flow in homogeneously and isotropically perfused tissue. *Physiological and Behavioral Temperature Regulation,* J. D. Hardy et al., eds., 1971, pp. 272–280.

Reid, M. H. and Mackay, R. S.: Application of new technique for system identification to dolphin temperature regulation and rabbit intraocular pressure control. *Medical and Biological Engineering,* Vol. 6, 1968, pp. 269–290.

Seagrave, R. C. and Miller, N. C.: A model for human thermoregulation. *Proceedings of the 24th Annual Conference on Engineering in Medicine and Biology,* Vol. 13, 1971, p. 56.

Smith, P. E.: Analog simulation of the physiological responses of men working in hot environments. *Proceedings of the San Diego Symposium for Biochemical Engineering,* 1962, pp. 117–125.

Smith, P. E. and James, E. W.: Human responses to heat stress-simulation by analog computer. *Archives of Environmental Health,* Vol. 9, 1964, pp. 332–342.

Spotila, J. R., Lommen, P. W., Bakken, G. S., and Gates, D. M.: A mathematical model for body temperatures of large reptiles: Implications for dinosaur ecology. *The American Naturalist,* Vol. 107, 1973.

Stolwijk, J. A. J.: A mathematical model of physiological temperature regulation in man. John B. Pierce Foundation Laboratory, Report No. NAS 9-9531.

Stolwijk, J. A. J.: A mathematical model of physiological temperature regulation in man. NASA CR-1855, 1971.

Stolwijk, J. A. J.: Mathematical model of thermoregulation. *Physiological and Behavioral Temperature Regulation,* J. D. Hardy, et al., eds., 1971, pp. 703–721.

Stolwijk, J. A. J. and Hardy, J. D.: Skin and subcutaneous temperature changes during exposure to intense thermal radiation. *Thermal Problems in Aerospace Medicine,* J. D. Hardy, ed., 1968, pp. 31–46.

Stow, R. W. and Schieve, J. F.: Measurement of blood flow in minute volumes of specific tissues in man. *J. Appl. Physio.,* Vol. 14, 1959, pp. 215–224.

Taylor, C. L.: Heat transfer applications in the human body. *Mech. Eng.,* Vol. 77, 1955, pp. 511–513.

Wessling, F. C. and Blackshear, P. L.: The thermal properties of human blood during the freezing process. *Trans. ASME, J. Heat Transfer,* Vol. 95, 1973, pp. 246–249.

Westland, R. A.: A biothermal analog. M.Sc. Thesis, University of California at Los Angeles, 1958.

Whittow, G. C., ed.: *Comparative Physiology of Thermoregulation,* Vol. II–Mammals, Academic Press, New York, 1971.

Winton, H. J. and Linebarger, R. N.: Digital simulation of human temperature control. *Proceedings of the Summer Computer Simulation Conference,* Vol. 2, 1970, pp. 830–836.

Wissler, E. H.: Steady-state temperature distribution in man. *J. Appl. Physiol.,* Vol. 16, 1961, pp. 734–740.

Wissler, E. H.: An analysis of factors affecting temperature levels in the nude human. *Temperature–Its Measurement and Control in Science and Industry,* J. D. Hardy, ed., Vol. 3, part 3, 1962, pp. 603–611.

Wissler, E. H.: A mathematical model of the human thermal system. *Bulletin of Mathematical Biophysics,* Vol. 26, 1964, pp. 147–166.

Wissler, E. H.: A mathematical model of the human thermal system. *Chem. Eng. Progress Symp. Series,* Vol. 62, 1966, pp. 66–78.

Wissler, E. H.: The use of finite difference techniques in simulating the human thermal system. *Physiological and Behavioral Temperature Regulation,* J. D. Hardy et al., eds., 1971, pp. 367–388.

Woodcock, A. H. and Breckenridge, J. R.: A model description of thermal exchange for the nude man in hot environments. *Ergonomics,* Vol. 8, 1971, pp. 223–235.

Wyndham, C. H. and Atkins, A. R.: An approach to the solution of the human biothermal problem with the aid of an analogue computer. *Proceedings of the 3rd International Conference on Medical Electronics,* 1960, pp. 32–38.

Wyndham, C. H. and Atkins, A. R.: A physiological scheme and mathematical model of temperature regulation in man. *Pflügers Arch.,* Vol. 303, 1968, pp. 14–30.

Yamamoto, W. S. and Raub, W. F.: Models of the regulation of external respiration in mammals: Problems and Promises. *Computers and Biomedical Engineering,* Vol. 1, 1967, pp. 65–104.

ACKNOWLEDGEMENT

The manuscript was typed by Mrs. S. Rosenthal. Her highly expert work is gratefully appreciated. Figures were prepared by Mrs. A. Rosen, whose help is also appreciated.

4

Tracer Methods in the Circulation

Herbert Weinstein

Department of Chemical Engineering
and Center for Biomedical Engineering
Illinois Institute of Technology
Chicago, Illinois, United States

and

Milorad P. Duduković

Department of Chemical Engineering
Washington University
St. Louis, Missouri, United States

CONTENTS

1 INTRODUCTION

1.1 General

The use of tracers to obtain information on the circulation dates back to Hering (73, 74), who in 1829 injected dyes into blood to determine blood circulation times. Stewart introduced the concept of using tracers to obtain blood flows in 1894 (172–174), and these ideas were subsequently developed and refined by Hamilton and coworkers (70,71). More recently, tracer techniques have been the subject of intensive development and are accepted as an extremely useful tool in studying complex biological systems. Tracer methods are used today for the determination of blood flow patterns, blood volume, and regional and local flow rates. At the same time they can be used successfully in measurements of global, gross effects of the transcapillary exchange processes. Use of the tracer method has especially increased with the development of radioactive tracers, and a large number of papers appeared in the last few decades discussing the various aspects of their use.

The results obtained by tracer methods do not yield values of the physical properties of the system but rather, a set of coefficients. These have to be

Partial support of this work by the National Science Foundation under grant ENG-74-13520 is gratefully acknowledged.

interpreted on the basis of a mathematical model in order to extract an estimate of the physical properties. How good an estimate is depends on the model chosen and the restrictions incorporated in its formulation. The parameters obtained in such a way are useful estimates of the physical quantities, especially at the present time, since this is the only way to assess their value. Mathematical models have another advantage in that, once the parameters are established, they can be used in predicting the behavior of the system.

It is the objective of this paper to present an overview of the basic principles of the tracer method and review, in particular, the use of the method in studies of blood circulation and transcapillary exchange processes. It is hoped that the material presented and the critical appraisal of the power and limitations of the method will create broader appreciation of the method, improve communication between workers in various fields, and enable experts from different disciplines to enter readily into this particular area of study. Although tracer techniques and mathematical models applied to the study of blood circulation and capillary exchange will be reviewed, the objective is not so much to cover in detail all the papers published in the area (an almost impossible task), but rather to demonstrate through the presentation of selected studies the various trends in this important research area.

1.2 Stimulus-response, the tracer method

In recent years, with the general availability of high-speed, large memory computers, mathematical modeling has probably become the single most important aspect of the design and analysis of engineering systems. It has also gained increased acceptance in previously rather descriptive disciplines of biological science. Mathematical modeling consists of formulating mathematical relationships which describe the behavior of the actual system such that the dependent and independent variables and parameters of the model are directly related to the physical and chemical quantities in the real system. The mathematical model can then be used to replace the original system in investigating a number of aspects of the system. In model building, one attempts to deduce the structure of an equation on the basis of observations of the real system. For some simpler systems it is sometimes possible to derive model equations starting from basic principles. More often, due to the complexity of the system, it is not feasible to represent it by fundamental equations. It is also difficult to establish values of necessary parameters due to our incapability of observing the system behavior under ideal conditions. In this situation it is necessary to devise the model of the system on the basis of observations of the system alone. An extremely useful tool in this type of analysis is the stimulus-response method which consists in perturbing the system in a prescribed way and observing the behavior (response, output) of the system to such perturbations. From a set of stimulus-response (input-output) data it is possible not only to establish a mathematical model for the system, but to also evaluate some of the necessary parameters.

The tracer method is a technique for observing the system, which usually consists of a population of specific entities such as molecules, by a process of labeling. The perturbation of the system (stimulus) is performed by introduction of a representative set of elements of the labeled specie (tracer, indicator), which are specifically marked in some way. This set may be considered a subset of a larger set of elements of the system (traced specie). From the movement, distribution, or behavior of the labeled specie (response) one generally infers the behavior of the traced specie and of the system itself.

1.3 Treatment of biological systems

Physiological functions in highly organized mammalian organisms consist of a series of complex metabolic processes. These processes, as has long been realized, are supported by blood circulation. Its complicated network of pathways, consisting of large and minute blood vessels, supplies oxygen and nutrients through various transport mechanisms to the living cells and removes their metabolic wastes. In order to properly assess the effects of many drugs used in chemotherapy and anesthesiology, to understand the functioning of body organs, and to properly diagnose pathological conditions, it is necessary to obtain quantitative information about the blood circulation and related exchange processes. At the present time it is almost impossible to measure the effect of these processes on a microscopic level. Important parameters such as local blood velocities, active blood volumes, permeability of capillary walls, and diffusivities in living tissue have not been measured directly. The physical properties of the system and its structure remain largely unknown, which makes any estimation of these parameters very difficult. At the same time the theory of non-Newtonian flow in vessels with viscoelastic walls subject to pulsations, seepage through the walls, vessel contractions and branchings, is in its infancy, and understanding the basic driving forces in membrane exchange is hardly complete. Faced with the extreme complexity of the system, insufficient theoretical knowledge, and lack of direct experimental methods, researchers have tried to employ indirect techniques for estimation of otherwise inaccessible parameters. We are concerned here with the portion of this work which employs tracer methods.

A biological system consists of a set of molecular species which can be considered as a continuum. The tracer is usually a different molecular specie from those normally present. Since the tracer is used to obtain information on flow patterns, diffusion fluxes, and such, it should have the same characteristics as the carrier fluid molecules or traced specie, and should be subject to the same external forces. If these conditions are met, then the average tracer molecule will behave like an average carrier fluid molecule. However, the only molecules and atoms that will exactly imitate the behavior of the population being traced are those that are identical with the population, and are thus incapable of separate identification. Thus, a perfect tracer does not exist. In each particular case, a tracer can be selected carefully so that useful data are obtained. The results of the experiments are interpreted in terms of idealized conceptual models in which labeled and nonlabeled species are assumed to behave identically. This often causes only negligible errors, due mainly to the difference in the diffusion fluxes of the tracer and traced specie.

1.4 Ways of using tracer information

The three most common tracer tests are pulse, step, or periodic injection of tracer. Various types of system variables may be examined as a response to such a stimulus. When properties of the circulation are studied, tracer molecules may be injected into the circulation and their reappearance at the injection site, or some other site, may be detected. This information yields estimates of blood flow rate and active blood volumes in circulation.

On the other hand, a tracer substance may be injected intravenously or intramuscularly in order to investigate rates of transcapillary exchange. Particular molecular species undergoing various metabolic processes may also be tagged. Detection of tagged specie in various parts of the organism and in metabolic wastes, yields information about the dynamics of species distribution through the organism, and provides an estimate of metabolic rates.

Thus, tracer methods may be successfully employed in research as well as in clinical diagnostic techniques. Interpretation of results obtained in such a way is presented in the next sections.

2 BASIC PRINCIPLES OF THE TRACER METHOD

2.1 General

Tracers have been used in the analysis of many types of systems. Categorization of these systems may be made on many different bases. We can arbitrarily separate systems into the categories of static or dynamic, open or closed, and single or multiphase. However, the categories are obviously not mutually exclusive, since a system with any number of phases must be either static or dynamic. Furthermore, it is the tracer measurement technique itself which determines to some extent whether the system is static or dynamic, open or closed, and single or multiphase. The equilibration method for obtaining blood volume is a static measurement in that it considers the circulation to be a static pool of blood. The Stewart-Hamilton measurement for cardiac output is a dynamic measurement on a portion of the same circulation. A tracer used in the circulation which decays rapidly gives an open system measurement on part of a closed system. Tracers such as cardiogreen, which is used in the Stewart-Hamilton method, are plasma tags, and the blood may have to be considered as a multiphase fluid when using them. Oxygen, on the other hand, is present in both cells and plasma and for its use as a tracer the blood could be considered as a single phase fluid. We will present basic principles of the tracer method for a dynamic, homogeneous system first, and then discuss modifications for applications to the other categories. The following discussion is kept brief and includes the minimum amount of material possible. The reader wishing to obtain a good background in the fundamentals is urged to read the articles of Meier and Zierler (115), Zierler (194), and a chemical reaction engineering text such as Levenspiel (103). The following discussion draws heavily on these references.

2.2 System restrictions

We must limit our discussions to systems which have the properites of stationarity and linearity, and which are conservative with respect to tracer and carrier fluid. We may relax a restriction for a special case but analysis of that case must then be specific to the situation. We will now examine these restrictions individually. Stationarity is the property of a system in a steady state. The inlet and outlet conditions are constant with time, as well as the properties of the system itself. Included in what we call the inlet and outlet conditions are the probability functions concerning behavior of the fluid in the system which we will develop below. Linearity with respect to tracer means that response of the system to successive tracer inputs is just additive; there is no interaction between inputs (8). The final restriction is self explanatory; neither tracer nor carrier fluid is consumed or produced in the system. This restriction is, however, the one most often violated in biological systems.

2.3 Conservation of mass laws

We can develop the basic notions of tracer analysis by performing elementary mass balances on the tracer and carrier fluid, respectively, in a vessel in an open system. We define an open system as one in which no part of the exit stream from the vessel reenters the vessel for a second passage. A closed system is one in which all of the exit stream reenters the vessel and new fluid is never added to the system. A recirculating system is one in which a portion of the exit stream is recycled to the entrance, and fresh fluid in an amount equal to that portion of the exit stream which leaves the system, is added to the recycled portion at the vessel entrance. Consider a vessel of volume V, through which there is a steady volumetric flow rate of F. The fluid is incompressible and inert, and there is a single exit to and a single exit from the vessel. With these constraints, the system will be stationary (steady state) and conservative. A bolus or mass, M_i, of tracer is injected into the carrier fluid in an ideal fashion as it enters the vessel. An ideal injection means the tracer enters the system over an infinitesimal time period. The tracer itself must behave physically in the vessel exactly as does the carrier fluid. The amount of tracer leaving the system between t and $t + dt$ is then

$$FC\,(t)\,dt$$

where $C\,(t)$ is the concentration at the exit. Now after a very large amount of time has passed after injection of tracer, the tracer is completely washed out of the vessel and a material balance just equates input and output of tracer

$$\int_0^\infty M_i \delta\,(t)\,dt = \int_0^\infty FC(t)\,dt$$

or

$$\frac{M_i}{F} = \int_0^\infty C(t)\,dt \tag{2.1}$$

This is our mass balance on tracer. We can construct from this balance a probability density of residence or transit times

$$E(t) = \frac{C(t)}{M_i/F} \tag{2.2}$$

This is often called a residence or transit time distribution (RTD) or exit age distribution, although in probability "language" it is not a distribution. $E(t)dt$ is the fraction of fluid which will spend a time between $t - dt/2$ and $t + dt/2$ in the vessel. The probability distribution of residence times is simply

$$G(t) = \int_0^t E(\tau)\,d\tau \tag{2.3}$$

The mass balance on tracer forms the basis for the calculation of cardiac output with the indicator dilution technique. If we know the mass of tracer injected, M_i, and can measure the concentration of tracer as a function of time, then the flow rate is

$$F = \frac{M_i}{\int_0^\infty C(t)\,dt} \tag{2.4}$$

The mass balance on carrier fluid also yields an important result. We can write this balance by first considering that fluid which has a residence time between t and $t + dt$ enters the vessel and leaves the vessel with rate $FE(t)dt$. The volume (volumetric flow rate × time) of fluid in the vessel with this range of residence times is simply $t(FE(t)dt)$, since this fluid flows through the vessel with a well defined superficial velocity and without interacting with other fluid according to our linear description. The total volume of fluid actively flowing in the vessel which equals the vessel volume is the sum of all these differential volumes

$$V = \int_0^\infty tFE(t)\, dt$$

or

$$\bar{t} = \frac{V}{F} = \int_0^\infty tE(t)\, dt$$

These two mass balances are the zeroth and first moments of the probability density function, $E(t)$. We have shown that normalization of the zeroth moment provides a measure of flow rate, and we can show that the first moment provides a measure of system volume–the flow volume between the injection and sampling sites.

$$V = F\int_0^\infty tE(t)\, dt = F\frac{\int_0^\infty tC(t)\, dt}{\int_0^\infty C(t)\, dt} \tag{2.5}$$

The residence time distribution contains all the information that we can get from a tracer measurement except for the normalization factor which gives us M_i/F. If we measure tracer response in arbitrary units, say counts per minute for a radioactive tracer, M, and cannot relate the measurement to actual tracer concentration, we can still obtain the $E(t)$ curve

$$E(t) = \frac{M(t)}{\int_0^\infty M(t)\, dt} \tag{2.6}$$

provided a bolus or pulse injection of tracer was made. However, we lose the ability to calculate flow.

A related problem is that of constant infusion or step function input of tracer. If we, at time $t = 0$, add tracer continuously to the inlet stream to the system, we then obtain at the outlet of the system a tracer concentration curve which starts at zero and assymptotically approaches the tracer concentration in the inlet stream, C_i. Since the integral of a Dirac delta function is a unit step function,

$$\int_{-\infty}^{\infty} \delta(t)\, dt = U(t) \tag{2.7}$$

where $U(t) = \begin{matrix} 0, t < 0 \\ 1, t < 0 \end{matrix}$

then the response of the system* to an input $C_0U(t)$ is the integral of $E(t)$

$$C_i \int_0^t E(\tau)\, d\tau = C_i G(t) \tag{2.8}$$

and we have lost the information available in the normalization of the tracer response to bolus injection. We can, however, obtain V/F, the perfusion rate from a constant infusion measurement.

$$\frac{dG(t)}{dt} = E(t)$$

so

$$\frac{V}{F} = \int_0^{\infty} tdG \tag{2.9}$$

So far we have worked with the zeroth and first moments of $E(t)$. There are in fact, an infinite number of moments of $E(t)$, and often it is better to represent the function by its moments in comparing models with experimental data. We can write in general

$$\mu_n = \int_0^{\infty} t^n E(t)\, dt \tag{2.10}$$

*See next section for a discussion of convolution.

where μ_n is the nth moment of $E(t)$ about the origin and n takes integral values from zero to infinity. Now the set of functions, $1, t, t^2, t^3, \ldots$ or $t^n, n = 0, 1, 2, \ldots$, is a complete set of functions which can be used over any finite interval to uniformly approximate $E(t)$ as long as it is a continuous function.* Since all

$$\mu_n \approx \int_0^Y t^n E(t)\, dt$$

where Y can be as large as we like–large enough so that $E(t)$ is undistinguishable from zero with respect to any concentration measuring device or such that

$$1 - \int_0^Y E(t)\, dt = \delta$$

where δ is as small as we wish. The set of moments μ_n is then a complete description of the function $E(t)$.

An alternate derivation of this result may be done with Laplace transforms. Taking the transform of the moment definition

$$\bar{\mu}_n = (-1)^n \frac{d^n}{ds^n} \bar{E}(0) \tag{2.11}$$

where s is the transform variable and super bar denotes the Laplace transform. A Taylor series expansion of $\bar{E}(s)$ about $s = 0$ gives

$$\bar{E}(s) = \sum_{n=0}^{\infty} \frac{1}{n!} \frac{d^n}{ds^n} [\bar{E}(0)] s^n \tag{2.12}$$

The coefficients of the Taylor series expansion are directly related to the transform of the moments and so the transformed moments completely describe the transform of $E(t)$. Again there is a technical problem with the region of convergence since we are concerned with the semi-infinite region $s \geqslant 0$. However, this problem has no practical significance. We really need only that $\bar{E}(s)$ is continuous and have continuous derivatives.

There are practical problems in the use of moments which usually restrict our use to just the first few. Each successive moment weights the long time response more heavily. However, the long time response, or tail of the curve, is usually of very small amplitude. When the response diminishes to the noise level of the instrumentation used in recording it, recording is usually ended. We have considerable error, then, at very low values of response and the weighting in the

*Weierstrass's Approximation Theorem (41).

moment calculation amplifies these errors. Moments greater than the third are usually never used for this reason. The first three moments, and often only the first two, are taken as a complete description of the response. Variance, or spread about the mean, is obtained from the second and first moments

$$\sigma^2 = \int_0^\infty (t - \mu_1)^2 E(t)\, dt = \mu_2 - \mu_1{}^2 \tag{2.13}$$

and is often used as a measure of tracer dispersion. The third moment of the function is related to the skewness.

2.4 Closed and recirculating systems

In order to examine closed and recirculating systems we must use the property of linearity. The principle of superposition implies the response of a linear system to successive inputs will simply be additive. This can be formalized with the convolution integral. If $E(t)$ is the response of a system to a unit impulse input, then the response of the system to a succession of arbitrary inputs is

$$C_o(t) = \int_0^t E(t - \tau) C_i(\tau)\, d\tau \tag{2.14}$$

where $C_i(t)$ is the function describing the summation of input signals and $C_0(t)$ is the output function or system response.

In dealing with closed or recirculating systems we make use of this integral by writing that the output function from the vessel makes up the appropriate part of the inlet function in the integral. A complete discussion of these systems follows in a later section.

It is interesting to examine how convolution can be done by using the moments of the curves. If we have in general

$$C_o(t) = E(t) * C_i(t)$$

where * denotes convolution, and take the Laplace transform of the equation we obtain

$$\bar{C}_o(s) = \bar{E}(s) \cdot \bar{C}_i(s)$$

The normalization is done by setting s to zero

$$\bar{C}_o(0) = \bar{C}_i(0)$$

since $E(t)$ is normalized ($\bar{E}(0) = 1$). We now normalize $\bar{C}_0(s)$ and $\bar{C}_i(s)$

$$\bar{c}_o(s) = \frac{\bar{C}_o(s)}{\bar{C}_o(0)} \qquad \bar{c}_i(s) = \frac{\bar{C}_i(s)}{\bar{C}_i(0)}$$

and differentiate and evaluate at $s = 0$

$$\frac{d\bar{c}(0)}{ds} = 1 \cdot \frac{d\bar{E}(0)}{ds} + 1 \cdot \frac{d\bar{c}_i(0)}{ds}$$

or

$$\mu_{1_o} = \mu_{1_E} + \mu_{1_i} \tag{2.15}$$

which says that first moments are added. We can differentiate again to obtain

$$\mu_{2_o} = \mu_{2_i} - 2\mu_{1_E}\mu_{1_i} + \mu_{2_E} \tag{2.16}$$

By repeated differentiation we can obtain each moment of $C_0(t)$ in terms of moments of the same order and smaller of the $C_i(t)$ and $E(t)$ curves. From the moments we can then construct the series approximation of $C_0(t)$. More importantly we can do the inverse, or deconvolution in this manner. We can solve the moment equations for the moments of $E(t)$ if we know $C_0(t)$ and $C_i(t)$ and construct $E(t)$ from them. The problem with this method of deconvolution is in the accuracy with which we can obtain the moments of the experimental curves.

2.5 Static systems

The dynamic system has been treated above. That it is a dynamic system which we have treated is clear since we are measuring tracer concentration as a function of time in order to get our probability density function of transit times. A good example of a static measurement technique is the equilibration method for determining blood volume. In this method, a known mass of tracer is injected into the blood stream and allowed to become uniformly dispersed. Then a sample of blood is taken and the concentration of tracer determined. The volume of dilution is simply

$$V = \frac{M_i}{C_t} \tag{2.17}$$

where V is the volume of dilution, M_i the mass of tracer injected, and C_t the tracer concentration at equilibrium (100). The problem associated with this and other static techniques in living systems is that the system itself is dynamic irrespective of the measurement technique. The system acts on the tracer, as it does on all materials, with a time constant which usually is small enough to affect the tracer concentration in a static method. Static methods usually have long measurement time associated with them compared to dynamic measurements. These actions by the system typically include purging of tracer and diffusion of tracer into extravascular fluid. Corrections to the static measurement are then made which attempt to correct for system dynamics, such as extrapolation of tracer concentration to zero time. We shall only consider dynamic techniques in the following material.

2.6 Single and multiphase systems

Blood is a two phase fluid being made up of plasma and cells. Tracers usually are confined to either the plasma volume or the cell volume. Since it is held that cells move with a velocity greater than that of plasma, errors can be introduced by considering the blood to be homogeneous, as is usually done. The errors come about because it is usually assumed that the tracer concentration measurement yields a volume average rather than a flow average. Concentration measurements made on samples which have been collected should be interpreted as flow average whereas measurements made directly on the flowing blood, such as in an optical densitometer, reflect more closely a volume averaging. The effect of the inhomogeneity of blood has been treated by Meier and Zierler (115) for the case of a flow averaged concentration measurement. The magnitude of the error in assuming that the system is homogeneous is shown to be small. We will confine ourselves hereafter to this homogeneous assumption.

3 TRACERS AND SYSTEM IDENTIFICATION

3.1 Basic concepts of biological modeling

A mathematical model represents a mathematical interpretation of the observed behavior of a system and can be used as a replacement for the system in testing the system's behavior under various conditions. Only in very special cases will a mathematical model be able to represent all details of the system. In general, the model is a description of the most important and relevant observable features of the system. Because of the difficulties involved, modeling is an adaptive procedure in which information gained from each attempt is used to discriminate between competing models and improve the next try. Mathematical models of biological systems serve two basic purposes: (a) simulation of real biological systems-simulation models, (b) parameter evaluation of biological systems-parameter estimation models.

3.2 Simulation models

Whenever the physical parameters of the system are known (through independent experimental measurements) and when a sufficient amount of data relating the causes and effects are available, i.e., stimulus-response of the system, and/or the physical laws governing the behavior of the system are reasonably well understood—simulation models may be developed to quantify relationships between causes and effects.

Once mathematical description of the functioning of the system is formulated, it should be tested against available experimental data. If agreement is good, the simulation model may be used to

1. study the behavior of the system under normal conditions,
2. study the effect on the system of a change in certain parameters (which in a real system may be achieved by therapeutic or surgical methods),
3. design and plan new experiments.

Simulation models should be only as accurate as the available experimental techniques. In formulating them, many secondary effects have to be neglected and should be incorporated at a later stage when better experimental data become available. A schematic of the development of a simulation model is presented in Figure 1.

Figure 1 Development of simulation models (schematic).

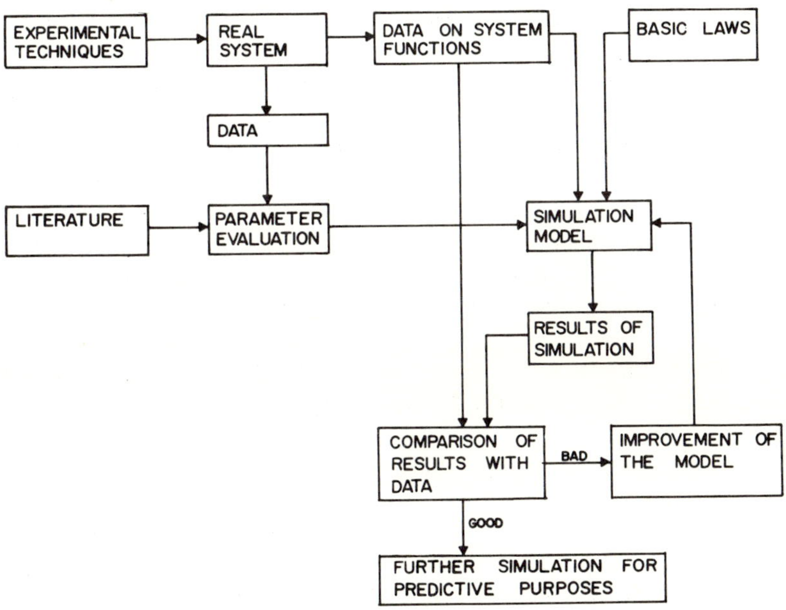

Simulation models may be formulated to describe pulmonary circulation (91), arterial blood flow (7), or drug distributions in mammalian organisms (20). The above mentioned models are just a few examples of a successful use of simulation models.

3.3 Parameter estimation models

In complex biological systems it is often impossible to measure certain physical parameters directly, especially as their true values cannot be determined independently *in vitro* but have to be estimated *in vivo*. The real system is then subjected to a rigorously controlled input (stimulus), and the output (response) of the system is recorded. It is then necessary to simulate the relationship between the input and the output from the formulated model and the input-output data in order to infer the values of pertinent parameters. If the main form of the model can be specified from basic principles of physics, chemistry, and biology, then the only unknowns are the coefficients of the model which can be estimated from data. These are true parameter estimation models. Due to the complexity of biological systems, it is often impossible to formulate a unique model. Statistically designed additional experiments, performed under such conditions that discrepancy between competing models is maximized, usually lead to a selection of the "best" model. Further experimentation planned under such conditions as to minimize a desired error criterion in parameter space will provide good estimates of system parameters.

When the system under consideration is extremely complex and, due to a multiplicity of competing processes, it is not clear which phenomena contribute most significantly to system response, one is faced with the problem of system identification. Pairs of input-output data are used to determine the transition matrix of the system, i.e., to determine the input-output relationship. On the basis of this information it is often possible to establish a model for system description and proceed with parameter evaluation. The theory of system identification emerges from the work in linear filtering and control, and is based on the developments of Kalman (86), Ho and Kalman (76), Bruni et al. (24), and Rossen and Lapidus (143). To the authors' knowledge, methods of system identification have not as yet been applied to tracer studies of biological systems.

Parameter estimation models have been used extensively to interpret data obtained by tracer methods. These models

1. provide a first estimate of otherwise unmeasurable parameters of biological systems,
2. indicate which type of experiments should be performed in order to distinguish between competing models, and therefore,
3. lead to better understanding of the biological systems.

In the preceding discussion we have divided biological mathematical models according to their main objective into simulation models and parameter estimation models. This division is a somewhat arbitrary one, as it can be argued that once the relationships between causes and effects are expressible in mathematical terms, every model becomes a simulation model. However, the above mentioned division can always be made according to the immediate objective of the model. A truly reliable simulation model should simulate a function of the system under certain conditions without adjusting parameters, as they should have been determined by independent experiments. These models are reliable and can be extrapolated to predict behavior of the system under unknown conditions (provided that these changes do not cause a variation of parameter values). Parameter estimation models hopefully simulate the system behavior, but actually are just a sophisticated version of curve fitting. Successful matching of the output of the model with the collected output data which is achieved by adjusting parameters, unfortunately does not immediately imply that the model "simulates" the function of the system, and further studies are necessary before the model can be accepted as a simulation one.

The main point is that, although one may choose to look at all models as being the same without distinguishing between the two categories, in biological modeling the distinction has to be made between whether the parameters were evaluated by independent measurements or whether through adjustment and matching of the model and system response. In the latter case the results predicted by the model subjected to new conditions should always be viewed with great caution.

As we have already mentioned, the classification of mathematical models is a difficult and somewhat arbitrary task. However, there are a number of common features which can be used to group the types of models. We can distinguish first of all between deterministic and stochastic models. In deterministic models the values of variables and parameters have definite fixed values, while stochastic models admit elements of randomness in the mathematical description. Although stochastic models are a closer description of most biological systems, often the use of deterministic models is simple and leads to good results. The basic principles of the tracer method can be established by deterministic and stochastic methods and lead to equivalent conclusions.

Biological models may also be classified according to their mathematical formulation into two groups: (a) lumped parameter models which are represented by ordinary differential equations and algebraic equations, with time as the only independent variable; (b) distributed parameter models which are represented by partial and ordinary differential equations with time and space being independent variables. It is worth pointing out that the terms lumped and distributed parameter models are misnomers as the variables are lumped or distributed as well as the parameters. Depending on whether the variables are continuous or may only take discrete values, one may also distinguish between continuous and discrete models.

3.4 Lumped parameter models

Due to the great organizational complexity of live organisms it is extremely difficult to obtain a precise geometrical description of the system under study. This is true whether it is the whole organism or only some of it parts (organs) that are being investigated. This is the main reason why lumped parameter models have gained wide acceptance in studies of biological systems. Faced with very complex or even unknown geometries, lumped parameter models are often the only choice left for interpretation of stimulus response data. A biological system under study, whether it is a whole organism or some of its functional parts, is often viewed as an assortment of pools or compartments. Compartment analysis is based on the assumption that specific pools can be identified and that the discharge of tracer can therefore be described by exponential equations. In engineering terminology the lumped parameter models of biological systems, which appear under the names of compartmental models in the biomedical literature, are simply representation of the system as a network of well mixed regions of constant volume which exchange matter among them at given, fixed rates. The system is then described with a set of first order differential equations explicit in derivatives. In engineering analysis the hypothesis of a well mixed region is often used in modeling; however, these regions are in general bounded by precisely defined boundaries. In modeling of biological systems a compartment may be identical to a geometrically well defined region enclosed by natural anatomic boundaries such as the heart, kidney, lungs, liver, etc. On the other hand, for some purposes of modeling a distinct compartment may be taken as all of a certain type of tissue, such as muscle or fat, without regard for its geometrical distribution. To some extent the selection of compartments depends upon the application at hand, and so is arbitrary. As mentioned above, often a compartment is distinguished by its exchange rate with the vascular system and not by its anatomic identity. Regions which are anatomically and functionally distinct may constitute a single compartment.

A compartment may be a physical compartment such as all the space within the cardiovascular system of an experimental animal. It may be a chemical compartment such as water, bicarbonate ion in an aqueous system containing carbon dioxide, cholesterol, or fatty tissue. Physiologists who use the tracer method frequently employ the word *pool* in place of *compartment*, and the two words may be taken as synonyms. Compartment size may signify a physical volume, such as the total number of liters of body water, but the basic quantity is the amount, not the volume of distribution. In most physical systems the compartment size is fixed and independent of the tracer used in determining it. In biological systems it can be dependent on the tracer, i.e., the distribution space of Na (24) in a certain organ (comprising a compartment) may not be equal to the distribution space of thocyanate, and thus a different compartment size may be determined depending on the tracer used.

Compartmental analysis is a popular tool among physiologists in interpretation of tracer data. The system under study is represented with a set of interconnected compartments and it is assumed that the tracer is injected rapidly into one compartment. By monitoring tracer concentration in some compartments one may infer the pool size of various compartments as well as transfer rates among compartments. Several texts have been published on this subject. Sheppard introduced the basic principles of compartment analysis and presented the required mathematical background in his book *Basic Princicples of the Tracer Method* as early as 1962 (157). Recently Steele in his book *Tracer Probes in Steady State Systems* (167) covered exclusively the applications of compartmental analysis to biological systems, while Shipley and Clark in their text *Tracer Methods for in-Vivo Kinetics* (162) also dwell mainly on lumped parameter models. Middleman (116) in one of the chapters of his recent book *Transport Phenomena in the Cardiovascular System* also presented a review of compartmental models.

The mathematical concepts of compartmental analysis are very simple. A simple component mass balance is applied to each compartment resulting in a set of ordinary differential equations relating the rate of change of a property within each compartment with the rates of input and output. Thus in a system of N compartments the mass balance on specie t (tracer) on the nth compartment yields an equation of the form

$$\frac{d({}^tC_n V_n)}{dt} = \sum_{\substack{i=1 \\ i \neq n}}^{N} {}^t r_{in} - \sum_{\substack{j=1 \\ j \neq n}}^{N} {}^t r_{nj} \tag{3.1}$$

where ${}^tC_n$ is the concentration of tracer in compartment n, V_n is the volume of the compartment, ${}^tr_{in}$ is the input rate of specie t from compartment i into compartment n, and ${}^tr_{nj}$ is the output rate of specie t from compartment n into compartment j. From further consideration of the system it can be determined which compartments may exchange matter between them. The resulting set of N equations may readily be solved analytically or numerically. Matrix notation allows for an elegant presentation of results, but in general, use of a large number of compartments is avoided in physiological tracer studies.

Tracer studies are frequently used in determining pool size of certain compartments within the body such as extracellular fluid, intracellular fluid, plasma, fat, muscle tissue, and in evaluating regional blood flow rates and transcapillary exchange rates. Simple lumped parameter models consisting of only a single compartment or two have proven to be very useful in estimating the parameters of the system. However, as experiments are performed mainly by physiologists, terminology varies from that used by engineers and this often presents difficulty in communication between experts in various areas. A few selected examples of compartmental modeling are presented below together with terminologies used in physiology and engineering.

Let us consider a well mixed region of volume V, which may consist of more than one phase. The phases within V are always in equilibrium, and density and volume of the system are constant in time. There is only one input and output, consisting of a single phase, to the system. This describes a single compartment with one inlet and outlet. A basic material balance on chemical specie j yields Eq. (3.2)

$$F(C_{jo} - C_j) = V \frac{d\bar{C}_j}{dt} \tag{3.2}$$

where F is the flow rate, C_{jo}, C_j concentration of specie j in the inlet flow and outlet flow, respectively; $\bar{C}_j$ is the volume averaged concentration within the compartment. The specie mass balance represented by Eq. (3.2) is often referred to as "Fick's principle" in the physiological literature. If Eq. (3.2) is applied to an organ, then C_{jo} is the arterial concentration, C_a, and the outlet concentration, C_j, is the venous concentration, C_v, which, according to the assumption of compartmental analysis, is always in equilibrium with the tissue concentration C.

$$C_v = \lambda C \tag{3.3}$$

where λ is the partition coefficient between blood and tissue. In most organs the blood volume is only a few percent of the tissue volume. Accordingly, Eq. (3.2) is rewritten as

$$\frac{V}{\lambda F} \frac{dC_v}{dt} + C_v = C_a \tag{3.4}$$

The response of the single compartment described by Eq. (3.4) to a step input in arterial concentration C_a is given by

$$C_v = C_a(1 - e^{-kt}) = \lambda C \tag{3.5}$$

$$\text{where } k = \frac{F\lambda}{V} \tag{3.6}$$

The situation just described gives the response of a compartment during "uptake," and by monitoring either C or C_v, the "uptake" curve may be obtained. The constant k, given by Eq. (3.6), is known as the "turnover constant" or fractional replacement rate (equivalent to fractional turnover rate). The unfortunate term "rate constant" is also often used in the medical literature. The turnover rate is simply the total steady state quantity of traced substance (carrier fluid) moving through the pool per unit time and is given by F, or kV/λ. The turnover time is the time required for that quantity which is exactly equal to pool size to move in and out. Thus the turnover time is equal to the mean residence time and is given by $1/k$ or $V/F\lambda$. Clearly, when the partition coefficient λ is unity, there is no difficulty in accepting the results of a single compartment model. If the response to a sudden injection of tracer is monitored, one obtains the "clearance" curve

$$C_v = \lambda C(0)e^{-kt} = \lambda C \tag{3.7}$$

Often in the medical literature the term "clearance" stands for the turnover rate of certain materials from various organs and is taken to represent an intrinsic diagnostic physiological parameter.

Kety and Schmidt (89, 90) introduced the use of inert gas uptake and clearance for the measurement of local blood flow.

Nitrous oxide and labeled krypton and xenon are among the most used gases in this technique. This represents a viable alternative to dye dilution. Although dye detection is often simpler, radioactive gases are soluble in air and effectively eliminated from the body in the lungs, thus circumventing the problem of tracer recirculation and allowing straightforward interpretation of collected data. Aukland et al. (3) applied a single compartment model in analysis of the data obtained by using hydrogen as a tracer in the study of the blood flow in myocardium. The data followed single exponential decay on a semi-log plot and the calculated coronary blood flow agreed well with the direct measurements of the venous outflow. The success of the simple model for myocardium is due to several factors. Myocardium is a highly homogeneous region, hydrogen equilibration is very rapid and the partition coefficient was found to be equal to one.

In general the correct differential equation for a single compartment model should read:

$$\left(V_B + \frac{V_T}{\lambda}\right)\frac{1}{F}\frac{dC_v}{dt} = C_a - C_v \tag{3.8}$$

and the simplified expression Eq. (3.4) is correct only when $V_B \ll V_T/\lambda$, which is the case when the organ does not contain much blood and the partition coefficient λ is larger than, or equal to, one.

Under pathological conditions, if one of the branches of a major coronary artery becomes occluded, a region of myocardium becomes "ischemic" (oxygen deprived). If this region is large enough it may behave as a separate compartment and the monoexponential clearance curve would change. Linder (109) investigated this condition by monitoring Kr^{85} clearance curves. While a healthy myocardium exhibited a monoexponential curve corresponding to a flow rate of 126 ml/min/100 g tissue, the curve that resulted after occlusion of the right branch of the anterior descending artery showed a two compartment character. As it was suspected that a collateral circulation to the ischemic region was established, the model of two compartments in parallel was used. This leads to the following form of the clearance curve.

$$C_v = f_1 C_1(0)e^{-k_1 t} + f_2 C_2(0)e^{-k_2 t} \tag{3.9}$$

where subscripts 1 and 2 denote regions 1 and 2 and f_1, f_2 stand for the fractions of volumetric flow in regions 1 and 2 as presented in Figure 2.

Equation (3.9) describes the superposition of two exponential curves. When the turnover constants are of different orders of magnitude a resolution of the curve into two portions may be readily accomplished. A straight line may be drawn at large times which corresponds to the component with the small turnover constant. By subtraction the other exponential curve may be obtained. From the intercepts at $t = 0$, if the partition coefficients λ_1 and λ_2 are known, an estimate of f_1 and f_2 may be obtained. The slopes of two monoexponential decays yield the values of the turnover constants. By such analysis Linder found that the fast (still healthy) compartment of the myocardium had a flow rate of 115 ml/min/100 gm while the slow compartment was perfused at a rate of only 8 ml/min/100 gm. It was also found that the "ischemic" region volume accounted for 22% of the total tissue volume of the heart.

A similar two parallel compartment model was applied by Haggendal, et al. (69) in determination of cerebral flow rates using radioactive Kr^{85} in saline as tracer. A two compartmental clearance curve was obtained and it was concluded that gray matter and white matter in the brain are separately perfused and that specific flow rate (flow rate per unit volume of tissue) through gray matter is

Figure 2 Two compartments in parallel model.

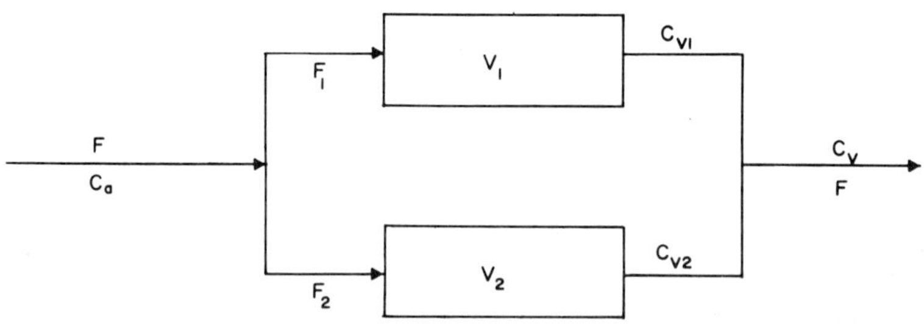

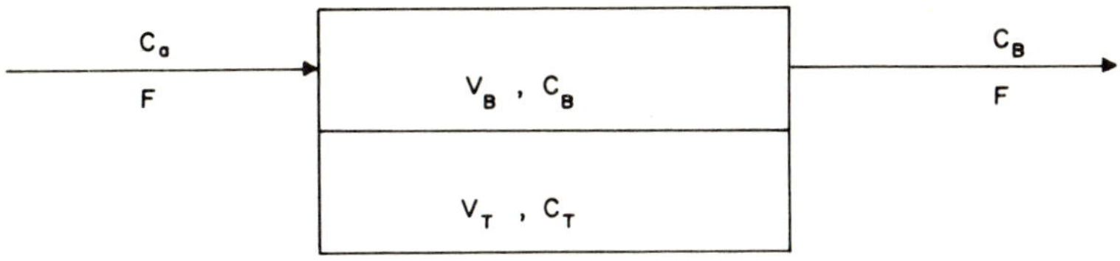

Figure 3 A two compartment model for blood-tissue exchange.

five times faster than through white matter. The results were expressed in terms of half time which is determined for each monoexponential portion of the curve as the time required for saturation concentration to be reduced by two.

$$t_{1/2} = \frac{0.693}{k} \tag{3.10}$$

The difficulty with compartmental analysis is that the clearance curve can always be fitted by a sum of exponential terms and the correct number of compartments is very difficult to determine. Even if the correct number of compartments is used, one has to rely on physiological knowledge of the system in order to determine whether the compartments are arranged in parallel, in series, or if there is a resistance barrier between them. We have illustrated the applicability of two compartment models in interpretation of tracer studies on myocardium and brain in determining regional blood flow rates. In the same manner, compartmental analysis may be applied in interpretation of trans-capillary exchange. A capillary bed in an organ is assumed to consist of two phases, blood and tissue. The solute passes between blood and tissue at a finite rate determined by some transfer resistance. The driving force for solute transfer is directly proportional to the concentration difference in the two phases–compartments, while the transfer resistance is inversely proportional to capillary permeability and capillary surface area available for transfer. Thus the model may be represented with two well mixed regions as shown in Figure 3, and described by the following set of equations.

$$V_B \frac{dC_B}{dt} = F(C_a - C_B) - PS(C_B - \lambda C_T) \tag{3.11}$$

$$V_T \frac{dC_T}{dt} = PS(C_B - \lambda C_T) \tag{3.12}$$

The solution for the outlet blood concentration is readily found. In the case of a clearance curve it is given by

$$\frac{C_B}{C_a(0)} = A_1 \exp\left(\frac{m_1 Ft}{V_B}\right) + (1 - A_1) \exp\left(\frac{m_2 Ft}{V_B}\right) \tag{3.13}$$

$$\text{where } A_1 = \frac{m_1 + \dfrac{\lambda V_B}{V_T}\dfrac{PS}{F} + \dfrac{PS}{F}}{m_1 - m_2} \tag{3.14}$$

$$m_{1,2} = -\frac{1}{2}\left(1 + \frac{\lambda V_B}{V_T}\frac{PS}{F} + \frac{PS}{F}\right) \pm \frac{1}{2}\left[\left(1 + \frac{\lambda V_B}{V_T}\frac{PS}{F} + \frac{PS}{F}\right)^2 - 4\frac{\lambda V_B}{V_T}\frac{PS}{F}\right]^{1/2} \tag{3.15}$$

The tissue region clearance curve may also be obtained and it can be shown that tissue desaturates more slowly than blood volume. The effect of the partition coefficient λ, blood to tissue volume ratio, V_B/V_T, and permeability number, PS/F, on the half time of the clearance curve may be determined. The larger the V_B/V_T ratio and the larger the PS/F, the sooner the monoexponential half time value is reached. The permeability number, PS/F, which is called the Stanton number in chemical engineering literature, has greater influence on the value of half time. This dimensionless quantity represents the ratio of material transported by mass transfer to the amount transferred by flow. For values PS/F larger than 10, for any value of V_B/V_T ratio, a monoexponential half time corresponding to a clearance curve from a single compartment is observed. By matching the experimental curve with the response of the model, the parameters of the system may be evaluated.

It is of importance to point out that parameters such as regional blood flow rate and permeability number could have been evaluated without the use of compartmental models by direct application of the basic principles of the tracer method as described in the previous section. Thus, from the moments of the exit age distribution curves one could, in principle, extract equivalent information without the use of compartmental models. Lassen and Serjsen (99) discussed advantages and disadvantages of multicompartmental modeling as applied to tracer curves.

Compartmental models have found their place in clinical applications and in complex modeling of the entire mammalian cardiovascular system. A study of cholesterol metabolism and the effects of cholesterol reducing agents in man (130, 146) was performed by applying a simple two compartment model (blood and tissue) to clearance curves of cholesterol. A multicompartmental model has been applied successfully in simulating the dynamics and equilibrium of drug distributions throughout the body thus developing new methods for chemotherapy and pharmacokinetics. At the present level of understanding of organ structure and function, and with the scarcity of reliable experimental data on diffusion coefficients and membrane transport, it is entirely appropriate to use lumped parameter models which minimize the need for detailed information while still being capable of simulating the phenomena of interest. Bischoff and Brown (20) and Bischoff (19) were particularly successful in developing a drug distribution model. A simplified flow chart of the circulation was presented as in Figure 4. Each large compartment represents an organ and consists of four well mixed homogeneous compartments: arterial capillary bed, venous capillary bed, interstitial fluid and intracellular fluid. A representative organ is shown in Figure 5. The model for the representative organ above is

$$V_{B_1}\frac{dC_{B_1}}{dt} = qC_1 - K_1A_1(C_{B_1} - C_I) - qC_{B_1} \tag{3.16}$$

$$V_{B_2}\frac{dC_{B_2}}{dt} = qC_{B_1} - qC_{B_2} - K_2A_2(C_{B_2} - C_I) \tag{3.17}$$

$$V_I\frac{dC_I}{dt} = K_1A_1(C_{B_1} - C_I) + K_2A_2(C_{B_2} - C_I) - K_3A_3(C_I - C_c) \tag{3.18}$$

$$V_c\frac{dC_c}{dt} = K_3A_3(C_I - C_c) \tag{3.19}$$

Each organ (group of organs) represented in Figure 5 is modeled in the way presented above with the exception of the right and left heart, which are

Figure 4 A schematic of mammalian circulation.

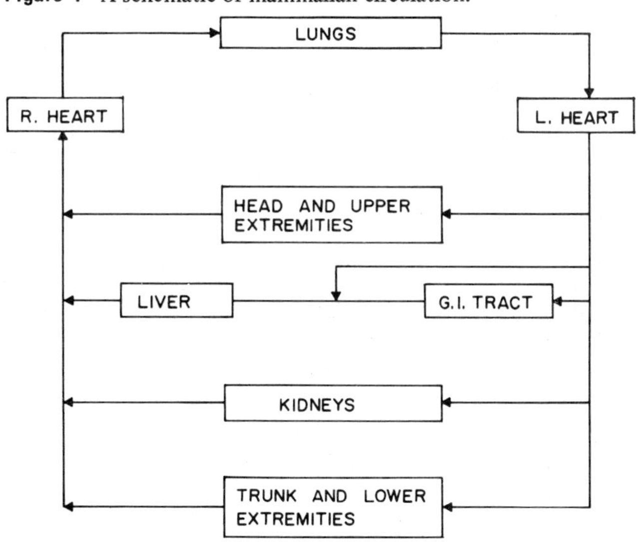

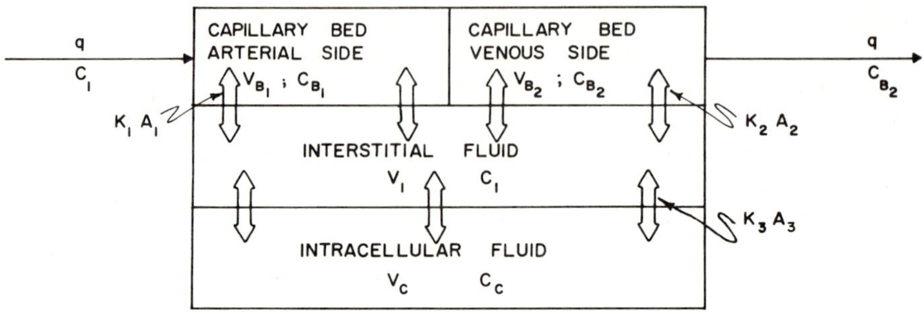

Figure 5 A representative organ used in Bischoff & Brown's model (20).

Figure 6 Final form of multiorgan model of Bischoff (19).

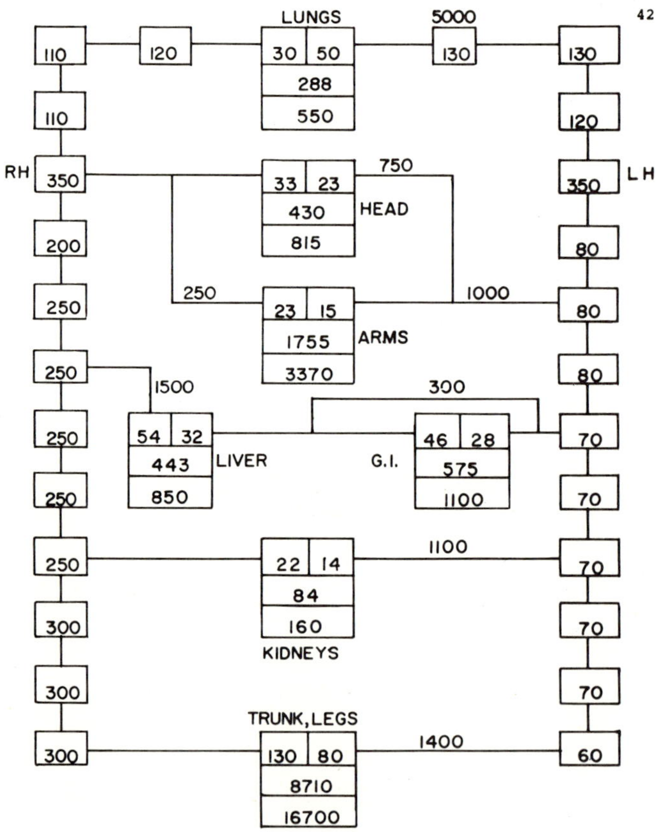

assumed to be well mixed single compartments. The blood pathways between organs are approximated by a few tanks in series. This has the effect of smoothing concentration profiles and is found to be preferable to the assumption of a time lag.

The model consists of over 50 first order linear differential equations explicit in derivatives with given initial conditions and can be readily solved by any marching procedure on a digital computer. The complete schematic of the model is presented in Figure 6. Numbers within the boxes are volumes in cubic centimeters, while numbers above the lines are flow rates in cubic centimeters per minute.

Parameters of the system, i.e., volumes, flow rates, permeability, area values in different regions were assumed on the basis of available physiological data, and are representative for healthy males.

The results were compared to some experimental data and show good agreement. The model clearly indicates that concentration-time profiles in different parts of the body will have different shapes. More details about potential use of the model in chemotherapy may be found in the literature (19, 20).

In summary, one can say that lumped parameter models have found extensive application to biological problems through compartmental analysis. These models have been successfully used as parameter estimation models in evaluation of regional blood flow rates and as simulation models in drug distribution studies. While compartmental analysis has a sound theoretical and mathematical basis, the resolution of experimental data still involves subjective interpretation, and results of analyses of this type should be viewed with caution.

3.5 Distributed parameter models

The transport of material from the blood to and across capillary walls, through the extracellular fluid and into intracellular fluid has received appreciable attention in the medical literature. Tracer methods are often employed in determining transcapillary rates, and a localized or microscopic approach is often taken in the interpretation of obtained results. This leads to distributed parameter model described by partial differential equations with time and space as independent variables. In spite of the great complexity of the living tissue and capillary pathways, many distributed parameter models were developed to explain the phenomena of capillary exchange and for interpretation of tracer data.

The first model for capillary exchange was developed by Krogh (93) in 1919, at the outset of his classical studies on the capillary circulation in skeletal muscles. This model represents a first important application of the theory of diffusion to physiological processes. Krogh considered capillaries to run in parallel arrays in the muscle, and assumed that each blood capillary of radius R_c supplied a coaxial cylinder of muscular tissue of diameter equal to the average distance $2R$ between open capillaries in the muscle. Being interested in the transport of oxygen, he assumed there was no barrier at the capillary wall, the partial pressure of oxygen at the wall was constant, and there was no net flux of oxygen out of the tissue cylinder, as the transport from the capillary is balanced at that surface with the transport from surrounding capillaries. The rate of metabolism of oxygen by muscle tissue was assumed to be independent of the oxygen concentration. The steady state exchange was considered, and axial diffusion neglected.

$$D\left(\frac{d^2p}{dR^2} + \frac{1}{R}\frac{dp}{dR}\right) = m \tag{3.20}$$

Boundary conditions are:

$$p = p_c \quad \text{at} \quad R = R_c$$

$$\frac{dp}{dR} = 0 \quad \text{at} \quad R = R_T$$

The solution obtained by Krogh is:

$$p = p_c + \frac{m}{D}\left(\frac{R^2 - R_c^2}{4} - \frac{R^2}{2}\ln\frac{R}{R_c}\right) \tag{3.21}$$

Hill (75), in 1928, modeled the diffusion of oxygen and lactic acid through tissues. He solved the equation for (a) steady state diffusion of oxygen into an active sheet of muscle where oxygen is consumed, (b) lactic acid diffusion out of the sheet, and (c) the simultaneous diffusion of oxygen and lactic acid.

Among the early works, those of Starling (166), Chambers and Zweifach (27), and Zweifach (198), are worth mentioning. They considered the absorption of fluids from the connective tissue spaces and the role of capillary permeability. Morales and Smith (118) also wrote extensively on capillary exchange.

In the fifties many works appeared based on Krogh's cylinder arrangement. Roughton (144) examined a more general form of Krogh's problem, treating a reversible chemical reaction of the diffusing specie with indiffusible substance in the surrounding tissue plus zero order consumption.

Schmidt (149, 150) developed a mathematical theory for exchange of substances between blood in capillaries, the homogeneous tissue, and the extracellular space. He set up partial differential equations describing the phenomena of filtration, plasma flow, capillary permeability, and radial diffusion in extracullar space. The resulting Laplace transforms were so complex that he did not attempt any complete inversion, but tried to find approximate solutions to some limiting cases. Finally, the mean concentration in the capillary, the concentration at the capillary wall, the mean concentration in extracellular space, and venous concentration were related to the concentration in the arterial blood. Numerical solutions for steady and unsteady states were obtained with arbitrarily estimated parameters. It appears that both the diffusion coefficient of the extracellular matrix and the permeability coefficient of the capillary wall are important parameters in the distribution of substances in tissue. Cotlove (38) applied Schmidt's model in evaluating the volume of distribution of inulin and sucrose in tissues. It appeared that distribution of inulin is diffusion-limited while the distribution of small molecules such as glucose is mainly flow-limited.

Sangren and Sheppard (147) considered a single capillary with uniform velocity and finite permeability at the wall, surrounded by a homogeneous, well mixed, interstitial fluid.

$$\frac{\partial C_T}{\partial t} = k'(C - C_T) \tag{3.22}$$

$$\frac{\partial C}{\partial t} = k(C_T - C) - u\frac{\partial C}{\partial z} \tag{3.23}$$

with the boundary conditions

$$C_T(t = 0) = 0 \tag{3.24}$$

$$C(z, 0) = a(z)H(z) \tag{3.25}$$

The solution was obtained by Laplace transforms, but it seems their inversion contains an error which was later corrected by Sheppard (157). Landahl (97) extended this model to allow for the consumption of the diffusing substance by first order reaction in the surrounding tissue.

Thews (178) gave a comprehensive mathematical treatment of Krogh's generalized problem, including the effects of axial diffusion, in the form

$$D\left(\frac{\partial^2 C}{\partial R^2} + \frac{1}{R}\frac{\partial C}{\partial R} + \frac{\partial^2 C}{\partial Z^2}\right) = \frac{\partial C}{\partial t} + m \tag{3.26}$$

Separation of variables and Laplace transformation were applied to obtain the solutions.

Blum (18) extended Thews' work, allowing for different diffusivities in the axial and radial direction. He considered first order, zero order, and Michaelis-Menten type reaction in the tissue, and included capillary permeability in the boundary conditions. Only the steady state problem was considered. Numerical results obtained with properly chosen biological parameters showed that axial diffusion might have a considerable effect on concentration profiles, especially in heterogeneous tissues where $D_Z > D_R$.

Bellman, Jacquez, and Kalaba (12), in attempting to give a quantitative approach to the distribution and uptake of drugs in organs, have set up differential equations similar to Schmidt's. They failed to find the unsteady state solutions and the short time behavior which they say is a "difficult problem analytically." Nevertheless, reformulating their distributed parameter model into a compartmental one, they founded the compartmental theory for treatment of drug uptake, which was successfully used later by Bischoff and Brown (20). Hudson and Carter (78) presented a mathematical analysis of the factors affecting distribution of oxygen in living tissue. Krogh's equation was applied to hexagonal symmetry, and the effects on the oxygen supply and its consumption rate of a variable diffusion coefficient, a capillary concentration profile, and mixing in extracellular fluid were considered.

Reneau, Bruley, and Knisely (136), extending the approach of Opitz and Schneider (123), presented a study of oxygen transport and consumption in the brain, and developed a very complex numerical method and computer program to solve their model. They included the axial and radial diffusion term in the capillary equation and the nonlinear function representing the erythrocyte-plasma oxygen equilibrium. The tissue equation included radial and axial diffusion terms, plus a zero order consumption term. Alternate direction numerical techniques were used to solve the problem, but the time for a single complete solution required ten hours on an IBM 7094 digital computer. Even in such a complex model, many simplifying assumptions had to be made. Papper and Kitz (126), in their text, *Uptake and Distribution of Anesthetic Agents,* described some of the simple distributed parameter models of Kety, Perl, and Landahl.

Many of the distributed parameter models which were developed for purposes of simulation of transcapillary exchange are used in interpretation of tracer data and in estimation of the parameters and rates of various transport processes. It is desired to introduce the same specific terminology through the discussion of some simple models, while the more extensive review of the methods of tracer analysis of transcapillary exchange will be presented in a later section. One of the simplest distributed parameter models for studying transcapillary exchange in skeletal muscle was employed by Renkin (138). The tissue region is assumed to be well mixed, and very large, relative to the blood region. As the washout of the tracer occurs by blood flow, it was assumed that tissue concentration C_T is uniform and independent of time for some finite period of time. The blood space, i.e., the capillary, is assumed to be linearly distributed and the concentration of the tracer in the blood varies with axial position along the capillary. Such a model which implies a distinct inlet and outlet position in space might be appropriate for treatment of the exchange of materials from a capillary bed into the surrounding tissue. A steady state material balance on an element of the capillary yields the following differential equation

$$F \frac{dC}{dZ} = -PO(C - \lambda C_T) \tag{3.27}$$

integrating the above equation from the capillary entrance, $Z = 0$, to the capillary end, $Z = \mathrm{L}$, and taking $S = AL$, one obtains

$$\frac{C_V - \lambda C_T}{C_a - \lambda C_T} = e^{-PS/F} \tag{3.28}$$

or presented in a different form

$$\frac{C_a - C_V}{C_a - \lambda C_T} = 1 - e^{-PS/F} \tag{3.29}$$

The term on the left hand side represents the ratio of the actual change in concentration across the blood region to the maximum possible change in concentration. This ratio is called "equilibration fraction" or "extraction factor" in the biomedical literature.

$$E = \frac{C_a - C_V}{C_a - \lambda C_T} \tag{3.30}$$

The product of the extraction factor and flow rate is called "capillary clearance"

$$Cl = FE = F(1 - e^{-PS/F}) \tag{3.31}$$

Thus a plot of $1 - C/F$ versus $\log F$ produces a straight line and enables the determination of the product of permeability and surface area, *PS*. The above method was used by Renkin in evaluation of Ru^{86} clearance curves from skeletal muscle. The effect of various physiological conditions produced by vasoactive drugs on *PS* was examined. Middleman (116) presented an order of magnitude argument showing that hydrodynamic factors in capillary tissue exchange are not transfer limiting. This implies that the major effect of vasoactive drugs is to increase the area available for transfer by opening more capillaries or increasing the permeability of capillary membranes. Renkin's assumption of the constant tissue concentration seems to hold well in some tissues, but cannot be used universally in all organs.

A model for a well mixed, but finite, tissue region with a linearly distributed capillary bed was developed by Johnson and Wilson (84). The equations describing this model are

$$\pi R_o^2 \frac{\partial C}{\partial t} + \pi R_o^2 U \frac{\partial C}{\partial Z} = 2\pi R_o P(\lambda C_T - C) \tag{3.32}$$

$$V_T \frac{\partial C_T}{\partial t} = \int_0^L 2\pi R_o P(C - \lambda C_T)\, dZ \tag{3.33}$$

$$\text{where } U = \frac{F}{\pi R_o^2} \tag{3.34}$$

The solution to a step input to the above set of equations has the form of an infinite series of exponentials. However, for large times, i.e., for $Ft/V_B \gg 1$, the tissue uptake curve has the following asymptotic form

$$\frac{C_T(t)}{C_a(0)} = 1 - \exp - \frac{[1 - \exp(-PS/F)]Ft}{V_T/\lambda} \tag{3.35}$$

and the half time for tissue saturation is given by

$$t_{1/2} = \frac{0.693\, V_T}{(1 - e^{-PS/F})F\lambda} \tag{3.36}$$

For very small transfer resistance, i.e., very large permeability number, $PS/F \gg 1$, the half time reduces to

$$t_{1/2} = \frac{0.693\, V_T}{F\lambda} \tag{3.37}$$

which is the half time of a monoexponential decay. Thus, under these conditions, the uptake curve may be well represented by a single compartment model as high rates of transfer make the tissue and blood region uniform in composition. The washout of the tracer is now entirely determined by the flow conditions; this is called a "flow limited" response. On the other hand, if transport rates are very slow, $PS/F \ll 1$, the expression for half time becomes

$$t_{1/2} = \frac{0.693\, V_T}{PS\lambda} \tag{3.38}$$

and washout of the tracer from the tissue is entirely "transfer limited."

The concepts of flow and transfer limited exchange were introduced by Goresky et al. (68), using different models than the one presented here. It was desired to introduce the concepts in this section while the detailed review of Goresky's papers will be presented in Section 5.

It is of interest to note that a distinction between lumped and distributed parameter models is not a rigid one and the diffusion equation describing the continuum can be readily developed as the limit of compartmental models. An especially elegant analysis, and proof of equivalence, was presented using the probabilistic approach by Shinnar and Naor (161).

3.6 The influence of tracer information on modeling

Modeling of biological systems and tracer methods used in physiology and medicine are entirely interrelated. Many models of biological functions have been formulated in order to simulate the behavior of the system under normal and pathological conditions. Often the parameters of these models cannot be evaluated by independent experiments *in vitro,* and tracer methods are employed *in vivo* for parameter estimation. Additional modeling is then required to interpret the results obtained by tracer analysis. Thus, models used for interpretation of tracer results enable us to estimate the parameters of the system and to build more sophisticated and more reliable simulation models. Often the information provided by the tracer method will lead to a proper choice of a model for a given organ or biological function. Modeling of tracer responses of various biological systems provides parametric classification of normal physiological states. By extending these models, diagnostic classification of abnormal states may be accomplished. For example, under normal physiological conditions the clearance curve of various highly soluble gases in myocardium exhibits a distinctive monoexponential decay. However, under pathological conditions a two compartment behavior is detected. Encouraged with the success of compartmental models in estimation of local flow rates and volumes, and with the usefulness of distributed parameter models in interpretation of transcapillary exchange tracer data, many combinations of these models were extended for simulation purposes. The compartmental, lumped parameter models found extensive use in simulation of drug distributions and became an important tool in pharmacokinetics and chemotherapy (19, 20). Various distributed parameter models which emerged from tracer studies of capillary exchange were successfully used in simulation of organ functins. Kuhn and Ramel (94), Marumo et al. (112), Lever (102), and Pinter and Shohet (131) applied simple distributed parameter models in simulating the exchange processes in the nephron, Henle's loop, and vasa recta of the kidney, while Jacquez et al. (82) and Palatt et al. (124) provided an excellent example of application of compartmental modeling to the description of the function of a very complex organ. Gas exchange in the lungs was investigated by Forster et al.

(55), and Milhorn and Pulley (117). Thews (178), Opitz and Shneider (123), Blum (18), and Reneau et al. (136) applied distributed parameter models to studies of oxygen transport in the brain. Only a distributed parameter model is capable of predicting oxygen deficiencies in certain regions under such conditions that the average compartmental concentration is still above the critical value. Tracer studies of the placenta gave rise to heat exchanger type models developed by Faber (52) and Rankin and Peterson (134).

Tracer methods have also been employed in determining complicated metabolic pathways and evaluating metabolic rates and rate constants. The success of modeling in this area led to attempts in simulating complex regulatory control functions of biological systems. Gatewood et al. (60) modeled the complex blood-glucose regulation and resolved the coupling of glucose-insulin interaction. His typical insulin infusion responses demonstrated that periodic input tests and frequency response analysis may be used in both normal subjects and diabetics. In a related study of glucose hormone dynamics, Young et al. (191) showed that Gatewood's model applies also to the dynamics of serum growth hormone. This study also recorded response of the system to single injections of insulin, which in a sense serves in the tracer role.

In summary, one can say that the influence of tracer information on modeling of biological systems is twofold. First of all, many models have been developed for interpretation of tracer results and were extended later for various simulation purposes. On the other hand, the information gathered by tracer techniques on the behavior of various organs and the quantitative estimates of many parameters that emerged from tracer studies made formulation of more sophisticated models possible.

4 TRACER ANALYSIS OF THE CIRCULATION

4.1 Cardiac output and central circulation volume

The Stewart-Hamilton method for obtaining cardiac output from a tracer or indicator dilution curve is the most extensively used tracer technique involving the circulation. Stewart (172–174) began the development of the method about 80 years ago, and Hamilton and coworkers extended and refined the technique (70, 71). The theoretical foundation for the work rests on the material balances for tracer and carrier fluid previously discussed. The equations derived from the material balances cannot, however, be applied directly to the output tracer concentration curve. The major problem involved with the development of this technique lay in rendering the measured curve suitable for use with the theoretical equations.

The central circulation is made up of the lungs and all or some of the heart chambers. The four heart chambers and the pulmonary artery upstream of the bifurcation are the only parts of the normal circulation through which the entire cardiac output passes in a single vessel. We will call such a point when all the cardiac output flows through a single cross section a nodal point. Any tracer method for obtaining cardiac output must then include at least one of these regions or nodal points between the injection and sampling site in order to effect the dilution of tracer in the entire blood stream. The injection site is usually at either one of the vena cavae or the pulmonary artery. Sampling is done at a location in either the aorta or a larger artery such as the brachial or femoral.

The recirculation of the traced blood is fast enough so that the central circulation response cannot be considered an open system. Some of the tracer reappears at the sampling site before all the tracer can pass the sampling site for the first time. This recirculation of tracer must be corrected in order to obtain the open system response or first passage time distribution of the central circulation. Fortunately, the tracer concentration curve shows a short time delay of several seconds, a sharp rise to a peak concentration, then a falloff toward the baseline of usually more than 50% of the peak concentration before the effects of recirculation are seen. The Stewart-Hamilton technique employs an exponential extrapolation of the descending portion of the curve down to the baseline to obtain a suitable approximation of the primary response. Many other methods have been advanced for obtaining the primary response from the measured curve which is commonly referred to as the indicator dilution curve. These methods are based on empirical and semitheoretical models. They have been the subject of many review articles, most of which are reported in the reviews of Harris and Newman (72) and Wingard et al. (189). There has also been considerable effort expended in inferring more than flow and volume for these models, especially from models of a compartmental nature. However, these attempts have not been notably successful.

Two of the models which have been proposed to fit the entire primary peak of the curve have also been applied to regional and whole circulation studies. Bassingthwaite et al. (9) introduced the lagged normal density curve which is the convolution of an exponential decay with a Gaussian distribution. This combination is made to include both effects of a diffusion type mixing process and a mixing tank or compartmental mixing. The random, or diffusion, type mixing is given by the Gaussian distribution

$$h_1(t) = \frac{1}{\sigma(2\pi)^{1/2}} \exp - \frac{1}{2}\left(\frac{t - t_c}{\sigma}\right)^2 \tag{4.1}$$

where t_c is the mean and σ the standard deviation. The mixing chamber effect is given by

$$h_2(t) = \frac{1}{\tau} e^{-t/\tau} \tag{4.2}$$

where τ is the time constant; V, chamber volume divided by F, the flow. The convolution of the two equations yields

$$E(t) = \frac{1}{\sigma(2\pi)^{1/2}} \exp\left[-\frac{1}{2}\left(\frac{t - t_c}{\sigma}\right)^2\right] - \tau \frac{dE(t)}{dt} \tag{4.3}$$

The three parameters give much flexibility in curve fitting. The authors used this equation to fit curves and interpreted changes in flow rate and injection sites with the parameters of the model. The model was also used in fitting the results of a transfer function analysis of a specific portion of the circulation (40).

Schlossmacher, et al. (148) applied the perfect mixers in series model of the chemical engineering literature to fitting veno-arterial dilution curves. They inverted the stimulus-response equation

$$C(t - t_a) = \frac{M_i}{F} \frac{1}{\tau_i(n-1)!}\left(\frac{t - t_a}{\tau_i}\right)^n e^{-(t-t_a)/\tau_i} \tag{4.4}$$

where t_a is the time of first appearance of tracer, and τ is the volume of a single mixer divided by F, for the values of $n = 3, 4, 5, 6$. All the curves they analyzed could be treated by these values of n. A table was presented with constants for each value of n which enabled the calculation of zeroth, first and second moments from curve width and peak height measurements. Because the moments of the curve are so easy to obtain with this method

$$\begin{aligned} \mu_1 &= n\tau_i \\ \mu_2 &= \left(\frac{1}{n} + 1\right)\mu_1^2 \end{aligned} \tag{4.5}$$

it is readily used in analyses requiring convolution or deconvolution.

The perfect mixers in series model yields a Poisson distribution and is a special set of the gamma variates. The authors were able to show that most of the empirical curve fitting equations for flow and volume in the literature fit this model and that little was lost in not using the more general gamma variate formulation.

The tracers used in cardiac output studies include optical, radioactive, and thermal types. In the last case, an impulse of thermal energy is used as the tracer, and temperature is measured at the sampling site. The measurement of tracer concentration usually introduces an error or distortion in the response since in most cases a sample is withdrawn through a sampling catheter. This error does not affect the cardiac output calculation but does affect the volume calculation. The cardiac output is obtained from

$$\text{C. O.} = F = \frac{M_i}{\int_0^\infty C_d(t)\,dt} = \frac{M_i}{\int_0^\infty C(t)\,dt} \tag{4.6}$$

where C_d is the measured concentration including the distortion and $C(t)$ is the actual concentration at the sampling site. Volume is obtained from

$$\frac{V}{F} = \frac{\int_0^\infty tC(t)\,dt}{\int_0^\infty C(t)\,dt} = \frac{\int_0^\infty tC_d(t)\,dt}{\int_0^\infty C_d(t)\,dt} - \frac{V_S}{F} \tag{4.7}$$

where V_S is the sampling system volume. A complete discussion of this distortion problem is given by Oleksy et al. (122) who used a moment analysis to remove the effects of distortion. The preceding equation is derived in the section on convolution and is

$$u_0 \text{ (central circulation)} = u_1 \text{ (distorted curve)} - u_1 \text{ (injection and sampling system)}$$

They measured experimentally the response of the sampling system and calculated its moments. They also calculated the variance of the undistorted response curve and from it were able to construct the undistorted primary response curve using the perfect mixers in series model of Schlossmacher et al. (148).

4.2 Regional flows and volumes

Many studies have been undertaken in which tracer methods have been used to obtain a measure of flow and volume in an organ or of a region of the circulation. At least in theory, if a tracer injection can be made over the inlet flow to the system to be analyzed and samples of tracer in a representative outlet stream can be obtained, flow and volume can be calculated. When these conditions are met, and recirculation is not fast enough to interfere with the recording of the primary response curve, flow and volume can be calculated directly from the tracer and carrier fluid material balances. In the more typical case when recirculation does occur, an exponential extrapolation of the Stewart-Hamilton type ordinarily does not yield a suitable approximation of the primary curve. For these cases we can use the linearity property of the tracer method. If we have a region, V, perfused by a flow, F, and we have for the concentration of tracer at the inlet just upstream of tracer, C_r, then C_i, the inlet concentration is

$$FC_i = FC_r + M_i \delta(t) \tag{4.8}$$

where $M_i\ \delta(t)$ is the bolus injection of tracer. $C_i(t)$ is the inlet concentration of tracer to the region and represents the combination of tracer injection and recirculation. The outlet concentration from the region is C_0 where

$$C_o(t) = \int_0^t P(t - \tau) C_i(\tau)\, d\tau \tag{4.9}$$

where $P(t)$ is the primary response curve we are trying to obtain. Since $P(t)$ is in the integrand of this convolution integral, we must deconvolute to obtain it and we must know C_r. A variation of this idea is to inject at a convenient point upstream of the region and sample at the inlet and outlet of the region since it is as easy to find C_i as it is to find C_r. Then we measure C_i and C_0 and can deconvolute to obtain P. This method has been used by a number of investigators. There is considerable difficulty in performing the deconvolution operation in an accurate manner, however. Small inaccuracies in the operation will make for large errors in the estimation of important parameters. Both C_i and C_0 tend to equal equilibrium values so that the tail of $P(t)$ is where the largest errors result. The basic information in $P(t)$ is V/F, the perfusion rate and

$$\frac{V}{F} = \int_0^\infty tP(t)\, dt \tag{4.10}$$

so that errors in the tail of $P(t)$ are emphasized by the weighting. Since $P(t)$ is a normalized function, the flow F cannot be evaluated from it. Coulam et al. (39, 40) have called this type of analysis "transfer function analysis" and have used it to analyze coronary, renal, and other circulations. They employed Fourier series descriptions of the upstream and downstream curves and then determined the Fourier series representation of the transfer function. The time domain transfer function is the primary response or first passage time distribution.

The input and output curves, recorded through identical sampling systems so there was no net effect of sampling system distortion, were expanded in Fourier series. If $f(t)$ and $g(t)$ are the upstream and downstream curves, respectively, and $F(n)$ and $G(n)$ the frequency domain representations, then

$$H(n) = \frac{G(n)}{F(n)} \tag{4.11}$$

is the frequency domain representation of $E(t)$ and

$$\phi(n)_H = \phi(n)_G - \phi(n)_F \tag{4.12}$$

the corresponding phase angle. In order to deal with the tail of the curve, the $E(t)$ was approximated by the lagged normal density distribution discussed earlier. Transfer functions for the coronary and renal circulations in dogs were evaluated.

In further work along these lines, Bassingthwaite and coworkers (7, 8, 91) used a model for the transfer function as an aid in deconvolution and determined transfer functions in the human leg, pulmonary vascular bed, and aorta. They correlated tracer dispersion with flow rate to obtain some physical insight on dispersion in the circulation. They found dispersion of indicator to be relatively small in degree and the characteristics of dispersion did not vary widely over a wide range of flow rates.

A great deal of work has been done on measurement of organ perfusion rates by the uptake and clearance methods as well as by bolus injection. Uptake and clearance are equivalent concepts; the former is the response to step increase in tracer inlet concentration and the latter response to a step decrease. These methods usually rely on a compartmental modeling approach and are discussed in detail in Section 3.4.

One aspect of this subject, not covered elsewhere in this chapter, is external monitoring of radioactive tracers. This technique has become of increasing interest in recent years. Zierler (195) developed the basic notions of this technique after Kety (88) proposed it. Kety assumed a single compartment model for his analysis and for this special case, concentration of tracer in the system is exactly the same as exit concentration. With this idealization, one obtains the exit age distribution directly from the external sampling with very little difference in analytical procedures between this case and the case of system exit sampling. For the general case with bolus injection, Zierler shows that a material balance on tracer yields

$$M(t) = M_i - \int_0^t FC(\tau)\, d\tau \tag{4.13}$$

where τ is a dummy variable, $M(t)$ is the mass of tracer in the system at any time, M_i is the mass of tracer injected, F is the flow rate, and $C(t)$ the exit concentration. Normalizing we obtain

$$m(t) = \frac{M(t)}{M_i} = 1 - \int_0^t E(\tau)\, d\tau \tag{4.14}$$

where $E(t)$ is the exit age distribution. On the left hand side we have a mass ratio, so if we begin counting radioactivity at the time of injection, we need not convert from counts per unit time to mass because the conversion factors will cancel. $m(t)$ is the instrument reading normalized on the initial reading. Integration of this expressions gives

$$\begin{aligned}\text{Area under normalized measured curve} &= \int_0^\infty m(t)\,dt = \int_0^\infty dt\left[1 - \int_0^t E(\tau)\,d\tau\right] \\ &= \frac{V}{F} \end{aligned} \tag{4.15}$$

This integration is equivalent to that of the material balance on carrier fluid developed in Section 4.2.

The problem with using this technique is that is is quite often impossible to inject directly into the inlet to the system of interest. Whenever the injection function to the system is not known, say due to injection well upstream of the inlet, the normalization factor is not known. This normalization factor, the instrument reading with all the tracer in the system, is necessary for correct evaluation of V/F if $E(t)$ is anything other than an exponential decay. Further, for unknown injection function, the concentration at the exit is

$$C_{\text{exit}}(t) = E(t)^* C_{\text{injection}}(t) \tag{4.16}$$

and the first moment of the normalized exit concentration curve will be equal to the sum of $E(t)$ and of the normalized injection function. Then we would also have this error included in the above V/F calculation.

Another application of tracers to obtain information on a region of the circulation is that of finding ventricular volumes. Shaffer (152) developed a technique based on constant infusion of indicator. Concentration was measured with enough resolution to determine stroke-by-stroke concentration in the exit stream. He derived an expression for end diastolic volume, EDV, in terms of the concentration in the exit stream after the nth beat, C_n, the equilibrium concentration, $C_{\max}$, and the stroke volume, SV.

$$EDV = \frac{SV}{(C_n - C_{n-1})/(C_{\max} - C_{n-1})} \tag{4.17}$$

This expression is based on a material balance over a single compartment with pulsed flow. SV was calculated from cardiac output and heart beat rate. Cardiac output was obtained from $C_{\max}$ and the mass rate of tracer infusion, I.

$$SU = \frac{I/C_{\max}}{\text{heart beat rate}} \tag{4.18}$$

Tracer was infused directly into the ventricle, so assumption of instantaneous mixing implied by the single compartment model leads to some error. Shaffer discusses the types and magnitudes of errors that can be introduced, and also reviews the literature on this topic up until 1964.

Ishii and MacIntyre (81), in a more recent paper, used a radioactive tracer with external monitoring to determine heart chamber volumes. They used scintillation cameras to record sequential radioisotope images and determine a set of counts per unit time curves for several regions of the circulation. They used a single compartment model for each chamber and measured the counts from only a portion of the chamber. The injection was made into the right antecubital vein so there was considerable dispersion of the tracer before it reached the heart. This can lead to errors in external sampling techniques, as discussed above. Measured heart chamber volumes were compared to those obtained from inserting modeling clay into cavities during autopsy, and agreement was fair.

4.3 Whole circulation tracer analysis

There have been several attempts to obtain first passage time distributions for the entire circulation in order to calculate whole body circulating blood volumes. Such a measurement technique would be useful in diagnosis and research for many obvious reasons. Hering's initial attempt to measure "circulation time" is the real beginning of this work and investigators developing the veno-arterial indicator dilution method were aware that the whole circulation response was implicit in the second peak of the tracer concentration curve. This peak is in fact the response of the entire circulation convoluted upon the first passage time distribution of the central circulation.

The work more directly related to whole circulation response curves was begun by Stephenson (169, 170). He published a classic paper in which he treated the recirculation peak of the indicator dilution curve including the whole body response by formulating the problem in terms of a Volterra integral equation of the second kind. However, he did not choose the same nodal point for both injection and sampling, and had problems of normalization. This work was extended with respect to the theory by Meier and Zierler (115), and Zierler (194) who, in his chapter in the *Handbook of Physiology*, discusses much of the fundamental theory of recirculation. His discussions cover the foundations of the work on transfer function analysis reported above and circulating blood volumes reported below.

Katz and Shinnar (87), using renewal theory were also able to derive a Volterra integral equation of the second kind to describe the recirculation problem and used for the normalization, the equilibration concentration of tracer. This normalization procedure leads to the same inaccuracies discussed in Section 2.5. Sheppard et al. (159) removed the effects of recirculation from a veno-arterial tracer curve by making use of a nodal point, whole body tracer curve. They also formulated the problem in terms of a Volterra integral equation of the second kind. Unfortunately, they did not make use of this information to analyze the whole body curves.

A major problem in the analysis of closed, recirculating systems using tracer methods has been pointed out by Naor, Shinnar, and Katz (119). They have shown that both the flow rate and the first passage time distribution cannot be extracted from the tracer response of a single injection even when concentration histories are taken at both the injection site and another arbitrary site simultaneously. Their demonstration is based on the viewpoint of applied probability theory. It is possible, however, to get around this theoretical limitation by the use of a model such as the Stewart-Hamilton technique or the perfect mixers in series method (148) for evaluation of the flow rate. Naor, Shinnar, and Katz develop, instead, a procedure for estimating a "maximum feasible flow" which is completely model-free.

There has been skepticism expressed concerning the use of a recirculation peak of a veno-arterial indicator dilution curve to derive a primary response curve of the entire circulation (121). This implies that a nodal point to nodal point curve of the entire circulation also cannot be used in this manner. It was suggested, a priori, that recirculation of tracer via faster pathways might occur two, or even three, times before the "whole body" curve form would be established, and that all of the tracer traversing some of the longer pathways might not return to the sampling site during the sampling period. A series of curves was obtained (Figure 7) by means of injection of indicator into the aortic root (Figure 7A) and bifurcation (Figure 7D), and into various major branches of the aorta, i.e., carotid (Figure 7C), renal (Figure 7B), and mesenteric arteries (Figure 7F), and a branch to the body wall (Figure 7E), with pulmonary artery sampling (153). The time of first appearance of indicator from all of these sites showed a modest spread. Each of the curves in Figure 7 was then adjusted to the same sensitivity of recording and multiplied by an assumed fraction of cardiac output traversing the organ or region represented by the dye curve, so as to account for all of the injected dye. The result is shown in Figure 8. The striking correspondence between the aortic root and renal artery curves suggests that indicator returning from the renal artery is an important component of the early part of the aortic root to pulmonary artery curve. As the renal artery curve declines, the aortic bifurcation and carotid artery curves rise; as these first three curves decline, the curves representing passage of indicator through the slower or longer pathways begin their rise. There was no definite evidence, at least in the greater parts of the curves, that their contribution to the aortic root to pulmonary artery curve resulted in multiple peaks or multimodality, such that one or more of the peaks would appear much after the first, and possibly be lost because of insufficient record length. These observations thus failed to yield any evidence that, beyond producing more dispersion of indicator, the greater

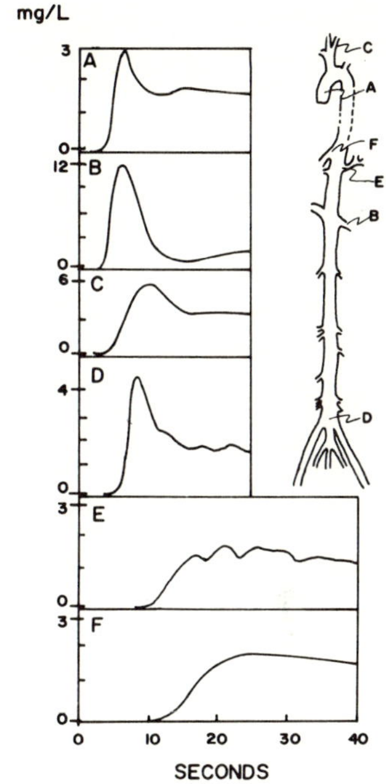

Figure 7 Tracer response curves with injection in various systemic arterial locations.

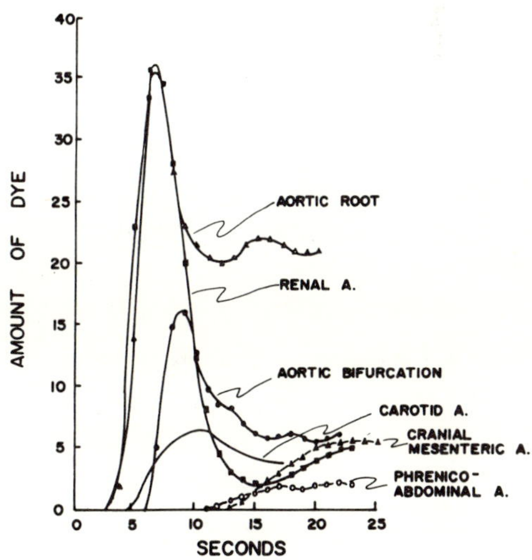

Figure 8 Composite of normalized tracer response curves with injection in various systemic arterial locations.

circulation differs in any material way from the lesser. Subsequently, experiments in which indicator was injected into the femoral artery with sampling at the femoral vein, and injected in the femoral vein and sampled in the pulmonary artery, failed to confirm earlier observations (121) that tracer returned to the central circulation from the lower extremities so late as not to contribute to the primary recirculation peak (loss of conservation). Further, a study done elsewhere using cardiopulmonary bypass to delay systemic recirculation (61) revealed no evidence of multimodality with a wide spread in peaks in a considerable portion of the whole body dye curve which would otherwise be hidden by recirculation.

Weinstein and associates (185, 186) presented a formulation and solution of the general problem of tracer response of recycle systems of which the "closed" circulation is a special case. They were able to apply their analysis to the problem of hemorrhagic shock and showed the effects of shock on circulating blood volume in three experimental animals. This formulation is presented here because of its fundamental nature in treating recirculation. The entire circulation is considered to be a linear dispersion unit with closed recycle loop representing the nodal point at which injection and sampling is made, as shown in Figure 9. A material balance on tracer at the injection site gives

$$C_i(t) + C_a(t) = C_S(t) \tag{4.19}$$

Here, $C_a(t)$ is the concentration of tracer at the outlet, and $C_S(t)$ the concentration of tracer at the inlet of the dispersing unit. $C_i(t)$ is the concentration of tracer at the injection site that would result from the tracer injection into a tracer-free system with no recirculation of tracer. If it is assumed that there is no tracer in the system before time $t = 0$, and that the dispersion unit is linear, stationary, and continuous, then

$$C_S(t)^*P(t) = C_a(t) \tag{4.20}$$

where * denotes convolution and $P(t)$ is the distribution of first passage times. Taking $C_i(t)$ to be an impulse injection, then $C_i(t) = \beta_0\delta(t)$ where $\beta_0 = M_i/F$, the normalization factor for the problem. Rearrangement of these two equations yields the equation

Figure 9 Circulation as a linear dispersion unit.

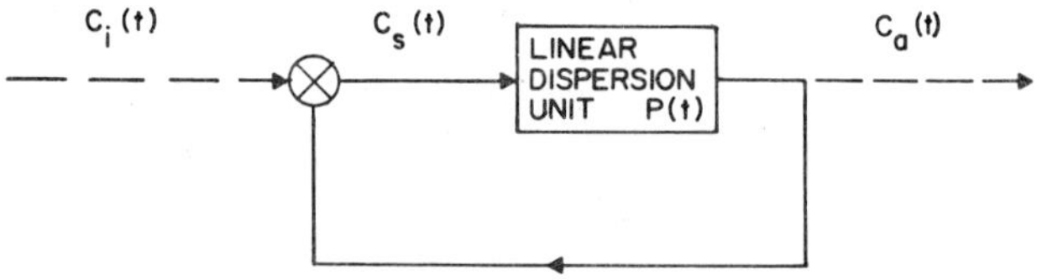

$$P(t) + Q_a(t)^*P(t) = Q_a(t) \tag{4.21}$$

where $Q_a = C_a/\beta_0$, the normalized tracer concentration at the sampling site. This integral equation is a Volterra type, second kind, and is readily inverted in the real time domain with the Liouville-Neumann method. The authors removed the distorting effect with a moment analysis and reported curves of $P(t)$ for both control and shock cases for the three dogs. The normalization factor, β_0, was obtained by a separate measurement of tracer concentration in the aorta. Since the nodal point used for injection and sampling was the pulmonary artery, the aorta curves were of the typical Stewart-Hamilton type. The method of Schlossmacher et al. (148) was used to obtain F from these curves since it is less affected by early recirculation than the Stewart-Hamilton method. The result obtained showed good agreement with measured blood loss in the three cases.

The general method of analysis presented by these authors can be extended, at least in theory, to a general analysis of the circulation which would include series and parallel paths. The formulation of the problem in terms of the Volterra integral equation of the second kind by manipulation of the material balance and convolution equations is the important key since this type of equation may be inverted readily by the Liouville-Neumann method in real times (41). Reasonable accuracy can be maintained and the inherent problems of numerical deconvolution are avoided.

5 TRACER ANALYSIS OF TRANSCAPILLARY EXCHANGE

5.1 General

The early realization that oxygen is needed for all life processes, and that blood is the main supplier of oxygen to the living tissues of the body, aroused increased interest in transcapillary exchange and molecular processes in and around the living cells. The exchange of matter that takes place in heterogeneous capillary beds in organs such as the brain and heart is of great complexity and occurs on the microscale. In order to better understand the physiological functions of many organs it is necessary to gain a better insight into these exchange processes. The pioneering work of Krogh (92, 93) introduced the concept of mathematical modeling of capillary exchange. In order to formulate a proper mathematical representation of such phenomena many important parameters such as the geometrical configuration of the capillary beds, the dimensions of the capillaries, the type of flow in the capillaries and the capillary membrane structure have to be determined. The nature of the driving forces for the exchange of matter, the inner volume of the capillary bed, the volume of the interstitial fluid around it, total volume of the organ, metabolic rates, diffusion, and capillary permeability have been intriguing subjects of experimental and theoretical studies. Many of these quantities are difficult to determine by direct observation. Therefore, various tracer stimulus-response methods have been devised and a number of models proposed for interpretation of the results.

In this section we review the mathematical models developed to explain the phenomena related to transcapillary exchange and the experimental results used in evaluating the process parameters. An attempt is made to provide enough general theory to relate the various models into a unified, comprehensive treatment of this subject. A descriptive section precedes this treatment to illustrate the system and to develop the terminology. The section concludes with a review of work in which tracer studies were used to obtain values for the parameters of transcapillary exchange in various organs.

5.2 Microcirculation

Microcirculation has been defined as "the ultimate portion of the cardiovascular system which is generally accepted as being concerned with the transfer of gases and nutrients and the removal of metabolic waste products" (188). Briefly, microcirculation can be described as blood flow through small vessels at the capillary level. The whole arteriovenous pathway can be roughly described in the following manner (4): "The arteries branch freely to form a primary network. Smaller arteries branch off to form a secondary network from which arterioles branch at more or less regular intervals. Terminal arterioles are the final arterial ramification, the branching of which continues as nonmuscular capillary vessels." Therefore, a capillary can be considered to be a small endothelial tube, about 8μ in diameter, with wall thickness of 1μ (4), devoid of smooth muscle and having a minimal amount of supporting elements. The capillary network unites into venules, these into veins to reproduce a pattern similar to that on the arterial side. Basic patterns of microcirculatory beds remain the same (188), although differences appear with the structural organization of the tissues. A good review of this subject is presented by Zweifach (197, 198). In almost all capillary networks in organ tissues one can find arcade connections, gradual divestment of spiral smooth muscle cells along terminal arterioles to form capillaries, irregular branching, and effective physiological shunts formed by arteriovenous capillaries. Every tissue is specific in the number of capillaries per unit area, their length and shape. In muscles, one can expect more capillaries arranged in parallel, running along the muscle fibers. They are distributed more radially in the spleen and much more randomly in the lungs and miocardium. The microscopic components of the vascular tree vary considerably in their distribution, size, wall thickness, and even the simplest true capillaries are structurally different in the various tissues and within the same tissue. The size of capillaries can be determined. Intaglietta and Zwerfach (80) reported that distribution of capillary lengths and radii in rabbit omentum is almost gaussian with little spread around the mean. Krogh (92, 93) found experimentally, by observation, that there are 1,350 capillaries/mm^2 in horse muscle and 2,630/mm^2 in dog muscle. By now it is well known that the number of active capillaries open to blood flow varies according to the tissue needs. Kuprianov and Kozlov (95) recently investigated the capillary bed structure in the mesentery.

Honig and his colleagues (77) measured the number of open capillaries by direct enumeration in the resting rat gracilis muscle. While the total number of capillaries was found to be 2,500–3,000/mm^2, only 240/mm^2 appeared to be open to flow in a resting muscle. Capillary length varied from 100μ to 700μ, and some of the parallel capillaries were found only 5μ apart. Countercurrent flow was identified in some closely spaced capillaries. Renkin (138) showed by his experiment with Rb-extraction that the total blood flow varies throughout the capillaries along the skeletal muscle. The principal adaptive mechanisms for this variation is the change in number of open capillaries.

The capillary network is of tremendous complexity and variety. When it is needed to interpret the experimental results obtained on such pathways, it is customary to single out a typical capillary for analysis. How well a single capillary represents such a complicated network is a question that has not yet been answered. However, single capillary models give more insight in the complicated processes that occur in the microcirculation and are a step forward toward more elaborate multicapillary models. By use of modern experimental techniques it is becoming possible to determine the capillary size distribution in different tissues and select a capillary which represents the mean of the distribution for analysis. A cross section of a typical capillary is presented in Figure 10.

In order to understand how matter exchanges across the capillary walls, information on the type of blood flow in the capillaries is needed. The flow was once considered to be fully developed laminar flow, but it has proved to be far more complex. Although the Reynolds number is certainly in the laminar region for a capillary, the flow itself is far from being laminar. First of all, blood is not a homogeneous fluid, as the red cells can always be distinguished from plasma. The flow may be pulsatile, and this problem has received much attention in the study of blood rheology. Contributions by Attinger (2), Wayland (184), Fung (59), and Goldsmith (62) are some of many worth mentioning. In the microcirculation, where the diameter of the capillaries approaches that of the red blood cell, the flow of plasma and erythrocytes deviates completely from homogeneous laminar flow. Bloch (17) has photographed the passage of red cells through the capillaries, and it appears that the cell occupies all the capillary cross section. Lighthill (108) formulated a model of compressible pellet movement through narrow, fluid filled, elastic tubes to simulate the mechanics of red cell motion through very narrow capillaries. His work was extended by Fitz-Gerald (54). Bugliarello (25) presented another comprehensive study on the subject.

Landis and Pappenheimer (98) and Pappenheimer (125) presented excellent surveys on the exchange of substances through capillary walls and

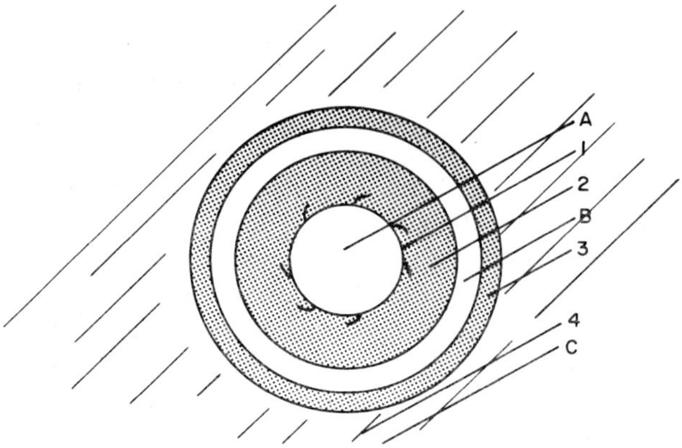

A - BLOOD 8 - 10μ

1 - ENDOCAPILLARY LAYER 50-100 Å

2 - ENDOTHELIUM 1000-2000 Å

B - SUBENDOTHELIAL SPACE 200 - 500 Å

3 - BASEMENT MEMBRANE 600 - 750 Å

C - INTERSTICIAL SPACE 40 - 50μ

4 - GROUND SUBSTANCE 40 - 50 μ

Figure 10 Cross section of a capillary.

capillary permeability. They confirmed that the fraction of total blood flow passing through true nutrient capillaries is subject to wide variation according to metabolic demands of the tissue, or to hemodynamic demands of the organism as a whole. Filtration due to differences in hydrostatic and osmotic pressure takes place across the capillary walls, and the extravascular circulation caused by it is important for removal of very large protein molecules. Metabolic exchanges, however, occur due to diffusion gradients and are entirely independent of the fluid movement due to filtration. The driving force for the diffusion is normally maintained by tissue metabolism. The mechanisms of organic membrane transport are very complicated; the fundamentals and a review have been given by Stein (168). Experimental measurements are possible in vitro; the techniques have been summed up by Passow and Stampfli (127), but are very hard to perform *in vivo.* As the diffusion coefficient and wall thickness are mostly unknown, the concept of capillary permeability is often used. The term *permeability* here, unfortunately, does not have a unique identity, since one must account for differences between the permeability to water, salts, proteins, etc. According to Bennett et al. (14), the capillaries in liver might have virtually infinite permeability to most substrates; those in brain and muscle, with a continuous membrane without fenestrations or pores, might display a finite permeability, and even zero permeability to many substrates. From the engineering point of view, permeability is a lumped transfer coefficient which describes the transport close to the capillary wall and through the membrane itself. Thus permeability may be dependent on the hydrodynamics of flow and on the nature and thickness of capillary membranes.

5.3 Modeling of the microcirculation

It is reported by Landis and Pappenheimer (98) that the essential factors influencing the rate and amount of capillary exchange are: distribution and rate of flow of capillary blood, volume of extravascular distribution compartments, permeability of capillary walls, diffusion coefficients in the interstitial tissue, permeability of cell membranes, and rates of reaction in the cells. They also indicate that the rate of exchange across capillary walls for gases is much greater than the rate of arrival by blood flow and, therefore, capillary permeability is not hindering. For all the other substances, permeability at the wall might play a role in the exchange process.

In spite of the great complexity of the living tissues and capillary pathways, many models have been developed to explain the phenomena of capillary exchange. A representative set of these was presented in Section 3.5.

The need to experimentally determine some of the microcirculation parameters led to the extension of the Stewart-Hamilton technique, which treated intravascular tracers to the treatment of extravascular indicators. Models were developed to infer permeability, clearance, volume of distribution of the diffusible substances, and blood flow rates.

Various tracer techniques have been devised and used in the study of microcirculation. The two most common ones are: single injection multiple tracer method, and the tissue clearance and uptake method. In the single injection multiple tracer technique a number of tracers are injected simultaneously and instantaneously into a major artery. Some of the tracers remain confined to the blood volume ("non-diffusible" tracers), the others diffuse across the capillary membranes into the interstices and eventually return to the blood (diffusible tracers). The sampling is done in a major vein. From the data obtained, regional blood flow rates, blood and interstitial volumes, capillary permeability, and other quantities are determined. Tissue clearance or uptake methods rely on the injection of diffusible tracers into the tissue. Exit blood concentration and tissue concentration are monitored in order to evaluate the pertinent parameters. Both methods rely heavily on mathematical models for data interpretation.

Kety (88) developed one of the first models for diffusible tracers which has had considerable use because of its simplicity. He considered a capillary surrounded by a well mixed tissue, i.e., no diffusion gradients are allowed in the tissue. The change of the total amount of diffusible tracer in the region (tissue and capillary) at any time is

$$\frac{dM}{dt_a} = F(C_a - C_v) \tag{5.1}$$

where $M = \bar{C}V_R = \lambda C_v V_R$ (5.2)

The concentration in the region is assumed to be the product of the measured venous concentration and the average partition coefficient. An instantaneous injection of tracer, M_i, is assumed by Kety and then the solution becomes simply

$$C_v(t_a) = \frac{M_i}{\lambda V_R} e^{-(F/\lambda V_R)t_a} \tag{5.3}$$

With all the assumptions made, differential Eq. (5.1), and the solution given by Eq. (5.3), describe a behavior of a perfectly mixed tank. A diffusible indicator can behave in the manner described by Eq. (5.3) as a single "washout curve" with no appearance lag, only if there is instantaneous equilibration of the material, both along the capillary and in the lateral direction. As discussed previously, Kety's model may be used only for highly soluble tracers in the regions where the ratio of the tissue and blood volume is much larger than one, and where the partition coefficient is equal to or larger than one.

Renkin (138, 139), and Crone (42), were interested in the "capillary diffusion capacity," which is the product of permeability P with exchange surface area of the capillary walls, S. They developed a formula

$$PS = F \ln \frac{1}{1 - E} \tag{5.4}$$

where $E = \dfrac{C_a - C_v}{C_a - \lambda C_T}$ (5.5)

is known as fractional extraction.

It is interesting to note that the problem they really solved is a steady state solution to the following equation

$$F \frac{dC}{dZ} = -PO(C - \lambda C_T) \tag{5.6}$$

$$C = C_a \quad \text{at} \quad Z = O \tag{5.7}$$

Then calling $C(Z = L) = C_v$ and taking $0 \cdot L = S$, one gets

$$C_v - \lambda C_T = (C_a - \lambda C_T)e^{-(PS/F)} \tag{5.8}$$

From Eq. (5.8) it is easy to develop the formula, Eq. (5.4). Renkin (138, 139) used this formula in evaluating the permeability in skeletal muscle perfused at a constant rate; Crone (42) determined the permeabilities in the brain, liver, lungs, kidney, and limbs using a single injection of a diffusible and a nondiffusible tracer and computed the extraction using Eq. (5.5) with $C_a = C_{\text{nondif.}}$, $C_v = C_{\text{dif.}}$.

It is apparent that this model is not complete, since only one capillary was considered at steady state with no return of material from the extravascular space. This description of an organ or a capillary network may be adequate only under certain conditions of low blood to tissue ratio. Similar simplified models are used in determining permeability and regional blood flow rates. Some of the experimental works have been done by Freis et al. (56) on the forearm, Chinnard and associates (29-33) on different organs, and Conn (35) on the heart.

Goresky and coworkers (65-68) were very successful in the early 60's in developing a model for determining blood volume in liver sinusoids and extravascular volumes of distribution of the diffusible tracers. Multiple tracer studies of the hepatic circulation of the dog were carried out using labeled red cells, albumin, inulin, sucrose, sodium, urea, water, and T-1824 (an inert optical tracer). Calculated volumes of distribution for extracellular substances diminished rapidly with increase in the molecular weight of the substances. This indicates the decrease of the rate of diffusion in the tissue, but nevertheless, all the results were obtained on the basis of a flow-limited linear two compartment model, assuming rapid, practically instantaneous, radial diffusion in the extracellular space. Goresky indicates when diffusion takes place slowly in relation to the blood flow, the rate of transcapillary passage will become limited chiefly by diffusion—"diffusion limited" response. In contrast, where diffusion perpendicular to flow takes place rapidly in comparison to the blood flow rate, the rate of transfer becomes chiefly limited by the rate of perfusion—"flow limited" response. He believes that the latter case applies to the exchange of most substances in the liver. Nevertheless, Goresky neglects axial diffusion by comparison to convection.

Goresky (65) considered a two compartment system with the following restrictions:

1. Flow is continued to one compartment.
2. Axial diffusion is neglected in both compartments.
3. The sinusoid capillary wall is not considered to be a phase boundary of any consequence for dissolved material.

4. Diffusion in the radial direction is assumed to be so rapid that concentrations in both compartments are equal.
5. Flow is assumed to be bolus or plug flow.

With these assumptions, the following differential equation is obtained

$$\frac{1+\gamma}{U}\frac{\partial C}{\partial t_a} + \frac{\partial C}{\partial Z} = 0 \tag{5.9}$$

$$\text{where } \gamma = \frac{A_B}{A_A} = \frac{\text{cross sectional area of tissue}}{\text{cross sectional area of capillary}} \tag{5.10}$$

It is easy to see that Eq. (5.9) represents unsteady state plug flow with velocity $U/(1+\gamma)$. If the initial concentration profile is $a(Z)$, as assumed by Goresky, then the solution is the well known delayed wave

$$C(Z, t_a) = a\left(Z - \frac{Ut_a}{1+\gamma}\right) \tag{5.11}$$

If there is no accessible extravascular space, $\gamma = 0$, the solution becomes

$$C(Z, t_a) = a(Z - Ut_a) \tag{5.12}$$

While the nondiffusible concentration travels like a wave with velocity U, the diffusible one is delayed and travels at a rate of $U/(1+\gamma)$. This results in a lag in the appearance of the second wave at the outflow and in the proportionate delay in all parts of the concentration wave. Goresky then treats a number of capillaries of different lengths, but some of the equations used are dimensionally incorrect. The red cell label was the earliest to appear in the hepatic venous outflow; and the concentration of injected cells rose to the highest and earliest peak and decayed most rapidly. These curves were taken as the standard nondiffusible tracer curves. The time for corresponding points on the diffusible tracer curves was increased by the factor of $(1+\gamma)$, and the corresponding concentrations were reduced by the factor of $1/(1+\gamma)$. Using these relationships, Goresky evaluated the factor $1/(1+\gamma)$ from the peak concentrations and the zero time for exchange from the times of appearance of the peak concentrations. The extravascular volume of distribution for each substance was found by multiplying γ with the sinusoidal plasma volume. The superpositon of the adjusted dilution curves upon the red cell dilution curve showed very good agreement. Goresky concludes that a flow-limited model closely represents reality in liver sinusoids.

Goresky (66) extended his flow-limited model allowing for a first order consumption of the tracer (sulfobromophthalein) in the liver. The model then becomes a description of a plug flow reactor with first order kinetics at unsteady state

$$\frac{\partial C}{\partial Z} + \frac{1+\gamma}{U}\frac{\partial C}{\partial t_a} + \frac{k\gamma C}{U} = 0 \tag{5.13}$$

with

$$\begin{aligned} C(Z, 0) &= a(Z) \\ C(0, t_a) &= 0 \end{aligned} \tag{5.14}$$

Goresky obtained the familiar solution of the damped wave in the form

$$C(Z, t_a) = e^{-(k\gamma t_a/1+\gamma)} a\left(Z - \frac{Ut}{1+\gamma}\right) \tag{5.15}$$

Goresky (67) also studied the distortion of the tracer response curves due to noninstantaneous injection and sampling system distortion. The same problem was considered and solved by Oleksy, Weinstein, and Shaffer (122).

Perl (128) developed a simplified model considering a slab of tissue exposed to arterial blood concentration on one face and venous blood concentration on the other.

Martin and Yudilevich (111) developed an integral equation description for the simultaneously obtained indicator dilution curves of two tracers, one confined to plasma, and another diffusible through the capillary barrier. Assuming that fractional extraction E is the same for all transit times in a homogeneous capillary bed, they arrived at the following relationship between the nondiffusible tracer concentration, C_{nd}, and diffusible tracer concentration, C_d

$$C_d(t_a) = (1-E)C_{nd}(t_a) + E\int_0^{t_a} C_{nd}(\tau)\phi(t_a - \tau)\, d\tau \tag{5.16}$$

where $\phi(\tau)d\tau$ is defined as the probability that a given particle will return to the intravascular space between times τ and $\tau + d\tau$ after its extraction from the blood. The function $\phi(\tau)$ therefore represents the backflux of tracer from the extravascular compartment into the vascular shape. Based on the assumption that the appearance times are the same for both indicators, and taking this as zero time, they obtained the fractional extraction,

$$E = 1 - \lim_{t\to 0} \frac{C_d(t)}{C_{nd}(t)} = 1 - \lim_{t\to 0} \frac{C_d'(t)}{C_{nd}'(t)} = 1 - \lim_{s\to\infty} \frac{\bar{C}_d(s)}{\bar{C}_{nd}(s)} \tag{5.17}$$

One should be able obtain $\phi(\tau)$ from Eq. (5.16) using Laplace transforms or series expansion. However, in interpreting the data, Martin and Yudilevich used the compartmental model and $\phi(\tau)$ was substituted with its expected value

$$\phi(\tau) = \beta \exp(-\beta\tau) \tag{5.18}$$

where $\beta = \dfrac{PS}{F}$

From convenient plots the fractional turnover rate, β, was found. Therefore, in the interpretation of data, assumption of equal plasma and interstitial concentrations was implicitly made.

At a symposium on "blood flow and exchange at microcirculatory level" in 1965, Zierler (196) presented a very good review on tracer dilution techniques, including diffusible tracers. He proposed the interstitial volume be calculated using the following formula

$$V_T = F(\bar{t}_{a,d} - \bar{t}_{a,nd}) \tag{5.19}$$

where $\bar{t}_{ad}$ and $\bar{t}_{and}$ are the mean residence time of the diffusible and nondiffusible tracers, respectively. He represented the vascular bed as a connection of four distributing systems: (a) arteries, (b) arteriovenous shunts, (c) capillaries, and (d) veins. Both the capillaries and arteriovenous shunts are considered arranged in parallel between arteries and veins. Furthermore, Zierler used the convolution principle to lump together the arterial and venous response into the response of large vessels. With these assumptions he arrived at the simplified picture of the vascular bed schematically represented in Figure 11.

The observed unit response of the whole system is given as

$$E(t_a) = x\int_0^{t_a} E_1(\tau)E_3(t_a - \tau)\,d\tau + (1 - x)\int_0^{t_a} \cdot E_2(\tau)E_3(t_a - \tau)\,d\tau \tag{5.20}$$

Figure 11 Schematic of the microcirculation following Zierler (194).

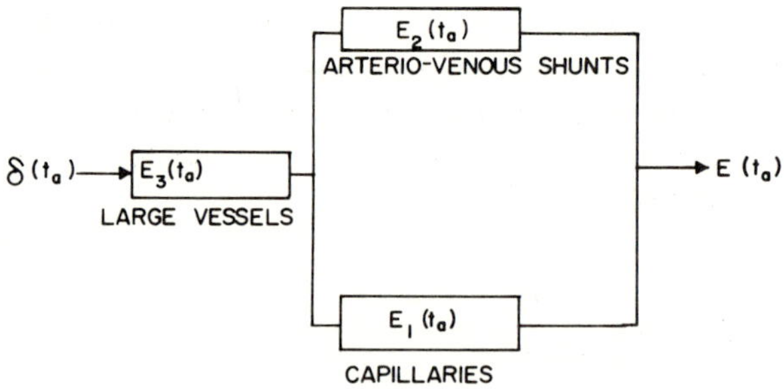

where x is the fraction that passes through the capillaries. The flow rate through the entire system, and the volume of the entire system, could be obtained by the classical methods reviewed earlier in this section. If one would be able to find a tracer that diffused completely through the capillary walls during one passage time, the obtained residence time distribution would be

$$E_4(t_a) = (1 - x)\int_0^{t_a} E_2(\tau)E_3(t_a - \tau)\, d \tag{5.21}$$

Subtracting $E_4(t)$ from $E(t)$ one would get $E_5(t)$ representing the residence time distribution for large vessels and capillaries in series.

Based on the investigation of transport of nonelectrolytes in an isolated perfused heart, Johnson and Wilson (85) developed the following model for capillary exchange. A capillary penetrates a region in which the extracellular concentration of a substance is homogeneous with respect to its spatial distribution. Within the capillary, concentration is a function of distance along the capillary and time. Flow in the capillary is assumed to be plug flow and radial variations are ignored. The model leads to two coupled equations

$$\pi R_0^2 \frac{\partial C}{\partial t_a} + \pi R_0^2 U \frac{\partial C}{\partial Z} = 2\pi R_0 P(C_T - C) \tag{5.22}$$

$$V_T \frac{\partial C_T}{\partial t_a} = \int_0^L 2\pi R_0 P(C - C_T)\, dz \tag{5.23}$$

A step change in concentration is considered through the boundary and initial conditions

$$C(0, t_a) = H(t_a), C(Z, 0) = 0, C_T(t_a = 0) = 0 \tag{5.24}$$

Using the dimensionless variables $z = Z/L$ and $t = Ut_a/L$ two characteristic parameters appear in the equation. One of them is

$$\gamma = \frac{V_T}{\pi R_0^2 L} = \frac{V_T}{V_c} = \frac{A_T}{A_c} \tag{5.25}$$

which is familiar from Goresky's calculations. The second one is

$$\beta = \frac{2PL}{R_0 u} = \frac{2\pi R_0 PL}{\pi R_0^2 u} = \frac{PS}{F} \tag{5.26}$$

The solution was obtained in the Laplace transform plane. However, inversion of the Laplace transforms was obtained only for large s, yielding the early time solutions. These are quite interesting as they predict that there will be appearance of transport susbstance in the capillary at a point beyond the convected material because of an extracellular space shunt due to the instantaneous diffusion postulated by the model. The characteristic time required for the concentration in the tissue space to relax to its new equilibrium value is of the order of γ/β if $\beta < 1$, or γ if $\beta > 1$. In any case, the characteristic time is of order much greater than one, so solutions for larger values of time are needed. Asymptotic solutions were obtained, and from these it was concluded that the rate constant for tissue buildup was

$$\frac{F}{V_T}\,[1 - e^{-(PS/F)}] \tag{5.27}$$

The model is of value as it clearly indicates the importance of the permeability to blood flow ratio on the intracapillary concentration profile. For high values of β a flow-limited pattern is obtained where γ determines the kinetics of distribution. For low values of β a transfer limited pattern is obtained. It is also important to notice the negligible utilization of a large section of the capillary area for highly permeable susbstances.

Perl and Chinard (129) proposed a model for interpretation of extravascular indicator dilution experiments in dog kidney. One of the inexplicable features in the kidney includes the experimental fact that appearance times and peak value times for highly diffusible extravascular indicators (tritium, krypton) are earlier than or coincidental with those for vascular indicators (T-1824 labeled albumin). The model has the mathematical form of the time dependent Fick's diffusion equation with a convective term. The indicator is assumed to enter the blood tissue region through vascular sources, to equilibrate instantaneously radially, and to be transported to vascular sinks by simultaneous diffusion and convection at a more macroscopic distance level. Finite permeability barriers are not allowed in the model, except for the limiting case of zero permeability. The solution containing the Peclet number, $\mathrm{Pe} = UL/D$, is obtained and the following cases are studied:

1. Pe $\to \infty$ axial diffusion is negligible compared to convection, and the equations reduce to the "delayed wave" model.
2. Pe ≈ 1, diffusion is comparable to convection; the amount of tracer is instantaneously dispersed from $x = 0$ to $x = L$, but the main portion travels with a reduced velocity approximating that of case 1.
3. Pe $\to 0$, convection is small relative to diffusion; the indicator equilibrates not only radially, but also axially and gets "washed out" by the vascular fluid. This is the flow-limited case proposed by Kety (88).

Levitt (105) critically examined the validity of the early extraction method (E_0 method) in estimating capillary permeability. The modified fractional extraction is defined as

$$E = 1 - \frac{C_d(t_a)}{C_{nd}(t_a)} \tag{5.28}$$

where the concentrations of the diffusible and nondiffusible tracers are measured in the venous effluent. Often the curves are extrapolated to zero appearance time and it is assumed that

$$E_o = E(t_a = 0) = 1 - e^{-(PS/F)} \tag{5.29}$$

Levitt examined the response to a concentration square wave input of finite, although small, duration. The response curves were found to be functions of two parameters $\beta = PS/F$, the permeability to flow ratio, and $1/\gamma = A_c/A_T$, the ratio of capillary to tissue cross-sectional area. The nondiffusible indicator traveled through the capillary without any change in shape according to the model. Its appearance time at the exit was equal to one capillary traversal time. The curves for the diffusible indicator can be described as an initial peak with the same appearance time as the nondiffusible indicator, followed by a tail. With increased values of β, the main portion of the tracer appears in the tail, and the peak is shifted from the origin. At very large values of β the peaks become higher, while at infinite β the delayed wave model by Goresky is obtained. It was shown that the E_0 method would accurately estimate capillary permeability only for organs with all capillaries identical, provided that γ is greater than three. Otherwise, the method underestimates the actual permeability. The square wave input of tracer used by Levitt probably represents real experiments better than does the Dirac δ-function.

Many experimental results have been analyzed on the basis of one or more of the previously mentioned models. The clearance of isotope from skeletal muscle was investigated by Hyman and Leventhall (79); blood flow and fluid compartment sizes in the dog heart were examined by Bailey and Harvey (5); and pulmonary spaces in lungs were evaluated by Taylor et al. (177). Bassingthwaighte and his colleagues (10) produced an extensive experimental study on blood flow and distribution volumes for several substances in the heart.

A study on capillary exchange modeling by Goresky, Ziegler, and Bach (68) appeared recently. In the first part of the study the authors examined a modified Sangren-Sheppard model (147), which we discussed earlier in Section 3.5, and which Levitt used in his work (105). The modification was accomplished by allowing for an active transport at the capillary wall which results in different permeabilities for the efflux and influx. The equations considered are

$$\frac{\partial C}{\partial t_a} + U\frac{\partial C}{\partial Z} + \gamma\frac{\partial C_T}{\partial t_a} = 0 \tag{5.30}$$

$$\frac{\partial C_T}{\partial t_a} - k_1 C + k_2 C_T = 0 \tag{5.31}$$

where the turnover constants are defined as follows

$$k_1 = \frac{P_1 S}{V_T} \qquad k_2 = \frac{P_2 S}{\lambda V_T} \tag{5.32}$$

where λ is the partition coefficient for the diffusible substance in extravascular tissue relative to that in the reference vascular fluid at equilibrium. The above equations were solved by others for $k_1 = k_2$. Goresky and his colleagues presented the solutions for many special cases: (a) $k_1 = k_2 = 0$ corresponding to the reference, nondiffusible substance, (b) $k_2 = 0$, k_1 finite, equivalent to Crone's model, (c) k_1 and k_2 finite similar to Sheppard's model, and (d) k_1 and k_2 infinite corresponding to Goresky's delayed wave model. In case (c) for finite values of k_1 and k_2, they presented an infinite series solution useful for small values of peremeability, and a Bessel function solution valid for large permeabilities. The outflow profiles from a single capillary for different values of permeability were calculated.

In the second part of their study, Goresky, Ziegler, and Bach attempted to deal with the problem of multicapillary modeling. They examined the following three cases: (a) time spent in large vessels was assumed constant and the nondiffusible tracer curve was assumed to result solely from varying capillary transit times; (b) time spent in each capillary was assumed constant and the reference tracer curve was assumed to result from varying times spent in the large vessels; (c) the combination of the previous two cases was considered, and it was shown that they were particular extremes of this more general case. The concentration curves of the diffusible tracers were obtained using some common probability density functions and the response of a single capillary.

Lee and Fronek (101) examined a single capillary model with finite radial diffusion in the tissue. The response of a model to a specifically constructed input function was considered in order to determine the exchanges in the extraction factor E with time. It is shown that the exchange is governed by three parameters: the permeability number β which is equivalent to the Stanton number; the Sherwood number, PR/D; and the ratio of characteristic diffusion and convection times, $D\bar{t}_m/R^2$.

5.4 General single capillary model and limiting cases

Almost all the models of microcirculation mentioned earlier treat a single, representative capillary and the adjacent tissue compartment. Additional simplifying assumptions are made in deriving the equations of momentum and mass transport. How well a single capillary model represents a complicated network in the microcirculation is a question that has not yet been answered. However, single capillary models do provide insight into the complicated processes that occur.

Capillary networks are different in various tissues. This makes an exact analysis very difficult. However, it is likely that each capillary supplies mainly, and possibly exclusively, the surrounding tissue. Therefore a representative analysis can consider the response of a single capillary, surrounded by a tissue compartment, to a sudden injection of tracer. This response can be considered to be typical for the whole capillary network, especially for capillaries arranged in parallel, or it can be combined with the capillary distribution branches and the response of large blood vessels to yield a model of an organ Even a model of a single cylindrical capillary of finite length with a concentric cylinder of tissue would be extremely complex and intractable if one were to include all the terms involved.

A reasonably general model was developed by Dudukovic (48) and presented by Dudukovic et al. (49, 50). A brief description of the model follows. A single average capillary is assumed to be of cylindrical shape with length L between 100μ and 800μ, and a diameter, $2R_0$, between 8μ and 10μ. A cylinder of tissue of radius R_1, 20μ to 200μ, surrounds the capillary. It has been shown by analysis (23, 136) that the velocity profile in the capillary does not influence the capillary and tissue concentration much. Hence, plug flow of plasma is assumed. The red cells move like slugs through a capillary and occupy its entire cross section. Secondary flow cells are induced between each successive pair of blood cells, and mixing of plasma is assumed to be good in the radial direction. The diffusion flux in the axial direction is neglected in the capillary, as is the presence of red cells. Plasma is considered to be a homogeneous fluid. The membrane of the capillary wall is considered to offer a finite resistance to passage of molecules. The partition coefficient of unit is assumed throughout the study.

With these assumptions the material balance for the capillary elements yields the following differential equation

$$\pi R_0^2 \frac{\partial C}{\partial t_a} = -\pi R_0^2 \frac{\partial C}{\partial Z} - 2\pi R_0 P(C - C_T|_{R=R_0}) \tag{5.33}$$

The tissue is considered as slightly anisotropic, with different properties in the axial and radial direction. Only diffusion fluxes are considered, as it has been shown that filtration and osmotic processes do not influence diffusion exchange to any appreciable extent (98, 125). The material balance on an element of tissue gives the following equation

$$\frac{\partial C_T}{\partial t_a} = D_Z \frac{\partial^2 C_T}{\partial Z^2} + \frac{D_r}{R} \frac{\partial}{\partial R}\left(R \frac{\partial C_T}{\partial R}\right) \tag{5.34}$$

As the model is to be applied to tracer experiment, the initial conditions are

$$\begin{aligned} C(Z, 0) &= 0 \\ C_T(Z, R, 0) &= 0 \end{aligned} \tag{5.35}$$

A boundary condition at the capillary wall is obtained from the continuity of the flux at the wall

$$- D_r \left.\frac{\partial C_T}{\partial R}\right|_{R=R_0} = P(C - C_T|_{R=R_0}) \tag{5.36}$$

If it is assumed that the capillaries are symmetrically arranged in a tissue cross section, each of them completely supplying its own cylinder of tissue with nutrients, then due to symmetry, a condition of zero flux can be prescribed at the outer boundary of the tissue cylinder

$$\left.\frac{\partial C_T}{\partial R}\right|_{R=R_1} = 0 \tag{5.37}$$

However, this boundary condition requires special attention. The tissue cylinder is only a model and is by no means insulated. Therefore, the conditon of zero flux at the outer boundary implies that the cylinder is in equilibrium with the rest of the tissue, or with other tissue cylinders, for every axial distance and at any time. Nevertheless, this boundary condition is a good approximation, as it is very likely that each capillary is indeed mainly supplying the neighboring tissue. If possible, a more general boundary condition should be applied at R_1, as indicated by the following equation

$$\left.\frac{\partial C_T}{\partial R}\right|_{R=R_1} = \phi(R_1, Z, t_a) \tag{5.38}$$

However, it is not presently possible to determine the function $\phi(R_1, Z, t_a)$. The conditions of zero flux in the axial direction, at $Z = 0$ and $Z = L$, are imposed on the tissue concentration.

The condition at $Z = 0$, for the concentration in the capillary is the function of time that describes the injection of the diffusible tracer. In order to find the transfer function of the model it is assumed that the input is well represented by the Dirac δ-function

$$C(0^-, t_a) = \frac{M_i}{\pi R_0^2 U}\,\delta(t_a) = C_0 \bar{t}_a \delta(t_a) \tag{5.39}$$

These equations are reduced to a dimensionless form with a following change of variables

$$u = \frac{C}{C_0} \qquad v = \frac{C_T}{C_0} \qquad \text{and}$$

$$t = \frac{t_a U}{L} \qquad r = \frac{R}{R_1} \qquad z = \frac{Z}{L}$$

The following dimensionless groups appear in the dimensionless equations

$$\text{Pe}_z = \frac{UL}{D_z} \quad \text{— Peclet number in the axial direction}$$

$$\text{Pe}_r = \frac{UR_1}{D_r} \quad \text{— Peclet number in the radial direction}$$

$$\text{Sh} = \frac{PR_1}{D_r} \quad \text{— Sherwood number}$$

$$\beta = \frac{PS}{F} = \frac{b\text{Sh}}{\text{Pe}_r} \quad \text{— permeability number equivalent to Stanton number}$$

$$b = \frac{2L}{R_0} \qquad l = \frac{L}{R_1} \qquad \rho = \frac{R_0}{R_1} \quad \text{— characteristic length ratios}$$

The Eqs. (5.33) to (5.39) now read as follows:

Governing equations

$$\frac{\partial u}{\partial t} + \frac{\partial u}{\partial z} = \beta(v|_\rho - u) \tag{5.40}$$

$$\frac{\partial v}{\partial t} = \frac{1}{\text{Pe}_z}\frac{\partial^2 v}{\partial z^2} + \frac{l}{\text{Pe}_r}\frac{1}{r}\frac{\partial}{\partial r}\left(r\frac{\partial v}{\partial r}\right) \tag{5.41}$$

Initial conditions

$$u(0, z) = 0 \tag{5.42}$$

$$v(0, z, r) = 0 \tag{5.43}$$

Boundary conditions

$$u(t, 0) = \delta(t) \tag{5.44}$$

$$\left.\frac{\partial v}{\partial z}\right|_0 = \left.\frac{\partial v}{\partial z}\right|_1 = 0 \tag{5.45}$$

$$\left.\frac{\partial v}{\partial r}\right|_1 = 0 \tag{5.46}$$

$$-\left.\frac{\partial v}{\partial r}\right|_\rho = \mathrm{Sh}(u - v|_\rho) \tag{5.47}$$

The problem is to solve the coupled system of equations, Eq. (5.40) to Eq. (5.47) for $u(t, z)$ and $v(t, z, r)$.

The formal solution of Eq. (5.40) is

$$u(t, z) = e^{-\beta z}\delta(t - z) + \beta \int_0^z v(t - z + z', z', \rho) e^{-\beta(z-z')} H(t - z + z') dz' \tag{5.48}$$

where $H(x)$ is the Heaviside unit step function.

The tissue concentration v still remains to be evaluated in order to obtain a complete solution for u from Eq. (5.48). Equations (5.41), (5.43), (5.45), (5.46), and (5.47) define the problem for v entirely, the only input to the system coming from the boundary condition Eq. (5.47). If it is assumed for the moment that $u(t, z)$ is a known function, then the problem for $v(t,z,r)$ can be solved using the appropriate Green's function (48).

The solution then has the form

$$v(t, z, r) = \frac{\beta\rho^2}{2} \int_0^t \int_0^1 u(t', z') G(t - t', z, r; z', \rho)\, dz'\, dt' \tag{5.49}$$

An integral equation for the tissue concentration at the boundary, $v(t,z,\rho)$, is obtained by substituting Eq. (5.48) into Eq. (5.49)

$$
\begin{aligned}
v(t,z,\rho) &= \frac{\beta\rho^2}{2}\int_0^t dt'\int_0^1 G(t-t',z\rho;z',\rho)\Bigg[e^{-\beta z'}\delta(t'-z') \\
&\quad + \beta\int_0^{z'} v(t'-z'+z'',z'',\rho)e^{-\beta(z'-z'')}H(t'-z'+z'')\,dz''\Bigg]dz' \\
&= \frac{\beta\rho^2}{2}\Bigg[H(1-t)\int_0^t e^{-\beta t'}G(t-t',z,\rho;t',\rho)\,dt' \\
&\quad + H(t-1)\int_0^1 e^{-\beta t'}G(t-t',z,\rho;t',\rho)\,dt'\Bigg] + \frac{\beta^2\rho^2}{2}\int_0^t dt'\int_0^1 \\
&\quad \cdot G(t-t',z,\rho;z',\rho)\,dz'\int_0^{z'} v(t'-z'+z'',z'',\rho)e^{-\beta(z'-z'')} \\
&\quad \cdot H(t'-z'+z'')\,dz'' \qquad (5.50)
\end{aligned}
$$

This together with Eq. (5.48) for $u(t,z)$ provides the formal solution to the system.

The existence and uniqueness of solution was established and the convergence of the iteration scheme was demonstrated (51). A few leading terms of the iteration scheme were evaluated analytically (51). Although a numerical method for computing the exit capillary concentration, $u(t,1)$, may be readily devised, the resulting expression is extremely cumbersome to use as it involves multiple infinite series. Evaluation of system parameters by matching such an expression to experimentally determined data would be extremely difficult.

This general single capillary model may be used, however, to assess various limiting cases and determine the influence of various modes of radial and axial transport on the shape of the response. Many simplified cases can be derived from the general one assuming limiting values of the system parameters Pe_z, Pe_r, Sh, β. The Stanton number, β, can take values from zero to infinity. Zero value implies that the capillary wall is impermeable to the tracer under consideration, i.e., it characterizes the tracer as a nondiffusible one. Infinite value of β indicates that the capillary wall offers practically no resistance to transfer of the particular tracer, i.e., one deals with a highly diffusible tracer.

The values of Peclet number in the axial direction, Pe_z, also lie in the interval from zero to infinity. A value which approaches zero indicates that the diffusive transport in the axial direction in the tissue is much faster than the convective transport in the capillary. A limiting value of infinity indicates that the transport due to diffusion is negligible with respect to that due to convection. Peclet number in the radial direction, Pe_r, may vary from zero to very large, but finite, values. A very small value implies that the radial equilibration in the tissue due to diffusion occurs much faster than the convective transport in the capillary. Large values of Pe_r indicate that radial diffusion in the tissue is slow compared to convection in the capillary.

Sherwood number, Sh, varies from zero to very large values. A zero value implies zero permeability, i.e., a case of nondiffusible tracer. Large values of Sh indicate that diffusion in the radial direction in the tissue is hindering the transport, while small values imply that permeability of the capillary wall determines the overall rate of transport.

Assuming limiting values of β, Pe_z, Pe_r, twelve simplified cases presented in Table 1 are considered. It is possible to obtain a closed form solution in the Laplace transform plane for all the cases but one (48). These solutions can be inverted numerically into the time domain. Although the concentration in the capillary, $u(t,z)$, and tissue concentration, $v(t,z,r)$, are obtained, it is only of major interest to evaluate the capillary concentration at the exit $u(t,1)$, or its transform $\bar{u}(s,1)$ which in a sense represents the transfer function of the model. The concentration at the outlet $u(t,1)$ can then be compared to experimentally obtained data in order to evaluate the remaining parameters of the system.

Solutions of the limiting cases can be compared with one another in order to gain insight into the bounds of the solution of the general case. The governing differential equations for all 12 limiting cases are readily obtained through material balances. Some of them can also be obtained as zero order perturbations of the general case for limiting values of parameters.

Dudukovic et al. (51) considered all 12 limiting cases and evaluated their transfer functions explicitly. Approximate solutions for small and large tissues were evaluated for most of the cases (48) by expanding the transfer function for large and small s, respectively. The first four moments of the solutions are calculated for all cases. The parameters of the system may be estimated by either transforming the experimental data and comparing them to the model response

Table 1 The Limiting Cases

Case	Permeability term (β)	Axial diffusivity term (Pe_z)	Radial diffusivity term (Pe_r)
1	zero	–	–
2	infinite	zero	zero
3	infinite	zero	finite
4	infinite	infinite	zero
5	infinite	infinite	finite
6	infinite	finite	zero
7	infinite	finite	finite
8	finite	infinite	zero
9	finite	infinite	finite
10	finite	zero	zero
11	finite	zero	finite
12	finite	finite	zero

Table 2 The Moments of the Solutions to the Limiting Case

Zeroth moment: $M_0 = 1$

First moment: $M_1 = \mu = \frac{1}{\rho^2} = \frac{V_{\text{total}}}{V_{\text{capillary}}}$

Variance: $\sigma^2 = M_2 - M_1^2$

$$\sigma^2 = a_1 \cdot \frac{1}{\rho^4} + a_2 \cdot \frac{\text{Pe}_R}{\rho^2}\left[\frac{\rho^4 - 1}{4} - \frac{(\rho^2 - 1)^2}{2} - \ln \rho\right]$$

$$+ a_3 \cdot \frac{2(1-\rho^2)}{\rho^6 \text{Pe}_z}\left[1 - \frac{(1-\rho^2)}{\rho^2 \text{Pe}_z}(1 - e^{-(\rho^2 \text{Pe}_z/1-\rho^2)})\right]$$

$$+ a_4 \cdot \frac{2(1-\rho^2)^2}{\beta\rho^4} + a_5\left[\frac{1 + e^{-\beta}}{\rho^4(1 - e^{-\beta})} - \frac{2(2-\rho^2)}{\beta\rho^2}\right]$$

Case number	a_1	a_2	a_3	a_4	a_5
1	0	0	0	0	0
2	1	0	0	0	0
3	1	1	0	0	0
4	0	0	0	0	0
5	0	1	0	0	0
6	0	0	1	0	0
7			not determined		
8	0	0	0	1	0
9	0	1	0	1	0
10	0	0	0	0	1
11	0	1	0	0	1
12			Equation (C) not applicable		

in the Laplace transform plane, or by matching the moments of the solution to the experimental data. When the system parameters are known the solutions in the time domain can be obtained by numerical inversion of the Laplace transform, or by a series of associated Laguerre polynomials.

The usefulness of the moments of the transfer function in evaluating system parameters was emphasized by Dudukovic (48). The first moments and the variances for all the limiting cases are presented in Table 2. A similar relationship was developed for the skewness (48). The variations of moments with system parameters are discussed in detail in the original study (48) where it was shown that variance and skewness may have a wide range of values from zero to values much larger than those of a perfectly mixed compartment. The analysis of the moments provides several parameters of the system. The volume of the blood and tissue space may be determined from the zeroth and first moment of the nondiffusible and diffusible tracer concentration curves at the capillary exit as established by Zierler (196). In order to evaluate diffusivities in the tissue space, and the permeability of the capillary wall, a specific model is needed for the interpretation of the higher moments.

It is of interest to point out some properties of the variance of the diffusible tracer curve. It can be seen from Table 2 that in the absence of axial diffusion, the variance is a sum of radial resistances due to permeability and radial diffusivity. Whenever one of these becomes infinite the corresponding resistance term vanishes. However, when axial diffusion proceeds at either finite or infinite rate, the radial resistances are weighted by terms which depend on axial diffusivities in a nonlinear fashion.

The properties of the responses for various limiting cases is discussed in detail by Dudukovic (48). In general, two types of response curves are obtained for a diffusible tracer. Whenever finite or infinite axial diffusivity is assumed, as in cases 2, 3, 6, 7, 10, 11, and 12, a theoretical zero appearance time is possible as the tracer may "bypass" the fluid and appear at the exit faster by diffusion through the tissue. The other group of models exhibit theoretical appearance times equal to the capillary mean transit time (equal to one on the dimensionless time scale). These are models 1, 4, 5, 8, and 9, which are based on the assumption of zero axial diffusivity in the tissue.

It is of interest to examine briefly the responses of various limiting cases. Case 1 with zero permeability represents the response of a single capillary to sudden injection of nondiffusible tracer. With the assumptions used, the response is a delta function appearing at mean capillary transit time (dimensionless time of one) as presented in Figure 12. Case 2, with infinite permeability

Figure 12 Capillary exit concentration for Case 1.

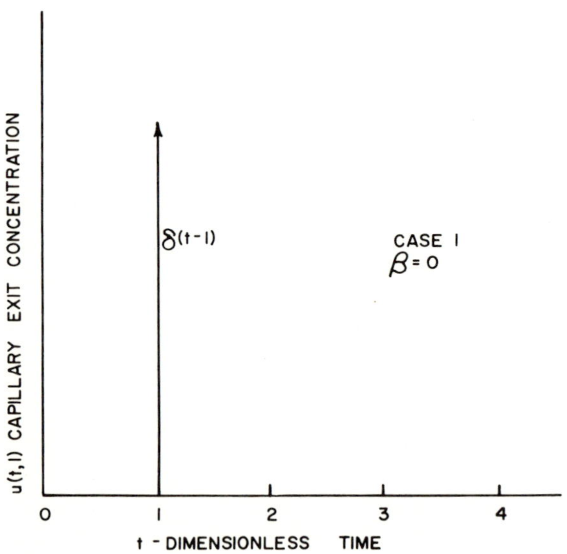

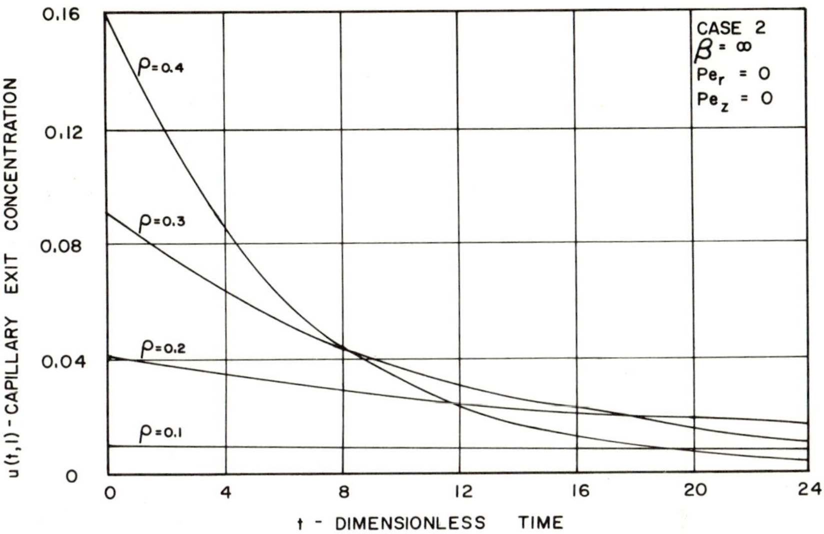

Figure 13 Capillary exit concentration for Case 2.

and infinitely fast axial and radial equilibration in the tissue is equivalent to Kety's model (88), and its response is presented in Figure 13 for various capillary volume to total volume ratios, ρ. The response of Case 3, which is based on assumptions of infinite permeability, axial diffusivity, and finite radial Peclet number, is presented in Figure 14 for various volume ratios, ρ, and different capillary mean transit time to characteristic diffusion time ratios, γ. These curves exhibit very long tails which decay slower than Kety's exponential. For very large diffusion times, and fairly large capillary to tissue volume ratios, the variance may be more than a hundred times greater than the variance for a perfectly mixed compartment. Naturally, the approach to Kety's model is achieved at high radial diffusivities especially for small values of ρ.

The response of Case 4, which is based on assumptions of no transport barrier at the capillary wall, and infinitely fast equilibration in the radial direction in the tissue with negligible axial diffusion, corresponds to the "delayed wave" model of Goresky (65). The response curves are delta functions at the mean transit time as shown in Figure 15.

The response curves for Case 5, which are based on assumptions of no barrier at the capillary wall, finite radial diffusivity, and zero axial diffusivity in the tissue, are presented in Figure 16. The curves for large diffusion time to capillary transit time ratios ($\gamma < 0.01$) rise very sharply to a high peak at the capillary transit time and are followed by a slowly decaying tail. The variance is larger than for a decaying exponential. With increasing γ, the peak of the curve moves to the right, toward the mean. For very large values of γ, Goresky's "delayed wave" model and the response of Case 4 is approached.

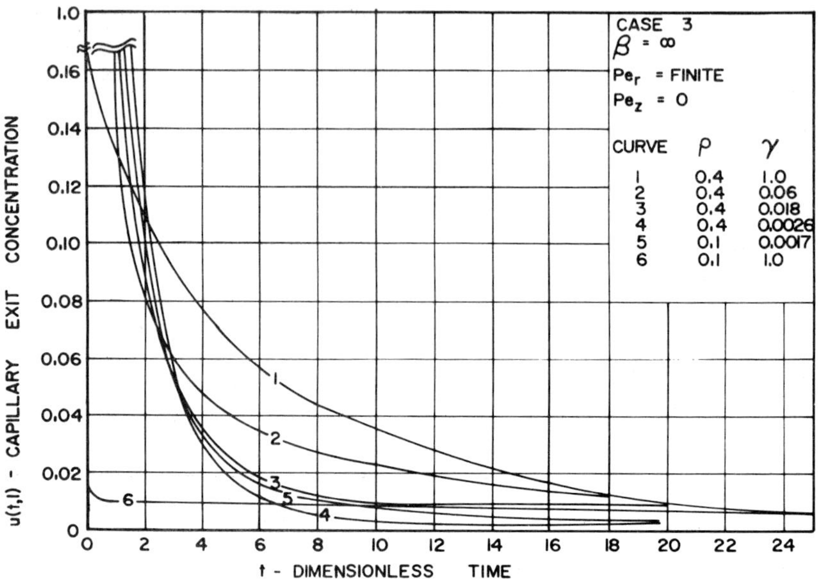

Figure 14 Capillary exit concentration for Case 3.

Figure 15 Capillary exit concentration for Case 4.

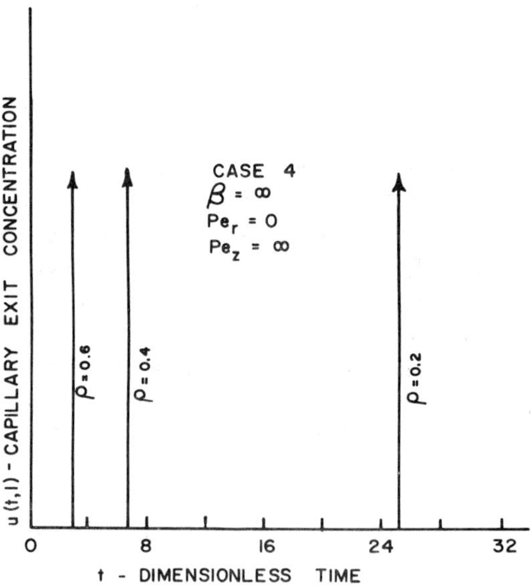

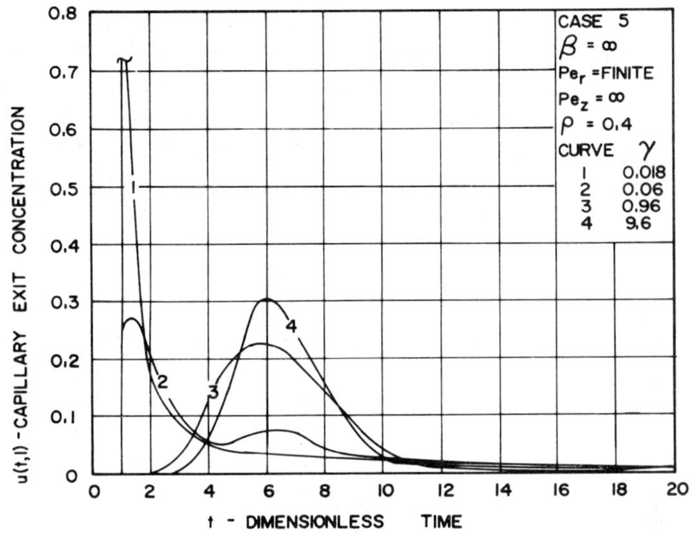

Figure 16 Capillary exit concentration for Case 5.

The response for Case 6, which is based on the assumptions of no barrier to transport at the capillary wall, and infinitely fast radial equilibration with finite axial diffusion in the tissue, is presented in Figure 17. This case is equivalent to the model of Perl and Chinard (129). For Peclet numbers less than

Figure 17 Capillary exit concentration for Case 6.

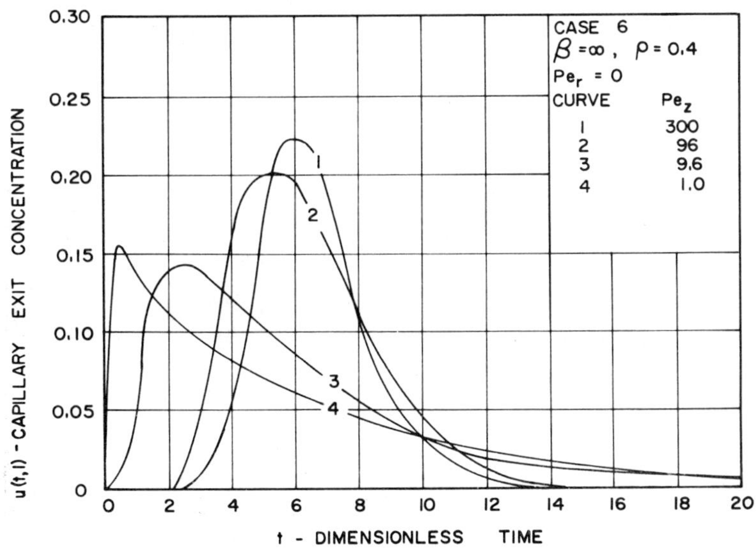

one, Kety's washout curve is asymptotically approached, while for large Peclet numbers (Pe > 300) the delayed wave model of Goresky is obtained in the limit.

The evaluation of the response for Case 7, which is based on infinite capillary permeability, and finite axial and radial diffusivity in the tissue, is as difficult to obtain as the solution to the general single capillary model. Some approximations to the solution are reported in the original study (48).

The response curves for Case 8, which incorporates the assumption of finite resistance to transport at the capillary wall, and instantaneous radial equilibration in the tisse with zero axial diffusivity, are presented in Figure 18. This limiting case is equivalent to the model of Sangren and Sheppard (147). A fraction $e^{-\beta}$ of tracer, which increases with decrease of permeability number β, appears at the outlet at the mean capillary transit time. The rest of the tracer appears as a very long, low intensity tail for small permeability numbers, or as a distinguishable curve with the peak close to the mean transit time of the diffusible tracer for large permeability numbers. For extremely large values of β the delayed wave model of Goresky is approached.

The response curves for Case 9, which, besides finite permeability and zero axial diffusion, allows for finite radial resistance in the tissue, combine the features of Case 5 and Case 8. They are presented in Figure 19. For high values of permeability number β, Case 5 is approached, while for high capillary transit time to diffusion time ratios, γ, Case 8 is closely matched.

The response for Case 10, which assumes finite permeability and infinitely fast equilibration in the axial and radial direction, is given in Figure 20. This limiting case corresponds to the model of Johnson and Wilson (84). The

Figure 18 Capillary exit concentration for Case 8.

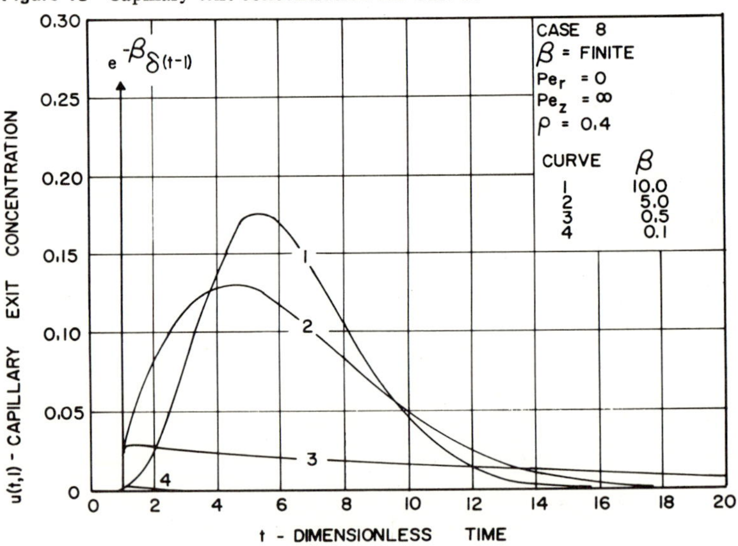

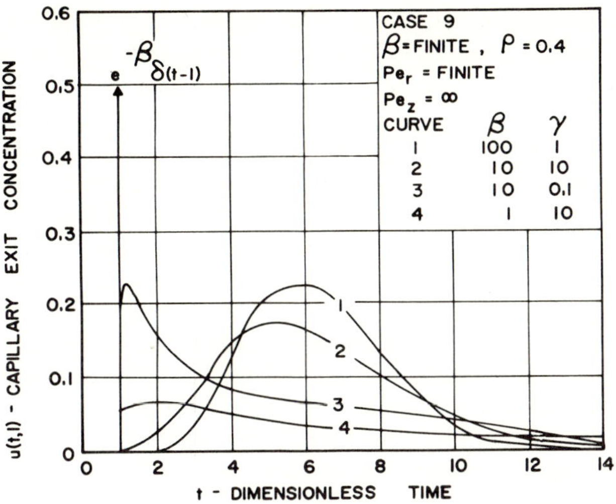

Figure 19 Capillary exit concentration for Case 9.

response curves have two main characteristics. The appearance time is always at the origin, due to instantaneous diffusion in the axial direction, and there is a weighted delta function at the mean capillary transit time. The fraction of tracer that appears at the capillary transit time, i.e., the fraction that has never left the blood volume, is given by $e^{-\beta}$. Thus this fraction varies from almost one for $\beta < 0.1$,

Figure 20 Capillary exit concentration for Case 10.

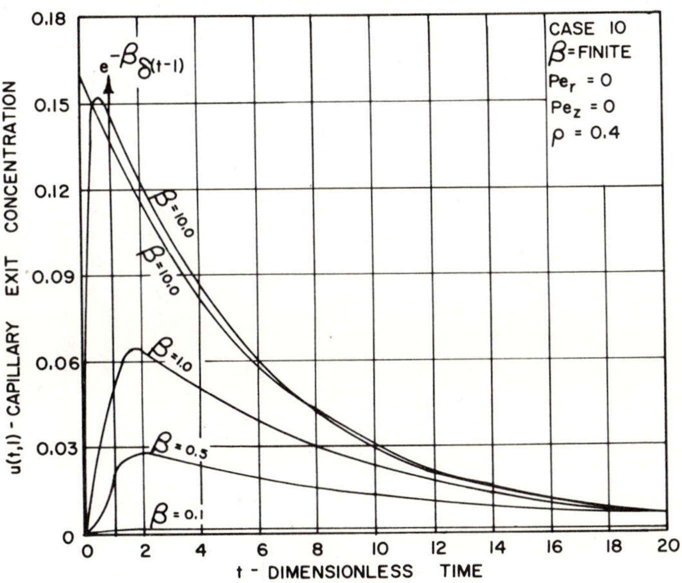

to practically zero for $\beta > 10$. At very high Stanton (permeability) numbers, β, Kety's monoexponential decay is observed. For low permeability numbers, β, there is no difference between Case 8 and Case 10 except in the appearance time. Thus for low permeabilities, axial diffusion is not overly important.

The response curves for Case 11, which is based on the assumption of finite permeability and radial diffusivity, and infinitely fast axial equilibration in the tissue, are given in Figure 21. This case combines the features of Case 3, and Case 10. The monoexponential decay of Kety is approached only if both permeability and radial diffusivity are sufficiently large, i.e., β and γ both very large. If β is large, and γ is small, the response of Case 3 is approached; if γ is large and β is not, the response is equivalent to the one considered in Case 10. Only for intermediate values of β ($1 < \beta < 10$), and small values of γ ($\gamma < 1$), do the response curves for Case 11 differ from those previously discussed.

The limiting Case 12 is based on the assumption that the capillary wall offers a finite resistance to solute transport, and radial equilibration in the tissue is instantaneous, while diffusion rate in the axial direction is comparable to the convective transport in the capillary. For small values of permeability ($\beta < 1$) this model gives the same response as Case 10 for all values of axial Peclet numbers of interest ($0.01 < \rho e_z \leqslant 1{,}000$). For intermediate values of permeability ($1 < \beta < 10$), Case 12 still behaves as Case 10 which indicates that over a large range of permeabilities, the axial diffusivity may range from small to very large without visible effect. For very large values of permeability number ($\beta > 10$) the response takes the shape of Case 6, i.e., of the convective model of Perl and Chinard (129).

Figure 21 Capillary exit concentration for Case 11.

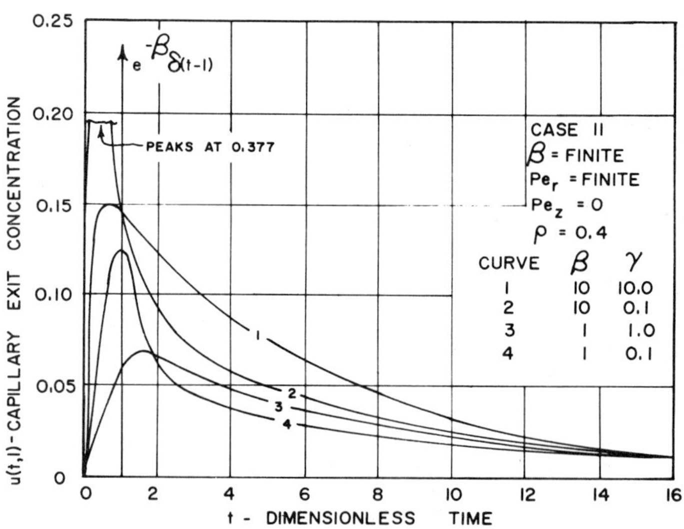

It is apparent from the above analysis that many models used as limiting cases to a general single capillary model may have the same values of the first three or four moments for certain values of parameters. Thus it is difficult to discriminate among various models on the basis of matching moments of the model response to the experimentally evaluated ones. More sophisticated analysis in discrimination procedure is not warranted at the present time for two main reasons. First, the experimental data should not be matched on the basis of a single capillary model because, in the limiting case of zero permeability, the model predicts a delta function response. Secondly, the quality of experimental data is rather poor, so sophisticated statistical analysis cannot be applied in discriminating between competing models. However, single capillary models are useful for understanding the significance and contribution of various transport processes. They are also an important starting point for multicapillary and whole organ modeling.

5.5 Tracer information and multicapillary modeling

The single capillary model is treated in a number of ways in the preceding section. The general distributed parameter model and several limiting cases are considered. All these models are useful in providing more information about the sensitivity of the response curves to certain system parameters. However, all of them represent the nondiffusible tracer response with a delta function at the mean capillary transit time. Therefore, these models cannot be applied directly to the experimentally determined curves for the nondiffusible and diffusible tracers in estimating system parameters. Although it is necessary to provide more information about the influence of system parameters on tracer response curves, which the models discussed do achieve, the ultimate goal is to evaluate the system parameters when the responses of a capillary network, or of an organ, to a sudden injection of both nondiffusible and diffusible tracer are given. This task is very hard to accomplish because of two factors. The capillary size distribution, and the arrangement of capillaries in the organ are too often unknown. The response of larger blood vessels, i.e., venules and arterioles, cannot be distinguished from the capillary network response, as mentioned earlier by Zierler (196). Evaluation of system parameters has to therefore be based on further modeling, whole organ modeling, and more simplifying assumptions.

In order to circumvent the problem of detailed organ modeling, compartmental models were used in interpreting clearance curves from various organs (42, 88, 138, 139). Due to unknown geometrical configurations and properties of capillary beds in various tissues, these models represent a good start in estimating the order of magnitudes of various parameters. Schmidt (149, 150), in his modeling of capillary exchange, attempted to include the presence of a capillary bed and to account for the interactions between capillaries. Goresky et al. (68) also tried to derive theoretical response curves from single capillary beds where capillaries were arranged in parallel. Levitt (106) attempted to account for the presence of other capillaries by modifying the single capillary model, corresponding to the limiting Case 5 treated in the previous section, by allowing some exchange with a well mixed region at the outer surface of the Krogh tissue cylinder. Dudukovic (48) showed that tracer response curves for a capillary network can be obtained on the basis of several models, each of which has some characteristic limitations. Dudukovic et al. (50) distributed these models in three categories:

1. modified single capillary models,
2. multicompartmental analysis and tanks in series models,
3. combination of the response of large vessels with capillary distributions and single capillary response.

5.5.1 Modified single capillary models The single capillary models considered in the previous section are based on the assumption of plug flow in the capillary. This simplifying assumption can be justified and reduces the complexity of the governing partial differential equations. However, there always exists a small, but finite, dispersion in the capillary, and the generalized axial dispersion model appears to be a good way to represent the capillary. The governing equation for the capillary then becomes

$$\frac{\partial u}{\partial t} + \frac{\partial u}{\partial z} = \frac{1}{\mathrm{Pe}_c}\frac{\partial^2 u}{\partial z^2} - \beta(u - v|_\rho) \tag{5.51}$$

where u, v, t, and z are dimensionless variables defined at the beginning of Section 5, while β is the permeability number. The capillary Peclet number, Pe_c, is defined by the following equation

$$\mathrm{Pe}_c = \frac{UL}{D_c} \tag{5.52}$$

where D_c is the axial dispersion coefficient in the capillary.

The solution for the capillary concentration can still be expressed by an integral involving the appropriate Green's function and tissue concentration at the boundary. This system of equations is even harder to solve than the one considered previously for the single capillary model. However, the dispersion Eq. (5.51) for the capillary can be used in some simplified limiting cases of the single capillary model. It is possible to find the moments of the solutions for these models. In this way, modified single capillary models are obtained. Their main feature is that both the nondiffusible tracer response curve, and the diffusible tracer response curve , can be matched adjusting the parameters of the model. These models have two or more parameters, at least one of which can be determined from the nondiffusible tracer response curve, and the others from the diffusible tracer response curve.

The modified single capillary model can be used as a whole organ model if the following assumptions are justified. First, the larger vessels do not contribute significantly to the dispersion of the nondiffusible tracer response curve, and their only contribution appears in the length of appearance time. Second, the capillary network is uniform and a representative capillary can be selected. The first assumption is very difficult to check, and the contribution of the branching of large vessels to tracer dispersion is not easily assessed. The second assumption can be checked experimentally by a microscopic analysis of the tissue considered.

If these assumptions are satisfied, the modified single capillary models may be used as whole organ models. In doing so, an estimate of the volume of large blood vessels may be obtained from the appearance time of the nondiffusible tracer curve. An approximate dispersion number ($a = 1/\mathrm{Pe}_c$) may be obtained from the variance of the nondiffusible tracer curve. The origin should be shifted to the appearance time in calculating this variance. The dispersion number can be taken to account for the dispersion of tracer caused by the capillary bed. From the diffusible tracer curve the volume of distribution and resistance to transfer may be determined. Dudukovic et al. (50) point out the usefulness of moments in estimating the parameters of the system described by a modified single capillary model. They also show that different models may be constructed depending on whether infinite or finite radial equilibration is assumed in the tissue, and whether assumption of negligible, or infinitely fast, axial transport in the tissue is made. It is shown that transfer functions may be obtained for all models and the first few moments are evaluated. When infinitely fast radial equilibration and zero axial diffusion is assumed, the moments are simple and convenient to use. The mean and the variance of the exit age distribution of the nondiffusible tracer become

$$\mu_{nd} = \bar{t}_c = \frac{V_c}{F} \tag{5.53}$$

$$\sigma_{nd}^2 = \{2a - 2a^2 [1 - e^{-(1/a)}]\}\bar{t}_c^2 \tag{5.54}$$

while the mean and the variance of the diffusible tracer curve can be represented in the following way

$$\mu_d = \frac{\bar{t}_c}{\rho^2} = \frac{V_c + V_T}{F} \tag{5.55}$$

$$\sigma_d^2 = \frac{\mu_d^2 \cdot \sigma_{nd}^2}{\mu_{nd}^2} + \frac{2}{\beta}(\mu_d - \mu_{nd})^2 \tag{5.56}$$

$$\sigma_d^2 = \frac{(V_c + V_T)^2 \cdot \sigma_{nd}^2}{V_c^2} + \frac{2}{F^2 \beta} \cdot V_T^2 \tag{5.57}$$

This suggests the use of the zeroth and first moment of the nondiffusible tracer response curve from an organ for evaluation of the local blood flow rate and blood volume. The first moment of the diffusible tracer response curve then yields the information about the tissue volume. Permeability number, β, of the organ can be evaluated using Eq. (5.56), and the calculated means and variances of the exit age distributions. Capillary Peclet number, Pe_c, can be obtained from the variance of the nondiffusible tracer response curve.

When infinitely fast axial equilibration in the tissue is assumed, the moments of the nondiffusible tracer exit age distribution are the same as indicated earlier. The zeroth and first moment of the diffusible tracer exit age distribution are the same as before, while the second moment is given by a fairly cumbersome expression

$$M_2 = \left[\frac{2}{\rho^4} \frac{1}{1 - T(a, \beta)} - \frac{2(2 - \rho^2)}{\beta \rho^2}\right] \bar{t}_c^{\,2} \tag{5.58}$$

where parameter T is

$$T(a, \beta) = \frac{\sqrt{1 + 4a\beta}\, e^{(1/2a)}}{(1 + 2a\beta) \sinh\left(\dfrac{\sqrt{1 + 4a\beta}}{2a}\right) + \sqrt{1 + 4a\beta}\, \cosh\left(\dfrac{\sqrt{1 + 4a\beta}}{2a}\right)} \tag{5.59}$$

By applying these two models to experimental data it was observed that the same order of magnitude was obtained for the permeability number, β, from each, and the values obtained with the second method tend to be up to 25% higher (48).

5.5.2 Tanks in series models The tanks in series model, as shown by Dudukovic et al. (50), is applicable to whole organ modeling. This model is of particular value when the tissue cannot be considered homogeneous, but has to be divided into several compartments. (Multicompartmental analysis is an application of the tanks in series model.) If it can be assumed that large blood vessels do not contribute to tracer dispersion, and the tracer function of the large vessels is $e^{-t_L s}$, then variance of the exit age distribution curve for the nondiffusible tracer determines the number of tanks, n, by which the capillary, or the organ blood volume, has to be approximated. By changing the adjustable parameters: axial diffusivity number, $h = n/\mathrm{Pe}_z$, radial diffusivity number, $K = 2m\gamma/(1 - P)P$, and permeability number, β, the exit age distribution for the diffusible tracer can be matched. Trial and error, or various regression methods, can be used in the matching procedure. Due to the large number of independent parameters, it is certain that the exit age distribution can be matched. It remains doubtful, however, what significance to attribute to the parameters determined in such a way, unless they can be checked experimentally in an independent manner. The moments of the tanks in series model for any larger number of tanks in the radial direction, m, are impossible to find analytically. Numerical procedures, as described by Dudukovic (48), can be applied.

The value of the tanks in series model lies in the fact that it is fairly simple to understand, visualize, and apply. This model immediately yields, not only capillary exit concentration, but also tissue concentrations in different elements, close to or far away from the capillary. If the capillary network configuration within a tissue can be determined, this lumped parameter model can be modified to better represent the real system by taking into consideration spatial capillary variations.

5.5.3 Combined models Another group of models is constructed (48, 50) taking into consideration the combined response of the large vessels and capillary network, and following the work of Goresky et al. (68). It is assumed that tracer is instantaneously injected in a large blood vessel. A certain amount of dispersion occurs in this vessel before it branches into a set of capillaries. Each of the capillaries exchanges the tracer with the adjacent tissue cylinder of radius, R_T. The capillaries have different radii, R_c, and lengths, L. If constant uniform velocity is assumed in all capillaries, the capillary network can be represented by a probability density function. This function ϕ_n can be defined in the following way

$$\phi_n\left(t_c, \frac{1}{\rho^2}\right) dt_c\, d\left(\frac{1}{\rho^2}\right) = \begin{array}{l}\text{fractions of the capillaries that}\\ \text{have transit time between } t_c\\ \text{and } t_c + dt_c \text{ and are surrounded}\\ \text{by a tissue cylinder of}\\ \text{cross-sectional area between}\end{array}$$

$$\frac{1}{\rho^2} \quad \text{and} \quad \frac{1}{\rho^2} + d\left(\frac{1}{\rho^2}\right) \tag{5.60}$$

Here t_c is the transit time, and $\rho = R_c/R_T$. It is assumed, furthermore, that capillary transit time is independent of the capillary and tissue radii. Then the probability density, ϕ_n, can be represented as a product of two densities

$$\phi_n\left(t_c, \frac{1}{\rho^2}\right) = \phi_1(t_c) \cdot \phi_2\left(\frac{1}{\rho^2}\right) \tag{5.61}$$

The response curve for the exit age distribution of a nondiffusible tracer can be represented as the convolution of the large vessel response and capillary network response

$$E_{nd}(t_a) = \int_0^{t_a} E_{LV}(t_a - \tau) E_{nd}{}^{nt}(\tau)\, d\tau \tag{5.62}$$

The capillary network response consists of the response of a single capillary of transit time t_c weighted with the appropriate fraction of capillaries with transit time t_c and summed over all capillaries

$$E_{nd}{}^{nt}(t_a) = \int_0^{\infty} E_{nd}{}^{c}(t_a, t_c)\phi_1(t_c)\, dt_c \tag{5.63}$$

Combining Eq. (5.62) and Eq. (5.63), the exit age distribution for a nondiffusible tracer becomes

$$E_{nd}(t_a) = \int_0^{t_a} E_{LV}(t_a - \tau)\, d\tau \int_0^{\infty} E_{nd}{}^{c}(\tau, t_c)\phi_1(t_c)\, dt_c \tag{5.64}$$

The exit age distribution for a diffusible tracer can be written, on the basis of a similar argument, in the following manner

$$E_d(t_a)$$
$$= \int_0^{t_a} E_{LV}(t_a - \tau)\, d\tau \int_0^{\infty} \phi_1(t_c)\, dt_c \int_1^{\infty} E_d{}^{c}\left(\tau, t_c, \frac{1}{\rho^2}\right) \phi_2\left(\frac{1}{\rho^2}\right) d\left(\frac{1}{\rho^2}\right) \tag{5.65}$$

The exit age distributions can be constructed if the response of large vessels, E_{LV}, the capillary distributions $\phi_1(t_c)$, $\phi_2(1/\rho^2)$, and the single capillary response are known. The response of the large vessels, E_{LV}, cannot be determined experimentally in an independent way. The distribution of capillary lengths can be evaluated by microscopic analysis of the tissue under consideration. With the assumption of uniform velocity, this corresponds to the distribution of capillary transit times. A gamma probability density can be used to approximate the capillary transit time distribution, $\phi_1(t_c)$. Variations in diameters of the capillaries can also be determined. With the assumption of constant tissue cylinder radius, this yields the density function $\phi_2(1/\rho^2)$. An exponential distribution is convenient to approximate $\phi_2(1/\rho^2)$. A model for the response of a single capillary has to be used. In this way the effect of large vessel response, of single capillary models, and of capillary distributions upon whole organ response, can be analyzed.

Whenever the exit age distributions are experimentally determined, the difficult problem of parameter estimation arises. Given the exit age distributions for a nondiffusible and diffusible tracer from a whole organ, the parameters of the system have to be evaluated. In order to extract the transfer function of a single capillary from a diffusible tracer injection, it is necessary to make assumptions, either about large vessel response, or single capillary response, to a nondiffusible tracer. The capillary distributions, $\phi_1(t_c)$, $\phi_2(1/\rho^2)$, or at least their first few moments, still have to be determined by independent analysis.

Information on the capillary distributions, $\phi_1(t_c)$, $\phi_2(1/\rho^2)$, is almost always unavailable, and response of the large vessels cannot be extracted from the exit age distribution of the nondiffusible tracer. The interpretation of the tracer curves can proceed in two ways. First, a single representative capillary, with the transfer function $e^{-\bar{t}_c s}$, can be assumed. In this case, the shape of the nondiffusible tracer curve is attributed to dispersions in large vessels. Second, response of large vessels can be neglected, and the whole nondiffusible tracer curve attributed to the distribution of capillary transit times. Estimating the parameters from the diffusible tracer curve in these two ways should give a range in which the true values lie. Dudukovic et al. (48, 50) combined these two methods with various models for a single capillary response. After detailed analysis they show that the lower bound for the permeability number β can always be found from the first two moments of the nondiffusible and diffusible tracer curve. The lower bound on the permeability number β is obtained when $\sigma_{\phi 1}{}^2$ is zero (all capillaries are assumed to have identical transit times), i.e., when the whole variance of the nondiffusible tracer is attributed to dispersion in the large blood vessels. The lower limit on permeability may be calculated from the following simple formula

$$\beta = \frac{2(\mu_d - \mu_{nd})^2}{\sigma_d{}^2 - \sigma_{nd}{}^2} \tag{5.66}$$

It is shown that consistent with the lower bound on β is the following extremum for the extraction factor

$$E = 1 - e^{-\beta} \tag{5.67}$$

Equations (5.66) and (5.67) can be used successfully in chemotherapy in estimating the average drug concentration in a particular tissue. The minimum value of the average drug concentration in the tissue as a function of time (averaged over tissue volume) is given by

$$\bar{C}_T(t_a) = \frac{E}{(\mu_d - \mu_{nd})F}\, C_{nd}(t_a) \tag{5.68}$$

Although the absolute lower bound for β emerges from the equations, a meaningful upper limit is difficult to assess. Unfortunately, along the whole locus of points given by Eq. (5.19), the value of β is infinite.

$$\bar{t}_c = \frac{\mu_d - \mu_{nd}}{\sigma_d{}^2 - \sigma_{nd}{}^2}\left[\sigma_{\phi_1}{}^2 + \sqrt{(\sigma_{\phi_1}{}^2)^2 + (\sigma_d{}^2 - \sigma_{nd}{}^2)\sigma_{\phi_1}{}^2}\right] \tag{5.69}$$

Without further information, either on mean capillary transit time, $\bar{t}_c$, or on variance in capillary transit times, $\sigma^2_{\phi_1}$, it is impossible to evaluate a meaningful upper limit on β.

This shows that the other method for combined models, which attributes the whole variance of the nondiffusible tracer to the distribution of capillary transit times and approximates the response of large vessels with pure time delay equal to the appearance time, does not provide an absolute upper bound on β. Although this method yields much larger values for β than the lower bound, it fails to estimate the upper bound on β due to the nonlinearity of the problem with respect to β.

In the original study (48), after moment analysis of the problem, it is established that the system of equations is always undetermined, and there is no unique solution for β. The permeability number can only be obtained if additional assumptions about structural properties of the system, such as mean capillary transit time, distribution of capillary lengths and radii, and capillary spacing, are made. Therefore, any sophisticated procedure of matching the entire curves for diffusible and nondiffusible tracers is fruitless, as any reasonable model with adjustable parameters will fit the data. Two more relations, or independent measurements of structural properties, are absolutely necessary.

Multicapillary modeling, although still in its infancy, is an important element to better understanding the complexities of transcapillary exchange. These early models are useful in indicating the path to be followed, and in determining additional experiments that have to be done. At the same time they serve as a useful comparison and check for simplified models based on compartmental analysis, or on single capillary response.

5.6 Tracer analysis of the kidney, brain, myocardium, liver, skeletal muscle

The applications of tracer techniques in evaluating important parameters of various organs have already been mentioned. It is of interest to point out that different tracer techniques are being used not only for research purposes, but also in newly developed diagnostic methods. Organ function has received considerable attention in the medical literature, and it is impossible to cover all of the papers that are related to the applications of the tracer method.

The kidney, with its complicated structure, has received a lot of attention. Chinard et al. (33), in performing tracer studies on various organs of a dog, found that simple models cannot be used to describe the response of the kidney. Models devised to infer such parameters as permeability from slopes of outflow patterns in various organs other than the kidney (42, 111), clearance flow per unit volume from downslopes in the lung (120, 133), or from relative appearance times and shapes of entire outflow patterns in the liver (64, 67), are found to be inapplicable to the kidney. The experimentally observed, inexplicable features in the kidney are: appearance and modal, or peak value, times for highly diffusible extravascular indications such as krypton, are earlier than those for vascular indicators, such as T-182_4 labeled albumin. Chinard and Enns (30, 31) and Levy and Sanceda (107) tried to explain these phenomena by hypothesizing a diffusion bypass. Perl and Chinard (129) formulated a distributed parameter model of simultaneous axial diffusion in the tissue and convection in the capillary, ignoring the permeability at the capillary wall (a reasonably good assumption for highly diffusible krypton). The model was capable of predicting earlier appearance times for highly diffusible indicators, as explained in an earlier section. Slotkoff et al. (163) used radioactive microspheres to determine the flow distribution in the kidney. An intensive review of this technique was presented by Wagner et al. (183). The use of the method is based on the assumptions that microspheres are distributed in a pattern equivalent to blood flow distribution, and all of the microspheres are trapped in the capillaries in a single passage. Then the flow per unit tissue, *TF*, is determined by the formula

$$TF = \frac{CO x TC}{Q}$$

where Q is the total radioactivity injected, TC is tissue radioactivity per unit weight of tissue, and CO is cardiac output. Slotkoff et al. (163) scanned the various portions of the kidney in order to determine the flow distribution under normal and pathological conditions. Thorburn et al. (179) tried to relate various portions of the tracer curves to responses of different anatomical zones. Recently, Dell et al. (45) applied diffusible tracers such as ^{133}Xe for diagnostic purposes. A computer program was devised, using an iterative least squares technique, to fit the experimental Xenon washout data by the sum of two, three, or four exponentials. It was assumed that experimental errors follow a Poisson distribution. A comparment, i.e., an exponential term, was declared nonsignificant if it did not significantly reduce the least squares error about the fitted line as judged by an F test. They showed that normal kidney is best fitted by four compartments. Eighteen out of thirty-eight studies in patients with essential hypertension yielded a four compartment curve with significant reduction in flow in one compartment. All patients with congestive heart failure had three compartment curves, while a typical response of patients with oliguric renal failure was a two compartment curve.

The microcirculation in the brain has also been a subject of intensive study. Hagendal et al. (69) used ^{85}Kr in saline to determine cerebral flow rate. He interpreted the data by assuming grey and white matter circulation to be arranged in parallel. Opitz and Schneider (123) showed how the Krogh arrangement of tissue and capillaries could be extended to cerebral grey matter. Oxygen transport in the brain received considerable attention and was studied by Roughton (145) and Reneau et al. (136). Tschetter et al. (181, 182) used radioactive microspheres to determine regional cerebral blood flows. Sundt and Waltz (176) studied the cortical blood flow, and microcirculation, before, during, and after temporary occlusion of the middle cerebral artery of squirrel monkeys by a ^{85}Kr washout curve. Recently, Broderson et al. (22), investigated the diffusion bypass of xenon in the brain circulation. The patterns of the microcirculation in the brain, as well as the kidney, are not entirely understood, and many parameters such as cerebral blood flow and capillary permeability, under normal and pathological conditions, have not yet been established.

Many investigators studied the microcirculation of the myocardium. Bassingthwaighte and his colleagues (10) applied various simple methods in analyzing the response to an intra-arterial injection of ^{133}Xe and antipyrine ^{125}I. The washout curves were flow limited. Ziegler and Goresky (192, 193) studied the uptake of ^{86}Rb. They also determined the volumes of distribution of ^{14}C-sucrose, ^{125}I-albumin, and inulin, and found the response curves to be diffusion limited. The response curves to water, antipyrine, and sodium were flow limited as predicted by Bassingthwaighte et al. (10). Blood flow rates, volumes of distribution of several tracers, and apparent transfer coefficients were determined.

Mason et al. (113) applied tracer methods in estimating the functional hepatic mass in normal dogs. Bradley et al. (21) used tracers to estimate the hepatic blood flow in man as early as 1945. Goresky (64, 66) and Goresky and Silverman (67) studied extensively the volume of distribution and capillary permeability in liver sinusoids by employing T-1824 as a nondiffusible tracer, and urea, albumin, inulin, and sucrose as diffusible tracers. He formulated this "delayed wave" model to explain the diffusible tracer curves, and inferred that the capillary wall is not a barrier for transport of the materials used as tracers. Goresky (66) also studied the rate of uptake of sulfobromophthalein in the liver.

Transcapillary exchange in skeletal muscle has been studied extensively in order to assess the resistance at the capillary wall to various substances of small and large molecular size. Muscle peremeabilities to EDTA and inulin were studied by Trap-Jensen and Lassen (180). Renkin (137–139), and Renkin and Rossel (140, 141), performed extensive studies on transport of potassium, and investigated the effects of different types of vasodilator mechanisms on transcapillary exchange of diffusible matter in skeletal muscle. Sheehan and

Renkin (154) studied the resistance to transport of various components in a perfused gracilis dog muscle. Comparing the tracer curves for potassium, rubidium, and sodium, they found the largest resistance at the capillary wall to sodium transport. They attributed the resistance to potassium and rubidium transport in part to the capillary wall, and in part to the muscle cell walls in the tissue. Chien (28) tried to explain the observed biphasic changes in permeability with flow rate by the Shunt circulation hypothesis. Lassen and Sejrsen (99) investigated the applicability of various interpretations of the tracer curves obtained in studies of gastrocnemium muscle.

The studies mentioned above are only a small portion of the work being done. They may serve as an indication of the research performed, and as a starting point in studying the tracer method or its applicability to a particular organ. It is interesting to note that in spite of physiological differences in various tissues there are many common factors. There seem to be practically no barriers to transport of gases in most of the tissues studied, while the movement of larger molecules is always hindered by some resistance at the capillary wall. The product of permeability and surface area for exchange per unit flow rate, i.e., the permeability number β, seems to be of the same order of magnitude for similar tracers for all organs studied (48). Further improvement in tracer techniques and the models required for their interpretation will lead to more accurate estimation of many parameters, and to more reliable diagnostic techniques and circulation models.

6 MATHEMATICAL MODELS

6.1 General

It has been more than adequately demonstrated here that tracer analysis and mathematical modeling are intimately related. The mathematical methods involved in solution of the models are therefore an important subject of this chapter. Models of many types—deterministic and stochastic, lumped parameter and distributed parameter—have been used which require an understanding of ordinary and partial differential equations, integral equations, and the elements of probability theory and statistics. The widespread application of compartmental modeling brought into the area of biomedical sciences the use of ordinary differential equations and elementary matrix theory. The necessary mathematical background for successful interpretation of compartmental models, familiar to engineers, has been reviewed by several experts in the area (157, 162, 166) and will not be presented here. Elements of probability theory related to concepts of residence time distributions are also familiar to engineers. Some of these concepts have been reviewed in Section 2 of this chapter. Additional information may be obtained from studies of probability theory (53), excellent reviews on the use of residence time distributions in chemical engineering (104), and physiology and medicine (194).

The emphasis in this section is on three topics which have not been extensively reviewed in the context of tracer techniques. One is a brief review of the types of integral equations that emerge in mathematical modeling of many biological systems. The second topic is parameter estimation by matching models and experimental data. This is one of the primary goals of tracer analysis. Finally, some typical sources of error in applications of the tracer method will be presented.

6.2 Integral equations

When modeling a biological system on which a tracer experiment has been made, we look for a unique characterization of the transport of molecules from one part of the system to another. A solution can be attempted by applying the principle of mass conservation, and by equating the rate of accumulation of particles of a particular species in an arbitrary volume element of the system to the algebraic sum of the input, output, and generation rate. This leads to a set of partial differential equations describing the complete distributed parameter system. If one is mainly interested in quantities of a certain species in various regions of the system, rather than in their complete spatial distribution, an approximation to the original system of equations can be made by representing the system with a set of well mixed regions, or compartments. This approximation is described by a system of ordinary differential equations, and is often used in biological modeling. Stephenson (169, 170) pointed out that the system of differential equations representing the compartmental model may be written as a system of integral equations.

Let us consider a simple linear biological system with a single influx, $f_{\text{in}}(t)$, and a single efflux, $f_{\text{out}}(t)$. If the impulse response of the system, which is assumed to be stationary, is given by $E(t)$, then the relationship between input and output can be readily described, with the convolution integral as

$$f_{\text{out}}(t) = \int_0^t f_{\text{in}}(\tau)E(t-\tau)\,d\tau \tag{6.1}$$

This simple principle of linear superposition was used by Stephenson in 1960 (170) to relate the input and output of a linear biological system. He also showed (171) that in a multiflux system, the i-th flux ϕ_i, is related to other fluxes ϕ_j by the following set of integral equations

$$\phi_i = \sum_j \int_0^t e_{ij}(t-\tau)\,\phi_j(\tau)\,d\tau + m_i(t) \tag{6.2}$$

where $m_i(t)$ is the generation term, while $e_{ij}(t-\tau)$ is the element of the transfer function matrix relating the contribution of the j-th flux to the i-th flux. Using Laplace transforms, Stephenson (171) reduced the system of Eq. (6.2) to a system of algebraic equations and applying matrix notation solved a few examples. Thus, through the work of Stephenson, the convolution theorem and Volterra integral equations of the first kind [Eq. (6.1)] found their place in the analysis of biological systems.

An integral equation formulation of a mathematical model has certain advantages over the equivalent system of differential equations. This became especially apparent when dealing with systems subject to recirculation or recycle. By 1948 Stephenson (169) developed an integral equation formulation to relate the instantaneous concentration of an indicator at one point in a circulation to all previous concentrations at a second point. The equation derived was equivalent to Eq. (4.21), which we shall continue here.

$$P(t) + Q_a(t)^*P(t) = Q_a(t) \tag{4.21}$$

We have obtained $Q_a(t)$ from $C_a(t)$, the measured concentration at a nodal point, and $\beta_0 = M_i/F$, the normalization factor.

Assuming that one can measure $C_a(t)$ and the β_0 without error, the primary response (i.e., the first passage response) $P(t)$ is to be determined.

Theoretically, both kinds of Volterra integral equations can be readily solved by Laplace transforms. Thus, if the analytical forms of $f_{\text{out}}(t)$ and $f_{\text{in}}(t)$ are given in Eq. (6.1), the solution for $E(t)$ may be obtained by

$$E(t) = L^{-1}\left[\frac{\bar{f}_{\text{out}}(s)}{\bar{f}_{\text{in}}(s)}\right] \tag{6.3}$$

Similarly, if $C_a(t)$ is given analytically, and β_0 measured independently, the solution for the primary response could be obtained by

$$P(t) = L^{-1}\left[\frac{\bar{Q}_{a}(s)}{1 + \bar{Q}_a(s)}\right] \tag{6.4}$$

Unfortunately, when applying tracer techniques, one deals with numerical data, and these formulae are not convenient to use. Taking Laplace transforms of numerical data, and numerically inverting Eqs. (6.3) and (6.4), is subject to gross inaccuracies. Several algorithms are available for numerical inversion of the Laplace transform (13, 36, 47), all of them being possibly subject to very large errors.

Fu (57), and Weinstein et al. (186), in their applications of the tracer method to evaluations of cardiac output, developed reliable numerical methods for solution of Volterra equations of the first and second kind. The relationship between the actually measured tracer concentration, $C_m(t)$, the impulsive response of the combined nonideal injection and sampling system, G_S, and the true physiological response, $C_a(t)$, is given by a Volterra integral equation of the first kind

$$C_m(t) = \int_0^t G_S(t-\tau)C_a(\tau)\,d\tau \tag{6.5}$$

Since $G_S(t)$ vanished for negative arguments, and using the change of variables $\xi = t - \tau$, Eq. (6.5) may be written as

$$C_m(t) = \int_0^\infty G_S(\xi)C_a(t-\xi)\,d\xi \tag{6.6}$$

Due to the closed circulatory nature of the physiological system, the tracer response curve approaches either a saturation concentration value, or a large time constant exponential decay, over a long period of time. Thus, either $C_a(t)$ and $C_m(t)$ are not absolutely integrable, or a record long enough to fully describe the tail is not taken. This was resolved by making the following transformations

$$C_m^*(t) = C_m(t+a) - C_m(t) \tag{6.7}$$

$$C_a^*(t) = C_a(t+a) - C_a(t) \tag{6.8}$$

where a is an arbitrary constant which is large enough. The shifted functions, $C_m{}^*(t)$ and $C_a{}^*(t)$, also satisfy integral Eq. (6.6) and they are absolutely integrable, hence their Fourier transforms exist. Then $C_a{}^*(t)$ is obtained by the following inverse Fourier transform

$$C_a^*(t) = \frac{1}{2\pi}\int_{-\infty}^{\infty} C_a^*(j\omega)e^{j\omega t}\,d\omega \tag{6.9}$$

where $C_a^*(j\omega) = \dfrac{C_m^*(j\omega)}{G_S(j\omega)}$ (6.10)

The Fourier transformation of the experimentally obtained $C_m{}^*(t)$ and $G_S(t)$, as well as the inversion of the calculated transform $C_a{}^*(j\omega)$, are carried out numerically. Fu (57) developed a simple numerical algorithm for this procedure. Weinstein et al. (186) developed an improved alternative numerical procedure for the solution of Volterra equations of the first kind. These programs can be made even more efficient by the use of the fast Fourier transform algorithm (37). A solution to the convolution integral, i.e., the deconvolution procedure, and solution of the Voleterra integral equation of the first kind may also be accomplished by application of the trapezoidal rule. The method has been discussed by Aseltine (1), and by Cuenod and Durling (43).

The Volterra integral equation of the second kind may be successfully solved by the Liouville-Neumann method (187), which is a procedure of successive integration and substitution. Fu (57) and Weinstein et al. (186) applied this method in evaluating the primary response $P(t)$ of the circulation from Eq. (4.21). Rearranging Eq. (4.21) we obtain

$$P(t) = Q(t) - \int_0^t Q(t-\tau)P(\tau)\,d\tau \tag{6.11}$$

and taking the first approximation to $P(t)$ to be $P_1(t) = Q(t)$, and substituting into Eq. (6.11) and peforming the integration, the next approximation is obtained. Repeating the iterative process, the following Neumann series is obtained

$$P(t) = \sum_{m=1}^{\infty} (-1)^{m-1} Q_m(t) \tag{6.12}$$

where $Q_m(t)$ is the m-th convolution of $Q(t)$ upon itself, i.e.,

$$Q_m(t) = \underbrace{Q(t)^*Q(t)^*Q(t)^* \ldots {}^*Q(t)}_{m \text{ times}} \tag{6.13}$$

where again, * denotes the operation of convolution. For continuous, bounded functions $Q(t)$ and $P(t)$ (discontinuities are allowed at a finite number of points), such that $|Q(t)| < M$ and $|P(t)| < N$, it can be readily shown that Eq. (6.12) converges as the m-th term is always bounded from above by the corresponding term of the convergent exponential series

$$|(-1)^{m-1}Q_m(t)| \leqslant \left| \int_0^t Q_{m-1}(t-\tau)Q(t)\,d\tau \right| \leqslant M^{m+1}\frac{t^m}{m!} \tag{6.14}$$

It follows that the reciprocal problem for $Q(t)$ has a solution in the form of the following Neumann series

$$Q(t) = \sum_{m=1}^{\infty} P_m(t) \tag{6.15}$$

where $P_m(t)$ is the m-th convolution of the primary response upon itself, and therefore, the response due to the m-th circulation. Equations (6.12) and (6.15) show that $P(t)$ and $Q(t)$ are resolvents of each other. The solution of Eq. (6.12) for the primary response $P(t)$ converges very rapidly, due to the fact that $Q(t)$ is zero for $t < t_a$ (t_a is appearance time). Only a finite number of terms of the Neumann series are required in calculating $P(t)$. In other words Eq. (6.12) may be written as

$$P(t) = \sum_{m=1}^{N} (-1)^{m-1} Q_m(t) \tag{6.16}$$

which is exact for $t < Nt_a$. However, for $t > Nt_a$, $P(t)$ only unmeasurably differs from zero and the contribution of that portion of the series may be neglected. Simple trapezoidal or Simpson's rules may be employed in evaluating $Q_m(t)$.

We may generalize from the above to say that the Neumann series method is an efficient method of solving Volterra integral equations of the second kind, which appear frequently in mathematical modeling of biological systems. Thus, for the general form of the integral equation

$$u(t) = f(t) + \lambda \int_0^t K(t-\tau) u(\tau)\, d\tau \tag{6.17}$$

where $u(t)$ is the unknown function, $f(t)$ is a given forcing function, $K(t)$ is the kernel, and λ a constant, the Neumann series solution takes the following form

$$u(t) = f(t) + \lambda f(t)^* \sum_{m=1}^{\infty} \lambda^{m-1} K_m(t) \tag{6.18}$$

The above series converges if the functions $u(t)$, $f(t)$, and $K(t)$ are sectionally continuous and bounded. The resolvent kernel is given by

$$S(t) = \sum_{m=1}^{\infty} \lambda^{m-1} K_m(t) \tag{6.19}$$

Substituting Eq. (6.19) into Eq. (6.18) one may view the resulting equation as an integral equation for $f(t)$. The solution by Neumann series is given by

$$f(t) = u(t) - \lambda u(t)^* \sum_{m=1}^{\infty} (-\lambda)^{m-1} S_m(t) \tag{6.20}$$

Comparing Eq. (6.16) to Eq. (6.17) one can readily establish the useful reciprocity relationship

$$K(t) = \sum_{m=1}^{\infty} (-1)^{m-1} S_m(t) \tag{6.21}$$

Integral equation formulations have also been used in modeling of transcapillary exchange. Martin and Yudelevich (111) arrived at a Volterra equation of the second kind by their probabilistic approach to transcapillary transport processes. Integral equations also appeared in the general capillary model formulation and multicapillary modeling, by using a deterministic approach (48).

6.3 Matching of models with experimental data

In order to estimate the parameters of the system from tracer experiments, we have to match the response of the mathematical model to the experimental data generated by the tracer studies. Tracer measurements in many physiological systems are subject to a considerable amount of fluctuation. This noise may often seriously obscure the informational content of the measurements. It becomes necessary to apply statistical methods to this information retrieval problem. It is often desired to not only obtain the best fit to data and determine from it the values of system parameters, but to also select the best model among competing models. Unfortunately, random errors in collecting data on physiological systems are often of such an order of magnitude that sophisticated data analysis is not warranted, as the information to be extracted cannot be better than the original data.

Parameter estimation from tracer data usually consists of three steps.

1. A model is hypothesized in the form of an integral or differential equation relating the output vector of the system $\mathbf{y}$ to the vector of independent variables of the system $\mathbf{x}$.
2. The vector $\mathbf{k}$ of parameters such as turnover constants and permeabilities is estimated on the basis of one of several procedures.
3. The significance of postulated variables is tested to find the minimum number of variables necessary to describe the process accurately.

Finally, discrimination between competing models that fit the data well may be accomplished on the basis of certain criteria, and by performing additional statistically designed experiments. The first step in parameter estimation, that of model building for a biological system, has already been discussed in this chapter. Here, we will review briefly the procedures available for parameter estimation by matching the response of the model to available data.

When the system contains only one or two unknown parameters, it is often possible to find a function of the output variable $F(y)$ which would yield a straight line when plotted against an appropriate function of the independent variable, $\phi(x)$. From the slope and intercept, the two unknown parameters can usually be determined. This simple graphical procedure is often used in the analysis of compartmental systems when the system may be approximated by one or two compartments (157, 162, 166).

The parameters of the system may be sought from the integrated expression relating the output of the system with the input, or from the original set of differential equations describing the system. The former path is more often taken in the analysis of biological systems. Thus, an integrated expression yields in general an m-dimensional vector of the response variables, $\mathbf{y}$, as a function of an n-dimensional vector of states, $\mathbf{x}$, and a p-dimensional vector of unknown parameters, $\mathbf{k}$.

$$\mathbf{y} = \mathbf{f}(\mathbf{x}, \mathbf{k}) \tag{6.22}$$

If, for example, we are studying transcapillary exchange with a diffusible tracer, and measuring only the outlet concentration of the capillary y, then $m = 1$, and the variable x is simply time t ($n = 1$). The vector $\mathbf{k}$, however, contains at least two unknown constants, the mean transit time and the permeability number, so that $p \geqslant 2$. Since p parameters have to be determined, one must perform at least p different experiments, i.e., at least p data points have to be taken. For each experiment, j, according to the predictions of the model

$$\mathbf{y}_j = \mathbf{f}(\mathbf{x}_j, \mathbf{k}) \qquad j = 1, 2, \ldots, M \tag{6.23}$$

where $M > p$ data points have been taken.

Because of errors in the measurements and inaccuracies in the model, it is impossible to hope for an exact fit as given by Eq. (6.23). Thus, actual recorded values of the output $\mathbf{y}_{aj}$ differ from the predicted values by the model $\mathbf{y}_j$

$$\mathbf{y}_{aj} = \mathbf{f}(\mathbf{x}_j, \mathbf{k}) + \boldsymbol{\epsilon}_j = \mathbf{y}_j + \boldsymbol{\epsilon}_j \tag{6.24}$$

where $\boldsymbol{\epsilon}_j$ is, in general, the vector of errors, or residuals, between the observations and the predicted dependent variable. In parameter estimation we then try to find a set of parameters, k, such that some scalar function of the errors $S(\mathbf{k})$, is minimized. Depending on the choice of $S(\mathbf{k})$ one may develop the following three methods: least squares estimation, maximum likelihood estimation, and Bayesian estimation.

The most commonly used procedure in analysis of biological systems is the least squares estimation of parameters. In general, the function $S(\mathbf{k})$, is defined as a weighted sum of squares of the errors:

$$S(\mathbf{k}) = \sum_{j=1}^{M} [\mathbf{y}_{aj} - \mathbf{f}(\mathbf{x}_j, \mathbf{k})]^T \mathbf{Q}_j [\mathbf{y}_{aj} - \mathbf{f}(\mathbf{x}_i, \mathbf{k})] \tag{6.25}$$

where $\mathbf{Q}_j$ is an m by m weighting matrix which is chosen in such a way as to reflect the knowledge of the relative precision of the various measurements. If $\mathbf{Q}_j = \mathbf{I}$, one obtains the unweighted least squares criterion. The values of parameters that minimize $S(\mathbf{k})$ may be denoted by $\hat{\mathbf{k}}$. If Eq. (6.25) for $S(\mathbf{k})$ is linear in parameters $\mathbf{k}$, then taking the derivatives of Eq. (6.25) with respect to each of the k's and setting them to zero results in p linear equations which can be readily solved for p parameters $\hat{\mathbf{k}}$. This is the linear least squares procedure, also known as multiple linear regression. Unfortunately, most often the parameters do not appear in a linear fashion in Eq. (6.31). This results in a nonlinear least squares method, and minimization of $S(\mathbf{k})$ in parameter space $\mathbf{k}$ is usually achieved in an iterative manner with the help of a computer. This minimization can be accomplished by direct search methods (132), or by application of either the Newton-Raphson technique (110) or the Gauss-Newton algorithm (6) in gradient methods. The maximum likelihood estimation is based on explicit assumptions concerning the form of the probability distribution of errors. Normal distributions are often used for this purpose. Once the form is assumed, any parameters appearing in the distribution function may be estimated along with the unknown parameters of the model. If the joint probability density function of all errors $\boldsymbol{\epsilon}$ in the observed variable is assumed to be $p(\boldsymbol{\epsilon},\boldsymbol{\psi})$, where $\boldsymbol{\psi}$ represents the set of unknown parameters of the distribution, then the likelihood function $L(\mathbf{k},\boldsymbol{\psi})$ may be formed

$$L(\mathbf{k},\boldsymbol{\psi}) = p[\mathbf{y}_a - \mathbf{f}(\mathbf{x}, \mathbf{k}), \boldsymbol{\psi}] \tag{6.26}$$

The maximum likelihood method simply consists in finding those values of $\mathbf{k}$ and $\boldsymbol{\psi}$ which maximize L. For reasons of convenience, $S_{ML} = \log L$ is usually maximized. Often some estimate of the range of parameter values may be made, a priori, and a relative probability density $p_0(\mathbf{k})$ may be established. Then, following the Bayesian estimation procedure, when the parameters are estimated on the basis of the collected data, the prior information is taken into account by multiplying the likelihood function by the prior distribution. In terms of the logarithms one has to maximize

$$S_B(\mathbf{k}, \boldsymbol{\psi}) = \log p[\mathbf{y}_a - \mathbf{f}(\mathbf{x}, \mathbf{k}), \boldsymbol{\psi}] + \log p_0(\mathbf{k}) \tag{6.27}$$

Sequential estimation, i.e., the reestimation of the parameters, after the results of each in a series of experiments, is an important application of the Bayesian estimation and extremely useful for on-line parameter estimation, regulation, and control. Further discussion of those methods, available algorithms for their application, and parameter estimation in ordinary and partial differential equations, is covered in detail in the recent text by Seinfeld and Lapidus (151).

In applications of tracer techniques to biological systems usually only one output tracer concentration is measured as a function of time, $y = f(t,\mathbf{k})$. If the differential equations describing the system are linear, which is almost always the case, then the transfer function of the system may be determined using Laplace or Fourier transforms. Parameter estimation criteria may be established in either the time or the frequency domain. Various criteria that may be used will be presented here for continuous data. If the data are available only at a set of R discrete times, analogous expressions may be developed replacing the integrals by finite sums.

An error criterion in the time domain with the weights $Q(t) = t^n$, is given by

$$S(\mathbf{k}) = \int_0^\infty [\bar{y}_a(t) - f(t,\mathbf{k})]\, t^n \, dt \tag{6.28}$$

This corresponds to matching the first p moments of the experimentally obtained curve, $y_a(t)$, with the moments of the theoretically determined response, $f(t,\mathbf{k})$. The resulting set of p equations can readily be solved for p parameters, k. The method of moments is simple to use; its drawback is, as mentioned previously, that the tail of the curve with its inherent inaccuracy carries a considerable weight in this particular error criterion. The unweighted least squares criterion in the time domain is given by

$$S(\mathbf{k}) = \int_0^\infty [y_a(t) - f(t,\mathbf{k})]^2 \, dt \tag{6.29}$$

By Parseval's theorem this criterion can be related to the following form in the frequency domain

$$S(\mathbf{k}) = \frac{1}{2\pi} \int_0^\infty [(\Delta R)^2 + (\Delta I)^2]\, d\omega \tag{6.30}$$

$$\begin{aligned} \text{where } \Delta R &= \operatorname{Re}[\bar{y}_a(\omega)] - \operatorname{Re}[\bar{f}(\omega,\mathbf{k})] \\ \Delta I &= \operatorname{Im}[\bar{y}_a(\omega)] - \operatorname{Im}[\bar{f}(\omega,\mathbf{k})] \end{aligned} \tag{6.31}$$

The minimization of S with respect to $\mathbf{k}$ should yield identical results using either Eq. (6.29) or Eq. (6.30).

An integral square error criterion may be formulated in the Laplace transform domain

$$S(\mathbf{k}) = \int_0^\infty [\bar{y}_a(s) - \bar{f}(s, \mathbf{k})]^2 \, ds \tag{6.32}$$

The procedures involved in moment matching, frequency response, and s-plane fitting were reviewed by Johnson et al. (85), and are summarized in Figure 22.

Each method has its advantages and disadvantages, and choice by each investigator is made according to the type of data he has available and his own preference. In biomedical work with tracers, the linear least squares method has been used extensively, as well as simple graphical procedures. More sophisticated procedures of nonlinear regression (160), and iterative minimization of the error criteria in parameter space (6), have also been used.

The field of parameter estimation is a very wide one, and in a review such as this, impossible to present outside the scope of basic concepts. More detailed reviews of available techniques for parameter evaluation are quoted below and may serve as an excellent starting point in selecting a desired method of parameter estimation in any particular study.

Berman (15) discussed the formulation and testing of linear compartmental models used for the interpretation of tracer kinetics. The choice of model, number of compartments, number of parameters, least squares fitting of the data, and judgment of fit are described. Cuenod and Sage (44), in an excellent review, compared the following parameter estimation and system identification procedures: numerical deconvolution, auto-correlation and cross-correlation,

Figure 22 Summary of procedures involved in moment matching, frequency response, and s-plane fitting (*based on Johnson et al.* [85]).

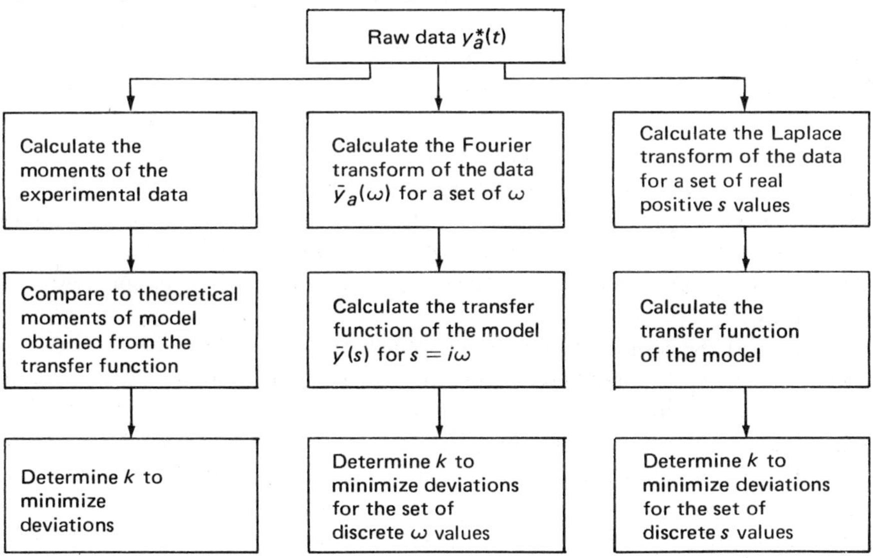

spectral density, sinusoidal response measurement, Wiener theory of nonlinear systems, gradient techniques, quasilinearization, and invariant imbedding. Bekey (11) presented a survey of a number of numerical techniques, including the search and gradient method, quasilinearization, and stochastic approximations, all of which are useful in parameter estimation. Mayne (114) discussed the applicability of regression analysis, or equivalently, Wiener-Kalman filtering in estimation of parameters, in many situations. Some simplified algorithms are presented. Berman and Weiss (16) developed a general purpose computer program (SAAM), designed to fit mathematical models to data by adjusting the parameters until a "best" fit is obtained. An open-end library of model types is incorporated within the program for routine use. Butcher and Cook (26) applied the popular fast Fourier transform (FFT) to system identification. Collins and Khatri (34) developed a method based on finite differences for determination of parameters in linear, nonlinear, and time varying distributed parameter systems described by partial differential equations. Smith (164) presented a method for estimating the Laplace transform of a dynamic system, given its input and output in sampled data form and corrupted by noise. Finally, Reid and Mackay (135), reviewed and discussed several methods used in biomedical engineering.

6.4 Sources of error and correction methods

The discrepancy between the tracer data recorded in an experiment and what a model predicts for the response is often considerable due to several factors. First of all, the model may not be a good representation of the system under study. Secondly, even if the basic form of the model is correct, it may be that the conditions of linearity and stationarity, which are usually assumed, were not even approximated during the experimental testing. Biological processes are stochastic in nature, and subject to random fluctuations which may be caused by many factors. The model can never include all of these factors and its representation of the system will be reasonable only if the effects of neglected quantities do not become important in the course of the experiment. It was mentioned earlier that a tracer is never a perfect one, i.e., there is always a difference in movement between the tracer and traced molecules. Finally, experimental techniques are often not rigorous enough, and errors are introduced by the method of tracer injection, sampling, and concentration evaluation.

Even if the model represents the system very well, the experimental data will differ from the predicted response due to both random and systematic errors. Random errors are caused by stochastic changes in the system during the experiments; systematic errors are introduced by improper experimental procedures such as tracer injection and detection. The statistical methods of data analysis will correct for the random errors which usually appear to have a Gaussian distribution. Some systematic errors can be corrected for, a priori, by modifying the tracer injection and detection method in use or, a posteriori, by making a necessary correction.

Gonzales-Fernandez (63) examined in detail the errors introduced by various methods of tracer injection and sampling. He correctly pointed out that Stephenson's (169) application of the theory of linear filters with lumped characteristics has limitations when applied to tracer studies. The main reason is, in many biological and physical systems, the dispersion of an indicator at every cross section is a function of space and time and cannot always be described by lumped characteristics. Both Stow (175), and Gonzales-Fernandez (63), noticed that correct flow estimates by the tracer method may be obtained only when either the point concentration $c(x,t)$, or velocity $V(x,t)$, is a function of time or space only. The mean flow concentration $\tilde{c}(t)$ at the surface S perpendicular to the main flow was defined as (where x is the coordinate along S)

$$\tilde{C}(t) = \frac{\int_S C(x,t)V(x,t)\,dA}{\int_S V(x,t)\,dA} = \frac{\int_S \phi(x,t)\,dA}{F(t)} \tag{6.33}$$

Then the amount of tracer collected between t_0 and $t_0 + T$, $I(T)$, is

$$I(T) = \int_{t_0}^{t_0+T} \int_S C(x,t)V(x,t)\,dA\,dt = \int_{t_0}^{t_0+T} \tilde{C}(t)F(t)\,dt \tag{6.34}$$

The time averaged mean flow concentration $\tilde{C}$ is given by

$$\tilde{C} = \frac{\int_{t_0}^{t_0+T} \tilde{C}(t)F(t)\,dt}{\int_{t_0}^{t_0+T} F(t)\,dt} \tag{6.35}$$

and the mean flow rate is

$$\tilde{F} = \frac{1}{T}\int_0^{t_0+T} F(t)\,dt \tag{6.36}$$

Then Eq. (6.34) becomes

$$\tilde{F} = \frac{I(T)}{\tilde{C}T} \tag{6.37}$$

Clearly, the mean flow rate cannot be determined from Eq. (6.37), as there is no way of evaluating $\widetilde{C}$.

If the point concentrations at S are not functions of position but of time only, $\widetilde{C}(t) = C(t)$, which can be realized experimentally by creating sufficient turbulence at the surface S, then the mean time concentration $\bar{C}$ is

$$\bar{C} = \frac{1}{T}\int_{t_0}^{t_0+T} C(t)\,dt \tag{6.38}$$

Using Eq. (6.34) and the mean value theorem of integral calculus, one can obtain an expression for an average flow rate F

$$\bar{F} = \frac{I(T)}{\bar{C}T} \tag{6.39}$$

It can be readily shown that if either $C(t)$ or $F(t)$ is constant, then $\bar{F} = \widetilde{F}$. If this is not the case then some errors are introduced by determining the mean flow rate by Eq. (6.39) which is used in tracer experiments. An analysis of errors was presented by Stow (175).

If the velocity profile is a function of position perpendicular to flow only, i.e., $V(x,t) = V(x)$, then

$$F = \frac{I(T)}{\int_{t_0}^{t_0+T} \widetilde{C}(t)\,dt} \tag{6.40}$$

However, $\widetilde{C}(t)$ in general cannot be measured, but the mean cross sectional concentration $\bar{\widetilde{C}}(t)$ at S can be obtained by transillumination, electrical conductivity, or radioactive counting, and is defined by

$$\bar{\widetilde{C}}(t) = \frac{\int_S C(x,t)\,dA}{\int_S dA} \tag{6.41}$$

which yields a different value for the flow rate

$$F_{\text{app}} = \frac{I(T)}{\int_{t_0}^{t_0+T} \bar{\widetilde{C}}(t)\,dt} \tag{6.42}$$

and, in general, $F_{\text{app}} \neq F$.

Clearly, whenever concentration and velocity can be separated, i.e., whenever each of them is a function of a single variable which is not the same one, then the correct formula for the flow rate can be obtained in terms of measurable quantities. Thus the mean flow rate is given by Eq. (6.39) where the concentration is either mean time concentration or mean cross-sectional concentration.

Gonzales-Fernandez (63) also analyzed the accuracy of determining the volume from the first moment of the tracer curve. He considered the example of a cylindrical tube of radius R in steady flow where the indicator is injected at the cross section $y = 0$, and detected at $y = L$. Two types of tracer injection and detection are possible. *Flow tagging* is achieved if for every time, t, the concentration at the injection plane, $y = 0$, is uniform in the whole cross section. This can be accomplished if, at the plane of indicator injection, a local turbulence is maintained for all times.

$$C(x, t) = \mu(t) \qquad \text{at} \qquad y = 0 \tag{6.43}$$

If the indicator is injected at any x on the cross section $y = 0$, at a uniform rate $\lambda(t)$, i.e., if the flux of the indicator is uniform at every infinitesimal area of the injection plane, then

$$C(x, t) = \frac{\lambda(t)}{V(x)} \tag{6.44}$$

and this describes *cross sectional tagging.* This type of tagging is accomplished if a slug of indicator is introduced uniformly over the injection plane by creating a temporary local turbulence. By allowing the flow to resume its previous pattern, the concentration of indicator at any point $(x, 0)$ in the injection plane will decrease inversely proportional to the velocity at that point $V(x)$. Thus cross sectional tagging is obtained.

At the detection plane, $y = L$, one may measure the mean flow concentration (mixing cup concentration) by sampling

$$\tilde{C}(t) = \frac{2\pi}{F} \int_0^R xC(x, t)v(x)\, dx \tag{6.45}$$

or the mean cross sectional concentration (by transillumination, for example)

$$\tilde{\bar{C}}(t) = \frac{2}{R^2} \int_0^R xC(x, t)\, dx \tag{6.46}$$

It was shown (37) that in all the cases of tagging and detection but one, the zeroth moment of the concentration curve would yield an accurate estimate of the flow rate. Only in the case of cross sectional tagging at the input site and mean cross sectional concentration measurement at the sampling site is the commonly used expression for determining the flow rate erroneous. However, the volume estimate obtained from the first moment of the tracer concentration curve is theoretically correct (for laminar flow) only in the case of flow tagging and mean flow concentration measurement.

In the same paper (63) the conditions under which the response of the whole system may be represented as convolution of the different portions of the system

$$G = G_1^* G_2^* G_3^* \dots {}^*G_5 \tag{6.47}$$

are also discussed.

It is also shown that under laminar flow conditions the response to a step input of tracer will not be given by the integral of the impulse response, if the former was injected by flow tagging, while the latter was introduced by cross sectioned tagging.

Many of the above developments are used in examining the errors in volume and flow measurements of the central circulation caused by the use of a sampling catheter. The same problem was considered by different investigators (46, 67, 96, 122, 156, 158). Further discussion of this problem appears in previous sections of this chapter.

NOTATION

a	$1/\mathrm{Pe}_c$
$a(Z)$	Arbitrary function used by Eq. (6.11)
A	Cross sectional area
$b = 2L/R_0$	Dimensionless ratio
C	Concentration of tracer
$\bar{C}$	Average concentration
C_a	Arterial concentration
C_d	Concentration of diffusible tracer
C_{nd}	Concentration of nondiffusible tracer
$C_0 = M_i/\pi R_0^2 L$	Initial concentration
C_T	Concentration of tracer in the tissue
C_V	Venous concentration
D	Diffusivity
D_c	Dispersion coefficient in the capillary
D_r	Radial diffusivity in the tissue
D_z	Axial diffusivity in the tissue

E	Extraction factor
$E(t)$	Exit age distribution
f	Fraction of flow
F	Flow rate
$G(x,x')$	Green's function
$G(t)$	Tracer probability distribution $\int_0^t E(\tau)d\tau$
$G_s(t)$	Injection and sampling system residence time distribution
h	Axial diffusivity number
$H(x)$	Heaviside's unit step function
I	Mass rate of tracer infusion
k	Mass transfer coefficient; time constant, PS/F; reaction rate constant
k'	Reaction rate constant
k_1, k_2	Turnover constants used in Eq. (5.32)
$K(t)$	Kernel of an integral equation
$l = L/R_1$	Characteristic dimensionless number
L	Capillary length
m	Metabolic rate constant; normalized mass of radioactive tracer
M	Mass of tracer in the system
M_i	Mass of tracer injected
O	Circumference
p	Partial pressure
$P(t)$	Primary response or first passage time distribution
P	Permeability
Pe	Peclet number
$\mathrm{Pe}_c = UL/D_c$	Capillary Peclet number
$\mathrm{Pe}_r = UR_1/D_r$	Radial Peclet number in the tissue
$\mathrm{Pe}_z = UL/D_z$	Axial Peclet number in the tissue
q	Flow rate
Q_a	Dimensionless concentration
r	Dimensionless radial coordinate
R	Radial coordinate
R_0	Radius of the capillary
R_1	Tissue cylinder radius; denotes region R_1
R_2	Denotes region R_2
s	Laplace transform variable
S	Surface area available for exchange
$\mathrm{Sh} = PR_1/D_r$	Sherwood number
t	Time; specie
t_a	Actual time; appearance time
$\bar{t}_{ad}$	Mean residence time of a diffusible tracer
$\bar{t}_{and}$	Mean residence time of a nondiffusible tracer
t_c	Capillary transit time

$\bar{t}_c$	Mean capillary transit time
t_L	Transit time in large vessels
$u = C/C_0$	Dimensionless capillary concentration
u	Velocity
U	Velocity
$U(t)$	Unit step function
$v = C_T/C_0$	Dimensionless tissue concentration
V	Volume
V_c	Capillary volume
V_T	Volume of the tissue
$z = z/L$	Dimensionless axial coordinate
Z	Axial coordinate

Greek symbols

$\beta = PS/F$	Permeability number
β_0	M_i/F_I normalization factor
γ	Characteristic number defined by Eq. (5.10)
$\delta(t)$	Dirac delta function
λ	Partition coefficient; parameter
μ	Moments of the exit age distribution
μ_d	Mean of the exit age distribution of a diffusible tracer
μ_{nd}	Mean of the exit age distribution of a nondiffusible tracer
$\rho = R_0/R_1$	Characteristic dimensionless number
σ	Standard deviation
σ^2	Variance of the curve
σ_d^2	Variance of the exit age distribution of the diffusible tracer
σ_{nd}^2	Variance of the exit age distribution of the nondiffusible tracer
τ	Time constant, V/F; dummy variable
ϕ	Probability density; function described by Eq. (5.16); phase angle

Other

$(\bar{\ })$	Laplace transform; average value
$(\tilde{\ })$	Average value
$(\ )*(\ )$	Denotes convolution
	All **boldface** symbols denote vector quantities

REFERENCES

1. Aseltine, J. A., *Transform Method in Linear System Analysis,* McGraw-Hill, New York, (1958).
2. Attinger, D. A., *Pulsatile Blood Flow,* McGraw-Hill, New York, (1964).
3. Aukland, K., Bower, B. F., and Berliner, R. W., "Measurement of Local Blood Flow with Hydrogen Gas," *Circ. Res.,* **14,** 164, (1964).
4. Bader, H., "The Anatomy and Physiology of the Muscle Wall," *Handbook of Physiology,* Section 2: *Circulation,* Vol. 2, W. F. Hamilton and P. Dow (eds.), American Physiological Society, Washington, D.C. (1963).
5. Bailey, L. E., and Harvey, S. C., "Effect of Ouabain on Cardiac ^{45}Ca Kinetics Measured by Indicator Dilution," *Am. J. Physiol.,* **216,** 123, (1969).
6. Bard, Y., "Comparison of Gradient Methods for the Solution of Nonlinear Parameter Estimation Problems," *SIAM J. Numer. Anal.,* **7,** 157, (1970).
7. Bassingthwaighte, J. B., "Plasma Indicator Dispersion in Arteries of the Human Leg," *Circ. Res.,* **19,** 332, (1966).
8. Bassingthwaighte, J. B., and Ackerman, F. H., "Mathematical Linearity of Circulatory Transport," *J. Appl. Physiol.,* **22,** 879, (1967).
9. Bassingthwaighte, J. B., Edwards, A. W. T., and Wood, E. H., "Areas of Dye-dilution Curves Sampled Simultaneously from Central and Peripheral Sites," *J. Appl. Physiol.,* **17,** 91, (1962).
10. Bassingthwaighte, J. B., Strandell, T., and Donald, D. E., "Estimation of Coronary Blood Flow by Washout of Diffusible Indicators," *Circ. Res.,* **23,** 259, (1968).
11. Bekey, G. A., "Identification of Dynamic Systems by Digital Computers," *Proc. Joint Automatic Control Conf.,* Boulder, Colo., (1969).
12. Bellman, R., Jacquez, J. A., and Kalaba, R., "Some Mathematical Aspects of Chemotherapy," *Bull. Math. Biophys.,* **22,** 185, (1960).
13. Bellman, R. E., Kalaba, R. E., and Lockett, J., *Numerical Inversion of the Laplace Transform,* American Elsevier Publishing Company, Inc., New York, (1966).
14. Bennett, H. S., Luft, J. H., and Hampton, J. C., "Morphological Classifications of Vertebrate Capillaries," *Am. J. Physiol.,* **196,** 381, (1959).
15. Berman, M., "The Formulation and Testing of Models," *Ann. New York Acad. Sci.,* **108,** 182, (1963).
16. Berman, M., and Weiss, M. F., *Users Manual for SAAM (Simulation, Analysis, and Modeling),* Public Health Service Publication No. 1703, National Institute of Arthritis and Metabolic Diseases–National Institute of Health, Bethesda, Md.
17. Bloch, E. M., "A Quantitative Study of the Hemodynamics in the Living Microvascular System," *Am. J. Anat.,* **110,** 125, (1962).
18. Blum, J. J., "Concentration Profiles in and around Capillaries," *Am. J. Physiol.,* **198,** 991, (1960).
19. Bischoff, K. B., in *Chemical Engineering in Medicine and Biology,* D. Hershey (ed.), Plenum Press, New York, p. 417, (1967).
20. Bischoff, K. B., and Brown, R. G., "Drug Distribution in Mammals," *Chem. Eng. Progress Symposium Series,* **62,** 33, (1966).
21. Bradley, S. E., Ingelfinger, F. J., Bradley, G. P., and Curry, J. J., "The Estimation of Hepatic Blood Flow in Man," *J. Clin. Invest.,* **24,** 890, (1945).
22. Broderson, P., Sejrsen, P., and Lassen, N. A., "Diffusion by-pass of Xenon in Brain Circulation," *Circ. Res.,* **32,** 363, (1973).
23. Bruley, D. F., and Knisely, M. H., "Hybrid Simulation-Oxygen Transport in the Microcirculation," *Chem. Eng. Symposium Series,* **66,** 22, (1970).
24. Bruni, C., Isidori, A., and Raberti, A., "A Method of Realization Based on the Moments of the Impulse Response Matrix," *IEEE Trans. Auto. Contr.,* **14,** 203, (1968).

25. Bugliarello, G., "Microcirculation Hemodynamics and other Biological Flow Problems," *Biomed. Fluid Mech. Symposium,* Fluids Engineering Division, Amer. Soc. of Mech. Engineers, Washington, D.C., (1966).
26. Butcher, W. E., and Cook, G. E., "Identification of Linear Sampled Data Systems by Transform Techniques," *IEEE Trans. Auto. Contr.,* **AC-14,** 582, (1969).
27. Chambers, R., and Zweifach, B. W., "Functional Activity of the Blood Capillary Bed with Special Reference to Visceral Tissue," *Ann. New York Acad. Sci.,* **46,** 683, (1947).
28. Chien, S., "A Theory for the Quantification of Transcapillary Exchange in the Presence of Shunt Flow," *Circ. Res.,* **29,** 173, (1971).
29. Chinard, F. P., "Evidence for a Diffusion By-pass of the Renal Cortical Vascular System," *Int. Conf. on Nephrology,* Washington, D.C., (1966).
30. Chinard, F. P., and Enns, T., "Transcapillary Pulmonary Exchange of Water in the Dog," *Am. J. Physiol.,* **178,** 197, (1954).
31. Chinard, F. P., and Enns, T., "Relative Renal Excretion Patterns of Sodium Ion, Chloride Ion, Urea, Water, and Glomerular Substances," *Am. J. Physiol.,* **182,** 247, (1955).
32. Chinard, F. P., Enns, T., and Nolan, M. F., "Indicator-Dilution Studies with Diffusible Indicators," *American Heart Association Monograph No. 4,* New York, (1962).
33. Chinard, F. P., Vosburgh, G. J., and Enns, T., "Transcapillary Exchange of Water and of Other Substances in Certain Organs of the Dog," *Am. J. Physiol.,* **183,** 221, (1955).
34. Collins, P. L., and Khatri, H. C., "Identification of Distributed Parameter Systems Using Finite Differences," *Trans. Am. Soc. Mech. Eng. (J. Basic Eng. Ser. D)* **91,** 239, (1969).
35. Conn, H. L., Jr., and Robertson, J. S., "Kinetics of Potassium Transfer in the Left Ventricle of the Intact Dog," *Am. J. Physiol.,* **181,** 319, (1955).
36. Cook, G. E., Dennan, E. D., and Carr, H. M., "Numerical Inversion of Laplace Transforms by the Laguerre-Gauss Quadrature Approximation," *IEEE Trans. Auto. Contr.,* **12,** 623, (1967).
37. Cooley, J. W., Lewis, P. A. W., and Welch, P. D., "The Finite Fourier Transform," *IEEE Trans. Audio Electrocoustics,* **AU-17** (2), 77, (1969).
38. Cotlove, E., "Mechanism and Extent of Distribution of Inulin and Sucrose in Chloride Space of Tissue," *Am. J. Physiol.,* **176,** 396, (1954).
39. Coulam, C. M., Warner, H. R., Marshall, H. W., and Bassingthwaighte, J. B., "A Steady-state Transfer Function Analysis of Portions of the Circulatory System Using Indicator Dilution Techniques," *Computers and Biomed. Res.,* **1,** 124, (1967).
40. Coulam, C. M., Warner, H. R., Wood, E. H., and Bassingthwaighte, J. B., "A Transfer Function Analysis of Coronary and Renal Circulation Calculated from Upstream and Downstream Indicator Dilution Curves," *Circ. Res.,* **19,** 879, (1966).
41. Courant, R., and Hilbert, D., *Methods of Mathematical Physics,* Vol. I, Interscience Publishers, Inc., New York, (1966).
42. Crone, C., "The Permeability of Capillaries in Various Organs as Determined by Use of Indicator Diffusion Method," *Acta Physiol. Scand.,* **58,** 292, (1963).
43. Cuenod, M., and Durling, A., *A Discrete-Time Approach for System Analysis,* Academic Press, New York, (1969).
44. Cuenod, M., and Sage, A. P., "Comparison of Some Methods Used for Process Identification," *Automatica,* **4,** 235, (1968).
45. Dell, R. B., Sciacca, R., Lieberman, K., Case, D. B., and Cannon, P. J., "A Weighted Least Squares Technique for the Analysis of Kinetic Data and its Application to the Study of Renal ^{133}Xenon Washout in Dogs and Man," *Circ. Res.,* **32,** 71, (1973).
46. Dow, P., "Effects of a Sampling System on Dye Curves," *Physiologist,* **1,** 17, (1958).

47. Dubner, H., and Abate, J., "Numerical Inversion of Laplace Transforms and the Finite Fourier Cosine Transform," *J. ACM,* **15,** 115, (1968).
48. Dudukovic, M. P., *Tracer Analysis of the Microcirculation,* PhD Thesis, Illinois Institute of Technology, Chicago, (1972).
49. Dudukovic, M. P., Weinstein, H., and Bernstein, B., *Tracer Analysis of the Microcirculation: General Single Capillary Model,* (in press).
50. Dudukovic, M. P., Weinstein, H., and Bernstein, B., *Tracer Analysis of the Microcirculation: Multicapillary Modeling,* (in press).
51. Dudukovic, M. P., Weinstein, H., Bernstein, B., and Shaffer, A. B., "Tracer Methods in the Analysis of the Microcirculation," *Proc. 26th Ann. Conf. on Eng. in Med. and Biol.,* Minneapolis, (1973).
52. Faber, J. J., "Application of the Theory of Heat Exchangers to the Transfer of Inert Materials in Placentas," *Circ. Res.,* **24,** 221, (1969).
53. Feller, W., *An Introduction to Probability Theory and Its Applications,* 3rd ed., John Wiley, New York, (1968).
54. Fitz-Gerald, J. M., "Mechanics of Red Cell Motion Through Very Narrow Capillaries," *Proc. Roy. Soc. B (London),* **174,** 193, (1969).
55. Forster, R. E., Roughton, F. J. W., Cander, L., Briscoe, W. A., and Kreuzer, F., "Apparent Pulmonary Diffusing Capacity for CO at Varying Alveolar O_2 Tensions," *J. Appl. Physiol.,* **11,** 277, (1957).
56. Freis, E. D., Higgins, T. F., and Morovitz, M. J., "Transcapillary Exchange Rates of Deuterium Oxide and Thiocyanate in the Forearm of Man," *J. Appl. Physiol.,* **5,** 526, (1952).
57. Fu, B., *Tracer Analysis of Recirculating Systems.* PhD Thesis, Illinois Institute of Technology, (1970).
58. Fu, B., Weinstein, H., Bernstein, B., and Shaffer, A. B., "Residence Time Distribution of Recycle Systems–Integral Equation Formulation," *Ind. & Eng. Chem. Process Design & Development,* **10,** 501, (1971).
59. Fung, Y. C., "Theoretical Considerations of the Elasticity of Red Cells and Small Blood Vessels," *Fed. Proceedings,* **25,** 1761, (1966).
60. Gatewood, L. C., Ackerman, E., Rosevear, J. W., Molnar, G. D., and Burns, T. W., "Test of a Mathematical Model of the Blood-Glucose Regulatory System," *Comp. Biomed. Res.,* **2,** 1, (1968).
61. Gianelli, S., Jr., Ayres, S. M., Vastola, J. W., Goldstone, R. A., and Buehler, M. E., "Indicator-dilution Curves Obtained Across the Systemic Circulation During Cardiopulmonary Bypass Perfusion," *Surgery,* **57,** 423, (1965).
62. Goldsmith, H. L., "Microscopic Flow Properties of Red Cells," *Fed. Proceedings,* **26,** 1813, (1967).
63. Gonzales-Fernandez, J. M., "Theory of the Measurement of the Dispersion of an Indicator in Indicator-Dilution Studies," *Circ. Res.,* **10,** 409, (1962).
64. Goresky, C. A., "Distribution of Substances in a Flow Limited Organ, the Liver," in *Compartments, Pools, and Spaces in Medical Physiology,* P. E. Bergner and C. C. Lushbauch (eds.), AEC Symposium No. 11, Clearinghouse for Federal Scientific and Technical Information, National Bureau of Standards, U.S. Department of Commerce, Springfield, Virginia, (1967).
65. Goresky, C. A., "A Linear Method for Determining Liver, Sinusoidal, and Extravascular Volumes," *Am. J. Physiol.,* **204** (4), 626, (1963).
66. Goresky, C. A., "Initial Distribution and Rate of Uptake of Sulfobromophthalein in Liver," *Am. J. Physiol.,* **207,** 13, (1964).
67. Goresky, C. A., and Silverman, M., "Effect of Correction of Catheter Distortion on Calculated Liver Sinusoidal Volumes," *Am. J. Physiol.,* **207,** 883, (1964).

68. Goresky, C. A., Ziegler, W. H., and Bach, G. G., "Capillary Exchange Modeling," *Circ. Res.,* **27**, 739, (1970).
69. Haggendal, E., Nilsson, N. J., and Norback, B., *Acta Physiol. Scand.,* **66**, Suppl. 258, 5, (1965).
70. Hamilton, W. F., Moore, J. W., Kinsman, J. M., and Spurling, R. G., "Simultaneous Determination of the Pulmonary and Systemic Circulation Times in Man and of a Figure Related to the Cardiac Output," *Am. J. Physiol.,* **84**, 338, (1928).
71. Hamilton, W. F., Moore, J. W., Kinsman, J. M., and Spurling, R. G., "Studies on the Circulation," *Am. J. Physiol.,* **99**, 534, (1932).
72. Harris, T. R., and Newman, E. V., "An Analysis of Mathematical Models of Circulatory Indicator-Dilution Curves," *J. Appl. Physiol.,* **28**, 840, (1970).
73. Hering, E., "Versuche die Schnelligkeit des Blutlaufs und der Absonderung Zu Bestimmen," *Ztschr. Physiol.,* **3**, 85, (1829).
74. Hering, E., "Versuche uber das Verhaltniss Zwischen der Zahl der Pulse und der Schnelligkeit des Blutlaufs," *Ztschr. Physiol.,* **5**, 58, (1833).
75. Hill, A. V., "Diffusion of Oxygen and Lactic Acid Through Tissue," *Proc. Roy. Soc. B,* **104**, 39, (1928).
76. Ho, B. L., and Kalman, R. E., "Effective Construction of Linear State-Variable Models from Input/Output Functions," *Regelungstechnik,* **14**, 545, (1966).
77. Honig, C. R., Frierson, J. L., and Patterson, J. L., "Comparison of Neural Controls of Resistance and Capillary Density in Resting Muscle, *Am. J. Physiol.,* **218**, (4), (1970).
78. Hudson, J. A., and Carter, D. B., "An Analysis of Factors Affecting Tissue Oxygen Tension," *Proc. Roy. Soc. B,* **161**, 247, (1964).
79. Hyman, C. H., and Leventhall, J., "Analysis of Clearance of Intra-arterially Administered Labels from Skeletal Muscle," *Am. J. Physiol.,* **203**, 1173, (1962).
80. Intaglietta, M., and Zweifach, B. W., "Geometrical Model of the Microvasculature of Rabbit Omentum from in-Vivo Measurements," *Circ. Res.,* **28**, 593, (1971).
81. Ishii, Y., and MacIntyre, W. J., "Measurement of Heart Chamber Volumes by Analysis of Dilution Curves Simultaneously Recorded by Scintillation Camera," *Circulation,* **44**, 37, (1971).
82. Jacquez, J. A., Carnahan, B., and Abbrecht, P. H., "A Model of the Renal Cortex and Medulla," *Math. Biosci.,* **1**, 227, (1967).
83. Johansson, B., Linder, E., and Seenan, T., "Collateral Blood Flow in the Myocardium of Dogs Measured with Krypton-85," *Acta Physiol. Scand.,* **62**, 263, (1964).
84. Johnson, J. A., and Wilson, T. A., "A Model for Capillary Exchange," *Am. J. Physiol.,* **210**, (6), 1299, (1966).
85. Johnson, J. L., Fan, L. T., and Wu, Y., "Comparison of Moments, S-Plane, and Frequency Response Methods for Analyzing Pulse Testing Data from Flow Systems," *Ind. Eng. Chem. Fund.,* **10**, 425, (1971).
86. Kalman, R. E., "Mathematical Description of Linear Dynamical Systems," *SIAM J. Control,* **1**, 52, (1963).
87. Katz, S., and Shinnar, R., "Parameters of Recirculation," *Proc. Ann. Conf. Eng. Biol. Med.,* **10**, 14.5, (1968).
88. Kety, S. S., "Measurement of Regional Circulation by the Local Clearance of Radioactive Sodium," *Am. Heart J.,* **38**, 321, (1949).
89. Kety, S. S., and Schmidt, C. F., "The Determination of Cerebral Blood Flow in Man by the Use of Nitrous Oxide in Low Concentrations," *Am. J. Physiol.,* **143**, 53, (1945).
90. Kety, S. S., and Schmidt, C. F., "The Nitrous Oxide Method for the Quantitative Determination of Cerebral Blood Flow in Man: Theory, Procedure, and Normal Values," *J. Clin. Invest.,* **27**, 476, (1948).
91. Knopp, T. J., and Bassingthwaighte, J. B., "Effect of Flow on Transpulmonary Circulatory Transport Functions," *J. Appl. Physiol.,* **27**, 36, (1969).

92. Krogh, A. S., *The Anatomy and Physiology of Capillaries.* Yale University Press, New Haven, (1929).
93. Krogh, A. S., "The Number and Distribution of Capillaries in Muscles with Calculations of the Oxygen Pressure Head Necessary for Supplying the Tissue," *J. Physiol.,* **52,** 409–415, (1919).
94. Kuhn, W., and Ramel, A., "Aktiver Salztransport als Moglecher (und Wahrscheinlicher) Einzeleffekt bei der Harnkonzentrierung in der Nierer," *Helv. Chem. Acta,* **42,** 628, (1959).
95. Kuprianov, V. V., and Kozlov, V. I., "Dynamic Structures of the Microvascular Bed," *Microvasc. Res.,* **3,** 22, (1971).
96. Lacy, W. W., Emanuel, R. W., and Newman, E. V., "Effect of the Sampling System on the Shape of Indicator Dilution Curves," *Circ. Res.,* **5,** 568, (1957).
97. Landahl, H. D., "Transient Phenomena in Capillary Exchange," *Bull. Math. Biophys.,* **16,** 55, (1954).
98. Landis, E. M., and Pappenheimer, J. R., "Exchange of Substances Through the Capillary Walls," *Handbook of Physiology,* Chapter 29, W. F. Hamilton and P. Dow (eds.), American Physiological Society, Washington, D.C., (1963).
99. Lassen, N. A., and Sejrsen, P., "Monoexponential Extrapolation of Tracer Clearance Curves in Kinetic Analysis," *Circ. Res.,* **29,** 76, (1971).
100. Lawson, H. C., "The Volume of Blood–a Critical Examination of Methods for its Measurement," *Handbook of Physiology,* Section 2: *Circulation,* Vol. 1, chapter 3, W. F. Hamilton and P. Dow (eds.), American Physiological Society, Washington, D.C., (1962).
101. Lee, J. S., and Fronek, A., "An Analysis of the Exchange of Indicators in Single Capillaries," *Microvasc. Res.,* **2,** 302, (1970).
102. Lever, A. F., The Vasa Recta and Countercurrent Multiplication," *Acta Med. Scand.,* **178,** Suppl. 434, (1965).
103. Levenspiel, O., *Chemical Reaction Engineering,* Wiley, New York, (1972) Chapter 9.
104. Levenspiel, O., and Bischoff, K. B., "Patterns of Flow in Chemical Process Vessels," *Advances Chem. Eng.,* **4,** 95, (1963).
105. Levitt, D. G., "Evaluation of the Early Extraction Method of Determining Capillary Permeability by Theoretical Capillary and Organ Models," *Circ. Res.,* **27,** 81, (1970).
106. Levitt, D. G., "Theoretical Model of Capillary Exchange Incorporating Interactions Between Capillaries," *Am. J. Physiol.,* **220,** 250, (1971).
107. Levy, M. N., and Sanceda, G., "Diffusion of Oxygen from Arterial to Venous Segments of Renal Capillaries," *Am. J. Physiol.,* **196,** 1336, (1959).
108. Lighthill, M. J., "Pressure Forcing of Tightly Fitting Pellets Along Fluid-Filled Elastic Tubes," *J. Fluid Mech.,* **34,** 113, (1968).
109. Linder, E., "Measurement of Normal and Collateral Blood Flow by Close-arterial and Intra-myocardial Injections," *Acta Physiol. Scand.,* **68,** Suppl. 272, 5, (1965).
110. Marquardt, D. W., "An Algorithm for Least Squares Estimation of Nonlinear Parameters," *J. Soc. Ind. Appl. Math.,* **11,** 431, (1963).
111. Martin, P., and Yudilevich, D., "A Theory for the Quantification of Transcapillary Exchange by Tracer-Dilution Curves," *Am. J. Physiol.,* **207** (1), 162, (1964).
112. Marumo, F., Yoshikawa, Y., and Koshikawa, S., "A Study on the Concentration Mechanism of the Renal Medulla by Mathematical Model," *Japan Circ. J.,* **31,** 1309, (1967).
113. Mason, M. F., Hanley, G., and Smith, A., "Application of the Saturation Principle to the Estimation of Functional Hepatic Mass in Normal Dogs," *Am. J. Physiol.,* **152,** 42, (1948).
114. Mayne, D. Q., "Parameter Estimation," *Automatica,* **3,** 245, (1966).

115. Meier, P., and Zierler, K. L., "On the Indicator-Dilution Method for Measurement of Blood Flow and Volume," *J. Appl. Physiol.,* **6,** 731, (1954).
116. Middleman, S., *Transport Phenomena in the Cardiovascular System,* John Wiley & Sons, New York, (1972).
117. Milhorn, H. P., Jr., and Pulley, P. E., Jr., "A Theoretical Study of Pulmonary Capillary Gas Exchange and Venous Admixture," *Biophysica J.,* **8,** 337, (1968).
118. Morales, M. F., and Smith, R. E., "On Theory of Blood Tissue Exchange of Inert Gases," *Bull. Math. Biophys.,* **10,** 191, (1948).
119. Naor, P., Shinnar, R., and Katz, S., "Indeterminacy in the Estimation of Flow Rate and Transport Functions from Tracer Experiments in Closed Circulations," *Int. J. Eng. Sci.,* **10,** 1153, (1972).
120. Newman, E. V., Merrell, M., Genecin, A., Monce, C., Milnor, W. R., and McKeever, W. P., "Dye Dilution Method for Describing the Central Circulation," *Circulation,* **4,** 735, (1951).
121. Nicholas, K. K., Warner, H. R., and Wood, E. H., "A Study of Dispersion of an Indicator in the Circulation," *Ann. New York Acad. Sci.,* **115,** 721, (1964).
122. Oleksy, S. J., Weinstein, H., and Shaffer, A. B., "Correction of Dye-Curve Distortion with the Use of Perfect Mixers in Series Model," *J. Appl. Physiol.,* **26,** 113, (1969).
123. Opitz, E., and Schneider, M., "Ergebnisse der Physiologie," *Biologischen Chemie und experimentallen Pharmakologie,* **46,** 126, (1950).
124. Palatt, P. J., Saidel, G. M., and Macklin, M., "Transport Processes in the Renal Cortex," *J. Theor. Biol.,* **29,** 251, (1970).
125. Pappenheimer, J. R., "Passage of Molecules through Capillary Walls," *Physiol. Rev.,* **33,** 387, (1953).
126. Papper, E. M., and Kitz, R. J., *Uptake and Distribution of Anesthetic Agents,* McGraw-Hill, New York, (1963).
127. Passow, H., and Stampfli, R., *Laboratory Techniques in Membrance Biophysics,* Springer Verlag, Berlin, (1969).
128. Perl, W., "Heat and Matter Distribution in Body Clearance Methods," *J. Theor. Biol.,* **2,** 201, (1962).
129. Perl, W., and Chinard, F. C., "A Convection-Diffusion Model of Indicator Transport Through an Organ," *Circ. Res.,* **22,** 273, (1968).
130. Perl, W., and Samuel, P., "Input-Output Analysis for Total Input Rate and Total Trace Mass of Body Cholesterol in Man," *Circ. Res.,* **25,** 191, (1969).
131. Pinter, G. G., and Shohet, J. L., "Origin of Sodium Concentration Profile in the Renal Medulla," *Nature,* **200,** 955, (1963).
132. Powell, M. J. D., "An Efficient Method for Finding the Minimum of a Function of Several Variables Without Calculating Derivatives," *Comp. J.,* **7,** 155, (1964).
133. Ramsey, L., Puckett, W., Jose, A., and Lacy, W., "Comparison of Slope and Mean Transit Time Volumes by Use of Diffusible and Nondiffusible Indicators," *Trans. Assoc. Am. Physicians,* **74,** 280, (1961).
134. Rankin, J. H. G., and Peterson, E. N., "Application of the Theory of Heat Exchangers to a Physiological Study of the Goat Placenta," *Circ. Res.,* **24,** 235, (1969).
135. Reid, M. H., and Mackay, R. S., "A Review and Discussion of Some Methods for System Identification and Modeling," *Med. Biol. Engin.,* **6,** 231–268, (1968).
136. Reneau, D. D., Jr., Bruley, D. F., and Knisely, M. H., "A Digital Simulation of Transient Oxygen Transport in Capillary-Tissue Systems," *AIChE J.,* **15** (6), 916, (1969).
137. Renkin, E. M., "Effects of Blood Flow on Diffusion Kinetics in Isolated Perfused Hindlegs of Cats," *Am. J. Physiol.,* **183,** 125, (1955).
138. Renkin, E. M., "Transport of Potassium-42 from Blood to Tissue in Isolated Mammalian Skeletal Muscles," *Am. J. Physiol.,* **197,** 1205, (1959).

139. Renkin, E. M., "Blood Flow and Transcapillary Exchange in Skeletal Muscle," *Physiol. Symp. Fed. Proc.*, **24**, 1092, (1965).
140. Renkin, E. M., and Rosell, S., "Influence of Sympathetic Adrenergic Vasoconstrictor Nerves on Transport of Diffusible Solutes from Blood to Tissues in Skeletal Muscle," *Acta Physiol. Scand.*, **54**, 223, (1962).
141. Renkin, E. M., and Rosell, S., "Effects of Different Types of Vasodilator Mechanisms on Vascular Tonus and on Transcapillary Exchange of Diffusible Material in Skeletal Muscle," *Acta Physiol. Scand.*, **54**, 241, (1962).
142. Rideout, V. C., and Katra, J. A., "Computer Simulation of the Pulmonary Circulation," *Simulation*, **14**, 239, (1969).
143. Rossen, R. H., and Lapidus, L., "Minimum Realizations and System Modeling," *AIChE J.*, **18**, 673, 881, (1972).
144. Roughton, F. J. W., "Diffusion and Chemical Reaction Velocity in Cylindrical and Spherical Systems of Physiological Interest," *Proc. Roy. Soc. B*, **140**, 203, (1952).
145. Roughton, F. J. W., "The Oxygen Equilibrium of Mammalian Hemoglobin in Oxygen," *Proc. Symp. New York Heart Assoc.*, 105–176, Little, Brown and Company, Boston, (1965).
146. Samuel, P., Holtzman, C. M., Meilman, E., and Perl, W., "Effect of Neomycin on Exchangeable Pools of Cholesterol in the Steady State," *J. Clin. Invest.*, **47**, 1806, (1968).
147. Sangren, W. C., and Sheppard, C. W., "A Mathematical Derivation of the Exchange of a Labeled Substance Between a Liquid Flowing in a Vessel and an External Compartment," *Bull. Math. Biophys.*, **15**, 387, (1953).
148. Schlossmacher, E. J., Weinstein, H., Lochaya, S., and Shaffer, A. B., "Perfect Mixers in Series Models for Fitting Veno-Arterial Indicator-Dilution Curves," *J. Appl. Physiol.*, **22**, 327, (1967).
149. Schmidt, G. W., "A Mathematical Model of Capillary Exchange as a Function of Tissue Structure," *Bull. Math. Biophys.*, **14**, 229, (1952).
150. Schmidt, G. W., "The Time Course of Capillary Exchange," *Bull. Math. Biophys.*, **15**, 447, (1953).
151. Seinfeld, J. H., and Lapidus, L., *Mathematical Methods in Chemical Engineering*, Volume 3, Process Modeling, Estimation, and Identification, Prentice Hall, Englewood Cliffs, N.J., (1974).
152. Shaffer, A. B., "Estimation of Ventricular Volumes by a Constant Infusion Indicator Dilution Technique," *Circ. Res.*, **15**, 168, (1964).
153. Shaffer, A. B., and Weinstein, H., personal communication.
154. Sheehan, R. M., and Renkin, E. M., "Capillary, Interstitial, and Cell Membrane Barriers to Blood-Tissue Transport of Potassium and Rubidium in Mammalian Skeletal Muscle," *Circ. Res.*, **30**, 588, (1972).
155. Sheppard, C. W., "Mathematical Considerations of Indicator Dilution Techniques," *Minn. Med.*, **37**, 93, (1954).
156. Sheppard, C. W., "Catheter Artifacts in Indicator Dilution Experiments," *Fed. Proceedings*, **16**, 118, (1957).
157. Sheppard, C. W., *Basic Principles of the Tracer Method*, John Wiley & Sons, New York, (1962).
158. Sheppard, C. W., Jones, M. P., and Couch, B. L., "Effect of Catheter Sampling on the Shape of Indicator Dilution Curves," *Circ. Res.*, **7**, 895, (1959).
159. Sheppard, C. W., Uffer, M. B., Merker, P. C., and Halikas, G., "Circulation and Recirculation of Injected Dye in the Anesthetized Dog," *J. Appl. Physiol.*, **25**, 610, (1968).
160. Sherwood, R. R., and Day, S. G., "Modeling of the Circulatory System Using Indicator Dilution Methods," *Proc. 26th Ann. Conf. on Eng. in Med. and Biol.*, September (1973).

161. Shinnar, R., and Naor, P., "Residence Time Distributions in Systems with Internal Reflux," *Chem. Eng. Sci.*, **22**, 1369, (1967).
162. Shipley, R. A., and Clark, R. E., *Tracer Methods for in-Vivo Kinetics,* Academic Press, New York, (1972).
163. Slotkoff, L. M., Logan, A., Jose, P., D'Avella, J., and Eisner, G. M., "Microsphere Measurement of Intrarenal Circulation of the Dog," *Circ. Res.*, **28**, 158, (1971).
164. Smith, F. W., "System Laplace-transform Estimation from Sampled Data," *IEEE Trans. Auto. Contr.*, **AC-13**, 37, (1968).
165. Snyder, M. F., and Rideout, V. C., "Computer Modeling of the Human Systemic Arterial Tree," *Biomech.*, **1**, 341, (1968).
166. Starling, E. H., "On the Absorption of Fluids from the Connective Tissue Spaces," *J. Physiol.*, **19**, 312, (1896).
167. Steele, R., *Tracer Probes in Steady State Systems,* Charles C Thomas, Springfield, Illinois, (1971).
168. Stein, W. D., *The Movement of Molecules Across Cell Membranes,* Academic Press, New York, (1967).
169. Stephenson, J. L., "Theory of Measurement of Blood Flow by the Dilution of an Indicator," *Bull. Math. Biophys.*, **10**, 117, (1948).
170. Stephenson, J. L., "Theory of Transport in Linear Biological Systems: I. Fundamental Integral Equation," *Bull. Math. Biophys.*, **22**, 1, (1960).
171. Stephenson, J. L., "Theory of Transport in Linear Biological Systems: II. Multiflux Problems," *Bull. Math. Biophys.*, **22**, 113, (1960).
172. Stewart, G. N., "Researches on the Circulation Time in Organs and on the Influences Which Affect It," *J. Physiol.*, London, **15**, 1, (1894).
173. Stewart, G. N., "Researches on the Circulation Time and on the Influences Which Affect It: IV. The Output of the Heart," *J. Physiol.*, London, **22**, 159, (1897-98).
174. Stewart, G. N., "The Output of the Heart in Dogs," *Am. J. Physiol.*, **57**, 27, (1921).
175. Stow, R. W., "Systematic Errors in Flow Determinations by the Fick Method," *Minn. Med.*, **37**, 30, (1959).
176. Sundt, T. M., and Waltz, A. G., "Cerebral Ischemia and Reactive Hyperemia," *Circ. Res.*, **28**, 426, (1971).
177. Taylor, A. E., Gaar, K., and Guyton, A. C., "Na^{24} Space, D_2O Space and Blood Volume in Isolated Dog Heart," *Am. J. Physiol.*, **211**, 66, (1966).
178. Thews, G., "Uber die Mathematische Behandlung Physiologischer Diffusionsprozesse in Zylinderformingen Objekten," *Acta Biotheoretica,* **10**, 105, (1953).
179. Thorburn, G. D., Kopald, H. H., Herd, J. A., Hollenberg, M., O'Morchoe, C. C., and Barger, A. C., "Intrarenal Distribution of Nutrient Blood Flow Determined with Kr^{85} in the Unanesthetized Dog," *Circ. Res.*, **13**, 290, (1963).
180. Trap-Jensen, J., and Lassen, N. A., "Restricted Diffusion in Skeletal Muscle Capillaries in Man," *Am. J. Physiol.*, **220**, 371, (1971).
181. Tschetter, T. H., Meyer, M. W., Klanen, A. C., and Resch, J. A., "Regional Cerebral Blood Flow Using Labeled Microspheres," *Paper ICPS,* Washington, D.C., Meeting August 25-31, (1968).
182. Tschetter, T. H., Klanen, A. C., and Resch, J. A., "Regional Cerebral Blood Flow in Dogs Using a Particle Distribution Method," *Proc. Soc. Exp. Biol. and Med.*, **131**, 1244, (1969).
183. Wagner, H. H., Rhodes, B. A., Sasaki, Y., and Ryna, J. P., "Studies of the Circulation with Radioactive Microspheres," *Invest. Radiol.*, **4** (6), 374, (1969).
184. Wayland, H., "Rheology and the Microcirculation," *Gastroenterology,* **52** (2), 342, (1967).

185. Weinstein, H., Fu, B., Bernstein, B., and Shaffer, A. B., "Residence Time Distributions in Recycle Systems with the Human Circulation as an Example," *Proc. Third Czech. Chem. Eng. Cong. (III CHISA)*, September 1-17, (1969).
186. Weinstein, H., Fu, B., Acosta, R., Bernstein, B., and Shaffer, A. B., "Estimation of Circulating Blood Volume from a Tracer Response Curve," *IEEE Trans. Biomed. Eng.*, **20**, 269, (1973).
187. Whittaker, E. T., and Watson, G. N., *A Course of Modern Analysis*, 4th Ed., Cambridge University Press, (1937).
188. Wiedeman, M. P., "Patterns of the Arteriovenous Pathways," *Handbook of Physiology*, Section 2: *Circulation*, Chapter 11, W. F. Hamilton and P. Dow (eds.), American Physiological Association, Washington, D.C., (1963).
189. Wingard, L. B., Jr., Chorbajian, T., and Galla, S. T., "Concepts of the Resident Time Distribution Applied to the Indicator Dilution Method," *J. Appl. Physiol.*, **33**, 264, (1972).
190. Yipintsoi, T., and Bassingthwaighte, J. B., "Circulatory Transport of Iodaantipyrine and Water in the Isolated Dog Heart," *Circ. Res.*, **27**, 461, (1970).
191. Young, D. R., Howard, J. C., Trigg, L. N., and Garcia, J. F., "A Preliminary Model of Serum Growth Hormone Response to Hypoglycemia in Man," *Comp. Biomed. Res.*, **3**, 74, (1970).
192. Ziegler, W. H., and Goresky, C. A., "Transcapillary Exchange in the Working Left Ventricle of the Dog," *Circ. Res.*, **29**, 181-207, (1971).
193. Ziegler, W. H., and Goresky, C. A., "Kinetics of Rubidium Uptake in the Working Dog Heart," *Circ. Res.*, **29**, 208-220, (1971).
194. Zierler, K. L., "Circulation Times and the Theory of Indicator-Dilution Methods for Determining Blood Flow and Volume," *Handbook of Physiology*, Section 2: *Circulation*, Vol. I, W. F. Hamilton and P. Dow (eds.), American Physiological Society, Washington, D.C., (1962).
195. Zierler, K. L., "Equations for Measuring Blood Flow by External Monitoring of Radioisotopes," *Circ. Res.*, **16**, 309, (1965).
196. Zierler, K. L., "Tracer-Dilution Techniques in the Study of Microvascular Behavior," *Physiology Symposium Fed. Proc.*, **24**, 1085, (1965).
197. Zweifach, B. W., "The Character and Distribution of the Blood Capillaries," *Anat. Res.*, **73**, 475, (1939).
198. Zweifach, B. W., "Peripheral Circulation," *Ann. Rev. Physiol.*, **10**, 225, (1948).

PART THREE

Mathematical Methods

5

Finite Difference Methods in Heat and Mass Transfer

J. Isenberg
J. H. Wiggins Company
Redondo Beach, California

G. de Vahl Davis
School of Mechanical and Industrial Engineering
University of New South Wales
Kensington, N.S.W., Australia

CONTENTS

1 INTRODUCTION, GENERAL EQUATIONS, AND FINITE DIFFERENCE OPERATORS

1.1 Introduction

In 1923, Lewis Fry Richardson published the first application of finite difference methods to a problem in fluid dynamics and heat and mass transfer: the problem of weather prediction. The calculations were performed by hand, during periods of rest from Richardson's duties as an ambulance driver for the French army during World War I. His results were wrong–largely because he was not aware of (and hence failed to satisfy) what is now called the stability condition. Nevertheless, his method embodies many of the basic principles being applied to similar problems 50 years later.

During this period, of course, new theories have been developed and new methods devised, and the availability of the high-speed digital computer has enabled these theories to be tested, and these methods to be implemented. Today it is possible to solve a wide range of problems of engineering significance by finite difference methods.

It is the purpose of this chapter to survey some of the methods available in the area of heat and mass transfer, and to describe several applications of these methods, so that the thermal analyst may be able to apply them to his own problems.

Since one should resort to numerical methods only for problems to which an analytical solution is either very difficult or impossible to obtain, convergence and stability proofs often cannot be found for these more complex–generally nonlinear–problems. Thus one is often forced to employ ad hoc approaches, with the numerical methods playing the role of an experimental tool. Indeed, the application of numerical methods as an art forms one of the interesting challenges in heat and mass transfer analysis. This work is composed in that spirit.

We are concerned here only with finite difference methods, although others–including finite element methods and methods based on Galerkin or similar techniques–are becoming popular. Recent numerical studies (e.g., Chen, 1972; Taylor and Hood, 1973) suggests that the finite element method has advantages when solving problems with complex boundaries, but is more costly in terms of computational effort. Galerkin methods (e.g., Catton et al., 1974) are very efficient but can only be used in simple geometries and, moreover, rely for their accuracy on the choice of suitable base functions. Finite difference methods are still in the most common use. Since they are highly suited to problems of conduction and convection and have been extensively applied to such problems, we have limited ourselves to these areas, and have not considered radiation heat transfer. Our presentation is based upon a consideration of laminar flow, although many, if not most, convection problems involved turbulent flow, and the reader will need to marry the techniques described here to the models of turbulence presented by Naot and Wolfshtein in a companion article in this volume. Further, selectivity has been imposed by restrictions of time and space, and by the interests and experience of the authors. This has resulted in the omission of such topics as compressible flow, boundary layer flow, and chemical reactions. The first two of these, in particular, would each warrant a review of this size to themselves.

The remainder of this section is devoted to a resume of the differential equations and boundary conditions relevant to the problems to be described, and to the definition of finite difference operators. The application of finite difference methods to the solution of parabolic and elliptic equations is discussed in Secs. 2 and 3. Additional areas of interest not usually found in standard texts, such as asymptotic expansions and stretched coordinates, are included. Section 4 treats parabolic-hyperbolic equations, an area the authors believe to have been neglected in the field of numerical methods. Where appropriate, reference is made in the text to practical implementations of the methods described. However, Sec. 5 is devoted exclusively to a discussion of several problems which have been solved by finite difference methods. Finally, Sec. 6 summarizes some of the more important recent developments and trends.

Attention has been largely (but not entirely) focused on methods employing the vorticity/vector potential (or stream function) form of the governing equations, rather than the primitive variable (velocity/pressure) form. This is because they are more suitable for convection heat transfer problems, especially when compressibility effects and material property variations are small and have been more commonly employed. However, the primitive variable approach has been used for incompressible, as well as for compressible, flow problems, despite certain difficulties that are encountered. Roache (1973) includes a discussion of both methods in his recent text on numerical fluid dynamics.

It has already been mentioned, and it is worth emphasizing, that the theoretical foundation of numerical analysis has failed to keep pace with the practical implementation of computational methods. In particular, it is generally impossible to obtain a theoretical estimate of the accuracy of a numerical solution of a problem involving the nonlinear equations of fluid flow. The practitioner is cautioned to use whatever means he can (such as mesh refinement, comparison with experimental results, comparison with approximate analytical solutions) to verify the quality of his results. Subject to such precautions, the authors believe that finite difference methods are now available that permit the achievement of reliable solutions to problems of significance in heat and mass transfer.

Those wishing to use the methods to be discussed in this article are advised to study the original papers referenced—particularly those describing the case studies—and to consult the literature of the subject. Both the heat transfer journals (such as *ASME Journal of Heat Transfer* and *International Journal of Heat and Mass Transfer*) and the computational journals (such as *Journal of Computational Physics, Computers and Fluids, Computer Methods in Applied Mechanics and Engineering, International Journal of Numerical Methods in Engineering*) regularly publish papers of relevance to the present topic.

1.2 The basic differential equations

The behavior of a fluid experiencing convection heat (or mass) transfer is described by a system of partial differential equations expressing the conservation of mass, momentum, and energy (or concentration). Conduction heat transfer is described by a simplified form of the energy equation. The derivation of these equations may be found in standard heat transfer texts.

The *continuity* (mass conservation) *equation* is

$$\frac{\partial \rho}{\partial t} = - \nabla \cdot \rho \mathbf{v} \tag{1.1}$$

or, in Cartesian coordinates,

$$\frac{\partial \rho}{\partial t} + v_x \frac{\partial \rho}{\partial x} + v_y \frac{\partial \rho}{\partial y} + v_z \frac{\partial \rho}{\partial z} = -\rho\left(\frac{\partial v_x}{\partial x} + \frac{\partial v_y}{\partial y} + \frac{\partial v_z}{\partial z}\right) \tag{1.1a}$$

Most heat transfer problems involving liquids and many of those involving gases can be analyzed as if the fluid were incompressible. In such cases (1.1) becomes

$$\nabla \cdot \mathbf{v} = \frac{\partial v_x}{\partial x} + \frac{\partial v_y}{\partial y} + \frac{\partial v_z}{\partial z} = 0 \tag{1.1b}$$

The form of the *momentum equation* to be considered is that applying to a fluid of constant viscosity that is "almost" incompressible (i.e., to a fluid of which the density variation may be neglected except in the buoyancy term of the momentum equation). In this form the equation, known as the Navier-Stokes equation, is

$$\rho \frac{D\mathbf{v}}{Dt} = - \nabla_p + \mu \nabla^2 \mathbf{v} + \rho \mathbf{g} \tag{1.2}$$

or, in Cartesian coordinates,

$$\rho\left(\frac{\partial v_x}{\partial t} + v_x \frac{\partial v_x}{\partial x} + v_y \frac{\partial v_x}{\partial y} + v_z \frac{\partial v_x}{\partial z}\right) = -\frac{\partial p}{\partial x} + \mu\left(\frac{\partial^2 v_x}{\partial x^2} + \frac{\partial^2 v_x}{\partial y^2} + \frac{\partial^2 v_x}{\partial z^2}\right) + \rho g_x \tag{1.2a}$$

with corresponding equations for the y and z components.

In many problems, fluid pressure is essentially constant. In others, fluid density is independent of temperature (i.e., the fluid is incompressible or "almost" incompressible, as defined above). In either case, and for constant thermal conductivity, the *energy equation* is

$$\rho c_p \frac{DT}{Dt} = k \nabla^2 T \tag{1.3}$$

or, in Cartesian coordinates,

$$\rho c_p\left(\frac{\partial T}{\partial t} + v_x \frac{\partial T}{\partial x} + v_y \frac{\partial T}{\partial y} + v_z \frac{\partial T}{\partial z}\right) = k\left(\frac{\partial^2 T}{\partial x^2} + \frac{\partial^2 T}{\partial y^2} + \frac{\partial^2 T}{\partial z^2}\right) \tag{1.3a}$$

The viscous dissipation term, which is seldom significant, has been omitted.

For fluids at rest, and for solids, $\mathbf{v} = 0$. From (1.3a) we obtain the *conduction equation*

$$\rho c_p \frac{\partial T}{\partial t} = k\left(\frac{\partial^2 T}{\partial x^2} + \frac{\partial^2 T}{\partial y^2} + \frac{\partial^2 T}{\partial z^2}\right) \tag{1.4}$$

Diffusion in a multicomponent system is described by equations analogous to the above, which need not be restated here. Similarly, the cylindrical and spherical forms of the equations will be omitted. For the purpose of explaining methods for their solution, the rectangular Cartesian forms will be sufficient.

1.3 Vorticity-vector potential formulation

Equations (1.2) are written in terms of what are sometimes called the primitive variables, viz., pressure and velocity. For an almost incompressible fluid, it is frequently convenient to recast these equations into a form involving the derived variables, vorticity and vector potential.

At this stage it is convenient to introduce the Boussinesq (1903) approximation appropriate to an almost incompressible fluid, viz., that the density is given by

$$\rho = \rho_0 [1 - \beta(T - T_0)] \tag{1.5}$$

in the buoyancy term of the momentum equations, and by $\rho = \rho_0$ elsewhere. Equation (1.2) then becomes

$$\frac{D\mathbf{v}}{Dt} = -\frac{\nabla P}{\rho_0} + \nu\nabla^2\mathbf{v} - \beta(T - T_0)\mathbf{g} \tag{1.6}$$

where P denotes the perturbation of static pressure from the hydrostatic value. Using the vector identity

$$(\nabla \times \mathbf{v})\mathbf{v} = \mathbf{v} \cdot \nabla\mathbf{v} - \frac{1}{2}\nabla\mathbf{v}^2$$

together with the continuity equation (1.1b), (1.6), and (1.3) may be written

$$\frac{\partial \mathbf{v}}{\partial t} = -(\nabla \times \mathbf{v}) \times \mathbf{v} - \frac{\nabla p}{\rho_0} + \nu\nabla^2\mathbf{v} - \beta(T - T_0)\mathbf{g} \tag{1.7}$$

$$\frac{\partial T}{\partial t} = -\nabla \cdot (\mathbf{v}T) + \alpha\nabla^2 T \tag{1.8}$$

where $\alpha = k/(\rho c_p)$ and p now denotes the total pressure $P + \frac{1}{2}\rho_0\mathbf{v}^2$.

Since $\mathbf{v}$ is solenoidal, i.e., $\nabla\cdot\mathbf{v} = 0$, it can be shown (Hirasaki and Hellums, 1970) that it is always possible to define a vector function $\boldsymbol{\psi}$ and a scalar function ϕ such that $\mathbf{v}$ is given by

$$\mathbf{v} = \nabla \times \boldsymbol{\psi} - \nabla \phi \tag{1.9}$$

$\boldsymbol{\psi}$ is solenoidal, i.e.,

$$\nabla \cdot \boldsymbol{\psi} = 0 \tag{1.10}$$

and ϕ satisfies the Laplace equation

$$\nabla^2 \phi = 0 \tag{1.11}$$

Introducing the vorticity vector $\boldsymbol{\zeta} = \nabla \times \mathbf{v}$, we have that

$$\begin{aligned} \boldsymbol{\zeta} &= \nabla \times \nabla \times \boldsymbol{\psi} && \text{(since } \nabla \times \nabla \phi = 0\text{)} \\ &= \nabla (\nabla \cdot \boldsymbol{\psi}) - \nabla^2 \boldsymbol{\psi} && \text{(vector identity)} \\ &= - \nabla^2 \boldsymbol{\psi} && \text{(by [1.10])} \end{aligned} \tag{1.12}$$

If we take the curl of (1.7) we finally obtain that

$$\frac{\partial \boldsymbol{\zeta}}{\partial t} = - \nabla \times \boldsymbol{\zeta} \times (\nabla \times \boldsymbol{\psi}) + \nu \nabla^2 \boldsymbol{\zeta} - \beta \nabla \times T\mathbf{g} \tag{1.13}$$

which, with (1.8), (1.11), and (1.12), constitutes a set of equations for T and the derived variables $\boldsymbol{\psi}$, ϕ, and $\boldsymbol{\zeta}$.

1.4 Two-dimensional flow

The equations of the previous section are appropriate to problems in three space dimensions and are thus (subject to the other restrictions noted) in their most general form. However, the cost, in terms of computer time, of obtaining solutions to these equations has been very high and it has only become practicable to solve fully three-dimensional problems in recent years. Among examples of such solutions are the works of Chorin (1968), Williams (1969), Caretto, Curr, and Spalding (1972), and Mallinson and de Vahl Davis (1973).

Fortunately, many problems in fluid dynamics and heat transfer can be treated as though they were two-dimensional, i.e., as though the dependent variables were functions of only two independent coordinates. There is a considerable practical advantage in making this simplification, as the amount of labor (i.e., computer time) required for the solution is thereby reduced by one or two orders of magnitude. However, as pointed out by Mallinson et al. (1973), the assumption of two-dimensional flow may not be valid even though it is very plausible, and care must be exercised in making it.

In two-dimensions (x and y) the vector potential becomes the familiar (scalar) stream function, defined by

$$v_x = \frac{\partial \psi}{\partial y} \qquad v_y = -\frac{\partial \psi}{\partial x} \tag{1.14}$$

which replaces (1.9). The scalar function has no counterpart in two-dimensional flow, and the only nonvanishing component of vorticity is that in the z-direction, defined by

$$\zeta = \frac{\partial v_y}{\partial x} - \frac{\partial v_x}{\partial y} \tag{1.15}$$

Thus, with the x-axis horizontal and the y-axis parallel to the gravity vector, the equations for ζ, ψ, and T become

$$\frac{\partial \zeta}{\partial t} = -\frac{\partial}{\partial x}(\zeta v_x) - \frac{\partial}{\partial y}(\zeta v_y) + \nu\left(\frac{\partial^2 \zeta}{\partial x^2} + \frac{\partial^2 \zeta}{\partial y^2}\right) - \beta g \frac{\partial T}{\partial x} \tag{1.16}$$

$$0 = \nabla^2 \psi + \zeta \tag{1.17}$$

$$\frac{\partial T}{\partial t} = -\frac{\partial}{\partial x}(T v_x) - \frac{\partial}{\partial y}(T v_y) + \alpha\left(\frac{\partial^2 T}{\partial x^2} + \frac{\partial^2 T}{\partial y^2}\right) \tag{1.18}$$

in which v_x and v_y are given by (1.14).

It should be noted that Eqs. (1.8), (1.16), and (1.18) are written in what is called the *conservative* form. This name arises from the fact that they imply the conservation of some property of the fluid. Consider, for example, the result of integrating (1.8) over a region R, say,

$$\int_R \frac{\partial T}{\partial t}\, dR = -\int_R \nabla \cdot (\mathbf{v}T)\, dR + \int_R \alpha \nabla^2 T\, dR$$

Noting that we may reverse the order of differentiation and integration on the left-hand side, and applying the divergence theorem to each term on the right-hand side, we obtain

$$\frac{\partial}{\partial t}\int_R T\, dR = -\int_B (\mathbf{v}T) \cdot \mathbf{n}\, ds + \alpha \int_B (\nabla T) \cdot \mathbf{n}\, ds$$

where B is the boundary of R, ds is a differential element of B, and $\mathbf{n}$ is the unit outward normal vector at the surface. In words: the rate of change of T in R is given by a balance between the convection of T across B and the diffusion of T across B. There is no source or sink of T in R; T is conserved. (Readers might prefer to multiply these equations throughout by the specific heat of the fluid, and regard them as expressing conservation of enthalpy.)

1.5 Boundary conditions

For the solution of (1.8), (1.11)-(1.13), or of (1.16)-(1.18), boundary conditions on the variables are required, which must be obtained from the specification of the particular problems.

Conditions for T may be readily formulated. They may be of the Dirichlet type, in which T is specified as a function of position (and, perhaps, time) on the boundary B of the solution domain, or the Neumann type, in which the normal gradient of T is specified on B. If heat is transferred between the solution region and its surroundings at a rate controlled by a specified heat transfer coefficient h, a combined condition

$$h(T_s - T_\infty) = k\left(\frac{\partial T}{\partial n}\right)_s \tag{1.19}$$

is obtained, where T_∞ is the temperature of the surroundings, n denotes the direction normal to the surface, s denotes conditions at the surface, and k is the thermal conductivity of the material in the solution region. Note that (1.19) applies even if this material is a fluid in motion, provided that a no-slip condition holds at B. Equation (1.19) may be written

$$aT + b\,\frac{\partial T}{\partial n} = c \tag{1.20}$$

With appropriate values for a, b, and c, (1.20) can be used to describe Dirichlet or Neumann conditions as well as the combined condition (1.19).

Boundary conditions on $\boldsymbol{\psi}$ and ϕ for (1.11) and (1.12) were proposed by Hirasaki and Hellums (1968, 1970) and were modified and implemented by Mallinson et al. (1973) in the case of a closed solution region; de Vahl Davis and Wolfshtein (1974) have applied them to an open system.

We consider only the commonest situation in which each boundary is normal to one of the coordinate directions. On the boundary which is, for example, normal to the z-axis, the conditions on ϕ and the three components of $\boldsymbol{\psi}$ are

$$\begin{gathered}\frac{\partial \phi}{\partial z} = -v_z \\ \psi_x = \psi_y = \frac{\partial \psi_z}{\partial z} = 0\end{gathered} \tag{1.21}$$

where v_z is the z component of velocity.

In two-dimensional flow ϕ disappears, and ψ reduces to a scalar which is constant along a streamline, and hence along any impermeable wall. This, together with the no-slip conditions, leads to

$$\psi = \text{constant} \qquad \frac{\partial \psi}{\partial x} = 0 \tag{1.22}$$

on a surface parallel to the y-axis, for example. In an open system, with flow across a boundary, conditions on the velocity components must be used in conjunction with (1.14) to derive suitable conditions on ψ.

Consider, for example, developing flow in the duct shown in Figure 1. Fluid enters at $x = 0$, at a total rate Q (per unit width in the z direction). The velocity distribution may be assumed to be uniform at $x = 0$, except at the walls $y = \pm L$. The velocity distribution shown in the figure is sometimes described as slug flow; the velocity at $x = 0$ will be $Q/2L$. At a sufficient distance x downstream, the flow will be fully developed. In the absence of thermal effects, the velocity distribution will be parabolic in laminar flow, hence v_x is known. In turbulent flow, or if there is heating or cooling at the walls, v_x will not be parabolic, but we may still be able to assume that the flow is fully developed in that $\partial v_x/\partial x = v_y = 0$. Thus the boundary conditions for v_x and v_y are

$$\begin{aligned}
&\text{at} \quad x = 0 \qquad v_x = \frac{Q}{2L} \qquad v_y = 0 \\
&\text{at} \quad x = X \qquad \frac{\partial v_x}{\partial x} = v_y = 0 \\
&\text{at} \quad y = \pm L \qquad v_x = v_y = 0
\end{aligned} \tag{1.23}$$

We now find the corresponding conditions for ψ.

The wall $y = -L$ is a streamline, hence ψ is a constant there; we may suppose, without loss of generality, that $\psi = 0$. Along $x = 0$, (1.14) tells us that

Figure 1 Flow in a duct.

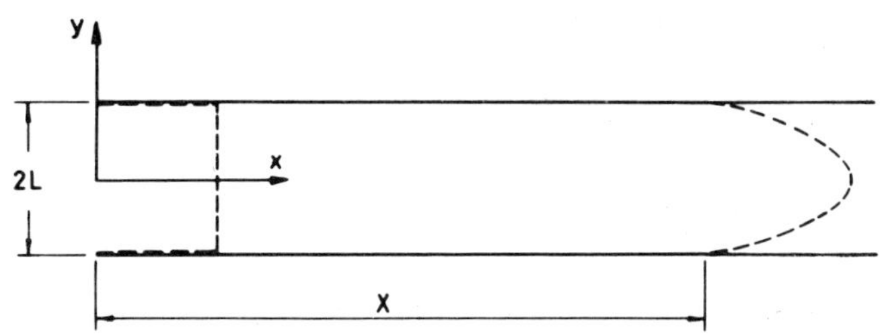

$$\psi = \int_{-L}^{y} v_x \, dy$$

and since $v_x = Q/2L$ = constant there, ψ varies linearly with y:

$$\psi = \frac{Q}{2L}(y + L)$$

and in particular at $y = L$, $\psi = Q$. Since the wall $y = +L$ is also a streamline, $\psi = Q$ along this wall. At $x = X$, ψ will vary in a manner depending upon $v_x(y)$, which in general is not known. But it is known–if the flow is fully developed–that $v_y = 0$. Hence, from (1.14), $\partial\psi/\partial x = 0$. Thus in terms of ψ the conditions (1.23) become

$$\begin{aligned}
&\text{at} \quad & x &= 0 \quad & \psi &= \frac{Q}{2L}(y + L) \\
&\text{at} \quad & x &= X \quad & \frac{\partial \psi}{\partial x} &= 0 \\
&\text{at} \quad & y &= -L \quad & \psi &= 0 \\
&\text{at} \quad & y &= L \quad & \psi &= Q
\end{aligned} \tag{1.24}$$

Boundary conditions for vorticity, needed for the solution of (1.13) or (1.16), are most readily derived from (1.12) or (1.17), respectively. The method is similar for two- and three-dimensional flow and will be described for the former.

Now

$$\zeta = \frac{\partial v_y}{\partial x} - \frac{\partial v_x}{\partial y}$$

But at the walls $y = L$, $v_y = 0$. Hence, $\partial v_y/\partial x = 0$, and

$$\zeta = -\frac{\partial v_x}{\partial y} = -\frac{\partial^2 \psi}{\partial y^2} \tag{1.25a}$$

At $x = 0$, v_x is constant. Hence, $\partial v_x/\partial y = 0$, and

$$\zeta = \frac{\partial v_y}{\partial x} = -\frac{\partial^2 \psi}{\partial x^2} \tag{1.25b}$$

At $x = X$, $v_y = 0$. Hence, $\partial v_y/\partial x = 0$ (fully developed flow), and

$$\zeta = -\frac{\partial v_x}{\partial y} = -\frac{\partial^2 \psi}{\partial y^2} \tag{1.25c}$$

The implementation of these boundary conditions in the numerical scheme will be described in a later section.

A more extensive discussion of boundary conditions than can be included here may be found in Roache (1972).

1.6 Nondimensional equations

The solution of the example of duct flow in the previous section is dependent on four quantities: Q, L, X, and the viscosity, ν, of the fluid. If heat transfer is included, and the energy equation must be solved, a fifth quantity α is required. Each of these must be specified before a numerical solution can be obtained, and the solution must be repeated entirely if any of them are varied.

The number of parameters can be reduced, and hence the range of applicability of a solution extended, if the equations and boundary conditions are recast in terms of nondimensional variables and parameters. The procedure is standard and straightforward, and the details will be omitted. However, it is important that this step be performed.

With (1.16)-(1.18), for example, the use of the nondimensional variables

$$x' \equiv \frac{x}{L} \qquad y' \equiv \frac{y}{L} \qquad v'_x \equiv \frac{2v_x L}{Q} \qquad v'_y \equiv \frac{2v_y L}{Q} \qquad t' \equiv \frac{Qt}{2L^2}$$

$$\psi' \equiv \frac{2\psi}{Q} \qquad \zeta' \equiv \frac{2\zeta L^2}{Q} \qquad T' \equiv \frac{T - T_1}{T_2 - T_1}$$

(in which $Q/2L$ has been used as a reference velocity and where T_1 and T_2 are two convenient reference temperatures) leads to

$$\frac{\partial \zeta}{\partial t} = -\frac{\partial}{\partial x}(\zeta v_x) - \frac{\partial}{\partial y}(\zeta v_y) + \frac{1}{\text{Re}}\nabla^2\zeta - \frac{\text{Ra}}{\text{Re}^2\text{Pr}}\frac{\partial T}{\partial x} \tag{1.26}$$

$$0 = \nabla^2\psi + \zeta \tag{1.27}$$

$$\frac{\partial T}{\partial t} = -\frac{\partial}{\partial x}(Tv_x) - \frac{\partial}{\partial y}(Tv_y) + \frac{1}{\text{RePr}}\nabla^2 T \tag{1.28}$$

in which the primes have immediately been discarded for clarity; ∇^2 denotes the Laplacian operator in terms of the *nondimensional* variables x and y; and

$$\text{Re} \equiv \frac{Q}{2\nu} \qquad \text{(a Reynolds number)}$$

$$\mathrm{Pr} \equiv \frac{\nu}{\alpha} \qquad \text{(the Prandtl number)}$$

and

$$\mathrm{Ra} \equiv \frac{\beta g \Delta T L^3}{\alpha \nu} \qquad \text{(the Rayleigh number)}$$

The boundary conditions may put into nondimensional form in a similar manner, in being necessary only to note that the size of the solution region is given by

$$0 \leqslant x \leqslant \frac{X}{L} \qquad -1 \leqslant y \leqslant 1$$

The quantity X/L is known as the aspect ratio.

To solve the flow problem, only two parameters–Re and X/L–are now required in place of the original four. One more–Pr–is required if the energy equation is to be solved, and a fourth–Ra–if natural convection effects are included. A significant generalization of the equations, and, hence, of their solution, has thus been achieved.

The set of nondimensional parameters is not unique. In particular, in problems for which a characteristic velocity ($Q/2L$ in the above example) is not available, i.e., in flows driven entirely by buoyancy, the choice of reference velocity is arbitrary. If we choose α/L for this purpose, then Re is replaced by 1/Pr and Eqs. (1.26) and (1.28) become

$$\frac{\partial \zeta}{\partial t} = -\frac{\partial}{\partial x}(\zeta v_x) - \frac{\partial}{\partial y}(\zeta v_y) + \mathrm{Pr}\,\nabla^2 \zeta - \mathrm{RaPr}\,\frac{\partial T}{\partial x} \tag{1.29}$$

$$\frac{\partial T}{\partial t} = -\frac{\partial}{\partial x}(T v_x) - \frac{\partial}{\partial y}(T v_y) + \nabla^2 T \tag{1.30}$$

1.7 Finite difference operators

Finite difference operators are a convenient notation for the methods to be discussed in this chapter. An extensive treatment of those operators can be found in Hildebrand (1968).

1.7.1 Notation and definition Let T be a dependent variable specified at discrete values of x. Let these discrete values of x be separated from each other by a constant interval, h. We shall adopt the notation $T_i = T(x_i)$ and $T_{i+1} = T(x_{i+1}) = T(x_i + h)$, where x_i is the value of x at point i.

The six basic finite difference operators are

Forward difference: $\Delta T_i = T_{i+1} - T_i$ (1.31)

Backward difference: $\nabla T_i = T_i - T_{i-1}$ (1.32)

Central difference: $\delta T_i = T_{i+½} - T_{i-½}$ (1.33)

Shift: $ET_i = T_{i+1}$ (1.34)

Derivative: $DT_i = \left(\frac{dT}{dx}\right)_i$ (1.35)

Average: $\mu T_i = \frac{T_{i-½} + T_{i+½}}{2}$ (1.36)

The operators can be shown to be associative, distributive, and commutative.

1.7.2 Taylor's series expansion T_{i+1} will now be expressed in terms of a Taylor's series expansion around T_i:

$$T_{i+1} = ET_i = T_i + \frac{hD}{1!} T_i + \frac{h^2 D^2}{2!} T_i + \cdots + \frac{h^N D^N}{N!} T_i + O(h^{N+1}) \tag{1.37}$$

where the last term denotes the order-of-magnitude of the discretization (or truncation) error resulting from termination of the series in the Nth term. This error can be expressed in two ways: firstly,

$$\frac{h^{N+1} D^{N+1}}{(N+1)!} T_{i+\theta} \qquad (0 \leqslant \theta \leqslant 1)$$

or, secondly, as the *leading* term of the truncation error,

$$\frac{h^{N+1} D^{N+1}}{(N+1)!} T_i$$

The Taylor's series expansion is valid when T possesses continuous derivations up to the Nth term. Indeed, the discretization error is zero if T is actually a polynomial of degree N or less. It is fortunate that most functions governing physical phenomena are expandable into a Taylor's series.

Equation (1.37) can be recast as

$$ET_i = \sum_{m=0}^{\infty} \frac{h^m D^m}{m!} T_i = e^{hD} T_i \tag{1.38}$$

yielding the following relation between the operators E and D:

$$E = e^h D \qquad \text{or} \qquad hD = \ln E \tag{1.39}$$

Using three of the previous formulae,

$$hD = \ln(\Delta + 1) = -\ln(1 - \nabla) = 2 \operatorname{arcsinh}\left(\frac{\delta}{2}\right) \tag{1.40}$$

and applying the series expansion for ln or arcsinh, one can readily express hD in terms of either forward or backward one-sided differences, as well as a central difference expansion that employs half-interval values. Since T is not given at half-interval points this last expansion clearly cannot be implemented. Instead, the relation

$$\mu = \left(1 + \frac{\delta^2}{4}\right)^{1/2} \tag{1.41}$$

is used to give

$$hD = 2\mu\left[\left(1 + \frac{\delta^2}{4}\right)^{-1/2} \operatorname{arcsinh}\left(\frac{\delta}{2}\right)\right] \tag{1.42}$$

in terms of the mean odd difference, $\mu\delta^m$. This equation is applied for odd derivatives while Eq. (1.40) is used for even derivatives.

1.7.3 First and second derivatives in terms of backward, forward, and central differences The following relations are useful for the approximations to be studied here.

For hDT_i

$$\Delta T_i - \frac{1}{2}\Delta^2 T_i + \frac{1}{3}\Delta^3 T_i - \cdots + (-1)^{m+1}\frac{1}{m}\Delta^m T_i + \cdots \tag{1.43a}$$

$$\nabla T_i + \frac{1}{2}\nabla^2 T_i + \frac{1}{3}\nabla^3 T_i + \cdots + \frac{1}{m}\nabla^m T_i + \cdots \tag{1.43b}$$

$$\mu\delta T_i - \frac{(1!)^2}{3!}\mu\delta^3 T_i + \cdots + (-1)^m \frac{(m!)^2}{(2m+1)!}\mu\delta^{2m+1} T_i + \cdots \tag{1.43c}$$

For $h^2 D^2 T_i$

$$\Delta^2 T_i - \Delta^3 T_i + \frac{11}{12}\Delta^4 T_i - \frac{5}{6}\Delta^5 T_i + \cdots \tag{1.44a}$$

$$\nabla^2 T_i + \nabla^3 T_i + \frac{11}{12}\nabla^4 T_i + \frac{5}{6}\nabla^5 T_i + \cdots \tag{1.44b}$$

$$\delta^2 T_i - \frac{1}{12}\delta^4 T_i + \frac{1}{90}\delta^6 T_i - \frac{1}{560}\delta^8 T_i + \cdots \tag{1.44c}$$

The leading error due to truncation of the series is equal to the first term neglected and is of the order h^N where N is the order of the neglected term. Thus, if, for example, the series of Eq. (1.44c) is truncated after the first term then the error is

$$\begin{aligned} -\frac{1}{12}\delta^4 T_i &= -\frac{1}{12}\delta^4 D^{-4} D^4 T_i \\ &= -\frac{h^4}{12}\frac{\delta^4}{[2\,\text{arcsinh}\,(\delta/2)]^4} D^4 T_i = O(h^4) \end{aligned} \tag{1.45}$$

h^4 accuracy for $D^2 T_i$ in Eq. (1.44c) can be obtained by applying the approximation

$$-\frac{1}{12}\delta^4 T_i = -\frac{1}{12}\delta^2 D^2 T_i + O(h^6) \tag{1.46}$$

which when substituted into Eq. (1.44c) yields

$$h^2 D^2 T_i = \frac{\delta^2 T_i}{1 + \dfrac{1}{12}\delta^2 T_i} + O(h^6) \tag{1.47}$$

Often when investigating the truncation error of a difference formula, one is interested in obtaining a series expansion for a difference operator in terms of the derivative operator, D. Some of the more important expansions are

$$\mu = \cosh\left(\frac{hD}{2}\right) = 1 + \frac{h^2 D^2}{2^2 2!} + \frac{h^4 D^4}{2^2 4!} + \frac{h^6 D^6}{2^2 6!} + \cdots \tag{1.48}$$

$$\delta = 2\sinh\left(\frac{hD}{2}\right) = hD + \frac{h^3 D^3}{2^2 3!} + \frac{h^5 D^5}{2^4 5!} + \cdots \tag{1.49}$$

$$\delta^2 = 4\sinh^2\left(\frac{hD}{2}\right) = 2\left(\frac{h^2 D^2}{2!} + \frac{h^4 D^4}{4!} + \frac{h^6 D^6}{6!} + \cdots\right) \tag{1.50}$$

$$\mu\delta = \sinh(hD) = hD + \frac{h^3D^3}{3!} + \frac{h^5D^5}{5!} + \cdots \tag{1.51}$$

$$\Delta = e^{hD} - 1 = hD + \frac{h^2D^2}{2!} + \frac{h^3D^3}{3!} + \cdots \tag{1.52}$$

$$\nabla = 1 - e^{-hD} = hD - \frac{h^2D^2}{2!} + \frac{h^3D^3}{3!} - \cdots \tag{1.53}$$

Thus the expression of (1.45), from (1.50) is of order h^4.

1.7.4 Extension to partial derivatives In applying finite differences to partial derivatives the position in space will be denoted by a separate index for each independent space variable, whereas a superscript will denote time. The increments will be denoted by the latter h with an index denoting the independent variable. Thus, for example,

$$T^{n+1}_{i-1,j,\,k+1}$$

denotes the temperature at the position in space $x - h_x$, y, $z + h_z$, and at time $t + h_t$. The indices i, j, and k as used here will always denote position in space and the superscript, n, will denote time.

Partial derivatives and differences will be denoted by the corresponding index. Thus,

$$D_x T^{n+1}_{i-1} = \left(\frac{\partial T}{\partial x}\right)_{x-h_x,\,t+h_t} \tag{1.54}$$

and

$$\Delta_t T_i^n = T_i^{n+1} - T_i^n \tag{1.55}$$

Clearly all the approximations developed in the previous section can be extended to partial derivatives.

2 PARABOLIC EQUATIONS

2.1 The diffusion equation in one space variable

Many of the concepts associated with the numerical solution of partial differential equations (PDE's) can be described by considering the simple example:

$$\frac{\partial T}{\partial t} = \frac{\partial^2 T}{\partial x^2} \tag{2.1}$$

which we suppose has been made nondimensional in such a way that $0 \leqslant T, x \leqslant 1$. Associated with (2.1) are the initial and boundary conditions

$$T(x, 0) = f(x) \qquad T(0, t) = g_1(t) \qquad T(T, t) = g_2(t) \tag{2.2}$$

and the solution is to be obtained in the semi-infinite rectangle shown in Figure 2.

We suppose that the solution region is covered by a rectangular grid or mesh, with mesh sizes of h_x and h_t, assumed for the present uniform in x and t, respectively. The solution is to be found at the nodes or mesh points. The notation T_i^n will be used to indicate the value of T at the ith mesh point along the x axis and at the nth time step,* i.e.,

$$T_i^n = T(ih_x, nh_t)$$

(For convenience in eventual programming, the origin is taken at $i = n = 0$.)

2.1.1 Explicit method The simplest finite difference approximation (FDA) to (2.1) is derived using a forward difference in time and a central difference in space, viz.,

*In some texts, a distinction in notation is sometimes made between the variables of the differential equation and the corresponding difference approximation. As the distinction is normally clear by context, the same notation will be used here for each.

Figure 2 Solution region and the finite difference mesh.

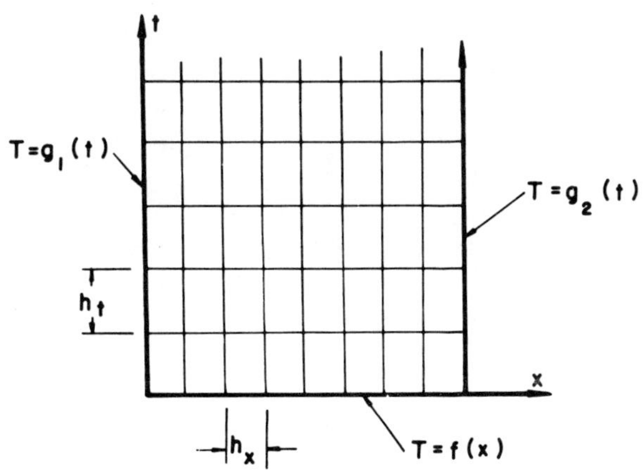

$$\frac{\Delta_t T}{h_t} = \frac{\delta_x{}^2 T}{h_x{}^2} + (h_t + h_x{}^2) \tag{2.3a}$$

and, neglecting the truncation error, we obtain the *explicit* formula

$$T_i^{n+1} = T_i^n + r(T_{i-1}^n - 2T_i^n + T_{i+1}^n) \tag{2.3b}$$

where $r = h_t/h_x{}^2$.

To illustrate the use of (2.3), suppose that (2.1) describes a heat conduction problem: a uniform rod is initially at a (nondimensional) temperature of zero, and at time zero each of its ends is suddenly heated to $T = 1$. Thus, $f(x) = 0, g_1(t) = g_2(t) = 1$.

We are immediately faced with a difficulty. In constructing a mathematical model of this problem, a singularity has been introduced which obviously does not exist in the physical situation: what are the values of $T(0, 0)$ and $T(1, 0)$? In practice we cannot change the temperature at these points "suddenly," i.e., instantaneously, but we have assumed that $T(0, 0-) = 0$ while $T(0, 0+) = 1$. The simplest method of dealing with this situation is to use the average value, 0.5, as the initial conditions at the ends. (A more accurate method using asymptotic expansions will be described later.)

Experience shows that at least ten intervals are required along the x axis in order to achieve reasonable accuracy. Thus, with $h_x = 0.1$ and $h_t = 0.005$, we obtain $r = 0.5$ and (2.3) assumes the simple form

$$T_i^{n+1} = \frac{1}{2}(T_{i-1}^n + T_{i+1}^n) \tag{2.4}$$

The reader may verify that (2.4) can be used to compute the values of T_i, $1 \leqslant i \leqslant 9$, for successive time steps. However, the solution progresses slowly. An attempt to accelerate the approach to steady state by the doubling the value of h_t to 0.01 will rapidly lead to values for T which are physically impossible. To discuss the reasons for this situation we must now introduce three important concepts associated with the finite difference solution of a partial differential equation.

2.1.2 Consistency, stability, and convergence Equation (2.3) was derived from (2.1) by replacing derivatives by finite difference approximations. We now carry out the reverse procedure, using the methods of Sec. 1.7.1, Eq. (2.3) is equivalent to

$$\frac{\Delta_t T}{h_t} = \frac{\delta_x{}^2 T}{h_x{}^2} \tag{2.5}$$

Since $\Delta = e^{hD} - 1$ and $\delta^2 = 4\sinh^2 hD/2$, (2.5) can be written

$$\frac{1}{h_t}\left(h_t D_t + \frac{h_t^2 D_t^2}{2!} + \cdots\right)T = \frac{1}{h_x^2}\left(h_x^2 D_x^2 + \frac{h_x^4 D_x^4}{12} + \cdots\right)T$$

or

$$\frac{\partial T}{\partial t} = \frac{\partial^2 T}{\partial x^2} - \left(\frac{h_t}{2}\frac{\partial^2 T}{\partial t^2} + \cdots\right) + \left(\frac{h_x^2}{12}\frac{\partial^4 T}{\partial x^4} + \cdots\right) \tag{2.6}$$

The two expressions in parentheses in (2.6) are the *truncation errors* of (2.3): they represent the difference between the FDA being solved and the PDE which it replaces. If h_t and h_x are reduced, these errors are also reduced and, in the limit h_t, h_x 0, the FDA approaches the PDE. This feature is known as *consistency:* and FDA is consistent with a PDE if, as h_t and h_x tend to zero, the truncation error of the FDA tends to zero and the FDA approaches the PDE. This is clearly a desirable feature of an FDA and is necessary for accuracy of the solution. However, it is not a sufficient condition: even though the FDA might approach a PDE as the step sizes tend to zero, it does not follow that the *solution* of the FDA approaches the *solution* of the PDE. It can be readily verified for the previous example that if h_t and h_x are reduced such that the ration h_t/h_x^2 retains the value of unity, the unrealistic results will persist.

This further condition is expressed in the concept of *convergence.* An FDA is convergent if, at each mesh point (x, t), the exact solution of the FDA approaches the solution of the PDE at the same point as the step sizes h_t and h_x tend to zero.

An FDA may be *conditionally* convergent. For example, it can be shown that (2.3) is convergent provided that $r \leqslant \frac{1}{2}$.

The third important feature of an FDA is its *stability,* concerned with the behavior of errors (introduced, typically, by round-off) in its solution as t (i.e., n) increases. An FDA is stable if the difference between its theoretical and its numerical solutions remains bounded at all space points i as n increases and h_t remains constant. Note that stability does not refer to the difference between the solutions of the FDA and the PDE; that is the province of convergence. Stability is concerned only with the FDA. Note also that the disastrous values resulting from the use of $h_t = 0.1$ (i.e., $r = 1$) in the previous subsection are *not* due to instability; these values can be computed exactly, i.e., without error, and are due to divergence (Smith, 1965).

However, there is an important link between consistency, convergence, and stability. Lax's equivalence theorem (see, for example, Isaacson and Keller, 1966) states that if an FDA to a well posed linear* initial value problem is

*Note the restriction to linear problems. A corresponding theorem for the more important class of non-linear problems has yet to be derived. Fortunately, experience suggests that the theorem is, nevertheless, usually valid for such problems as well.

consistent, then stability is a necessary and sufficient condition for convergence. This theorem is of considerable practical, as well as theoretical, importance. For while it is quite easy to demonstrate the consistency (or otherwise) of an FDA, and is generally also fairly easy to show its stability, it is often very difficult to establish convergence. Happily, this theorem relieves us of the need to do so, and it is sufficient to prove consistency and stability.

2.1.3 The von Neumann method of stability analysis The assemblage of FDA's for all mesh points i can be written in the general form

$$(T^{n+1}) = [A](T^n) + (B) \tag{2.7}$$

where the parentheses denote column vectors and square brackets denote a matrix known as the amplification matrix.

Suppose that an error is introduced into some or all the elements of (T^n), and denote the erroneous vector by (U^n). Then after one time step (U^{n+1}) will be given by

$$(U^{n+1}) = [A](U^n) + (B)$$

Denoting the vector of errors by (ϵ^n), we obtain

$$(\epsilon^{n+1}) = (T^{n+1}) - (U^{n+1}) = [A](\epsilon^n) \tag{2.8}$$

i.e., (ϵ^n) obeys the same FDA but with homogeneous boundary conditions. It is the behavior of (ϵ^n) that we wish to study. The FDA will be stable if (ϵ^n) remains bounded as $n \to \infty$.

In the von Neumann method of stability analysis, the initial error vector is expressed as a finite Fourier series

$$\epsilon_{\mathrm{i}} = \sum_m A_m e^{m\pi x} = \sum_m B_m e^{\mathrm{i}\beta_m i}$$

where $\mathrm{i} \equiv \sqrt{-1}$, with as many terms as there are elements in the error vector. Because we are considering only linear FDA's, it is sufficient to study the propagation of the error due to just a single term $e^{\beta i}$. The coefficient B is a constant and can be omitted.

Thus we shall seek a solution of the error FDA (2.8) in the form

$$\epsilon_{\mathrm{i}}^n = \xi^n e^{\mathrm{i}\beta i} \tag{2.9}$$

where the time dependence of (ϵ) is expressed in the term ξ^n. This will satisfy the initial condition at $t = 0$ and shows that (ϵ) will remain bounded if $|\epsilon| \leqslant 1$. The quantity ξ is called the amplification factor.

The error FDA corresponding to (2.3) is, by (2.8),

$$\epsilon_i^{n+1} = \epsilon_i^n + r(\epsilon_{i-1}^n - 2\epsilon_i^n + \epsilon_{i+1}^n)$$

Using (2.9), we obtain

$$\xi^{n+1} e^{i\beta i} = \xi^n e^{i\beta i} + r\xi^n (e^{i\beta(i-1)} - 2e^{i\beta i} + e^{i\beta(i+1)})$$

or

$$\xi = 1 - 2r + r(e^{-i\beta} + e^{i\beta}) = 1 - 2r + 2r\cos\beta = 1 - 4r\sin^2\left(\frac{\beta}{2}\right)$$

We require that $|\xi| \leqslant 1$, i.e., that

$$-1 \leqslant 1 - 4r\sin^2\left(\frac{\beta}{2}\right) \leqslant 1 \qquad \text{for all } \beta$$

The right hand inequality is satisfied for all $r > 0$; the left hand inequality requires

$$r\sin^2\left(\frac{\beta}{2}\right) \leqslant \frac{1}{2}$$

which is satisfied for all β if

$$r \leqslant \frac{1}{2} \tag{2.10}$$

Since in practice the situation $r = 0$ must be excluded (as this requires $h_t = 0$) we see that the explicit FDA (2.3) is stable if $0 < r \leqslant \frac{1}{2}$.

The von Neumann method when extended to the more general case of variable coefficients can only be applied locally. Mitchell (1969) points out that there is much numerical evidence indicating that a FDA for such cases will be stable if the von Neumann conditions is satisfied in the field, as though the coefficients were constant.

2.1.4 Du Fort-Frankel method An attempt to overcome the stability limit (2.10) of the simple explicit method is the Du Fort-Frankel method which uses the FDA

$$\frac{T_i^{n+1} - T_i^{n-1}}{2h_t} = \frac{T_{i-1}^n - (T_i^{n+1} + T_i^{n-1}) + T_{i+1}^n}{h_x^2} \tag{2.11}$$

It can readily be shown that (2.11) is equivalent to

$$\frac{\partial T}{\partial t} = \frac{\partial^2 T}{\partial x^2} - \left(\frac{h_t}{h_x}\right)^2 \frac{\partial^2 T}{\partial t^2} - \frac{h_t^2}{6}\frac{\partial^3 T}{\partial t^3} + \frac{h_x^2}{12}\frac{\partial^4 T}{\partial x^4} + \cdots$$

Thus, the method is consistent with (2.1)–with truncation errors going to zero–only if h_t goes to zero faster than h_x. If h_t, h_x go to zero in such a way that $h_t/h_x = c$ remains constant, the (2.11) is consistent not with (2.1), but with the hyperbolic equation

$$\frac{\partial T}{\partial t} + c^2 \frac{\partial^2 T}{\partial t^2} = \frac{\partial^2 T}{\partial x^2}$$

If (2.11) is to be used to solve (2.1), the condition $h_t \ll h_x$ must be satisfied.

A stability analysis of (2.11) leads to

$$\xi = \frac{2r \cos\beta \pm (1 - 4r^2 \sin^2\beta)^{1/2}}{1 + 2r}$$

If $r \leqslant \frac{1}{2}$, $|\xi| \leqslant 1$ for all β. If $r > \frac{1}{2}$,

$$\xi = \frac{2r \cos\beta \pm \mathrm{i}(4r^2 \sin^2\beta - 1)^{1/2}}{1 + 2r}$$

Hence

$$|\xi|^2 = \frac{4r^2 \cos^2\beta + 4r^2 \sin^2\beta - 1}{4r^2 + 4r + 1} = \frac{4r^2 - 1}{4r^2 + 4r + 1} \leqslant 1$$

Thus the Du Fort-Frankel method is stable for all positive r. Note that it is a three time level formula and requires an alternative starting procedure.

2.1.5 Implicit methods Each of the methods described so far is explicit. The FDA contains only one known value of T at the new time level–which can be calculated explicitly from known quantities. The methods are, therefore, easy to implement and require only a modest computational effort per time step. However, each of the two usable methods is restricted to small h_t: one because of a stability condition, and the other because of a consistency condition. This restriction can be removed, permitting the use of larger time steps, but only at a cost of greater computational effort per step.

Instead of approximating $\partial^2 T/\partial x^2$ at time n, we use a weighted average of approximations at the times n and $n + 1$:

$$\frac{T_i^{n+1} - T_i^n}{h_t} = \theta \left(\frac{T_{i-1}^{n+1} - 2T_i^{n+1} + T_{i+1}^{n+1}}{h_x^2} \right) + (1 - \theta) \left(\frac{T_{i-1}^n - 2T_i^n + T_{i+1}^n}{h_x^2} \right)$$

There are now three unknown values of T in the equation for each mesh point i, which can be rewritten

$$-\theta r T_{i-1}^{n+1} + (1 + 2\theta r) T_i^{n+1} - \theta r T_{i+1}^{n+1} = r(1 - \theta) T_{i-1}^n + [1 - 2r(1 - \theta)] T_i^n + r(1 - \theta) T_{i+1}^n \qquad (2.12)$$

If the weighting factor $\theta = 1$, the method is called fully implicit. If $\theta = \frac{1}{2}$, the method known as the Crank-Nicolson method is obtained. If $\theta = 0$, the explicit method (2.3) is recovered.

The stability of (2.12) may be investigated in the usual way, and it will be found that stability is conditional, requiring

$$0 < r \leqslant \frac{1}{2(1 - 2\theta)} \quad \text{for} \quad \theta \leqslant \frac{1}{2}$$

$$0 < r \quad \text{for} \quad \theta > \frac{1}{2}\,*$$

A study of the consistency of (2.12) shows that it is equivalent to

$$\frac{\partial T}{\partial t} = \frac{\partial^2 T}{\partial x^2} - \left(\frac{h_t}{2} \frac{\partial^2 T}{\partial t^2} + \frac{h_t^2}{6} \frac{\partial^3 T}{\partial t^3} + \cdots \right) + \left(\theta h_t \frac{\partial^3 T}{\partial t \partial x^2} + \frac{h_x}{12} \frac{\partial^4 t}{\partial x^4} + \frac{h_t h_x^2}{12} \frac{\partial^5 T}{\partial t \partial x^4} + \cdots \right)$$

$$= \frac{\partial^2 T}{\partial x^2} + O(h_t + h_x^2) \qquad (2.13)$$

and is thus consistent with (2.1).

For general values of θ, we see that (2.12) is only first-order accurate in time. But if the Crank-Nicolson value $\theta = \frac{1}{2}$ is selected, second-order accuracy is

*Fox (1962) has noted a trivial restriction for the Crank-Nicolson method requiring $r \leqslant 1/\sin(\pi h_x)$.

achieved. For since, from the differential equation (2.1) itself, $\partial/\partial t \equiv \partial^2/\partial x^2$, it follows that $\partial^2 T/\partial t^2 = \partial^3 T/\partial t \partial x^2$ and with $\theta = \frac{1}{2}$, the first terms

$$\frac{h_t}{2}\frac{\partial^2 T}{\partial t^2} \quad \text{and} \quad h_t \frac{\partial^3 T}{\partial t \partial x^2}$$

in each pair of parentheses in (2.13) will cancel each other, leaving a truncation error of $(h_t^2 + h_x^2)$.

The accuracy of (2.12) can be further improved by forcing all the terms of order $(h_t + h_x^2)$ to vanish, i.e., by forcing

$$\frac{h_t}{2}\frac{\partial^2 T}{\partial t^2} = h_t \frac{\partial^3 T}{\partial t \partial x^2} + \frac{h_x^2}{12}\frac{\partial^4 T}{\partial x^4} \tag{2.14}$$

Since $\partial/\partial t \equiv \partial^2/\partial x^2$, (2.14) can be written

$$\frac{h_t}{2}\frac{\partial^4 T}{\partial x^4} = \left(\theta h_t + \frac{h_x^2}{12}\right)\frac{\partial^4 T}{\partial x^4}$$

Hence, if θ and r are chosen to satisfy

$$\theta = \frac{1}{2} - \frac{1}{12r} \tag{2.15}$$

the method becomes of order $(h_t^2 + h_t h_x^2)$. It can be seen that (2.15) is compatible with the stability requirement (2.13), which can be written

$$\theta \geqslant \frac{1}{2} - \frac{1}{4r}$$

The implicit method is only advantageous if a value of r in excess of $\frac{1}{2}$ is employed. Equation (2.15) suggests that suitable values might be $r = 1$ with $\theta = \frac{5}{12}$ and $r = 2$ with $\theta = \frac{11}{24}$.

2.1.6 The Thomas algorithm for tri-diagonal systems Since (2.12) contains three unknowns, T_i^{n+1} cannot be computed independently of the values of T^{n+1} at neighboring mesh points. If (2.12) is written out for each of the M (say) mesh points, a tri-diagonal linear system of equations is obtained, of the general form

$$\begin{aligned} b_1 u_1 + c_1 u_2 &= d_1 \\ a_2 u_1 + b_2 u_2 + c_2 u_3 &= d_2 \\ \cdots\cdots\cdots\cdots\cdots &\cdots \\ a_i u_{i-1} + b_i u_i + c_i u_{i+1} &= d_i \\ \cdots\cdots\cdots\cdots\cdots &\cdots \\ a_M u_{M-1} + b_M u_M &= d_M \end{aligned} \tag{2.16}$$

where, in the present context, $u_1 = T_1^{n+1}$; $u_2 = T_2^{n+1}$ etc.

The system (2.16) is solved by an elimination procedure usually attributed to Thomas (1949). The method is basically a modified Gaussian elimination technique and can be found in basic texts (Smith, 1965).

The solution of a system of M equations by this method requires $5M$-4 operations (multiplications and divisions) and is clearly superior to the normal elimination methods which do not take advantage of the tri-diagonal nature of the system and require about $M^3/3$ operations.

2.1.7 Boundary conditions So far, attention has been limited to conduction problems, and to the situation in which the unknown function is specified on the boundaries: in the heat transfer context, the temperature is given there.

We now turn our attention to treatment of the mixed boundary condition

$$T + a\frac{\partial T}{\partial x} = b \tag{2.17}$$

at $x = 0$, say, where in general a and b may be functions of time. We have seen in Sec. 1 that special forms of (2.17) apply to heat flux and convective boundary conditions. The equation will be applied using Figure 3.

Since, in the notation of Figure 3, T_0 is not specified, it must be computed from the boundary condition, either alone or in conjunction with the FDA. The simplest procedure is to replace the derivative in (2.17) with a forward difference approximation $(T_1 - T_0)/h_x$, leading to

$$T_0 = \frac{aT_1/h_x - b}{a/h_x - 1} \tag{2.18}$$

At each time step, T_1^{n+1} is determined from the FDA in the usual way; T_0^{n+1} may then be computed from (2.18). However, the forward difference approximation to $\partial T/\partial x$ has a truncation error of $O(h_x)$, whereas the FDA is $O(h_x^2)$. The greater error at the wall can affect the quality of the whole solution.

A far better procedure is to use a central difference approximation to $\partial T/\partial x$ at the wall, which requires the introduction of the fictitious mesh point labeled "−1" in Figure 3, a distance h_x *outside* the wall. If this point existed, then $\partial T/\partial x$ could be found from

Figure 3 Fictitious mesh point outside a boundary.

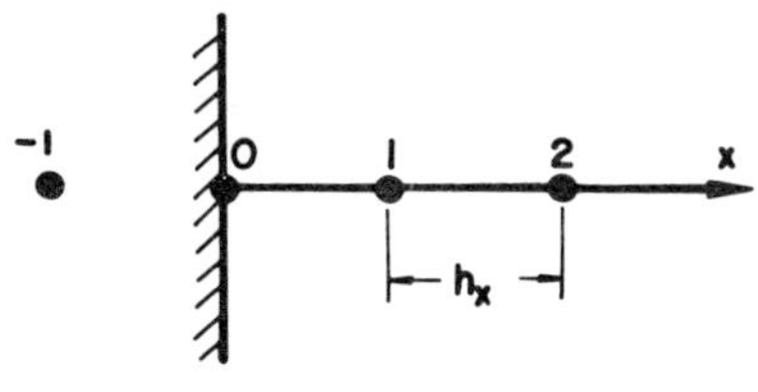

$$\left.\frac{\partial T}{\partial x}\right|_0 = \frac{T_1 - T_{-1}}{2h_x} \tag{2.19}$$

so that (3.27) becomes

$$T_{-1} = T_1 + \frac{2h_x T_0}{a} - \frac{2h_x b}{a} \tag{2.20}$$

The FDA is now applied at $i = 0$; this will call for a value of T_{-1}, for which (2.20) is used. For example, (2.12) at $i = 0$ is

$$-\theta r T_{-1}^{n+1} + (1 + 2\theta r)T_0^{n+1} - \theta r T_1^{n+1} = \text{terms in } T^n$$

which, when (2.20) is introduced, becomes

$$\left(1 + 2\theta r - \frac{\theta r h_x}{a}\right) T_0^{n+1} - 2\theta r T_1^{n+1} = -\frac{\theta r h_x b}{a} + \text{terms in } T^n \tag{2.21}$$

Since (2.19) is $O(h_x{}^2)$, the accuracy of the solution is not impaired. Moreover, (2.21) is compatible with the tri-diagonal system (2.12), so the solution procedure remains the same.

2.1.8 Example 1: Heat transfer in countercurrent plate heat exchangers Heat transfer in counter-flow parallel plate heat exchangers was investigated by Gutfinger et al. (1970). The governing equations in nondimensional form are

$$\text{Gz}_h v_{x_h} \frac{\partial T_h}{\partial x} = \frac{\partial^2 T_h}{\partial y^2} \tag{2.22a}$$

for the hot stream, and

$$\text{Gz}_c v_{x_c} \frac{\partial T_c}{\partial x} = \frac{\partial^2 T_c}{\partial y^2} \tag{2.22b}$$

for the cold stream. Here Gz represents the Graetz number

$$\text{Gz} \equiv \frac{a^2 v_{x_{\text{avg}}}}{\alpha L}$$

where a is the width of each channel, $v_{x_{\text{avg}}}$ is the average velocity, α is the thermal diffusivity, and L the length of the exchanger. The model is depicted in Figure 4. The boundary conditions are

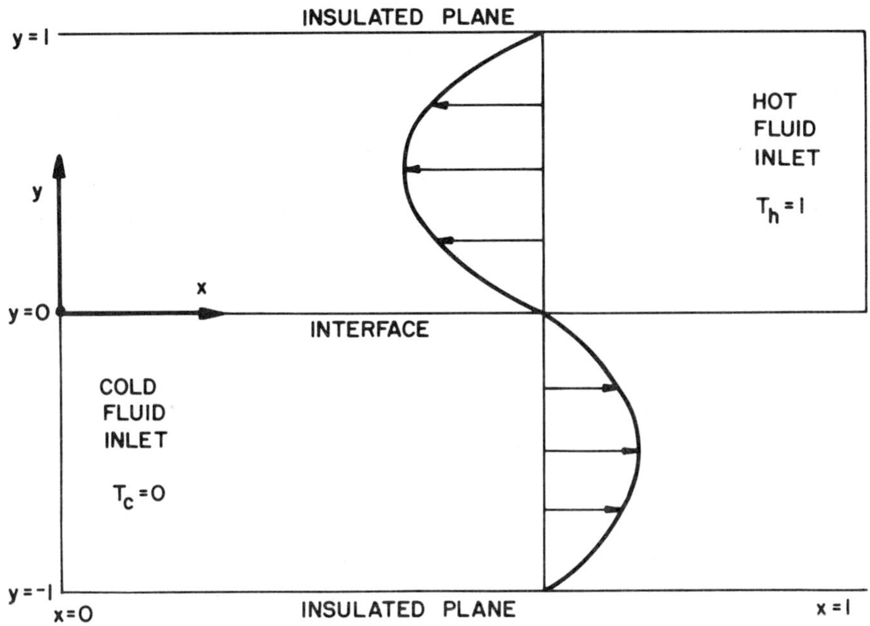

Figure 4 Schematic of heat transfer in countercurrent plate heat exchangers.

$$T_h(1, y) = 1 \qquad \frac{\partial T_h}{\partial y}(x, 1) = 0$$

for the hot stream, and

$$T_c(0, y) = 0 \qquad \frac{\partial T_c}{\partial y}(x - 1) = 0 \tag{2.23}$$

for the cold stream. At the interface between the two streams, there exist two conditions–continuity of temperature and continuity of heat flux–giving

$$T_h(x, 0) = T_c(x, 0) \tag{2.24a}$$

$$K \frac{\partial T_h}{\partial y}(x, 0) = \frac{\partial T_c}{\partial y}(x, 0) \tag{2.24b}$$

where K is the conductivity ratio k_h/k_c. Inherent in these two equations is the assumption that the thermal resistance of the interface is negligible.

The velocity profiles used were those for laminar flow. For the Newtonian case this gives

$$v_x = 6(|y| - y^2) \tag{2.25}$$

for each stream. The negative velocity profile for the hot stream implies that the solution here must be marched out in the negative axial direction from $x = 1$ to $x = 0$. Thus Eqs. (2.22) can be regarded as an extension of the simple heat equation (2.1) where the governing equation is now initial-valued in x instead of t, and contains a variable coefficient.

The system and boundary equations were solved numerically by assuming the temperature profile at the interface and successively correcting this profile by the following iterative sequence:

1. The temperature distribution in each stream was solved separately using the given interfacial profile as a boundary condition, and the FDA of Crank-Nicolson.
2. The interfacial profile was corrected so as to obey (2.24b) by using internal points computed by step 1.

The correction of step 2 was implemented by means of a 3-point approximation for the first derivative that is correct for one-sided differences to $O(h_y^2)$. For constant x at the interface this yields from Eq. (1.43b)

$$\frac{\partial T_c}{\partial y} = \frac{3T_0 - 4T_{-1} + T_{-2}}{2h_y} \tag{2.26a}$$

and from Eq. (1.43a)

$$K\frac{\partial T_h}{\partial y} = K\frac{-3T_0 + 4T_1 - T_2}{2h_y} \tag{2.26b}$$

where the adjacent points in the cold and hot streams are denoted by negative and positive indices, respectively. Equating the above two relations results in a corrected value of the temperature at each grid point on the interface:

$$T_0 = \frac{4(T_{-1} + KT_1) - (T_{-2} + KT_2)}{3(1 + K)} \tag{2.27}$$

where the values for the internal points appearing on the right-hand side are taken from the temperature distribution produced by step 1.

The sequence of iterations described above always converged. Additionally, convergence was accelerated by applying Aitken's ∇^2-process (Hildebrand, 1974) after every third iteration. For a relative iterative accuracy of 0.1% for all points on the interface, it was usually found that convergence was attained within 15 iterations, if the initial guess of the interfacial profile was linear.

2.1.9 Asymptotic expansions As noted earlier, in parabolic equations one often encounters singularities between the initial and boundary conditions. If we are willing to ignore the singularity and substitute an average value, then the solution will be inaccurate near the singularity. A better approach is to employ an asymptotic expansion near the singularity using simplifying assumptions. This usually results in a simple analytic expression for the solution which is valid near the singularity.

To illustrate this concept we shall take as our first example our familiar uniform rod with an initial temperature of zero and ends suddenly heated to $T = 1$. Due to symmetry, the area of interest is the region $0 < x \leqslant \frac{1}{2}$. For short times, only the region near $x = 0$ "senses" the temperature disturbance and from the perspective of this region, the inner core can be regarded as semi-infinite. Thus, for short times, $T(x, 0) = 0$, $T(0, t) = 1$, and $T(\infty, t) = 0$. With these conditions the heat equation (3.1) can be easily solved by introducing a similarity variable:

$$\eta \equiv \frac{x}{t^{1/2}} \tag{2.28}$$

which yields, in terms of the complementary error function,

$$T = \operatorname{erfc} \frac{\eta}{2} = \operatorname{erfc}\left(\frac{x}{2t^{1/2}}\right) \tag{2.29}$$

This solution will be valid for the original problem until the center, $x = \frac{1}{2}$ starts to experience a significant rise in temperature. At this point in time, (approximately $t = 0.01$) we revert to our finite difference approximation after having obtained an accurate solution for short times.

Usually, the application of asymptotic expansions results in a substantial simplification of the original problem. Consider heat transfer to a fluid in laminar flow within a circular duct—the classical Graetz problem. The governing equation in dimensionless form is

$$\frac{\partial^2 T}{\partial \omega^2} + \frac{1}{\omega}\frac{\partial T}{\partial \omega} = (1 - \omega)\frac{\partial T}{\partial z} \tag{2.30}$$

with $T(\omega, 0) = 0$ and $T(1, z) = 1$, where the wall of the tube corresponds to $\omega = 1$. Note that the boundary conditions are identical to those in the previous example.

In order to develop an asymptotic expansion for small z, we would tacitly assume that Eq. (2.30) is valid for this region. It is well known that the effect of axial conduction cannot be neglected for small z thus changing Eq. (2.30) into an elliptic PDE. We shall circumvent this difficulty by limiting ourselves to the mathematical parabolic PDE, for the purpose of demonstrating the asymptotic expansions.

For small z the region of interest is that adjacent to the wall, $\omega = 1$. Since the second term on the left-hand side of (2.30) represents curvature, for the foregoing region of interest this term is negligible in comparison to the first. Also, for the right-hand side,

$$(1 - \omega^2) = (1 + \omega)(1 - \omega) \cong 2(1 - \omega) \tag{2.31}$$

This is tantamount to assuming that the parabolic velocity profile becomes linear adjacent to the wall. (This assumption is basic to the Leveque solution.) The simplified governing equation is

$$2(1 - \omega)\frac{\partial T}{\partial z} = \frac{\partial^2 T}{\partial \omega^2} \tag{2.32}$$

with boundary conditions $T(\omega, 0) = 0$, $T(1, z) = 1$, and $T(-\infty, z) = 0$. The similarity transformation to be used here is

$$\eta \equiv \frac{1 - \omega}{z^{1/3}} \tag{2.33}$$

which yields

$$\frac{d^2 T}{d\eta^2} + \frac{2}{3}\eta^2 \frac{dT}{d\eta} = 0 \tag{2.34}$$

with boundary conditions, $T(0) = 1$ and $T(\infty) = 0$. The solution is

$$T = 1 - \frac{\int_0^{2\eta^3/9} \lambda^{-2/3} e^{-\lambda}\, d\lambda}{\Gamma\left(\frac{4}{3}\right)} \tag{2.35}$$

which is valid for a value of z so small, that only a small region adjacent to $\omega = 1$ is affected. From here the solution is marched out using boundary layer coordinates suggested by the similarity transformation. We transform to η-z coordinates where η is as defined above. When recast in these coordinates, Eq. (3.40) becomes

$$\frac{\partial^2 T}{\partial \eta^2} + \left(\frac{2 - z^{1/3}\eta}{3}\eta^2 - \frac{z^{1/3}}{1 - z^{1/3}\eta}\right)\frac{\partial T}{\partial \eta} = z(1 - z^{1/3}\eta)\eta \frac{\partial T}{\partial z} \tag{2.36}$$

Here η is a boundary layer coordinate since it expands with increasing z.

The technique here is to assume a value for $z_{b.l.}$, which defines the region in which the boundary layer coordinates can be employed. The outer boundary in η is then $z_{b.l.}^{-1/3}$. The solution is initiated for a very small value of z via Eq. (2.35). From this point, Eq. (3.46) is applied with boundary conditions $T(\eta, 0) = 0$, $T(0, z) = 1$, and $\partial T/\partial\eta\ (z_{b.l.}^{-1/3}, z) = 0$. The solution is marched out until $z = z_{b.l.}$. If the value of $T(z_{b.l.}^{-1/3},\ z_{b.l.})$–corresponding to $T(0, z_{b.l.})$ in ω-z coordinates–shows a barely perceptible rise from zero, then the assumed value of $z_{b.l.}$ is acceptable and the marching is resumed using Eq. (2.30) without interpolation. Otherwise, the value of $z_{b.l.}$ is adjusted and the process repeated.

While the method of asymptotic expansions yields a smooth transition from the region of the singularity, most problems can be solved without this method if one is willing to tolerate the early inaccuracies caused by the "shock." However, there are problems where the method *must* be applied. In a theoretical study of unsteady heat transfer to a draining film, Isenberg and Gutfinger (1973) obtained the following field equation:

$$\frac{1}{2}(x^2 - 2x + 4)\frac{\partial T}{\partial t} = \frac{\partial^2 T}{\partial x^2} - \frac{x}{4t}(x^2 - 3x + 2)\frac{\partial T}{\partial x} \tag{2.37}$$

with boundary conditions $T(x, 0) = 1$, $T(0, t) = 1$, and $T(1, t) = 0$. The appearance of t in the denominator of the last term induced large oscillations in the solution for small t, when conventional numerical techniques were employed. While the oscillations eventually died out, they still persisted within a portion of the region of interest, creating significant errors. The proper remedy was asymptotic expansions.

Noting that for the initial period (very small values of t) the region of temperature variation lies close to $x = 1$, this value of x was substituted into Eq. (3.27), obtaining

$$\frac{3}{2}\frac{\partial T}{\partial t} = \frac{\partial^2 T}{\partial x^2} - \frac{1-x}{4t}\frac{\partial T}{\partial x} \tag{2.38}$$

The similarity variable used was

$$\eta \equiv \frac{1-x}{t^{1/2}}$$

which resulted in the ordinary differential equation

$$\frac{d^2 T}{d\eta^2} + \eta\frac{dT}{d\eta} = 0 \tag{2.39}$$

with boundary conditions $T(0) = 0$ and $T(\infty) = 1$. The solution of Eq. (3.49) is

$$T = \operatorname{erf}\left(\frac{\eta}{2^{1/2}}\right) = \operatorname{erf}\left[\frac{1-x}{(2t)^{1/2}}\right] \tag{2.40}$$

The methodology here was the same as that for the prior example. Transforming to the η-t boundary layer coordinates gave

$$\frac{\partial^2 T}{\partial \eta^2} + \eta \frac{\partial T}{\partial \eta} = \frac{t}{2}(3 + \eta^2 t)\frac{\partial T}{\partial t} \tag{2.41}$$

which was employed for t less the assumed value of $t_{b.l.}$.

2.2 The transport equation in one space dimension

The methods and concepts discussed so far can be applied to more general situations. We now consider some problems associated with the diffusion and convection of a scalar field T through a fluid traveling with a constant velocity v_x in a direction parallel to the x axis, according to the equation

$$\frac{\partial T}{\partial t} + v_x \frac{\partial T}{\partial x} = \alpha \frac{\partial^2 T}{\partial x^2} \tag{2.42}$$

T can be regarded as temperature, vorticity, or concentration, with α the appropriate diffusion coefficient.

We first note that, with $v_x = 0$, (2.42) reverts to (2.1) and we have a pure diffusion problem. The foregoing analyses carry over directly, with appropriate allowance being made for the diffusion coefficient α. For example, the stability condition for the explicit method (2.10) becomes

$$r \leqslant \frac{1}{2\alpha}$$

2.2.1 Pure convection Consider now the *hyperbolic* convection problem

$$\frac{\partial T}{\partial t} + v_x \frac{\partial T}{\partial x} = 0 \tag{2.43}$$

The general solution of which is $T = f(x - v_x t)$, i.e., the initial distribution of T is translated along the x axis at a speed v_x without change in shape. Using a forward difference in time and a central difference in space, the FDA to (2.43) is

$$\frac{T_i^{n+1} - T_i^n}{h_t} + \frac{v_x}{2h_x}(T_{i+1}^n - T_{i-1}^n) = 0 \tag{2.44}$$

When stability analysis is applied to (2.44), it is found that, say,

$$\xi = 1 - \frac{v_x h_t}{h_x} \mathbf{i} \sin \beta = 1 - \lambda \mathbf{i} \sin \beta$$

where $\lambda \equiv v_x h_t / h_x$. Thus,

$$|\xi|^2 = 1 + \lambda^2 \sin^2 \beta \geqslant 1$$

and (2.44) is thus unconditionally unstable.

2.2.2 Upwind differences A method used to overcome this instability is due to Lelevier (Richtmyer and Morton, 1967) and is known as upwind differencing or the donor-cell technique. A one-sided difference is used for the convection term, the direction of the difference being "upwind," i.e., forward if $v_x < 0$ and backward if $v_x > 0$. (Although the present analysis is restricted to v_x = constant, the method is used in the general case as well.) The FDA's are

$$\frac{T_i^{n+1} - T_i^n}{h_t} + \frac{v_x}{h_x}(T_{i+1}^n - T_i^n) = 0 \qquad \text{if} \qquad v_x < 0$$

$$\frac{T_i^{n+1} - T_i^n}{h_t} + \frac{v_x}{h_x}(T_i^n - T_{i-1}^n) = 0 \qquad \text{if} \qquad v_x > 0$$

Consider the case $v_x > 0$. The amplification factor is found to be

$$\xi = (1 - \lambda + \lambda \cos \beta) - \mathbf{i}\lambda \sin \beta$$

Therefore,

$$|\xi|^2 = 1 - 2\lambda(1 - \lambda)(1 - \cos \beta) \tag{2.45}$$

When $\lambda < 1$, the second term on the right-hand side of (2.45) is always positive, and ξ has a maximum value of 1 when $\cos \beta = 1$. Thus the method is stable. However, it should be noted that (2.9) can also be regarded as one term in a Fourier decomposition of the solution of the FDA itself. Since $\xi < 1$ except when $\cos \beta = 1$, most terms in the solution will be attenuated as n increases, which should not occur in this situation. When $\lambda = 1$, $\xi = 1$ for all β, the method is just stable, and the wave form shape is retained for all time. When $\lambda > 1$, ξ^2 has a maximum of $(2 - 1)^2 > 1$ when $\cos \beta = -1$, and the procedure is unstable. Thus the stability criterion is

$$\lambda = \frac{v_x h_t}{h_x} \leqslant 1 \tag{2.46}$$

This equation is known as the Courant-Friedrichs-Lewy (CFL) condition, and applies generally to explicit FDA's to hyperbolic equations. In effect, it requires that the fluid be prevented from travelling more than one mesh distance in one time step.

2.2.3 Amplitude and phase errors The upwind difference approximation described in the previous subsection—and, indeed, many other approximations—lead to an amplification factor ξ which is generally different from unity and which is complex. The fact that $|\xi| < 1$ means, as noted above, that the amplitude of the computed solution will decrease with increase of time. It is thus desirable that $|\xi|$ should be as close to unity as possible.

A different type of error is introduced if ξ is complex. The analytical solution $T = f(x - v_x t)$ can alternatively be written as the sum of terms like

$$T_m e^{\mathrm{i}k_m(x-v_x t)} = T_m e^{\mathrm{i}(\phi - k_m v_x t)} \tag{2.47}$$

where T_m is the amplification factor associated with the component of wave number k_m. The quantity $(-k_m v_x t)$ is thus the phase shift at a point x after a time t in relation to the phase ϕ at $t = 0$. After one time step, therefore, the phase shift should be $(-k_m v_x h_t)$.

On the other hand, the numerical solution propagates according to

$$T[ih_x, (n+1)h_t] = T(ih_x, nh_t) \tag{2.48}$$

and writing $T(ih_x, nh_t)$ as

$$\sum_m B_m e^{\mathrm{i}k_m x} = \sum_m B_m e^{\mathrm{i}k_m ih_x} = \sum_m B_m e^{\mathrm{i}\beta_m i}$$

we have, for the term of wave number $k_m = \beta_m / h_x$,

$$T[ih_x, (n+1)h_t] = B_m \xi e^{\mathrm{i}\beta_m i} = AB_m e^{\mathrm{i}(\beta_m i + \theta)} \tag{2.49}$$

where we have written, in general, $\xi = Ae^{i\theta}$. Thus the numerical phase shift is θ per time step, which is to be compared with the actual phase shift

$$-k_m v_x h_t = -k_m \lambda h_x = -\lambda \beta_m \tag{2.50}$$

In the case of upwind differencing, for example,

$$\xi = (1 - \lambda + \lambda \cos \beta) - \mathrm{i}\lambda \sin \beta$$

so that

$$\theta = \tan^{-1}\left(\frac{-\lambda \sin \beta_m}{1 - \lambda + \lambda \cos \beta_m}\right) \tag{2.51}$$

For $\lambda = 1$, $\theta = -\beta_m$ and the numerical and analytical phase shifts are equal. For any other λ (< 1), θ is less than $(-\lambda\beta_m)$, by an amount which depends upon β_m. In other words, the various harmonics in the initial distribution of T will, in the numerical solution, be convected at different speeds and the distribution will become distorted with increase of time. This distortion is known as *phase error;* to minimize it, the imaginary part of ξ should be as small as possible.

Several finite difference schemes have been developed with the objective of eliminating or reducing phase error. Roberts and Weiss (1966) proposed a method which has a truncation error $O(h_t^2 + h_x^4)$. It is based on the use of a "staggered" mesh; i.e., the solution progresses in half time steps $h_t/2$ and involves the use of mesh points midway between the regular mesh points. The result is a method which, although costly in terms of computational effort, achieves a significantly lower phase error than the simpler conventional methods. The authors recommend its use both for inviscid problems and for viscous problems, provided that diffusion is small in comparison with convection.

Other higher order schemes have been introduced by Arakawa (1966), Crowley (1967), and Fromm (1968, 1971).

2.2.4 The central differences approximation to the transport equation We return now to a consideration of the full equation (2.42). Using a central difference formula for the convection term (which is unconditionally unstable in the absence of diffusion),the FDA

$$\frac{T_i^{n+1} - T_i^n}{h_t} + \frac{v_x}{2h_x}(T_{i+1}^n - T_{i-1}^n) = \frac{\alpha}{h_x^2}(T_{i-1}^n - 2T_i^n + T_{i+1}^n) \tag{2.52}$$

is obtained. The amplification factor is found to be

$$\xi = 1 - 2r(1 - \cos\beta) - i\lambda \sin\beta \tag{2.53}$$

where $r \equiv h_t\,\alpha/h_x^2$ and $\lambda \equiv v_x h_t/h_x$. The diffusion term in (2.52) is responsible for the second term on the right of (2.53); by reducing the real part of ξ below unity it exercises a stabilizing effect.

The magnitude of ξ is given by

$$|\xi|^2 = 1 + 4r^2 + 4r^2\cos^2\beta - 4r + 4r\cos\beta - 8r^2\cos\beta + \lambda^2\sin^2\beta \tag{2.54}$$

Now suppose that $\lambda \leqslant 2r$. Then

$$|\xi|^2 \leqslant 1 + 4r^2 + 4r^2 \cos^2\beta - 4r + 4r\cos\beta - 8r^2\cos\beta$$
$$+ 4r^2 \sin^2\beta \leqslant 1 - 4r(1 - 2r)(1 - \cos\beta)$$

Since $\cos\beta \leqslant 1$, $|\xi|^2$ will be less than or equal to unity provided $r \leqslant \frac{1}{2}$. Thus stability is assured if the two conditions:

$$\lambda \leqslant 2r \leqslant 1 \tag{2.55}$$

are satisfied. (Numerical experiments show that these conditions are also necessary but note that we have not proved this.)

From the definitions of λ and r it can be seen that the requirement $\lambda \leqslant 2r$ is equivalent to

$$\frac{v_x h_x}{\alpha} \leqslant 2$$

where $v_x h_x/\alpha$ can be interpreted as a Peclet number based on the cell size.

2.2.5 The upwind difference approximation to the transport equation When an upwind approximation is used for the convection term, the FDA is

$$\frac{T_i^{n+1} - T_i^n}{h_t} + \frac{v_x}{h_x}(T_i^n - T_{i-1}^n) = \frac{\alpha}{h_x^2}(T_{i-1}^n - 2T_i^n + T_{i+1}^n) \tag{2.56}$$

hence,

$$\xi = 1 - (\lambda + 2r)(1 - \cos\beta) - \mathrm{i}\lambda\sin\beta$$

and, after some manipulation,

$$|\xi|^2 = 1 - 2\lambda(1 - \lambda)(1 - \cos\beta) - 4r(1 - 2r)(1 - \cos\beta)$$
$$+ 4\lambda r(1 - \cos\beta)^2 - 4r^2 \sin\beta$$

Stability requires $|\xi|^2 \leqslant 1$, i.e.,

$$4\lambda r(1 - \cos\beta) \leqslant 2\lambda(1 - \lambda) + 4r(1 - 2r) + 4r^2(1 + \cos\beta) \tag{2.57}$$

The maximum value of the left hand side of (3.62) is $8\lambda r$, occuring when $\cos\beta = -1$ and hence the stability criterion is

$$8\lambda r \leqslant 2\lambda(1 - \lambda) + 4r(1 - 2r)$$

or

$$\lambda + 2r \leqslant 1 \tag{2.58}$$

For values of cos β other than -1, the left hand side of (2.57) is reduced by a factor between zero and unity, and the right hand side is increased by the term $4r^2 \sin^2 \beta(1 - \cos \beta)$ which is always positive. Thus, if (2.58) is satisfied, (2.57) will be satisfied for all β and hence stability is assured.

Comparison of (2.58) and (2.46) shows that the presence of the diffusion term in the upwind FDA has a destabilizing effect, in contrast to that found in the previous section for the central difference approximation.

2.3 A simplified notation

In the following sections, consideration will be given to problems involving two or three space variables. In order to render the notation more compact, the following convention will be adopted: the superscript (for time) and the subscripts (for position) will only be printed if the do *not* have the default values of *n, i, j,* and *k,* respectively. Thus,

$$T \equiv T^n_{i,j,k}$$

$$T^{n+1}_{i-1} \equiv T^{n+1}_{i-1,j,k}$$

and so on.

2.4 The transport equation in multi-dimensional space variables

2.4.1 Explicit method The analogue of (2.56) in two space variables is

$$\frac{\partial T}{\partial t} + v_x \frac{\partial T}{\partial x} + v_y \frac{\partial T}{\partial y} = \alpha\left(\frac{\partial^2 T}{\partial x^2} + \frac{\partial^2 T}{\partial y^2}\right) \tag{2.59}$$

With $v_x = v_y = 0$, (2.59) describes a pure diffusion problem, for which the explicit FDA is

$$T^{n+1} = T + r_x(T_{i-1} - 2T + T_{i+1}) + r_y(T_{j-1} - 2T + T_{j+1}) \tag{2.60}$$

where $r_x = h_t/h_x{}^2$, $r_y = h_t/h_y{}^2$. It can readily be verified that (2.60) is stable if

$$r_x + r_y \leqslant \frac{1}{2} \tag{2.61}$$

If, as is often the case, $h_x = h_y$, (2.61) becomes

$$r \leqslant \frac{1}{4}$$

a condition which is twice as restrictive on h_t as (2.10). In three space dimensions, it can similarly be shown that stability is ensured if

$$r_x + r_y + r_z \leqslant \frac{1}{2}$$

or

$$r \leqslant \frac{1}{6}$$

in the case of equal mesh sizes.

In the presence of the convection terms, and using central differences, the explicit FDA to (2.59) is

$$\begin{aligned} T^{n+1} &= T - \frac{\lambda_x}{2}(T_{i+1} - T_{i-1}) + \frac{\lambda_y}{2}(T_{j+1} - T_{j-1}) \\ &\quad + r_x(T_{i-1} - 2T + T_{i+1}) + r_y(T_{j-1} - 2T + T_{j+1}) \end{aligned} \tag{2.62}$$

Following an argument similar to that used in Sec. 2.2.4, and if we consider, for simplicity, the case in which $\lambda_x = \lambda_y = \lambda, r_x = r_y = r$, then it can be shown that (2.62) is stable if

$$\lambda \leqslant 2r \leqslant \frac{1}{2} \tag{2.63}$$

Using upwind differences for the convection term, the FDA for $v_x, v_y > 0$, is

$$\begin{aligned} T^{n+1} &= T - \lambda_x(T - T_{i-1}) - \lambda_y(T - T_{j-1}) \\ &\quad + r_x(T_{i-1} - 2T + T_{i+1}) + r_y(T_{j-1} - 2T + T_{j+1}) \end{aligned} \tag{2.64}$$

With $\lambda_x = \lambda_y = \lambda, r_x = r_y = r$, it can be shown that (2.64) is stable if

$$\lambda + 2r \leqslant \frac{1}{2} \tag{2.65}$$

Both (2.63) and (2.65) are twice as restrictive as the corresponding one-dimensional conditions. The extension to three space dimensions is obvious.

The Du Fort-Frankel method (Sec. 2.1.4) can be extended to two and three dimensions in an obvious manner. It is again found to be unconditionally stable, but the consistency condition $h_t \ll h_x$ remains, and the method must be used with caution.

2.4.2 Example 2: Time-dependent temperature distribution in flow past a sphere The transport equation for potential flow past axisymmetric bodies was numerically solved for the case of a step change in temperature (or concentration) of a sphere by Bentwich, Szwarcbaum, and Sideman (1965). An interesting feature of this work is the application of Boussinesq's transformation to facilitate the computation.

The governing equation in dimensionless form is expressed in cylindrical coordinates:

$$\frac{\partial T}{\partial t} + \mathrm{Pe}\left(v_\omega \frac{\partial T}{\partial \omega} + v_z \frac{\partial T}{\partial z}\right) = \frac{1}{\omega}\frac{\partial}{\partial \omega}\left(\omega \frac{\partial T}{\partial \omega}\right) + \frac{\partial^2 T}{\partial z^2} \tag{2.66}$$

where the Peclet number, $\mathrm{Pe} = W_\infty R/\alpha$, W_∞ is the rectilinear dimensional velocity at infinity, and R is the dimensional radius of the sphere. The boundary conditions are

$T = 1$ on the surface of the sphere ($r = 1$) (2.67a)

and $T = 0$ initially in the fluid (2.67b)

The velocity field given by v_ω and v_z is that for potential flow past a sphere. In terms of the dimensionless velocity potential and stream function, this is

$$\phi = \left(r + \frac{1}{2r^3}\right)\cos\theta \tag{2.68a}$$

$$\psi = \frac{1}{2}\left(r^2 - \frac{1}{r}\right)\sin^2\theta \tag{2.68b}$$

where r and θ are the spherical coordinates given by

$$r\cos\theta = z \qquad \text{and} \qquad r\sin\theta = \omega$$

Applying Boussinesq's transformation we transfer to ϕ-ψ space coordinates resulting in the following energy equation

$$\frac{\partial T}{\partial t} = (v_\omega{}^2 + v_z{}^2)\left(\frac{\partial^2 T}{\partial \phi^2} + r^2\sin^2\theta\,\frac{\partial^2 T}{\partial \psi^2}\right) + 2v_z\frac{\partial T}{\partial \psi} + \mathrm{Pe}(v_\omega{}^2 + v_z{}^2)\frac{\partial T}{\partial \phi} \tag{2.69}$$

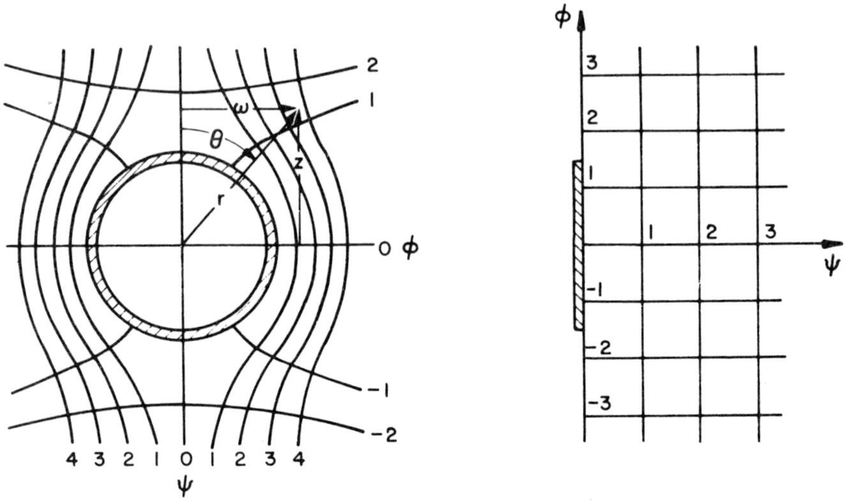

Figure 5 Coordinate system for temperature distribution in flow past a sphere.

The coordinate systems are shown in Figure 5.

After some manipulation one can readily obtain

$$v_\omega{}^2 + v_z{}^2 = \left(1 - \frac{1}{r^3}\right)^2 + \frac{3}{r^3}\left(1 - \frac{1}{4r^3}\right)\sin^2\theta \tag{2.70}$$

$$v_z = \frac{1}{2r^3}\,(2\cos^2\theta - \sin^2\theta) - 1 \tag{2.71}$$

Equation (2.69) was solved numerically by an explicit method. Let indices i, j, and n correspond to the ϕ, ψ, and t coordinates, respectively. The explicit FDA is

$$\frac{\Delta_t T^n_{i,j}}{h_t} = (v_\omega{}^2 + v_z{}^2)_{i,j}\,\frac{\delta_\phi{}^2 T^n_{i,j}}{h_\phi{}^2} + (r^2\sin^2\theta)_{i,j}\,\frac{\delta_\psi{}^2 T^n_{i,j}}{h_\psi{}^2}$$

$$+ \operatorname{Pe}(v_w{}^2 + v_z{}^2)_{i,j}\,\frac{\mu_\phi\delta_\phi T^n_{i,j}}{h_\phi} + 2(v_z)_{i,j}\,\frac{\mu_\psi\delta_\psi T^n_{i,j}}{h_\psi} \tag{2.72}$$

with $T^{n+1}_{i,j}$ as the only unknown. Along $\psi = 0$ for $|\phi| > \frac{3}{2}$, $\sin^2\theta = 0$ and the field equation becomes initial-valued in ψ. The above FDA was amended along this line in order to avoid the introduction of extraneous points. This was done by using a forward difference approximation for $\partial T/\partial\psi$, instead of the more accurate central difference formula. Since the original central difference formula possesses $h_\psi{}^2$ accuracy, the use of a 3-point, one-sided FDA with similar accuracy would have been warranted here.

Using von Neumann stability criterion, the following requirement was derived: *

$$(v_w{}^2 + v_z{}^2)_{i,j} h_t \left(\frac{1}{h_\phi{}^2} + \frac{1}{h_\psi{}^2} \right) \leqslant \frac{1}{2} \tag{2.73}$$

The contribution of this work is in advocating the use of ϕ-ψ coordinates that by definition are more naturally suited for specifying the pertinent region. As the Peclet number increases, the steady state thermal layer becomes thinner in the forward end of the sphere, and extends itself into a lengthening and thinning tail from the rear end. Had the problem been solved using r-θ coordinates, computer memory would have had to be allocated for irrelevant points outside this thermal boundary layer.

It is interesting to note that for large Pe (greater than 100) the velocity-potential directed diffusion $\partial^2 T/\partial\phi^2$ becomes negligible. The problem then becomes initial-valued in the ϕ direction and can be marched out from $\phi = 1.5$. When Pe becomes large enough then $\partial^2 T/\partial\phi^2$ will be vanishingly small so as to enable the central difference for $\partial T/\partial\phi$ to induce an instability similar to that of Richardson's method which has been mentioned earlier. This is probably why the investigators limited the Peclet number in their study to values less than 1000. Indeed, when the term $\partial^2 T/\partial\phi^2$ is dropped from Eq. (2.69), the equation takes on a different form (parabolic-hyperbolic) which will be discussed later.

The domain under study possesses infinite boundaries. Numerically, we must give finite values to these infinite boundaries. There are two methods for choosing these finite values. One method advocates the testing of successively larger values until the temperature field is no longer sensitive to the location of the boundaries. The decision whether to expand the boundary or not demands at least two calculations of the temperature field. The second method uses the zero flux condition $\partial T/\partial n = 0$ at the numerical "infinity" instead of $T = 0$. We then check the *boundary* points to determine if all their temperatures are less than a given small value ϵ. If this condition is not satisfied then the domain is expanded. This method is particularly superior to the first for time-dependent problems such as that described here. This is because the calculation confines itself to the pertinent portion of the domain and grid points are added only when their temperatures become significant, with the growth of the thermal boundary layer. It has been found (Wittke and Chao, 1967) that ϵ should be no larger than 0.001, however, the reader is advised to test various values of ϵ for his specific problem.

2.4.3 Crank-Nicolson method To overcome the stability restrictions of explicit methods, most workers have used an implicit method. For example, the

*Actually, as shown earlier, the convection terms impose an additional restriction.

Crank-Nicolson method described in Sec. 2.1.5 for one-dimensional conduction or diffusion can be extended to two and three dimensions:

$$\frac{T^{n+1} - T}{h_t} = \alpha\mu_t\left(\frac{\delta_x{}^2}{h_x{}^2} + \frac{\delta_y{}^2}{h_y{}^2} + \frac{\delta_z{}^2}{h_z{}^2}\right)T^{n+½} \tag{2.74}$$

Equation (2.74) is unconditionally stable, but requires the solution, at each time step, of a system of equations of order equal to the number of unknown values of T. Since this may be several hundred or more (depending, of course, upon the mesh size), and since the Thomas algorithm can no longer be used (the system not being tri-diagonal), the computational labor is considerable. Direct solution, by Gauss-Jordan elimination for example, is impractical, and iterative or relaxation methods must be employed.

Consider a two-dimensional problem in which, for simplicity, $h_x = h_y$. Then (2.74) at a point (i,j) can be written

$$rT_{i-1} + rT_{j-1} - (1 + 4r)T + rT_{j+1} + rT_{i+1} = S \tag{2.75}$$

where $r = \alpha h_t/2h_x{}^2$; all the values of T on the left hand side are at the "new" time level of $(n + 1)$, and the superscript has been omitted for clarity; S contains all the terms from (2.74) which are at the "old" time level of n.

To solve a system like (2.75) by relaxation, it is first written

$$T = \frac{1}{1 + 4r}(rT_{i-1} + rT_{j-1} + rT_{j+1} + rT_{i+1} - S) \tag{2.76}$$

At any time n, the value of S is known, and the remaining terms on the right-hand side of (2.76) can be first estimated by also taking values at time n. Successive application of (2.76) for all i and j for which T is unknown leads to a set of values for T of improved accuracy. Thus the iteration form of (2.76) is

$$T^{m+1} = \frac{1}{1 + 4r}(rT^m_{i-1} + rT^m_{j-1} + rT^m_{j+1} + rT^m_{i+1} - S) \tag{2.77}$$

where the superscript m denotes the iteration number. All values of T^m (2.77) are estimates of T^{n+1} at the appropriate position—estimates which improve in accuracy as m increases. The process is said to have converged* when successive estimates of T differ by less than some specified small quantity for all i and j.

The iterative solution of a system of linear equations like (2.75) will converge only if certain conditions on the coefficients are satisfied. One way of expressing these conditions is to say that the coefficient matrix must be

*Note that "convergence" here refers to the approach of an iterative procedure to a solution, in contrast to the meaning ascribed in Sec. 2.1.2. The double use of the word is unfortunate.

diagonally dominant. It can be shown that this requires the absolute value of the coefficient of T to be greater than, or equal to, the sum of the absolute values of the coefficients of the other unknowns in each equation, and to be strictly greater than the sum in at least one equation. In the case of (2.75) the requirement is

$$1 + 4r \geqslant r + r + r + r$$

which is true for any value of r. Thus (2.75) is unconditionally convergent.

The process defined by (2.77) is known as Jacobi iteration and is not very efficient, because use is not being made of the most up-to-date values. If the new values are computed in order of increasing i and j, with all values of j being used for a given i before the next value of i is taken, then $T^{m+1}_{i-1,j}$ and $T^{m+1}_{i,j-1}$ will have been computed by the time $T^{m+1}_{i,j}$ is being found. It will be more efficient, therefore, to replace (2.77) by:

$$T^{m+1} = \frac{1}{1+4r}\left(rT^{m+1}_{i-1} + rT^{m+1}_{j-1} + rT^{m}_{j+1} + rT^{m}_{i+1} - S\right) \qquad (2.78)$$

which is known as Gauss-Seidel iteration and which converges in about half the number of steps as Jacobi iteration.

A further improvement can be achieved. If T^m is added to and substracted from the right hand side of (2.78), it becomes

$$T^{m+1} = T^m + \frac{1}{1+4r}\left[rT^{m+1}_{i-1} + rT^{m+1}_{j-1} + rT^{m}_{j+1} + rT^{m}_{i+1} - (1+4r)T^m - S\right] \qquad (2.79)$$

The second term on the right hand side of (2.79) can be regarded as a correction to be added to T^m to make it more nearly correct. As the process converges, this term goes to zero. The improvement is achieved by "over-relaxing," i.e., by increasing the correction term by a factor ω somewhat greater than unity:

$$T^{m+1} = T^m + \frac{\omega}{1+4r}\left[rT^{m+1}_{i-1} + rT^{m+1}_{j-1} + rT^{m}_{j+1} + rT^{m}_{i+1} - (1+4r)T^m - S\right] \qquad (2.80)$$

This process is usually known as successive over-relaxation (SOR). It depends for its efficiency on finding optimum value of ω. In general situations, ω_{opt} can be estimated by performing a few test iterations (say five or ten) with different values of ω, each time starting from the same initial set of values of T, then continuing with the value that has yielded the most rapid rate of convergence. The rate of convergence can be estimated by calculating, at the end of each set of test iterations, the quantity $\Sigma_{i,j}|T^{m+1} - T^m|$.

It can be shown that, using ω_{opt}, SOR is faster than Gauss-Seidel iterations by a factor of the order of $2/h_x$. For small mesh sizes, therefore, it is worth some extra effort at the start to find ω_{opt}.

When the convection terms are added to (2.74), the situation changes. Using central difference approximations to these terms, the FDA can be written

$$\frac{T^{n+1} - T}{h_t} = S^{n+1} + S^n$$

$$\text{where } S^{n+1} = -\frac{v_x}{4h_x}(T_{i+1}^{n+1} - T_{i-1}^{n+1}) - \frac{v_y}{4h_y}(T_{j+1}^{n+1} - T_{j-1}^{n+1}) + \frac{T_{i-1}^{n+1} - 2T^{n+1} + T_{i+1}^{n+1}}{2h_x^2} + \frac{T_{j-1}^{n+1} - 2T^{n+1} + T_{j+1}^{n+1}}{2h_y^2}$$

and S^n is a similar expression in terms of T^n.

This can be written as

$$-(r_x + \lambda_x)T_{i-1}^{n+1} - (r_y + \lambda_y)T_{j-1}^{n+1} + (1 + 2r_x + 2r_y)T^{n+1} - (r_y - \lambda_y)T_{j+1}^{n+1} - (r_x - \lambda_x)T_{i+1}^{n+1} = R \quad (2.81)$$

$$\text{where } r_x = \frac{h_t}{2h_x^2} \qquad \lambda_x = \frac{v_x h_t}{4h_x}$$

$$r_y = \frac{h_t}{2h_y^2} \qquad \lambda_y = \frac{v_y h_t}{4h_y}$$

and R contains all the terms at time level n.

A number of cases can be distinguished, depending upon the relative sizes of the r's and λ's. For the sake of simplicity in the present discussion, suppose that $r_x = r_y = r$ and $\lambda_x = \lambda_y = \lambda$. Then, if $\lambda \leqslant r$, diagonal dominance of (2.81) is assumed without further restriction. However, if $\lambda > r$, diagonal dominance requires

$$\lambda \leqslant r + \frac{1}{4} \quad (2.82)$$

In other words, λ is permitted to exceed r, but only by an amount 0.25.

Experience shows that (2.82) is sufficient for the convergence of (2.81), but that it is not necessary. It can be violated to a limited extent (depending upon h_x and h_y) while still permitting the steady solution of (2.81) to be obtained at large values of n. However, the number of iterations required at each time step rises sharply, and the transient behavior of T oscillates unrealistically about the anticipated smooth transition from the initial to the final conditions.

Neglecting the term $\frac{1}{4}$, (2.82) implies that $v_x h_x/\alpha$ should be less than 2. If (2.81) is an FDA to the energy equation (with T denoting temperature), then α represents the thermal diffusivity and (2.82) requires that (approximately) the Peclet number based on mesh size should be less than 2.

If upwind differences are used to approximate the convection terms in the transport equation, it can easily be seen that (2.81) is replaced by

$$-(\lambda_x + r_x)T_{i-1}^{n+1} - (\lambda_y + r_y)T_{j-1}^{n+1} + (1 + 2r_x + 2r_y + \lambda_x + \lambda_y)T^{n+1} - r_y T_{j+1}^{n+1} - r_x T_{i+1}^{n+1} = R \quad (2.83)$$

where now $\lambda_x = v_x h_t/2h_x$, $\lambda_y = v_y h_t/2h_y$, and the r's are as above.

It can be seen that diagonal dominance is obtained for all (positive) values of r_x, r_y, λ_x, and λ_y. Thus (2.83) is unconditionally convergent.

2.4.4 False diffusion Since the use of upwind differencing leads to unconditional convergence of the iterative procedure for the solution of the FDA, one is led to ask why central differences are even considered. The answer lies in the accuracy of the solution, which is largely determined by a diffusion-like truncation error which is present in upwind difference approximations but not in central difference approximations.

Consider again the one-dimensional equation

$$\frac{\partial T}{\partial t} + v_x \frac{\partial T}{\partial x} = \alpha \frac{\partial^2 T}{\partial x^2} \quad (2.42)$$

and the corresponding upwind and central approximations

$$\frac{T^{n+1} - T}{h_t} + v_x \frac{T - T_{i-1}}{h_x} = \alpha \frac{T_{i-1} - 2T + T_{i+1}}{{h_x}^2} \quad (2.56)$$

and

$$\frac{T^{n+1} - T}{h_t} + v_x \frac{T_{i+1} - T_{i-1}}{2h_x} = \alpha \frac{T_{i-1} - 2T + T_{i+1}}{{h_x}^2} \quad (2.52)$$

respectively. (It is sufficient to treat explicit formulae; the results apply also to implicit methods.)

Applying the analysis of Sec. 2.1.2, we find that (2.56) and (2.52) are equivalent to

$$\frac{\partial T}{\partial t} + O(h_t) + v_x \frac{\partial T}{\partial x} - \frac{v_x h_x}{2} \frac{\partial^2 T}{\partial x^2} + O({h_x}^2) = \alpha \frac{\partial^2 T}{\partial x^2} + O({h_x}^2) \quad (2.84)$$

$$\frac{\partial T}{\partial t} + O(h_t) + v_x \frac{\partial T}{\partial x} + O(h_x{}^2) = \alpha \frac{\partial^2 T}{\partial x^2} + O(h_x{}^2) \tag{2.85}$$

respectively.

Equation (2.84) contains a term $(v_x h_x/2)(\partial^2 T/\partial x^2)$ which does not appear in (2.85) and which may be grouped with the diffusion term on the right-hand side to yield

$$\frac{\partial T}{\partial t} + v_x \frac{\partial T}{\partial x} = \left(\alpha + \frac{v_x h_x}{2}\right) \frac{\partial^2 T}{\partial x^2} + O(h_t + h_x{}^2)$$

At small values of α and large values of v_x, the false diffusion term $v_x h_x/2$ will dominate the true diffusion α unless h_x is reduced appropriately.

De Vahl Davis and Mallinson (1974) considered an analogous problem of forced convection in two dimensions characterized by a Reynolds number. They showed that the effects of false diffusion are greatest in regions where the fluid is moving at an angle of 45° to the two coordinate directions. For a nominal Reynolds number of 1000, and with a mesh size of $\frac{1}{30}$, the effective Re was as low as 240 in some regions. To ensure an effective Re $\geqslant$ 900 throughout the flow required a reduction of the mesh size to $\frac{1}{100}$: an excessive amount of computation for what is still only modest accuracy.

Thus, while the use of upwind differences has a beneficial effect on the stability of the solution, it entails a cost in terms of accuracy which may be considerable at large values of Re (i.e., small values of α).

2.4.5 Alternating direction implicit methods In order to keep the FDA tri-diagonal, and hence permit the use of the Thomas elimination algorithm, (2.74) can be replaced by the implicit equation

$$\frac{T^{n+1} - T}{h_t} = \frac{\delta_x{}^2 T^{n+1}}{h_x^2} + \frac{\delta_y{}^2 T^n}{h_y^2} \tag{2.86}$$

in which only the x-derivatives on the right-hand side have been computed at the new time level $(n + 1)$. Thus there are only three unknowns in each equation—T_{i-1}^{n+1}, T^{n+1}, and T_{i+1}^{n+1}—and the system can be readily solved.

The stability of (2.86) can be studied by the von Neumann method using, as a trial function for the solution of the error FDA,

$$\epsilon_{i,j}^n = \xi^n e^{i\beta i} e^{i\gamma j}$$

and it will be found that

$$\xi = \frac{1 - 4r \sin^2(\gamma/2)}{1 + 4r \sin^2(\beta/2)}$$

where $r = h_t/h_t^2$. For $\beta \to 0$ and $\gamma \to \pi$, $|\xi|$ can exceed unity for sufficiently large values of r. Thus the method is not unconditionally stable.

If (2.86) is replaced by

$$\frac{T^{n+1} - T}{h_t} = \frac{\delta_x^2 T^n}{h_x^2} + \frac{\delta_y^2 T^{n+1}}{h_y^2} \tag{2.87}$$

again leading to a tri-diagonal system, it will be found that

$$\xi = \frac{1 - 4r \sin^2(\gamma/2)}{1 + 4r \sin^2(\beta/2)}$$

and $|\xi| > 1$ under certain conditions.

However, if the two steps (2.86) and (2.87) are combined:

$$\frac{T^* - T}{h_t/2} = \frac{\delta_x^2 T^*}{h_x^2} + \frac{\delta_y^2 T^n}{h_y^2} \tag{2.88a}$$

$$\frac{T^{n+1} - T^*}{h_t/2} = \frac{\delta_x^2 T^*}{h_x^2} + \frac{\delta_y^2 T^{n+1}}{h_y^2} \tag{2.88b}$$

in which T^* is an intermediate variable, not given any physical significance and discarded as soon as (2.88a) is completed, then the stability of the double process is described by

$$\xi = \left[\frac{1 - 4r \sin^2(\beta/2)}{1 + 4r \sin^2(\gamma/2)}\right]\left[\frac{1 - 4r \sin^2(\gamma/2)}{1 + 4r \sin^2(\beta/2)}\right]$$

which is less than or equal to unity for all r, β, and γ. Two potentially unstable processes have been combined to yield one which is unconditionally stable.

T^* is computed at all mesh points using (2.88a); T^{n+1} is then found from (2.88b). Each step can be interpreted as advancing the solution over half a time step (although T^* is *not* regarded as part of the solution), and hence a time step of $(h_t/2)$ is used in each of (2.88).

The consistency of the combined process can be investigated as described above; the method will be found to be unconditionally consistent, with a truncation error $O(h_t^2 + h_x^2)$.

This method was introduced by Peaceman and Rachford (1955) and is known as an alternating direction implicit (ADI) method. Some modifications and extensions are described later in this section, and the next section shows how ADI methods can be used to solve elliptic equations. ADI methods are used very widely for the solution of parabolic and elliptic equations in both two and three dimensions.

Birkhoff and Varga (1959) have noted that ADI methods do not necessarily extend to the cases of variable coefficients nor to nonrectangular regions. Nonetheless, the variety of problems already solved by these methods indicates them to be far more flexible than the case for which the formal proof exists (constant coefficients along with a rectangular region).

2.4.6 Example 3: Collapse of vapor bubbles with translatory motion Wittke and Chao (1967) studied the heat transfer controlled collapse of single rising vapor bubbles. They assumed constant translational velocity, spherical bubbles, and potential flow for the liquid and arrived at the following dimensionless axisymmetric energy equation for the liquid phase

$$\frac{\partial T}{\partial t}+\left\{\frac{\mathrm{Pe}}{2}\mu\left[1-\frac{1}{\left(1+\frac{r}{R}\right)^3}\right]-\frac{2}{R\left(1+\frac{r}{R}\right)}-\dot{R}\left[1-\frac{1}{\left(1+\frac{r}{R}\right)^2}\right]\right\}\frac{\partial T}{\partial r}$$
$$+\left\{\frac{\mathrm{Pe}}{2}\frac{1-\mu^2}{R\left(1+\frac{r}{R}\right)}\left[1+\frac{1}{2\left(1+\frac{r}{R}\right)^3}\right]+\frac{2\mu}{R^2\left(1+\frac{r}{R}\right)^2}\right\}\frac{\partial T}{\partial \mu}$$
$$=\frac{\partial^2 T}{\partial r^2}+\frac{1-2\mu}{R^2\left(1+\frac{r}{R}\right)^2}\frac{\partial^2 T}{\partial \mu^2}\qquad (0<r\leqslant\infty,\ -1\leqslant\mu\leqslant 1,\ t>0)\tag{2.89}$$

where $\mu \equiv -\cos$, r is the dimensionless radial distance from the bubble wall, R is the dimensionless bubble radius, $\dot{R}$ is the rate of change of the bubble radius, and Pe is the Peclet number. The value of R starts initially from 1 and decreases to a final value of R_f, where R_f is a parameter greater or equal to zero. R is given by an energy balance at the bubble wall:

$$R=\frac{\mathrm{Ja}}{2[1-(1-c_\rho)(R_f/R)^3]}\int_{-1}^{1}\left(\frac{\partial T}{\partial r}\right)_{r=0}d\mu\tag{2.90}$$

where Ja (the Jakob number) and c_ρ are parameters of the problem. The noncondensables within the bubble are assumed to be uniformly distributed and the gaseous mixture assumed to be of uniform temperature. The presence of noncondensables causes a decrease of the vapor (and wall) temperature as collapse proceeds. By assuming a linear variation of vapor pressure with temperature and applying Boyle's law, the authors obtained the bubble temperature (and boundary condition):

$$T = 1 - \left(\frac{R_f}{R}\right)^3 \quad \text{at} \quad r = 0 \tag{2.91a}$$

The other boundary conditions used were

$$T = 0 \quad \text{at} \quad t = 0 \tag{2.91b}$$

and

$$\frac{\partial T}{\partial \mu} = 0 \quad \text{at} \quad \mu = \pm 1 \tag{2.91c}$$

This last boundary condition appears to be incorrect. It evolved from the condition that

$$\frac{\partial T}{\partial \theta} = 0 \quad \text{at} \quad \theta = 0, \pi$$

Now, by the transformation for μ,

$$\left(\frac{\partial T}{\partial \mu}\right)_{\mu=-1} = \lim_{\theta \to 0} \frac{\partial T/\partial \theta}{\sin \theta} = \left(\frac{\partial^2 T}{\partial \theta^2}\right)_{\theta=0} \tag{2.92a}$$

and similarly

$$\left(\frac{\partial T}{\partial \mu}\right)_{\mu=1} = -\left(\frac{\partial^2 T}{\partial \theta^2}\right)_{\theta=\pi} \tag{2.92b}$$

Since neither of the right-hand terms of Eqs. (2.92a) and (2.92b) are necessarily zero, the boundary conditions of Eq. (2.91c) are not applicable. The appropriate condition here evolves from the fact that the coefficient of $\partial^2 T/\partial \mu^2$ in Eq. (2.89) vanishes at $\mu = \pm 1$, thus rendering the field equation initial-valued in the μ direction along these two lines. A similar phenomenon arose in the work of Bentwich et al. (1965) which was discussed earlier.

The field equation was solved numerically by the Peaceman-Rachford, ADI method where central differences were used for the first and second derivatives in both r and μ. First, R given by Eq. (2.90), was evaluated by numerical integration (an analytical expression was used for the initial time step). Then R was computed by a forward difference approximation in time. The final step in this sequence was the numerical solution of the field equation. The total time step was adjusted so that R decreased in increments of approximately 0.01.

The boundary conditions applied at infinity were those of zero flux. Wittke and Chao used an expanding network which has been described earlier in this section. Their criterion for expansion was $\epsilon = 0.001$.

A comparison of theoretical and experimental results shows satisfactory agreement. This indicates that the solution was fairly insensitive to the boundary conditions used at $\mu = \pm 1$.

It is interesting to speculate about the benefits gained through the application of the previously discussed Boussinesq transformation (Bentwich et al., 1965) where the ϕ and ψ coordinates used would correspond to flow around a noncollapsing sphere. Mathematically, the transformation is plausible for the collapsing bubble; however, it may not be preferable for numerical computation. As noted in the previous example for the *noncollapsing* sphere, the ϕ-ψ system is superior to the r-θ system since the latter uses much more extraneous points in the grid. For the case of the collapsing sphere this argument holds only when the flow field induced by the translational velocity of the bubble dominates that induced by the radial collapse. When radial collapse dominates, the resulting temperature field will tend toward radial symmetry and computation in r-θ coordinates will be more appropriate. Most of the experimental runs of Wittke (1966) yielded parametric values that correspond to a translational-velocity dominated temperature field.

2.4.7 Locally one-dimensional methods Recently, Russian numerical analysts have developed a splitting known as "locally one-dimensional" methods (LOD). By these methods the conduction equation in two-dimensions is split into

$$\frac{1}{2}\frac{\partial T}{\partial t} = \frac{\partial^2 T}{\partial x^2} \tag{2.93a}$$

$$\frac{1}{2}\frac{\partial T}{\partial t} = \frac{\partial^2 T}{\partial y^2} \tag{2.93b}$$

while the conduction equation in three dimensions is split into

$$\frac{1}{3}\frac{\partial T}{\partial t} = \frac{\partial^2 T}{\partial x^2} \tag{2.94a}$$

$$\frac{1}{3}\frac{\partial T}{\partial t} = \frac{\partial^2 T}{\partial y^2} \tag{2.94b}$$

$$\frac{1}{3}\frac{\partial T}{\partial t} = \frac{\partial^2 T}{\partial z^2} \tag{2.94c}$$

This gives a one-dimensional equation to be solved in each stage of the solution.

These split equations can be solved by either explicit or implicit methods. Thus, by the explicit method (2.93) leads to

$$T^{n+1/2} = \frac{1}{2} r_x T_{i-1} + (1 - r_x)T + \frac{1}{2} r_x T_{i+1} \tag{2.95a}$$

and

$$T^{n+1} = \frac{1}{2} r_y T_{j-1}^{n+1/2} + (1 - r_y)T^{n+1/2} + \frac{1}{2} r_y T_{i+1}^{n+1/2} \tag{2.95b}$$

Similarly, applying the Crank-Nicolson equation gives

$$-\frac{1}{4} r_x T_{i-1}^{n+1/2} + \left(1 + \frac{1}{2} r_x\right) T^{n+1/2} - \frac{1}{4} r_x T_{i+1}^{n+1/2} = \frac{1}{4} r_x T_{i-1} + \left(1 - \frac{1}{2} r_x\right) T + \frac{1}{4} r_x T_{i+1} \tag{2.96a}$$

and

$$-\frac{1}{4} r_y T_{j-1}^{n+1} + \left(1 + \frac{1}{2} r_y\right) T^{n+1} - \frac{1}{4} r_y T_{j+1}^{n+1} = \frac{1}{4} r_y T_{j-1}^{n+1/2} + \left(1 - \frac{1}{2} r_y\right) T^{n+1/2} + \frac{1}{4} r_y T_{j+1}^{n+1/2} \tag{2.96b}$$

As in ADI methods no physical significance should be attached to the values at the $(n + \frac{1}{2})$ time level.

The stability condition for Eqs. (2.95) and (2.96) can be evaluated by the von Neumann method. Inserting $\xi^n e^{i\beta i} e^{i\gamma j}$ for $T_{i,j}{}^n$ in (2.95a) gives the requirement

$$-1 \leqslant \left[1 - 2r_x \sin^2\left(\frac{\gamma}{2}\right)\right]\left[1 - 2r_y \sin^2\left(\frac{\beta}{2}\right)\right] \leqslant 1 \tag{2.97}$$

For the case of $r_x = r_y = r$, the condition is

$$r \leqslant \frac{1}{2}$$

which is seen to be an improvement over the regular explicit restriction of $r \leqslant \frac{1}{4}$.

This is because Eq. (2.95) relates one point at the $(n+1)$ level to nine points at the n level and such formulae are known to possess better stability.

The system of Eq. (2.96) can be readily shown to have unconditional stability.

Applications of LOD methods are at present infrequent, since they are relatively new. However, their simple formulation, particularly in three dimensions, makes them appealing to the analyst.

2.4.8 Further ADI and LOD methods* We consider again the numerical solution of

$$\frac{\partial T}{\partial t} + \frac{\partial}{\partial x}(v_x T) + \frac{\partial}{\partial y}(v_y T) + \frac{\partial}{\partial z}(v_z T) = \alpha \nabla^2 T + S \tag{2.98}$$

After approximating the spatial derivatives by FDA's, the equation may be represented at each mesh point in the solution field by

$$\frac{\partial T}{\partial t} = (a_x + a_y + a_z)T + S_D \tag{2.99}$$

where a_x, a_y, and a_z are point operators representing the FDA's to the derivatives in the coordinate direction indicated by the subscript, and S_D is the FDA of S. Writing $(\mathbf{T})$ as the vector of mesh point values, the set of equations (2.99) may be written as

$$\frac{\partial}{\partial t}(\mathbf{T}) = [A_x + A_y + A_z](\mathbf{T}) + (S_D) \tag{2.100}$$

where A_x, A_y, and A_z are matrix operators formed from the point operators a_x, a_y, and a_z, and square brackets denote matrices.

We have seen that the explicit advancement of (2.98) by

$$(\mathbf{T})^{n+1} = (\mathbf{T})^n + h_t[A_x + A_y + A_z](\mathbf{T})^n + (S_D)^n$$

is straightforward, but is subject to a severe restriction on h_t if numerical instability is to be avoided.

As with equations in a single space variable, unconditional stability can be achieved by the use of an implicit relationship between $(\mathbf{T})^n$ and $(\mathbf{T})^{n+1}$, viz.

$$(\mathbf{T})^{n+1} = (\mathbf{T})^n + h_t[A_x + A_y + A_z](\mathbf{T})^{n+1} + (S_D)^n \tag{2.101}$$

Equation (2.101) is only first-order accurate in time, and second order accuracy can be achieved if the Crank-Nicolson equation

*This section has been adapted from Mallinson (1973).

$$(\mathbf{T})^{n+1} = (\mathbf{T})^n + \frac{h_t}{2}[A_x + A_y + A_z]((\mathbf{T})^n + (\mathbf{T})^{n+1}) + (S_D)^{n+½} \tag{2.102a}$$

is employed. This may be written

$$\left[I - \frac{h_t}{2}[A_x + A_y + A_z]\right](\mathbf{T})^{n+1} = \left[I + \frac{h_t}{2}[A_x + A_y + A_z]\right](\mathbf{T})^n + (S_D)^{n+½} \tag{2.102b}$$

(where I denotes the unit matrix) and requires either the use of iterative methods, as discussed above, or the inversion of $I - h_t/2\,(A_x + A_y + A_z)$ at each time step.* The former requires considerable computational effort and the latter, for any realistic number of mesh points, involves a prohibitive amount.

As noted earlier, an alternative is to use a series of subtime intervals over which the advancement is achieved by a partially implicit procedure, such that at the end of the time interval an approximation to (2.101) or (2.102) is obtained. It has been found (Douglas, 1961; Aziz, 1965) that schemes which are based on (2.102) are invariably more successful.

The ADI of Peaceman and Rachford, described earlier, was the earliest application of this idea. Using the present notation it gives for the two-step advancement

$$(\mathbf{T})^* = (\mathbf{T})^n + \frac{h_2}{2}A_x(\mathbf{T})^* + \frac{h_t}{2}A_y(\mathbf{T})^n + \frac{h_t}{2}(S_D)^{n+½} \tag{2.103a}$$

$$(\mathbf{T})^{n+1} = (\mathbf{T})^* + \frac{h_t}{2}A_x(\mathbf{T})^* + \frac{h_t}{2}A_y(\mathbf{T})^{n+1} + \frac{h_t}{2}(S_D)^{n+½} \tag{2.103b}$$

where $(\mathbf{T})^*$ denotes an intermediate set of values of $(\mathbf{T})$, apparently at the time $t + h_t/2$ but in fact having no physical significance.

$(\mathbf{T})^*$ can be eliminated from these two equations to yield

$$(\mathbf{T})^{n+1} = (\mathbf{T})^n + \frac{h_t}{2}[A_x + A_y]((\mathbf{T})^n + (\mathbf{T})^{n+1}) + (S_D)^{n+½} + \frac{h_t^2}{4}[A_xA_y]((\mathbf{T})^{n+1} - (\mathbf{T})^n) \tag{2.104}$$

*Note that since the elements of this matrix involve the velocity components they will, in general, vary from one time step to the next.

and the Peaceman-Rachford scheme is seen to be a perturbation of (2.102a).

The foregoing scheme cannot be extended to three dimensions, since the method becomes unstable. Successful extensions to three dimensions using ADI methods were achieved independently by Douglas (1961) and Brian (1961). As described by Douglas, advancement of $(\mathbf{T})$ over the time step h_t is achieved through the use of two intermediate field evaluations:

$$\left[I - \frac{h_t}{2} A_x\right](\mathbf{T})^* = \left[I + \frac{h_t}{2}\,[A_x + 2A_y + 2A_z]\right](\mathbf{T})^n + h_t(S_D)^n \tag{2.105a}$$

$$\left[I - \frac{h_t}{2} A_y\right](T)^{**} = (T)^* - \frac{h_t}{2} A_y (T)^n \tag{2.105b}$$

$$\left[I - \frac{h_t}{2} A_z\right](T)^{n+1} = (T)^{**} - \frac{h}{2} A_z (T)^n \tag{2.105c}$$

The procedure is not strictly an advancement over three time subintervals as would be a generalization of the Peaceman-Rachford scheme, but has been derived by a direct factorization of the Crank-Nicolson equation (2.102). Again, the intermediate variables $(T)^*$ and $(T)^{**}$ have no physical significance. This scheme was successfully used by Aziz (1965) in a problem of natural convection.

A simpler scheme, suggested by Samarskii (1963), was used by Chorin (1968) to solve the Navier-Stokes equations without Δp. It belongs to the family of the LOD methods. Using the present notation, the scheme may be written:

$$[I - h_t A_x](T)^* = (T)^n \tag{2.106a}$$

$$[I - h_t A_y](T)^{**} = (T)^* \tag{2.106b}$$

$$[I - h_t A_z](T)^{n+1} = (T)^{**} + h_t(S_D)^n \tag{2.106c}$$

and is attractive since less computation is required per time step than in the case of the Douglas scheme.

The elimination of $(T)^*$ and $(T)^{**}$ from (2.106a)-(2.106c) leads to

$$\begin{aligned}\frac{(T)^{n+1} - (T)^n}{h_t} &= [A_x + A_y + A_z](T)^{n+1} + (S_D)^n \\ &\quad - h_t[A_y A_z + A_z A_x + A_x A_y](T)^{n+1}\end{aligned} \tag{2.107}$$

As Chorin (1969) pointed out, the FDA is therefore correct to order h_t. However, for constant h_t, the error associated with the last term in (2.107) will persist into the steady state solution (if one exists) as $t \to \infty$.

An alternative, which does not suffer from this defect, was suggested by Samarskii and Andreev (1963). It is

$$[I - \theta h_t A_x](\mathbf{T})^* = [A_x + A_y + A_z](\mathbf{T})^n + (S_D)^n \tag{2.108a}$$

$$[I - \theta h_t A_y](\mathbf{T})^{**} = (\mathbf{T})^* \tag{2.108b}$$

$$[I - \theta h_t A_z](\mathbf{T})^{***} = (\mathbf{T})^{**} \tag{2.108c}$$

$$(T)^{n+1} = (\mathbf{T})^n + h_t(\mathbf{T})^{***} \tag{2.108d}$$

The intermediate variables may be eliminated to yield

$$\begin{aligned}\frac{T^{n+1} - T^n}{h_t} &= [A_x + A_y + A_z](1 - \theta)(\mathbf{T})^n \\ &+ [A_x + A_y + A_z]\theta(\mathbf{T})^{n+1} + (S_D)^n \\ &- \theta^2 h_t [A_y A_z + A_z A_x + A_x A_y]((\mathbf{T})^{n+1} - (\mathbf{T})^n) \\ &+ \theta^3 h_t^2 [A_x A_y A_z]((\mathbf{T})^{n+1} - (\mathbf{T})^n)\end{aligned} \tag{2.109}$$

For $\theta = \frac{1}{2}$, (2.109) is seen to be an extension of the Crank-Nicolson scheme (2.102). It will be noted that the terms of order h_t and $h_t{}^2$ vanish if a steady solution, in which $(\mathbf{T})^{n+1} = (\mathbf{T})^n$, is reached. It can readily be verified that (2.109), with $\theta = \frac{1}{2}$, is identical to the Douglas scheme (2.105). However, the former is easier to implement and is preferred to the latter.

2.4.9 Conservative FDA's It was pointed out in Sec. 1.4 that, by virtue of the continuity equation, the connection terms

$$v_x \frac{\partial T}{\partial x} + v_y \frac{\partial T}{\partial y}$$

in two dimensions can be written

$$\frac{\partial}{\partial x}(v_x T) + \frac{\partial}{\partial y}(v_y T)$$

This formulation implies the conservation of some fluid property (heat, momentum, etc., depending upon the significance of T).

If the conservative form of the differential equation is differenced, e.g., by writing

$$\frac{\partial}{\partial x}(v_x T) = \frac{(v_x)_{i+1} T_{i+1} - (v_x)_{i-1} T_{i-1}}{2h_x} + O(h_x{}^2) \tag{2.110}$$

etc., it can be shown that the FDA also possesses this conservative property, whereas the use of

$$v_x \frac{\partial T}{\partial x} = (v_x)_i \frac{T_{i+1} - T_{i-1}}{2h_x} + O(h_x^2) \tag{2.111}$$

does not. This can be verified by summing the FDA's for all mesh points (equivalent to integrating the differential equation over the solution region) and noting that the contribution to the flux of T at the internal mesh points cancel out for the first expression, but not for the second.

Intuitively, it would seem reasonable to suppose that if fluid properties are conserved in nature, then the FDA's should also have this property. Roache (1973) refers to some evidence that the use of conservative differencing may lead to some loss of accuracy, and points out that conservative schemes do not, of course, necessarily imply accuracy. (Conservative schemes, for example, may still be unstable.)

However, the great majority of evidence supports the use of conservative differencing, certainly for the class of problems considered in this review. We (and Roache) recommend its use.

2.5 Suggested literature

Vedamurthy and Sastri (1974) combined analyzed conductive and radiative heat transfer to the walls of fluidized bed combustors. The governing energy equation for the emulsion packet of layers of solid and gaseous elements on the wall was reduced to a nonlinear unsteady one-dimensional equation with a nonlinear boundary condition at the wall. The temperature field was solved by the Crank-Nicolson method.

Two other recent papers deal with linear models that are also solved by the Crank-Nicolson method. Tamir and Rachmilev (1974) presented results for the direct contact condensation of an immiscible vapor on a thin film which is generated as liquid spreads over a sphere. The governing energy equation is two-dimensional, but parabolic in the direction of flow along the surface of the sphere. Since the boundary condition on the wall of the sphere was of the mixed type, with a coefficient dependent on the initial-valued coordinate, the equation was solved numerically. Hwang and Sheu (1974) employed the Crank-Nicolson method in considering the effect of the radial velocity component on laminar forced convection in the entrance region of circular tubes. The flow field was approximated by an analytical solution due to Langhaar (1942).

An interesting model of heat transfer in a rod bundle with helical wrap spaces was presented by Graves and Catton (1973). A flow field was postulated and the resulting energy equation was parabolic in the direction of axial flow. The solution was subsequently marched out numerically in this direction.

3 ELLIPTIC EQUATIONS

In contrast to parabolic PDE's which are a class of propagation problems, elliptic PDE's are a type of equilibrium problems. In the former the boundary conditions are not specified over all the edges of the domain. In the latter all the boundary conditions must be given; otherwise the problem is improperly posed. Richardson (1925) gave a lucid contrast between the two types of problems by referring to parabolic equations as marching problems and elliptic equations as jury problems.

3.1 Laplace's equation in two variables

The most common type of elliptic equation is that of Laplace which in two dimensions is:

$$\frac{\partial^2 T}{\partial x^2} + \frac{\partial^2 T}{\partial y^2} = 0 \tag{3.1}$$

A rectangular network is superimposed on the field in which the intervals in each specific coordinate are equal, i.e., the rectangles are all equal. Applying central differences to a typical nodal point, *i,j*, yields

$$\frac{\delta_x^2 T_{i,j}}{h_x^2} + \frac{\delta_y^2 T_{i,j}}{h_y^2} + O(h_x^2 + h_y^2) = 0 \tag{3.2}$$

where the leading truncation error is

$$-\frac{1}{12} h_x^2 \left(\frac{\partial^4 T}{\partial x^4}\right)_{i,j} - \frac{1}{12} h_y^2 \left(\frac{\partial^4 T}{\partial y^4}\right)_{i,j}$$

Rearranging Eq. (3.2) gives

$$T_{i,j} = \frac{1}{2[1 + (h_x^2/h_y^2)]} \cdot \left(T_{i-1,j} + T_{i+1,j} + \frac{h_x^2}{h_y^2} T_{i,j-1} + \frac{h_x^2}{h_y^2} T_{i,j+1}\right) \tag{3.3}$$

which indicates that for a square grid the value at each nodal point in the FDA solution will be the average of the four neighboring points.

Since the set of equations described by Eq. (3.3) is sparse and diagonally dominant, the Gauss-Seidel method described earlier is also applicable here. Often accelerated convergence is obtained using successive over-relaxation (SOR). For a square grid the sequence takes the form

$$T_{i,j}^{m+1} = \frac{\omega}{2}\left(T_{i-1,j}^{m+1} + T_{i+1,j}^{m} + T_{i,j-1}^{m+1} + T_{i,j+1}^{m}\right) + (1-\omega)T_{i,j}^{m} \tag{3.4}$$

where the superscript m denotes the iteration and the spacewise advancement is in the directions of increasing i and j. The degree of over-relaxation is given by the value of ω which for Eq. (3.4) varies from one, corresponding to the Gauss-Seidel method, up to a maximum of two. Frankel (1950) determined the optimum for a rectangular domain as a function of network size, while Starling (1963) made a similar study for a cylinder.

3.1.1 Higher accuracy formulas As shown earlier h^4 accuracy can be obtained via the approximation

$$h^2D^2 = \frac{\delta^2}{1 + (1/12)\delta^2} \tag{1.47}$$

This yields a nine-point formula given by the equation

$$\left[\delta_x^2\left(1 + \frac{1}{12}\delta_y^2\right) + \frac{h_x^2}{h_y^2}\delta_y^2\left(1 + \frac{1}{12}\delta_x^2\right)\right]T_{i,j} = 0 \tag{3.5}$$

3.1.2 Richardson's h^m extrapolation method Richardson (1910) and Richardson and Gaunt (1927) introduced the technique of the deferred approach to the limit by extrapolation. Although the method has since then been implemented in numerical integration in the Romberg method, it appears to have been first applied to finite differences.

If we employ an h^m accurate FDA and solve a field for two different square mesh sizes, h_1 and h_2, then the truncation error at any point for each mesh size is $c_1 h_1{}^m$ and $c_2 h_2{}^m$, respectively. Hence c_1 and c_2 are not constants, but contain derivatives which have to be evaluated at unknown and different places in the two expressions. In extrapolating to the limit we assume that the mesh sizes h_1 and h_2 are sufficiently small so as to enable us to equate c_1 to c_2, yielding a better approximation given by

$$T = \frac{h_2{}^m T_1 - h_1{}^m T_2}{h_2{}^m - h_1{}^m} \tag{3.6}$$

where T_1 and T_2 are obtained from the FDA using h_1 and h_2, respectively.

3.2 Point and line iteration methods

The Gauss-Seidel and SOR methods are examples of a class known as *point* iteration methods. These methods yield an explicit correction for each point, i.e., the correction for the point is the only unknown in the equation.

Faster convergence is usually effected when *line* or *row* iteration methods are used. These methods result in an implicit correction, giving a system of equations that is relatively simple to solve. Usually the system is tri-diagonal, so the Thomas algorithm described earlier can be readily applied.

We shall illustrate the line iteration method by applying it to the equation

$$\frac{\partial^2 T}{\partial x^2} + \frac{\partial^2 T}{\partial y^2} = f(x, y) \tag{3.7}$$

with linear boundary conditions specified on a rectangular domain. The FDA for this equation is

$$\frac{\delta_x{}^2 T_{i,j}}{h_x^2} + \frac{\delta_y{}^2 T_{i,j}}{h_y^2} = f_{i,j} \tag{3.8}$$

We apply the method to each line in j in the direction of increasing j and obtain for the case of a square grid ($h_x = h_y = h$):

$$T_{i-1,j}^{m+1} - 4T_{i,j}^{m+1} + T_{i+1,j}^{m+1} = -T_{i,j-1}^{m+1} - T_{i,j+1}^{m} + h^2 f_{i,j} \tag{3.9}$$

$T_{i,j-1}^{m+1}$ is already known from the calculation for row j–1 and $T_{i,j+1}{}^{m}$ is known from the previous iteration. Applying this equation to all the unknown points in the j'th row yields a tri-diagonal system of equations. After solving this row we proceed to the $j + 1$ row and repeat the procedure. It can be shown that less iterations in m are required for convergence than in any of the point iteration methods.

3.3 Alternating direction implicit methods

Faster convergence is afforded by the use of ADI methods. For this implementation we seek the asymptotic limit to the parabolic equation

$$\frac{\partial T}{\partial t} = L(D_x{}^2, D_y{}^2, D_x, D_y, x, y, t)T \tag{3.10}$$

of Sec. 2.

The central idea here is to choose the optimum set of time intervals so as to reach steady-state with a minimum of time steps. Too large a time interval will introduce oscillations into the solution and increase the number of time steps required. A discussion of optimum time intervals (better known as "good parameter sequences") for the Peaceman-Rachford and Douglas-Rachford methods can be found in either Ames (1969) or Mitchell (1969).

3.4 Locally one-dimensional methods

Like the ADI method, the LOD methods also have a counterpart in elliptic equations. The literature on this application is sparse. In his book, Mitchell (1969) reported that no parameter sequences were as yet available for the LOD method.

It is obvious that both the ADI and LOD methods belong to the class of line iteration methods described earlier.

3.5 Irregular boundaries

When the boundary is irregular, i.e., it is not parallel to one of the coordinates, finite differences can still be employed, but care must be taken if h^2 accuracy is desired. The problem is especially complicated when derivative boundary conditions are encountered. The reader is referred to Ames (1969) for a full discussion along with background references.

3.6 Example 4: Effects of variable thermal properties on moving band source temperatures

Isenberg and Malkin (1975) have studied heat transfer to a semi-infinite medium with variable thermal properties from a moving band source. The computed temperature field is of interest in machining and sliding contact processes.

For the steady-state problem the governing equation in nondimensional form is

$$\frac{c_p}{c_{p_0}} \mathrm{Pe} \frac{\partial T}{\partial x} = \frac{\partial}{\partial x}\left(\frac{k}{k_0}\frac{\partial T}{\partial x}\right) + \frac{\partial}{\partial y}\left(\frac{k}{k_0}\frac{\partial T}{\partial y}\right) \tag{3.11}$$

where c_p is specific heat, k is thermal conductivity, Pe is a Peclet number, and the index 0 denotes properties at $T = 0$. The property ratios are given functions of the material. The boundary conditions are

$$\frac{\partial T}{\partial y} = -1 \qquad y = 0 \qquad -1 \leqslant x \leqslant 1 \tag{3.12a}$$

$$\frac{\partial T}{\partial y} = 0 \qquad y = 0 \qquad |x| > 1 \tag{3.12b}$$

$$T = 0\left(\text{or } \frac{\partial T}{\partial n} = 0\right) \qquad \text{at} \qquad x \to \pm\infty \qquad \text{or} \qquad y \to \infty \tag{3.12c}$$

The variable k/k_0 occurs frequently in variable property problems. In order to avoid calculating space derivatives for it, the Kirchhoff transformation is used. Defining

$$\Theta \equiv \frac{1}{k_0} \int_0^T k(T')\, dT' \tag{3.13}$$

yields the final form of the governing equation:

$$\mathrm{Pe}\, \frac{\partial \Theta}{\partial x} = \frac{\alpha}{\alpha_0} \left(\frac{\partial^2 \Theta}{\partial x^2} + \frac{\partial^2 \Theta}{\partial y^2} \right) \tag{3.14}$$

where α is the temperature-dependent thermal diffusivity. The boundary conditions remain the same with T replaced by Θ.

Since the domain is semi-infinite, a dynamic finite boundary similar to that of the previous example of Wittke and Chao was employed. The criterion for expansion was a boundary temperature exceeding 10^{-4}.

If $\Theta = 0$ is regarded as the "neutral" temperature then the only perturbation to this temperature is afforded by means of the boundary condition at $-1 \leqslant x \leqslant 1$ and $y = 0$ (the heat source), which constitutes a relatively small portion of the boundary. Problems of this type classically demand a relatively fine mesh (and large computer time) for the FDA in order to obtain good accuracy in the vicinity of the perturbation. There are several ways to reduce the number of grid points required. The obvious way is the use of unequal intervals (Young, 1962), but this would reduce the accuracy of the FDA from h^2 to h due to the approximation for the second derivative. Another way is to divide the region into a number of sub-domains with different mesh sizes; however, the programming here is usually formidable. A third way is to warp the coordinate system by a transformation that yields for the original system a fine grid in the region of interest, which gradually becomes courser in withdrawing from this region. The transformation generally gives a more complicated expression for the field equation.

The last device mentioned above was applied to the example under discussion. The problem was redefined in ξ-η coordinates according to the transformation:

$$x = \frac{\sinh(\xi)}{\sinh(1)} \tag{3.15a}$$

$$y = \sinh(\eta) \tag{3.15b}$$

By employing these coordinates, equal intervals in the ξ-η plane, when transformed to the x-y plane, yielded a fine grid in the region adjacent to the band source.

The Peaceman-Rachford ADI method was used to solve the temperature field. In each iteration, the value of α/α_0 at each point was computed from the temperature obtained from the previous iteration. A good initial guess was afforded by the analytical solution of Jaeger (1942). The temperature field, T, was calculated from the final values for Θ via the Kirchhoff transformation.

The analyst is cautioned against using Eq. (3.11) for large Peclet number. For this case the term $\partial^2\Theta/\partial x^2$ becomes vanishingly small and the field equation becomes parabolic in x and the more tractable form of the field equation makes it easier to solve using one of the methods of the previous section.

3.7 Suggested literature

The mass transfer and motion of solid particles suspended by a compressible fluid flowing in a rotating cascade of a turbomachine was solved numerically by Hussein and Tabakoff (1974). The governing combined continuity and momentum equation was formulated in two-dimensions for blade-to-blade stream surfaces of revolution and solved by the SOR method.

Two recent works suggest improved numerical methods for elliptic equations. Brazier (1974) developed an optimum SOR procedure for any domain or coefficient matrix. In the proposed method an individual relaxation factor is computed for each mesh point. Although for most cases, the rates of convergence were comparable to those for the conventional SOR method, some cases with nonsymmetrical coefficients, evolving from convection, yielded convergence rates several times faster. Raithby and Torrance (1974) have formulated weighted upwind differencing schemes for combined conduction and convection. In the direction treated the scheme is weighted upwind just enough to achieve stability.

4 PARABOLIC-HYPERBOLIC EQUATIONS*

In this section we address ourselves to PDE's of the form

$$\frac{\partial T}{\partial t} + v_x \frac{\partial T}{\partial x} + v_y \frac{\partial T}{\partial y} = \alpha \frac{\partial^2 T}{\partial y^2} \tag{4.1}$$

Equations of this type govern such phenomena as transient heat transfer for laminar forced convection in ducts (Siegel and Sparrow, 1959), transient heat transfer for laminar natural convection on a flat plate (Hellums and Churchill, 1961), condensation of rising bubbles in an immiscible medium (Isenberg and Sideman, 1970), and solidification of a glass-like film on a vertical wall (Isenberg and Gutfinger, 1975).

Equation (4.1) is initial-valued in both t and x, and boundary-valued in y. If either the first or second term on the left-hand side were to be dropped then

*This section has been adapted from Isenberg (1975).

the resulting equation would be parabolic (even without the term $v_y\ \partial T/\partial y$). By contrast, if the last term on the left-hand side and the term on the right would both be dropped, then the equation would be hyperbolic. For want of a better name, Eq. (4.1) will be referred to as a "parabolic-hyperbolic" equation. In this section we shall examines several FDA's for this equation.

4.1 Completely explicit methods

Hellums (1960) in his study of natural convection on a vertical plate proposed the following FDA for the parabolic-hyperbolic equation:

$$\frac{\Delta_t T^n_{i+1,j}}{h_t} + v^n_{x_{i+1,j}} \frac{\Delta_x T^n_{i,j}}{h_x} + v^n_{y_{i+1,j}} \frac{\Delta_y T^n_{i,j+1}}{h_y} = \alpha \frac{\delta_y^{\,2} T^n_{i+1,j}}{h_y^{\,2}} \qquad (v_x > 0, v_y < 0) \quad (4.2)$$

Here, again, upwind differences are used for the convective terms. The scheme is shown in Figure 6a.

In applying the von Neumann condition for stability, Hellums used the trial function

$$\epsilon^n_{i,j} = \xi^n e^{\mathrm{i}\beta i} e^{\mathrm{i}\gamma j} \qquad (4.3)$$

Figure 6 Schematics of explicit and partially implicit approximations for parabolic-hyperbolic equations.

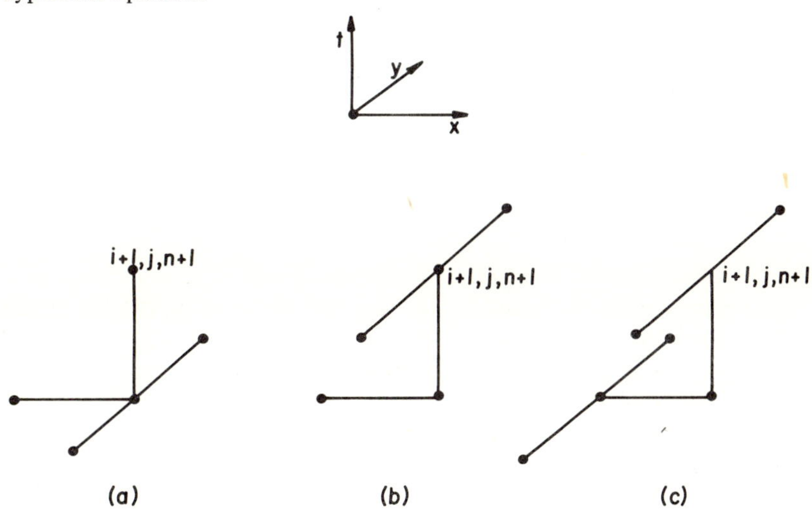

(see Sec. 2.4.5), and derived the following restriction:

$$h_t\left(\frac{v_x}{h_x} - \frac{v_y}{h_y} + \frac{2\alpha}{h_y^2}\right) \leqslant 1 \tag{4.4}$$

Note that this restriction reduces to that of Lelevier (Richtmyer and Morton, 1967) for the case of pure *convection* ($v_y = \alpha = 0$) described in Sec. 2.2.1. Thus the maximum time step varies from

$$\frac{1}{v_x/h_x} \quad \text{to} \quad \frac{1}{v_x/h_x - v_y/h_y + 2\alpha/h_y^2}$$

The truncation error of this scheme is $O(h_x + h_t + h_y)$. In attempting to reduce the error by using smaller increments, the restriction of the size of h_y imposed by (4.4) will severely limit the time step.

By contrast, using a central difference approximation for $v_y\ \partial T/\partial y$ gives the stability criteria

$$h_t\left(\frac{v_x}{h_x} + \frac{2\alpha}{h_y^2}\right) \leqslant 1 \tag{4.5a}$$

$$\text{and } |v_y| \leqslant \frac{2\alpha}{h_y} \tag{4.5b}$$

if v_y is small when compared to α then Eq. (4.5) is less restrictive than (4.4). Additionally, the new FDA has a truncation error of $O(h_x + h_t + h_y^2)$.

4.2 Partially implicit methods

As noted in the previous section, the least restriction on stability for the previous FDA's is

$$\frac{v_x h_t}{h_x} \leqslant 1 \tag{4.6}$$

which is always present due to the very nature of upwind differencing. However, it is seen from Eqs. (4.4) and (4.5a) that the presence of the increment, h_y, greatly constrains the time step. We shall now seek methods for allowing the stability condition for parabolic-hyperbolic equations to approach that of pure convection as given above.

For the sake of simplicity we shall drop the term $v_y\ \partial T/\partial y$. From the prior analysis of this term, the study of the following schemes can clearly be extended to account for its effect.

Equation (4.1) can be made implicit in the term on the right-hand side by evaluating the latter at the $t + h_t$ time level. The FDA is, using our simplified notation;

$$T_{i+1}^{n+1} - T_{i+1} + v_x \frac{h_t}{h_x}(T_{i+1} - T) = T_{i+1,j-1}^{n+1} - 2T_{i+1}^{n+1} + T_{i+1,j+1}^{n+1} \tag{4.7}$$

The scheme is shown in Figure 6b.

Substituting $\xi_{i,j}^n$ from Eq. (4.3) yields the amplification factor

$$\xi = \frac{\lambda_x(e^{-i\beta} - 1) + 1}{1 + 2r_y(1 - \cos\gamma)} \tag{4.8}$$

where $\lambda_x \equiv v_x h_t/h_x$ and $r_y \equiv \alpha h_t/h_y^2$, and by Eq. (4.6) $\lambda_x \leqslant 1$. Since $|\xi| \leqslant 1$,

$$|\lambda_x(e^{-i\beta} - 1) + 1| \leqslant |1 + 2r_y(1 - \cos\gamma)| \tag{4.9a}$$

or

$$1 + 2\lambda_x(\lambda_x - 1)(1 - \cos\beta) \leqslant [1 + 2r_y(1 - \cos\gamma)]^2 \tag{4.9b}$$

Now since $\lambda_x \leqslant 1$, the left-hand term can never be greater than one, while the right-hand term can never be less than one. Thus, the condition $\lambda_x \leqslant 1$ is sufficient to guarantee the stability of (4.7). As a result, this partially implicit FDA is unrestricted in the size of h_y.

It is interesting to note that as $\partial T/\partial x \to 0$, (4.7) reduces to the backward FDA for the unsteady heat conduction equation. An analogous reduction is obtained for $\partial T/\partial t \to 0$. This observation leads us to consider approximations that will reduce to Crank-Nicolson accuracy as $\partial T/\partial x \to 0$ or $\partial T/\partial t \to 0$. To affect this we shall now approximate the diffusion term by the average of that at the points i, j, n, and $i+1, j, n+1$, giving

$$T_{i+1}^{n+1} - T_{i+1} + v_x \frac{h_t}{h_x}(T_{i+1} - T)$$
$$= \frac{\alpha}{2}\frac{h_t}{h_y^2}(T_{i+1,j-1}^{n+1} - 2T_{i+1}^{n+1} + T_{i+1,j+1}^{n+1} + T_{j-1} - 2T - T_{j+1}) \tag{4.10}$$

The scheme is depicted in Figure 6c. The amplification factor for (4.10) is

$$\xi = \frac{\lambda_x(e^{-i\beta} - 1) - Se^{i\beta} + 1}{1 + S} \tag{4.11}$$

where $S \equiv (1 - \cos\gamma)\,\alpha h_x/h_y^{\,2}$.

We now impose the condition $\lambda_x \leqslant 1$. We require that

$$|\lambda_x(e^{i\beta} - 1) - Se^{i\beta} + 1| \leqslant 1 + S \tag{4.12a}$$

$$\text{or } (1 - \lambda_x)^2 - 2(1 - \lambda_x)(\lambda_x - S)\cos\beta + (\lambda_x - S)^2 \leqslant (1 + S)^2 \tag{4.12b}$$

Now, if $\lambda_x > S$, the maximum on the left side of (4.12b) corresponds to $\cos\beta = -1$, giving

$$(1 - S)^2 \leqslant (1 + S)^2 \tag{4.13}$$

which shows stability. When $\lambda_x \leqslant S$, the maximum is given by $\cos\beta = 1$ resulting in

$$(1 + S - 2\lambda_x)^2 \leqslant (1 + S)^2 \tag{4.14}$$

Since $\lambda_x \leqslant 1$ the left side of (4.14) has a maximum of $(1 + S)^2$. Thus the FDA of (4.10) is unrestricted in the size of h_y.

To test the consistency of Eqs. (4.7) and (4.10), we apply the methods of Sec. 2.1.2. In operator form (4.7) and (4.10) are recast as

$$\frac{\Delta_t T}{h_t} + v_x \frac{\Delta_x T}{h_x} = \alpha \frac{E_t \delta_y^{\,2} T}{h_y^{\,2}} \tag{4.15}$$

$$\text{and} \quad \frac{\Delta_t T}{h_t} T + v_x \frac{\Delta_x T}{h_x} = \frac{\alpha}{2h_y^{\,2}}\left(E_t \delta_y^{\,2} T + E_x^{-1} \delta_y^{\,2} T\right) \tag{4.16}$$

Equation (4.15) can be expanded into

$$\frac{1}{h_t}\left(h_t D_t + \frac{h_t^2 D_t^2}{2} + \cdots\right)T + \frac{v_x}{h_x}\left(h_x D_x - \frac{h_x^{\,2} D_x^{\,2}}{2} + \cdots\right)T$$
$$= \frac{\alpha}{h_y^{\,2}}\left(1 + h_t D_t + \frac{h_t^2 D_t^2}{2}\right)\left(h_y^{\,2} D_y^{\,2} + \frac{h_y^{\,4} D_y^{\,4}}{12} + \cdots\right)T \tag{4.17a}$$

or

$$\frac{\partial T}{\partial t} + v_x \frac{\partial T}{\partial x} = \alpha \frac{\partial^2 T}{\partial y^2} - \left(\frac{h_t}{2}\frac{\partial^2 T}{\partial t^2} + \cdots\right) + v_x\left(\frac{h_x}{2}\frac{\partial^2 T}{\partial x^2} - \cdots\right)$$
$$+ \alpha\left(\frac{h_y^2}{12}\frac{\partial^4 T}{\partial y^4} + h_t \frac{\partial^3 T}{\partial t \partial y^2} + \cdots\right) \tag{4.17b}$$

with a truncation error of $O(h_x + h_t + h_y^2)$.

For Eq. (4.16) we obtain

$$\frac{\partial T}{\partial t} + v_x \frac{\partial T}{\partial x} = \alpha \frac{\partial^2 T}{\partial y^2} - \left(\frac{h_t}{2}\frac{\partial^2 T}{\partial t^2} + \cdots\right) + v_x\left(\frac{h_x}{2}\frac{\partial^2 T}{\partial x^2} - \cdots\right)$$
$$+ \alpha\left(\frac{h_y^2}{12}\frac{\partial^4 T}{\partial y^4} + \frac{h_t}{2}\frac{\partial^3 T}{\partial t \partial y^2} - \frac{h_x}{2}\frac{\partial^3 T}{\partial x \partial y^2} + \cdots\right) \tag{4.18}$$

Therefore, both (4.15) and (4.16) are consistent with the PDE's that they represent.

4.3 Higher accuracy methods

The Lax-Wendroff formula (1960, 1964) is the best known FDA for first-order hyperbolic equations. It is developed by a Taylor's expansion around the point $i + 1, j, n + 1$. For the parabolic-hyperbolic equation this gives

$$T_{i+1}^{n+1} = E_t T_{i+1}^n = \left(1 + h_t D_t + \frac{h_t^2 D_t^2}{2} + \cdots\right) T_{i+1}^n$$
$$= \left[1 + h_t(-v_x D_x + \alpha D_y^2) + \frac{h_x^2}{2}(-v_x D_x + \alpha D_y^2)^2\right] T_{i+1}^n$$
$$= \left(1 - h_t v_x D_x + h_t \alpha D_y^2 + \frac{h_t^2}{2} v_x^2 D_x^2\right.$$
$$\left. - h_t^2 v_x \alpha D_x D_y^2 + \frac{h_t^2}{2}\alpha^2 D_y^4\right) T_{i+1}^n \tag{4.19}$$

Since our solution must be limited to low-order terms, we drop the last term on the right. This yields the FDA

$$\frac{\Delta T_{i+1}}{h_t} + v_x \frac{\mu_x \delta_x}{h_x} T_{i+1} = \alpha \frac{\delta_y^2}{h_y^2} T_{i+1} + \frac{h_t}{2} v_x^2 \frac{\delta_x^2}{h_x^2} T_{i+1}$$
$$- h_t v_x \alpha \frac{\mu_x \delta_x \delta_y^2}{h_x h_y^2} T_{i+1} \tag{4.20}$$

The scheme is shown in Figure 7a.

The inclusion of the last two terms has increased the accuracy of the approximation to $O(h_x^2 + h_t + h_y^2)$. Unfortunately, the scheme is unstable. The amplification factor is

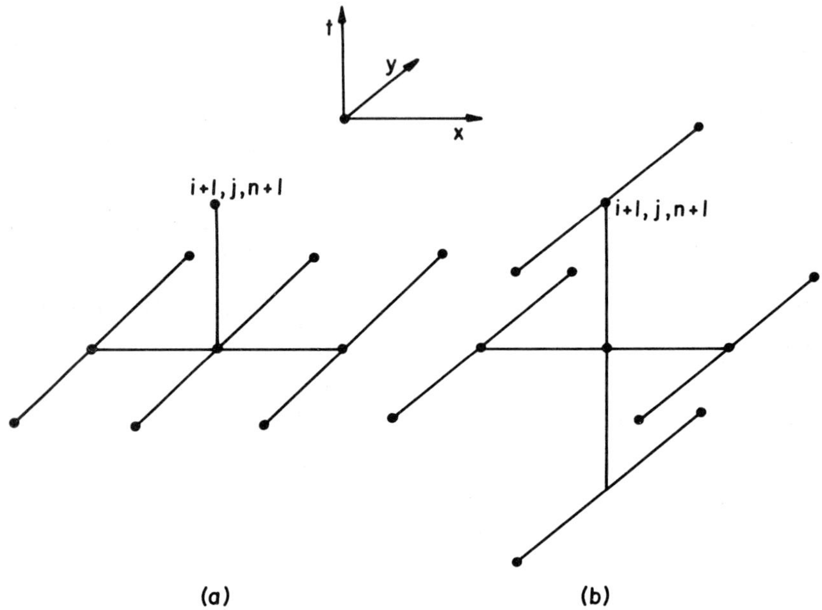

Figure 7 Higher accuracy approximations for parabolic–hyperbolic equations: (a) modified Lax-Wendroff (b) modified leap frog.

$$\xi = 1 - \lambda_x^2(1 - \cos\beta) - 2r_y(1 - \cos\gamma)$$
$$+ \mathrm{i}[\lambda_x r_y(1 - \cos\gamma) - \lambda_x] \sin\beta \tag{4.21}$$

Now, λ_x can have a maximum value of one. Therefore, we can choose $\lambda_x = 1$, $\cos\beta = \cos\gamma = -1$. This gives the condition $|\xi| \geqslant 1$ for any value of r_y.

A second high-accuracy formula for first-order hyperbolic equations, is the "leap frog" scheme, which for our formula becomes

$$\frac{\mu_t\delta_t}{h_t} T_{i+1} + v_x \frac{\mu_x\delta_x}{h_x} T_{i+1} = \frac{\alpha}{4h_y^2} \delta_y^2(T + T_{i+2} + T_{i+1}^{n+1} + T_{i+1}^{n-1}) \tag{4.22}$$

The method is depicted in Figure 7b. It is seen to be a three-level formula that requires equal intervals in each of the coordinates. The accuracy of the method is $O(h_x^2 + h_t^2 + h_y^2)$. An analysis of stability will readily show that for $\lambda_x \leqslant 1, r_y \leqslant \frac{1}{2}$.

4.4 Completely implicit methods

Since Eq. (4.1) is defined in the quarter plane $0 \leqslant x,\ t \leqslant \infty$, one can extend implicit FDA's for the hyperbolic PDE to the parabolic–hyperbolic equation.

One such FDA based on the formula of Wendroff (1960) is centered at $i+\frac{1}{2}$, $j, n+\frac{1}{2}$:

$$\frac{\mu_x \delta_t}{h_t} T_{i+1/2}^{n+1/2} + v_x \frac{\mu_t \delta_x}{h_x} T_{i+1/2}^{n+1/2} = \alpha \frac{\mu_x \mu_t \delta_y^2}{h_y^2} T_{i+1/2}^{n+1/2} \tag{4.23}$$

The FDA is shown in Figure 8a.

The FDA of (4.23) is described by von Rosenberg (1969). Numerically, the scheme appears to have been first implemented by Killeen (1966) in his study of the unsteady draining process. He observed that as the solution was marched out, erratic results were eventually generated by the model. Killeen's work was continued by Stack (1967), who employed the above FDA to solve both the momentum and energy equations of the nonisothermal draining problem. Here again, the solution was observed to break down.

Since the (4.23) is implicit in both the x and t advancement, to determine stability we must examine the error propagation in each direction. For the timewise propagation we insert our familiar

$$\epsilon_{i,j}^n = \xi^n e^{\mathrm{i}\beta i} e^{\mathrm{i}\gamma j}$$

into (4.23) and obtain

Figure 8 Completely implicit approximations for parabolic–hyperbolic equations: (a) modified Wendroff (b) modified backward.

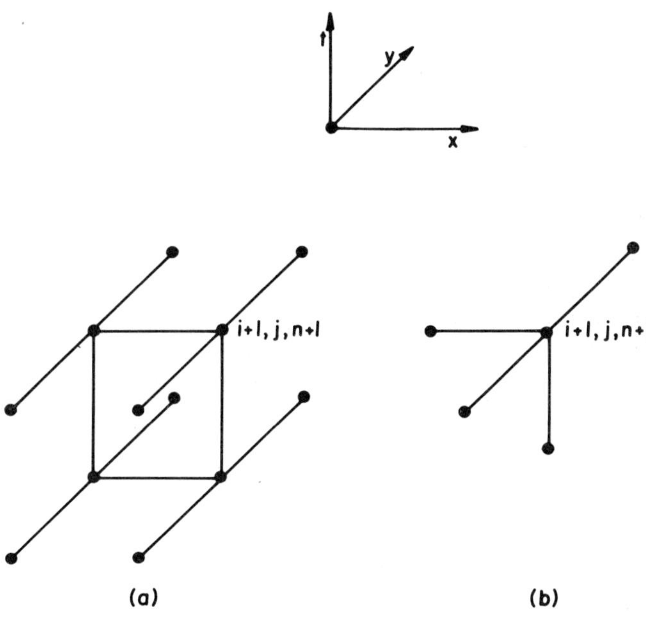

$$|\xi|^2 = \frac{[1 - r_y(1 - \cos\gamma)]^2 + \lambda_x^2 \tan^2\beta/2}{[1 + r_y(1 - \cos\gamma)]^2 + \lambda_x^2 \tan^2\beta/2} \tag{4.24}$$

which is always less than or equal to unity indicating that the method is unconditionally stable. A similar result is obtained for x propagation.

The authors' own experiments with the FDA of (4.23) indicates that it is sensitive to singularities on the boundary. Indeed, if the initial conditions on $x = 0$ and $t = 0$ are not identical at $x = t = 0$, then strong oscillations will be induced into the region, that attenuate very slowly. Since the problems studied by Killeen and Stack are coupled and nonlinear, oscillations may have developed that caused the solution to diverge. This is particularly plausible in the case of a singularity at $x = t = 0$, which physically will propagate a surface of discontinuity into the $x - t - y$ region.

A method that is particularly insensitive to singularities is the backward FDA:

$$\frac{\nabla_t}{h_t} T_{i+1}^{n+1} + v_x \frac{\nabla_x}{h_x} T_{i+1}^{n+1} = \frac{\delta_y^2}{h_y^2} T_{i+1}^{n+1} \tag{4.25}$$

which is depicted in Figure 8b. This FDA can be shown to be unconditionally stable.

To test the accuracy of (4.23) and (4.25), they were each used to solve the transient temperature distribution in laminar flow between parallel plates. The boundary conditions used were

$$\begin{aligned} T &= 1 \quad \text{at} \quad t = 0 \\ T &= 0 \quad \text{at} \quad x = 0 \\ T &= 1 \quad \text{at} \quad y = 0 \text{ and } y = 1 \end{aligned}$$

The solution generated by (4.23) oscillated unrealistically until steady-state was approached. Equation (4.25) gave a smooth transition to $t \to \infty$. Moreover, when the explicit FDA (4.2) was applied to this problem, the solution for the same mesh size in x and t as used for (4.25), with the constraint imposed on h_t by the stability condition for (4.2), showed excellent agreement with the solution generated by Eq. (4.25) with a maximum error less than 1%.

5 EXAMPLES OF SYSTEMS OF PARTIAL DIFFERENTIAL EQUATIONS

5.1 Coupled systems of equations

Convection heat transfer problems require, obviously, the solution of the energy equation which, for two-dimensional flow and constant properties, is Eq. (1.18).

In forced convection, the velocity components required in (1.18) can be obtained by solution of the equations of motion without consideration of thermal effects. However, in the presence of natural convection, or when the motion is generated entirely by buoyancy forces, the equations for v_x and v_y involve T, and the full system (1.16)-(1.18) must be solved simultaneously.

Even the equations of motion themselves, in either the primitive variable or the vorticity-stream function form, constitute a coupled system* and present a significant numerical problem.

In this section, a discussion will be given of a few examples of solutions of such systems which have appeared in the literature. The discussion must, because of space restrictions, be brief, and reference should be made to the original papers for full details. The references cited were not necessarily the first to use the particular techniques described. It is believed, however, that they provide illustrative examples of the methods discussed in preceding sections. They have also been chosen for their heat transfer interest, although almost no mention of results is made, attention here being directed to the methods used.

The principal difficulty in the solution arises from the non-linearity of the equations, and hence of the finite difference approximations. Methods for the solution of nonlinear systems of algebraic equations are known, but are not well developed, and instead a process of quasi-linearization and iteration is usually adopted.

The basis of a typical method is as follows. The equations to be solved can be written in the general form

$$\frac{\partial \zeta}{\partial t} = f_1(\zeta, v_x, v_{yy} T) \tag{5.1}$$

$$\nabla^2 \psi = -\zeta \tag{5.2}$$

$$\frac{\partial T}{\partial t} = f_2(T, v_x, v_y) \tag{5.3}$$

$$v_x = \frac{\partial \psi}{\partial y} \qquad v_y = -\frac{\partial \psi}{\partial x} \tag{5.4}$$

*The vorticity ϕ can be eliminated from (1.16) and (1.17) to yield (in the absence of the buoyancy term) a single equation for ψ:

$$\frac{\partial}{\partial t}(\nabla^2 \psi) + \frac{\partial(\nabla^2 \psi, \psi)}{\partial(x, y)} = \nu \nabla^4 \psi$$

However, an efficient procedure for the solution of this equation is not known and the second order system is almost invariably solved instead.

where v_x, v_y, and T are assumed to be constant over one time step, and (5.1) is solved for ζ at the end of the time step. The new value of ψ is then found from (5.2), following which new values of v_x and v_y are available from (5.4). Finally, (5.3) yields a new T. (The symbols ζ, ψ, etc., are here taken to refer to the matrix of values throughout the solution region; the word *solve* implies the use of one of the numerical processes discussed in Sec. 2 and 3.) This completes the advancement through one time step of ζ, ψ, and T and the cycle may be repeated until a steady state is reached (assuming that one exists).

The procedure is complicated by the need to evaluate boundary conditions for vorticity. At a stationary impermeable wall, for example, it was shown in Sec. 1.5 that the vorticity is given by

$$\zeta = -\frac{\partial^2 \psi}{\partial y^2} \tag{5.5}$$

where y is the direction normal to the wall. As originally suggested by Woods (1954), it is assumed that the variation of ζ with y is linear over the mesh interval adjacent to the wall, i.e., that

$$\zeta = \zeta_0 + \frac{y}{h_y}(\zeta_1 - \zeta_0) \tag{5.6}$$

where the subscripts 0 and 1 denote mesh points on the wall at a distance h_y from it. Upon insertion of (5.6) into (5.5), the latter may be integrated twice to yield

$$\zeta_0 = -\frac{3\psi_1}{h_y^2} - \frac{\zeta_1}{2} \tag{5.7}$$

Thus (5.7) enables boundary values for ζ to be found from the interior values of ζ and ψ after (5.1) and (5.2) have been solved.

However, the solution of (5.1) for new interior values of ζ was based on the assumption that the old boundary values of ζ applied over the whole of the time step. Wilkes and Churchill* (1966) found that they were unable to obtain solutions to a particular natural convection problem above a certain value of the Grashof number, and attributed their difficulty to the fact that the interior and boundary vorticities were "out of step." De Vahl Davis and Thomas (1969) confirmed this, and showed that a recomputation of the interior vorticity and stream function after calculation of an average boundary vorticity, over the time step–and better still, several iterations of the sequence (5.1), (5.2), (5.7)–allowed the Grashof number range to be extended to the accuracy limit imposed by truncation errors.

*They actually used a different formula for boundary vorticity, but the point is still valid.

A further discussion of the numerical implementation of boundary conditions may be found in Roache (1973).

Anyone who has tried to do so will know that one of the major problems in obtaining a numerical solution to systems such as (5.1)-(5.4) is the control of instabilities. These can arise from the nonlinearity of the equations, from the coupling between the equations via the buoyancy term, from the coupling due to boundary conditions, and from other sources. Despite the extensive literature on stability analysis (some of which has been included here) and the variety of methods available for performing this analysis, it is only possible to analyze the stability of linear systems of FDA's. This is accomplished by substituting the appropriate Fourier terms for each of the dependent variables and requiring that the eigenvalues of the resulting amplification matrix are all less than unity in modulus (Carnahan et al., 1969). However, the vast majority of systems encountered in physical problems are nonlinear, and reliance must ultimately be placed on numerical experiments.

5.2 Transient and steady state solutions

For some problems, it is necessary to obtain the true transient solution of the equations following some change in boundary conditions. In these cases, it is necessary to solve the set (5.1)-(5.4), with (5.7), as it stands involving what is essentially a triply-nested iterative procedure:

1. the innermost iteration, to solve the elliptic equation (5.2);
2. the iterative procedure just described to compute the boundary vorticity properly; and
3. the advancement through time of the full system.

In other problems only the final steady state may be of interest. In these cases, three approaches may be adopted. Firstly, the transient terms may be discarded, leading to a set of elliptic equations. This procedure is not often followed, since each of these equations must be solved iteratively, together with an outer iteration through the system as a whole, and this generally proves to be slower than the alternative approaches. Secondly, the transient terms may be retained and the system solved as described above, with only the solution as t being of interest. Finally, the method of the false transient (as described below) may be used, which yields only the steady solution–assuming it exists–but is much less demanding in computer time.

5.3 Combined free and forced convection in horizontal rectangular channels

Cheng and Hwang (1969) described a numerical method for obtaining flow and heat transfer results for steady fully developed laminar forced convection in horizontal rectangular channels with secondary flow caused by buoyancy forces. The wall temperature was assumed to vary linearly in the flow direction, and to be constant at any cross-section. The method they used would probably not be chosen today in its entirety, but their paper nevertheless contains useful illustrations of some of the techniques described in earlier sections.

The equations of motion and energy were written in nondimensional form as

$$v_x \frac{\partial}{\partial x} \nabla^2 \psi + v_y \frac{\partial}{\partial y} \nabla^2 \psi = \nabla^4 \psi - \frac{\mathrm{Ra}\mathrm{Re}}{\bar{v}_z} \frac{\partial T}{\partial x} \tag{5.8}$$

$$v_x \frac{\partial v_z}{\partial x} + v_y \frac{\partial v_z}{\partial y} = \nabla^2 v_z + 4 \tag{5.9}$$

$$\mathrm{Pr}\left(v_x \frac{\partial T}{\partial x} + v_y \frac{\partial T}{\partial y}\right) = \nabla^2 T - v_z \tag{5.10}$$

where v_x and v_y are the horizontal and vertical secondary velocities, v_z is the axial velocity (with an average value of $\bar{v}_z$) and ψ and T are the stream function and temperature, respectively. The flow is characterized by the Rayleigh, Reynolds, and Prandtl numbers, together with the cross-sectional dimensions of the channel (which appear in the boundary conditions).

These are the steady state equations; moreover (despite the remark in Sec. 5.1), the biharmonic form of the equation of motion was used, thus avoiding the difficulties associated with boundary values for vorticity. FDA's were constructed and then solved using the following procedure:

1. Equation (5.9) was solved for v_z by assuming v_x and v_y to be zero at all mesh points;
2. Equation (5.10) was solved for T with the same assumption, using the values of v_z found in (*i*).

The values of v_z and T thus found are those of the forced convection problem in the absence of buoyancy effects. They were then used as first estimates for the solution of the full problem.

3. Equation (5.8) was solved for ψ (the first time this was done $v_x = v_y = 0$);
4. v_x and v_y were computed from FDA's to $v_x = \partial\psi/\partial y, v_y = -\partial\psi/\partial x$;
5. Equation (5.9) and (5.10) were then used to obtain improved values of v_z and T, respectively.

The cycle from 3 to 5 was repeated until successive estimates of each variable differed by less than a specified small quantity.

In the present context, the results are of less interest than the method used to obtain them. The principle reasons for citing this work are its use of iteration to linearize the equations and of simplified equations to obtain initial estimates for the solution of the full equations.

5.4 Natural convection in a rectangular enclosure

Wilkes and Churchill (1966) were the first of a series of authors (Elder, 1966; de Vahl Davis, 1968; Rubel and Landis, 1969; Quon, 1972) to use numerical methods on this problem. A rectangular container of fluid was heated on one of its vertical walls and cooled on the opposite wall. The horizontal walls were insulated or varied linearly in temperature between the values of the hot and cold walls. The motion was assumed to be two-dimensional (an assumption subsequently shown by Mallinson and de Vahl Davis (1973) to be invalid except at the midplane of the cavity, although the effect of the motion in the third direction on the heat transfer is restricted to a region of about one cavity width from the end walls).

Their nondimensional equations were

$$\frac{\partial \zeta}{\partial t} + v_x \frac{\partial \zeta}{\partial x} + v_y \frac{\partial \zeta}{\partial y} = \frac{\text{Gr}}{2} \frac{\partial T}{\partial y} + \nabla^2 \zeta \tag{5.11}$$

$$\frac{\partial T}{\partial t} + v_x \frac{\partial T}{\partial x} + v_y \frac{\partial T}{\partial y} = \frac{1}{\text{Pr}} \nabla^2 T \tag{5.12}$$

$$\nabla^2 \psi = -\zeta \tag{5.13}$$

$$v_x = \frac{\partial \psi}{\partial y} \qquad v_y = -\frac{\partial \psi}{\partial x} \tag{5.14}$$

where ζ is the vorticity and the other variables are defined above. The parameters of the problem are the Grashof and Prandtl numbers and the height/width (aspect) ratio of the enclosure (which appears in the boundary conditions).

An ADI technique was used to advance (5.12) and then (5.11) through one time step. Equation (5.13) was solved by SOR, and new boundary values for vorticity computed. Finally, values of v_x and v_y were found from (5.14), thus completing the set of calculations for one time step.

The results yielded distributions of velocity and temperature in the fluid, and enabled a Nusselt number to be found as a function of Grashof and Prandtl numbers and the geometry of the cavity. As mentioned, later works have extended and improved upon these results. (Thomas and de Vahl Davis, 1971, also studied annular cavities.)

5.5 Convection in a rotating annulus

Williams (1969) discussed some of the problems of finding a satisfactory numerical method for integrating the full Navier-Stokes equations for viscous, incompressible, three-dimensional flow. His objective was to seek a method that would represent both the temporal and spatial distributions as faithfully as possible, using FDA's that resembled the true Navier-Stokes equations. Hence, he worked with the primitive variable equations.

He considered a fluid contained between two vertical coaxial cylinders and two horizontal planes. The container rotated about its axis at a constant rate. The fluid was thermally driven away from a state of solid rotation by holding the two cylindrical walls at different constant temperatures. The base was thermally insulated and the free upper surface (assumed horizontal) behaved the same way, owing to the presence of a lid (not in contact) inhibiting interaction with the overlying air.

The equations of motion, energy and continuity, were written in cylindrical coordinates:

$$\frac{Dv_\phi}{Dt} = -\frac{1}{\omega}\frac{\partial p}{\partial \phi} + \nu I(v_\phi) - \left(2\Omega + \frac{v_\phi}{\omega}\right)v_\omega \tag{5.15}$$

$$\frac{Dv_\omega}{Dt} = -\frac{\partial p}{\partial \omega} + \nu F(v_\omega) + \left(2\Omega + \frac{v_\phi}{\omega}\right)v_\phi \tag{5.16}$$

$$\frac{Dv_z}{Dt} = -\frac{\partial p}{.\partial z} + \nu H(v_z) + \beta g T \tag{5.17}$$

$$\frac{DT}{Dt} = \alpha \nabla^2 T \tag{5.18}$$

$$\frac{\partial v_\phi}{\partial \phi} + \frac{\partial}{\mu\omega}(\omega v_\omega) + \omega\frac{\partial v_z}{\partial y} = 0 \tag{5.19}$$

where v_ϕ, v_ω, and v_z are the velocity components in the circumferential, radial, and axial directions, respectively; p is the deviation of pressure from the hydrostatic value; Ω is the angular velocity of the container; and I, F, and H are functions containing the viscous terms (the details of which are omitted here). D/Dt denotes the substantial derivative:

$$\frac{Dq}{Dt} \equiv \frac{\partial q}{\partial t} + \frac{1}{\omega}\frac{\partial}{\partial \omega}(\omega v_\omega q) + \frac{1}{\omega}\frac{\partial}{\partial \phi}(v_\phi q) + \frac{\partial}{\partial z}(v_z q)$$

The system (5.15)-(5.19), together with suitable boundary conditions, are sufficient for the determination of the five unknowns, v_ϕ, v_ω, v_z, p, and T.

However, it was found convenient to replace (5.19) by an equation for p derived from (5.15)-(5.17) and (5.19). To obtain this, the equations of motion were written in vector form:

$$\frac{\partial \mathbf{v}}{\partial t} + \mathbf{v} \cdot \nabla \mathbf{v} + 2\Omega x \mathbf{v} = -\nabla p + \beta g T - \nu \nabla x \zeta \tag{5.20}$$

where $\boldsymbol{\zeta} = \nabla \mathbf{x} \mathbf{v}$ is the vorticity vector, and the continuity equation in divergence from

$$\mathfrak{D} \equiv \nabla \cdot \mathbf{v} = -\frac{1}{\rho} \frac{D\rho}{Dt} = 0 \tag{5.21}$$

The divergence of (5.20)

$$\frac{\partial}{\partial t} = -\left[\nabla^2 p - \beta g \frac{\partial T}{\partial z} - 2\Omega \zeta + \nabla \cdot (\mathbf{v} \cdot \nabla \mathbf{v})\right] \tag{5.22}$$

is an equation for the prediction of $\mathfrak{D}$. However, since $\mathfrak{D} = 0$, p must satisfy the right-hand side of (5.22) set to zero, i.e.,

$$\nabla^2 p = \nabla \cdot \mathbf{G} \tag{5.23}$$

where $\mathbf{G}$ contains the inertia, buoyancy, etc., terms from the equations of motion. The boundary conditions for (5.23) were obtained from the original equations of motion (5.15)-(5.17). They are that

$$\frac{\partial p}{\partial \omega} = G_\omega \quad \text{and} \quad \frac{\partial p}{\partial z} = G_z \tag{5.24}$$

on the $\omega =$ constant and $z =$ constant boundaries, respectively. The solution of (5.23) was facilitated by converting it to an equation with homogeneous boundary conditions. To do this, let p_0 be any reasonable function satisfying (5.24). Then $p^* = p - p_0$ satisfies

$$\nabla^2 p^* = G^* = \nabla \cdot G - \nabla^2 p_0 \tag{5.25}$$

with $\partial p/\partial \omega = \partial p/\partial z = 0$ on the boundary.

An interlacing mesh was used. The points p and T are located at the intersections of the surfaces $\omega =$ constant, $\phi =$ constant, and $z =$ constant, where the increments h_ω, h_ϕ, and h_z are uniform throughout the solution region. v_ϕ points are located on circumferential lines midway between p, T points; v_ω points are located on radial lines midway between p, T points; and v_z points

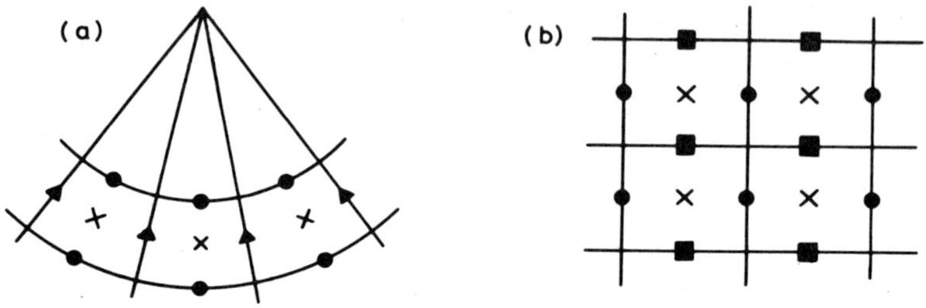

Figure 9 The mesh point distribution (a) on horizontal planes; (b) on vertical planes; x: p, T points; ▲: v_ϕ (circumferential velocity) points; ●: v_ω (radial velocity) points; ■: v_z (axial velocity) points.

are located on vertical lines of the p, T mesh, midway between points. This is illustrated in Figure 9. The discrete fluid element centered on p and bounded by v_ϕ, v_ω, v_z points forms the fundamental fluid element for which most concepts and properties such as mass conservation apply. The physical boundaries are placed so that they fall halfway between p points, i.e., on appropriate circumferential and horizontal lines in Figure 9.

Full details of the FDA's are given in the reference. As an example, the FDA to (5.17) is

$$\frac{\mu_t \delta_t v_z}{h_t} + \frac{1}{\omega}\,\delta_\omega \frac{\mu_z(\omega v_\omega)\cdot \mu_\omega v_z}{h_\omega} + \frac{1}{\omega}\,\delta_\phi \frac{\mu_z v_\phi \cdot \mu_\phi v_z}{h_\phi} + \delta_z \frac{\mu_z v_z \cdot \mu_z v_z}{h_z} = -\frac{\delta_z p}{h_z} + \nu H^*(v_z) + \beta g \mu_z T \quad (5.26)$$

where $H^*(v_z)$ denotes the appropriate FDA to $H(v_z)$.

Notice that (5.26) is a three time level formula, yielding an expression for v_z^{n+1}. The nonlinear, rotational, pressure, and buoyancy terms were evaluated at the central level n, while the diffusion terms were evaluated at $(n-1)$. The averaging is necessary to provide values at points where the variables are not explicitly defined. An improved degree of accuracy was obtained through use of such an interlacing mesh.

The variables v_ϕ, v_ω, v_z, and T were obtained from explicit FDA's of which (5.26) is an example. To obtain the finite difference analogue of (5.25), let the FDA's to the equations of motion be written

$$\frac{\delta_t \mu_t v_\omega}{h_t} = -\frac{\delta_\omega p}{h_\omega} + G_\omega$$

$$\frac{\delta_t \mu_t v_\omega}{h_t} = -\frac{(1/\omega)\delta_\phi p}{h_\phi} + G_\phi$$

$$\frac{\delta_t \mu_t v_z}{h_t} = -\frac{\delta_z p}{h_z} + G_z$$

where $\mathbf{G} \equiv (G_\omega, G_\phi, G_z)$ represents all the terms other than the transient $(\partial/\partial t)$ and pressure gradient terms. Substituting these into the FDA to (5.19), viz.

$$\frac{\delta_\phi v_\phi}{h_\phi} + \frac{\delta_\omega(\omega v_\omega)}{h_\omega} + \frac{\omega \delta_z v_z}{h_z} = 0$$

yields

$$\nabla^2 p = \nabla \cdot \mathbf{G} \tag{5.27}$$

where $\nabla^2 p \equiv \dfrac{1}{\omega}\, \delta_\omega \dfrac{\omega \delta_\omega p / h_\omega}{h_\omega} + \dfrac{1}{\omega^2} \dfrac{\delta_\phi^2 p}{h_\phi^2} + \dfrac{\delta_z^2 p}{h_z^2}$

$$\nabla \cdot G \equiv \frac{1}{\omega}\, \delta_\omega \frac{\omega G_\omega}{h_\omega} + \frac{1}{\omega} \frac{\delta_\phi G_\phi}{h_\phi} + \frac{\delta_z G_z}{h_z}$$

Solution of (5.27) each time step provided the values of p necessary to complete the marching process.

There is, however, a complication arising from the fact that inevitably there will be a small error—perhaps due only to round-off—in p. This creates an artificially nonzero divergence $\mathcal{D}$ which can lead to computational instability. This was overcome by using the round-off divergence at one step as a correction term to the forcing terms of (5.27), which was written

$$\nabla^2 p = \nabla \cdot \mathbf{G} - \frac{\delta_t \mu_t \mathcal{D}}{h_t} \tag{5.28}$$

since $\mathcal{D}^{n-1} \neq 0$, but $\mathcal{D}^{n+1}$ should be zero, (5.28) becomes

$$\nabla^2 p = \nabla \cdot \mathbf{G} + \frac{\mathcal{D}^{n-1}}{2h_t} \tag{5.29}$$

This stratagem was found to prevent the growth of $\mathcal{D}$ as time n increases.

Equation (5.29) was solved by expanding p into two finite series of harmonics in the ϕ and z directions, leading to a second-order ordinary difference equation in ω for the coefficients.

Extensive details are given by Williams in his paper of this procedure and many other computational features.

This was one of the first attempts to solve a fully three-dimensional problem. Using a UNIVAC 1108 computer, $(14)^3 = 2744$ mesh points could be handled in direct core storage, and the solution per time step required 2.00 seconds. With auxiliary drum storage, the mesh could be refined to 60 X 34 X 34 = 69260 points, for which the solution required 95.7 seconds per time step. Clearly, computations with such mesh are very time-consuming and expensive but the cost can be reduced with advances in technology and improvements in programming.

5.6 The method of the false transient

In the previous section, a method for solving the transient primitive variable equations was described. In this section, a method for obtaining the steady solution of the vorticity-vector potential equations is discussed. It was developed by Mallinson and de Vahl Davis (1973)* and has also been employed by de Socio, Sparrow, and Eckert (1973).

The method was based on the conversion of the elliptic equations (1.11) and (1.12) into parabolic equations by the addition of false transient terms; furthermore, the time derivatives were given modified coefficients. The equation set to be solved is, in nondimensional variables,

$$\frac{\mathrm{Pr}}{\alpha_\zeta}\frac{\partial \zeta}{\partial t} = -\nabla \times (\zeta \times \mathbf{v}) + \mathrm{Pr}\nabla^2\zeta - \mathrm{RaPr}(\nabla \times T\mathbf{g}) \tag{5.30}$$

$$\frac{1}{\alpha_\psi}\frac{\partial \psi}{\partial t} = \nabla^2\psi + \zeta \tag{5.31}$$

$$\frac{1}{\alpha_\phi}\frac{\partial \phi}{\partial t} = \nabla^2\phi \tag{5.32}$$

$$\frac{\partial T}{\partial t} = -\nabla \cdot (\mathbf{v}T) + \nabla^2 T \tag{5.33}$$

Equations (5.31) and (5.32) are no longer physically meaningful. However, as $t \to \infty$, all the transient terms decay to zero if a steady solution exists. Thus a false transient and a true steady solution are obtained.

The reason for the insertion of the transient terms into (5.31) and (5.32) is that they can be solved–i.e., advanced through one time step–with considerably less effort than that required for the solution of the corresponding elliptic equations. The inclusion of Pr on the left hand side of (5.30) has the effect of

*The method was developed in connection with a study of motion in a closed cavity. With no flow through the boundaries of the system, (1.11) and (1.21) show that $\phi = 0$ everywhere, and this reference therefore did not consider ϕ. The method has been extended to open systems by de Vahl Davis and Wolfshtein (1974).

adjusting the development of the vorticity transient in cases where $\Pr \neq 1$ in such a way that the total solution time becomes more or less independent of Pr.

The method is a generalization of the device mentioned by Dorodnicyn (1969) and employed in the artificial compressibility technique of Chorin (1967). In contrast with the latter, however, this method uses a purely artificial construction and can be applied to any set of elliptic equations.

The time advancement of (5.30)-(5.33) was achieved using the Samarskii-Andreev ADI scheme described in Sec. 2.4.8. It is unconditionally stable when applied to a single, linear equation, but when applied to the coupled, nonlinear system an upper limit on h_t exists. It is $0.8\ \{\min(h_x, h_y, h_z)\}^2$ when $\alpha_\zeta = \alpha_\psi = \alpha_\phi = 1$ in either two or three dimensions. For *any* values of Ra and Pr, solutions could be obtained if h_t was chosen to satisfy this condition. If the condition was violated, the instability could still be controlled if α_ζ (or, less effectively, α_ψ) was reduced. It was found that the total solution time was reduced by exceeding the stability limit on h_t in (5.31)-(5.33) by a factor of 5-10, provided h_t in (5.30) was reduced by a similar factor. The adjustments to h_t in the various equations were achieved by the α's.

One time step on an IBM 360/50 computer required about 5 seconds for $(11)^3 = 1331$ mesh points; steady state was achieved in the order of 100 time steps. Some results were given by Mallinson and de Vahl Davis (1973) for a three-dimensional natural convection problem and for the three-dimensional motion generated in a box by the sliding in its own plane, of one wall. The method was found to be one to two orders of magnitude faster than a conventional iterative procedure when applied to a two-dimensional natural convection problem. The disadvantage of the method (apart from the obvious fact that the transient solution is lost) is that a large storage capacity is required: there are twelve variables associated with each mesh point (three components each of $\boldsymbol{\zeta}$, $\boldsymbol{\psi}$, and $\mathbf{v}$; T; ϕ; and a work space for the ADI procedure).

5.7 The development of free convection between heated vertical plates

Bodoia and Osterle (1962) investigated the development of free convection in a viscous fluid between heated vertical plates. The fluid was assumed to enter the channel with flat profiles $v_x = 0$ and $T = 0$, and to obey the Boussinesq (1903) approximation for an almost incompressible fluid (see Sec. 1.3). The geometry of the problem is shown in Figure 10. The governing equations in nondimensional primitive variable form are:

$$\frac{\partial v_x}{\partial x} + \frac{\partial v_y}{\partial y} = 0 \tag{5.34}$$

$$v_x \frac{\partial v_x}{\partial x} + v_y \frac{\partial v_x}{\partial y} = \frac{\partial^2 v_x}{\partial y^2} - \frac{dp}{dx} + T \tag{5.35}$$

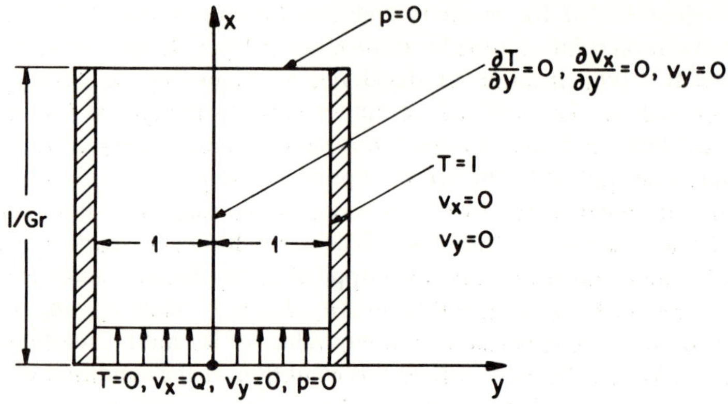

Figure 10 Schematic of free convection between parallel plates.

$$v_x \frac{\partial T}{\partial x} + v_y \frac{\partial T}{\partial y} = \frac{1}{\mathrm{Pr}} \frac{\partial^2 T}{\partial y^2} \tag{5.36}$$

where x and y are the vertical and transverse coordinates, respectively, while p represents the so-called x-dependent pressure defect (the perturbation of static pressure from the hydrostatic value). The dimensions of the pertinent region are width = 2 and height = 1/Gr, where Gr is a Grashof number. In order to completely define the problem, the following equation must be imposed for all x:

$$\int_0^1 v_x \, dy = Q \tag{5.37}$$

The boundary conditions are

$$v_x = Q \qquad v_y = T = 0 \qquad \text{at} \qquad x = 0 \qquad 0 \leqslant y \leqslant 1 \tag{5.38a}$$

$$\frac{\partial v_x}{\partial y} = v_y = \frac{\partial T}{\partial y} = 0 \qquad \text{at} \qquad y = 0 \qquad x \geqslant 0 \tag{5.38b}$$

$$v_x = v_y = 0 \qquad T = 1 \qquad \text{at} \qquad y = 1 \qquad x \geqslant 0 \tag{5.38c}$$

$$p = 0 \qquad \text{at} \qquad x = 0 \qquad \text{and} \qquad x = 1/\mathrm{Gr} \tag{5.38d}$$

The last boundary conditions specifies that the pressure defect p vanishes when the fluid emerges from the channel.

The coupled nonlinear system is hyperbolic in Eq. (5.34), parabolic with respect to v_x in (5.35) and parabolic with respect to T in (5.36). The solution was marched out in the x direction by dividing the region from $y = 0$ to $y = 1$ into n equal intervals and applying the following partially implicit system of FDA's:

$$\frac{\mu_y \delta_x}{h_x} v_{x_{i+\frac{1}{2}, j+\frac{1}{2}}} = \frac{\Delta_y}{h_y} v_{y_{i+1,j}} \tag{5.39}$$

$$v_{x_{i,j}} \frac{\Delta_x}{h_x} v_{x_{i,j}} + v_{y_{i,j}} \frac{\mu_y \delta_y}{h_y} v_{x_{i+1,j}} = \frac{\delta_y^2}{h_y^2} v_{x_{i+1,j}} - \frac{\Delta_x}{h_x} p_i + T_{i+1,j} \tag{5.40}$$

$$v_{x_{i,j}} \frac{\Delta_x}{h_x} T_{i,j} + v_{y_{i,j}} \frac{\mu_y \delta_y}{h_y} T_{i+1,j} = \frac{1}{\mathrm{Pr}} \frac{\delta_y^2}{h_y^2} T_{i+1,j} \tag{5.41}$$

and $$\sum_{j=1}^{n} v_{x_{i+1,j}} + \frac{1}{2} v_{x_{i+1,0}} = nQ \tag{5.42}$$

where (5.42) is the trapezoidal approximation to (5.37). Thus it is seen that the first FDA is uncoupled from the rest of the system. Moreover, the system has been linearized by using coefficient values of v_x and v_y from the i level in the FDA for the x-momentum and energy equations. For these two reasons the system of linear equations generated by the above FDA's while not tri-diagonal, possesses a relatively sparse coefficient matrix and was therefore solved from row i to $i + 1$ by a special form of the Gaussian elimination scheme. In this manner the solution was marched out in the x-direction until $p = 0$, which determined the end of the channel. The value of x at this level gave $1/\mathrm{Gr}$. Thus, while Pr and Gr were parameters of the problem, once Pr was chosen then the assumed value of Q dictated the resulting value for $1/\mathrm{Gr}$.

The technique outlined here has been extended for the same problem to three other possible boundary conditions. Miyatake and Fujii (1972, 1973) studied the case of one plate isothermal and the other plate insulated, and the case of unequal uniform plate temperatures. Miyatake et al. (1973) investigated the case of uniform flux at one plate with the other plate insulated. In all these three studies a fully implicit FDA was used with iteration at each level of x. At the start of each iteration the partically implicit FDA's of Bodoia and Osterle were used and the values obtained for the $x + h_x$ level were successively refined until convergence was obtained.

5.8 Solidification of a liquid film on a vertical wall

Isenberg and Gutfinger (1975) studied the mechanism of heat transfer to a falling glass-like film with temperature-dependent viscosity. Due to this dependency there exists a coupling between the equations of motion and energy.

The governing equations in nondimensional form in terms of the temperature T, stream function ψ, and film thickness F, are

$$\frac{\partial T}{\partial t} + \frac{1}{F}\frac{\partial \psi}{\partial y}\frac{\partial T}{\partial x} + \frac{1}{F}\frac{\partial (y\psi_s - \psi)}{\partial x}\frac{\partial T}{\partial y} = \frac{1}{F^2}\frac{\partial^2 T}{\partial y^2} \tag{5.43}$$

$$\frac{\partial^2 \psi}{\partial y^2} = F^3(1-y)\exp\left\{C\left[1 - \frac{1}{T + B(1-T)}\right]\right\} \tag{5.44}$$

and
$$\frac{\partial F}{\partial t} + \frac{\partial \psi_s}{\partial x} = 0 \tag{5.45}$$

where ψ_s is the value of ψ at $y = 1$. The boundary conditions for the system are:

$$\psi = \frac{\partial \psi}{\partial y} = 0 \quad \text{at} \quad y = 0 \tag{5.46a}$$

$$T = 1 \quad \text{at} \quad t = 0 \quad x > 0 \quad 0 \leqslant y \leqslant 1 \tag{5.46b}$$

$$T = 0 \quad \text{at} \quad x = 0 \quad t > 0 \quad 0 \leqslant y \leqslant 1 \tag{5.46c}$$

$$T = 0 \quad \text{at} \quad y = 0 \quad t > 0 \quad x > 0 \tag{5.46d}$$

$$\frac{\partial T}{\partial y} = 0 \quad \text{at} \quad y = 1 \quad t > 0 \quad x > 0 \tag{5.46e}$$

By the above boundary conditions the initially molten glass is cooled by the vertical wall ($y = 0$) maintained at $T = 0$, while the free interface ($y = 1$) is insulated. The parameters of the problem are B and C in Eq. (5.44).

Equation (5.43) is a nonlinear parabolic-hyperbolic equation in T, (5.44), effectively is a second order ODE in ψ, and (5.45) when recast in the form

$$\frac{\partial F}{\partial t} + \left(\frac{\partial \psi_s}{\partial F}\right)_t \frac{\partial F}{\partial x} = 0 \tag{5.47}$$

is seen to be a nonlinear first order hyperbolic equation in F.

The coupled system of (5.43)-(5.45) was solved iteratively. First the temperature at the line $(i+1, n+1)$ were calculated by the FDA scheme of Eq. (4.25) with the first derivative $\partial T/\partial y$ added in central difference form. The computed temperatures were then used to calculate ψ and F at the line $(i+1, n+1)$ via (5.44) and (5.45), and the iteration repeated until convergence was obtained.

5.9 Suggested literature

A number of works have appeared in the literature based on the original solution of free convection on a vertical plate (Hellums and Churchill) described in Sec. 4. Some have dealt with non-Newtonian fluids as exemplified by the studies of Fujii et al. (1973, 1974) of a Sutterby fluid for the cases of constant wall temperature and heat flux. The governing system of equations was solved by marching in the vertical direction with iterative correction between steps. A problem of greater complexity was the study of the steady laminar wake above a thin vertical isothermal plate cooled by natural convection, due to Hardwick and Levy (1973). The domain of interest was divided into two regions. The leading edge of the plate was solved by a *parabolic* set of equations, whose solution was applied to the *elliptic* set governing the trailing edge and the wake. The former region was solved by the Gauss-Seidel method while the latter was solved by applying the Dufort-Frankel method until steady state was attained. Unlike the study of Hellums and Churchill, these studies were only concerned with the steady-state solutions, hence none of the governing equations were parabolic–hyperbolic.

Systems of PDE's have also been solved for complex geometries. In an investigation of the effect of thermocapillary flows on heat transfer in film condensation, Cary and Mikic (1973) solved three coupled elliptic equations. The mathematical model consisted of a rectangle combined with a quadrant.

Ever since the pioneering work of Wilkes and Churchill (1966) on the ADI solution of two-dimensional free convection in a rectangular enclosure, numerous variations on the subject have appeared in the literature. Ozoe and Churchill (1972) extended the solution to non-Newtonian Ostwald de Waele and Ellis fluids. Greenspan and Shultz (1974) presented results for axisymmetric free convection in a vertical cylinder with localized heating from below. Jones (1974) analyzed the convective field generated by an exothermic chemical reaction in a gaseous medium within an infinitely long cylindrical enclosure. The reaction was represented by a distribution of heat sources with an Arrhenian temperature dependence and the flow field assumed to be laminar. Ozoe, Sayama and Churchill (1974) investigated free convection in an inclined square enclosure and their solution was subsequently extended to an inclined rectangular enclosure by Ozoe, Yamamoto, Sayama, and Churchill (1974). The example of Mallinson and de Vahl Davis (1973) of free convection in a box, cited earlier, constitutes an important numerical extension into the realm of three dimensions.

The phenomenon of axial corner flow has recently been numerically solved by two methods. Ghia and Davis (1974) obtained a set of four elliptic two-dimensional equations which were solved by the ADI method of Peaceman-Rachford. Their results are in good agreement with an earlier solution of Rubin and Grossman (1971) who employed the Gauss-Seidel method. The ADI method not only converged much faster, but exhibited better stability characteristics.

Several papers dealing with heat transfer and flow in curved tubes have appeared recently, motivated by the work of Dean (1928). The problem is interesting because of the secondary flow generated by the curvature, a phenomenon that enhances pressure drops and heat transfer rates. Some of the first numerical solutions were obtained by Mori and Nakayama (1965, 1967) for the laminar and turbulent regimes. Their solutions were marched out in the axial direction as a set of parabolic PDE's. More recently Miyazaki (1974) presented results for curved rectangular tubes rotating about an axis through the center of curvature of the tube. The governing set of elliptic equations were solved numerically by the SOR method.

Another subject of recent numerical research is the combined effects of forced and free convection in entrance regions. Ou et al. (1974) studied the case of large Prandtl number fluids in rectangular chambers. The governing equations are elliptic in both vorticity and stream function, and parabolic in temperature. The former were solved by the line iterative method, and the latter by the ADI method of Peaceman-Rachford. Yang and Yu (1974) presented results for magnetohydrodynamic flow between two parallel plates. The governing equations include both Jouleon effects and viscous dissipation and were solved iteratively in the vertical direction by the Crank-Nicolson method.

Bankston and Smith (1973) analyzed the steady laminar axisymmetric motion of an incompressible vapor in a heat pipe. The governing elliptic equations are elliptic in both the stream function and vorticity. In a companion paper, Rohani and Tien (1973) considered the two-dimensional heat and mass transfer in the vapor region of a heat pipe into which noncondensible gas has been introduced for the purpose of temperature control. They obtained four coupled elliptic equations with mass composition, temperature, stream function, and vorticity as the dependent variables. In a later work, Tien and Rohani (1974) also investigated the effects of vapor pressure drop on heat pipe performance. Three coupled elliptic equations, with stream function vorticity and enthalpy as the dependent variables, govern the phenomenon and were solved by the Gauss-Seidel method. These problems form part of a class of recirculating flows treated extensively by Gosman et al. (1969).

6 NEW DEVELOPMENTS AND TRENDS FOR FUTURE RESEARCH

Research continues on finite difference methods to achieve higher accuracy, faster convergence, and to reduce the computer memory required. The reader is best advised to study the latest literature appearing in the computer, and heat transfer journals to be advised of new developments, trends, and active fields of research. For this purpose, some of the more important recent studies have been summarized in this section.

Yalamanchili and Chu (1973) compared stability and oscillation characteristics of the finite difference, finite element, and weighted residual (Galerkin) methods for transient two-dimensional heat conduction. For both the stability and nonoscillation criteria, the finite element method was more restrictive than the Galerkin method, but less restrictive than the finite difference method. The latter result contrasts with an earlier study by Lemmon and Heaton (1969) for the one-dimensional case which showed the finite element method to be more restrictive than the finite difference method in nonoscillation characteristics and equally restrictive in stability. Obviously, more research is indicated on this subject.

An unconditionally stable second order accurate, implicit FDA was developed by Khosla and Rubin (1974) for parabolic equations. For the case of combined convection and conduction in a specific direction the limitation of a cell Peclet number of two for central differencing was overcome by a modified central difference. The scheme was proven to be stable for all step sizes, as exemplified by the author's application of the method to the nonlinear Burger's equation.

While the use of triangular networks have not been discussed here, they sometimes are used in cartesian coordinates in fitting irregular boundaries and in changing grid size (Dusinberre, 1961). In a recent paper, Mahata and McNary (1974) developed a similar scheme for axisymmetric problems in cylindrical coordinates. Four applications were given and the resulting accuracies were shown to be comparable to those obtained by both finite rectangular differencing and finite elements.

Singularities arise in heat transfer problems either from point sources or from sharp boundary corners. These singularities are conventionally treated by mesh refinement, which causes an increase in computer time and memory. Emery (1973) proposed a superior method that superimposes a closed-form analytical solution and a numerical solution in the vicinity of the singularity. As noted by Emery, the idea was first formulated mathematically by Motz (1947) in solving a Poisson equation for a sharp reentrant corner by finite differences. The technique has since been referred to as "singularity programming."

Numerous papers have appeared in the literature dealing with the moving boundary problem as typified by the diffusion of oxygen in absorbing tissue (Crank and Gupta, 1972a), melting or solidification (Boley, 1961) and freeze-drying (Greenfield, 1974). As noted by Gupta (1974) there is no exact analytical solution available for a *general* moving boundary problem, hence motivating various numerical methods to be proposed from time to time. (A good review of these methods is given by Gupta, and the reader is referred to his paper for more complete background.) Douglas and Gallie (1955) advocated dividing the entire region into equal intervals and adjusting the time step so as to move the boundary from node to node. Conversely, Murray and Landis (1959) proposed using a fixed number of intervals on either side of the moving boundary and adjusting the size of the intervals as the bounndary moves. Both of these methods can cause large inaccuracies or instability. Recently Crank and Gupta (1972b) employed cubic splines and polynomials to transfer the unequal interval from the vicinity of the moving boundary to the fixed boundary. The interpolations required by this method are avoided by the use of a Taylor's expansion in space and time, as proposed by Gupta (1974).

A class of problems of greater complexity is the *two-dimensional* moving boundary, where the position of the boundary is a function of two space variables. Early attempts to solve such problems (Dusinbere, 1949; Eyres et al., 1947) applied one energy equation to the entire region and artifically modified the specific heat in the vicinity of the moving boundary to account for the latent heat released. Other attempts have basically been variations of the one-dimensional methods mentioned above and have been more accurate than the one-region approach. Nonetheless, the latter methods require unequal and/or nonorthogonal mesh arrangements to cope with the moving boundary. Recently Kroeger and Ostrach (1974) used a sequence of conformal mappings to warp an arbitrary moving body such that all boundaries become coordinate lines in the transformed domain. The transformations must be carried out after each time step, but result in simple transformed equations and boundary conditions. The method was applied to solve the temperature and flow fields during solidification of a metal, which included the effects of free convection in the liquid region. The transformations are detailed by Kroeger (1972) in his thesis.

NOTATION

A	matrix operator
a	width of channel, also point operator
c_p	specific heat
c_ρ	density coefficient
D	derivative operator
$\mathfrak{D}$	divergence
E	shifting operator
F	film thickness
Gr	Grashof number
Gz	Graetz number
g	gravity
h	interval width. Also heat transfer coefficient
Ja	Jakob number
K	conductivity ratio
k	thermal conductivity
k_m	wave number
L	length
n	normal

Pe	Peclet number
Pr	Prandtl number
p	pressure. Also perturbation of static from hydrostatic pressure
Q	flow rate
R	dimensionless bubble radius
$\dot{R}$	rate of change of bubble radius
R_f	final dimensionless bubble radius
Ra	Rayleigh number
Re	Reynolds number
r	ratio, h_t/h^2, where h pertains either to x, y, or z. Also fraction of $\alpha h_t/h^2$. Also dimensionless radial distance from bubble wall
S	source
T	temperature
t	time
U	erroneous vector
v	velocity
x	coordinate
y	coordinate
z	coordinate

Greek symbols

α	thermal diffusivity. Also diffusion coefficient
β	coefficient associated with angle of Fourier term. Also coefficient of Boussinesq approximation
γ	coefficient associated with angle of Fourier term
Δ	forward difference operator
δ	central difference operator
ϵ	vector of errors. Also temperature at infinity
ζ	vorticity
η	similarity variable
Θ	temperature of Kirchhoff transformation
θ	value between zero and unity. Also angle in spherical coordinates. Also numerical phase shift.
λ	ratio, vh_t/h, where h is the mesh size associated with the direction of v. Also fraction of vh_t/h
μ	averaging operator. Also $-\cos\theta$. Also absolute viscosity
ν	kinematic viscosity
ξ	amplification factor
ρ	density
ϕ	scalar function for three-dimensional flow. Also velocity potential. Also phase angle
ψ	stream function for two-dimensional flow

ψ	vector function for three-dimensional flow
ω	radius in cylindrical coordinates. Also degree of over-relaxation
∇	backward difference operator. Also vector gradient

Subscripts

c	cold fluid
h	hot fluid
i,j,k	pertaining to space coordinates
0	T_0 or $T=0$
s	surface
t	time
x,y,z	space coordinates
θ,ω	space coordinates
ϕ,ψ	space coordinates (Boussinesq transformation)
∞	infinity

Superscripts

m	iteration
N	order
n	pertaining to time, t
$'$	nondimensional

BIBLIOGRAPHY

W. F. Ames, 1969, *Numerical Methods for Partial Differential Equations.* Nelson, London.

A. Arakawa, 1966. *J. Comp. Phys.*, vol. 1, p. 119.

K. Aziz, 1965, A numerical study of cellular convection. *Ph.D. Thesis,* Rice University, Houston, Texas.

C. A. Bankston and H. J. Smith, 1973. *ASME, J. Heat Transfer,* vol. 95, pp. 371–376.

M. Bentwich, G. Szwarcbaum, and S. Sideman, 1965. *Trans. ASME,* Paper 65-HT-38.

G. Birkhoff and R. S. Varga, 1959. *Trans. Am. Math. Soc.*, vol. 92, pp. 13–24.

J. R. Bodoia and J. F. Osterle, 1962. *Trans. ASME,* Ser. C., vol. 84, p. 40.

B. A. Boley, 1961. *J. Math. Phys.*, vol. 40, pp. 300–313.

Boussinesq, J., 1903. *Theorie Analytique de la Chaleur,* vol. 2, Gauthier-Villars, Paris.

P. H. Brazier, 1974. *Computer Methods in Appl. Mech. and Engng.*, vol. 3, pp. 335–347.

P. L. T. Brian, 1961. *AIChE J.*, vol. 7, p. 367.

L. S. Caretto, R. M. Curr, and D. B. Spalding, 1972. *Comp. Methods in Appl. Mech. and Engng.*, vol. 1, p. 39.

B. Carnahan, H. A. Luther, and J. O. Wilkes, 1969, *Applied Numerical Methods.* Wiley, New York.

J. D. Cary and B. B. Mikic, 1973. *ASME J. Heat Transfer,* vol. 95, pp. 21–24.

I. Catton, P. S. Ayyaswamy, and R. M. Clever, 1974. *Int. J. Heat and Mass Trans.*, vol. 17, p. 173.

R. T. Chen, 1972. *Phys. Fluids,* vol. 15, p. 2098.

K. C. Cheng and G. J. Hwang, 1969. *ASME J. Heat Transfer,* vol. 91, p. 59.

A. J. Chorin, 1967. *J. Comp. Phys.*, vol. 2, p. 12.
A. J. Chorin, 1968. *Math. Comp.*, vol. 22, p. 745.
A. J. Chorin, 1969. *Math. Comp.*, vol. 23, p. 341.
J. Crank and R. S. Gupta, 1972a. *J. Inst. Math. Applics.*, vol. 10, pp 19–33.
J. Crank and R. S. Gupta, 1972 b. *J. Inst. Math. Applics.*, vol. 10, pp. 296–304.
W. P. Crowley, 1967. *J. Comp. Phys.*, vol. 1, p. 471.
L. M. de Socio, E. M. Sparrow, and E. R. G. Eckert, 1973. *Computers and Fluids*, vol. 1, p. 273.
G. de Vahl Davis, 1968. *Int. J. Heat and Mass Transfer*, vol. 11, p. 1675.
G. de Vahl Davis and R. W. Thomas, 1969. *Phys. Fluids*, vol. 12, (suppl. II), p. 198.
G. de Vahl Davis and G. D. Mallinson, To be published.
G. de Vahl Davis and M. Wolfshtein, 1974. *Proc. 4th Int. Conf. Numerical Methods in Fluid Dynamics*, to be published.
W. R. Dean, 1928. *Phil. Mag. and J. of Sci.*, 7th. Ser., vol. 5, p. 673.
A. A. Dorodnicyn, 1969, *Proc. 4th Australian Comp. Conf.*, vol. 2, The Griffin Press, Nethey, S. Australia.
J. Douglas, Jr., 1961. *Num. Math.*, vol. 4, p. 41.
J. Douglas and T. M. Gallie, 1955. *Duke Math. J.*, vol. 22, pp. 557–570.
G. M. Dusinberre, 1949, *Numerical Analysis of Heat Flow*. McGraw-Hill, New York.
G. M. Dusinberre, 1961, *Heat Transfer Calculations by Finite Differences*. International Textbook Co., Scranton, Pa.
T. W. Elder, 1966. *J. Fluid Mech.*, vol. 24, p. 823.
A. F. Emery, 1973. *ASME J. Heat Transfer*, vol. 95, pp. 344–351.
N. R. Eyres, D. R. Hartree, J. Ingram, R. Jackson, R. J. Sargant, and J. B. Wagstaff, 1947. *Phil. Trans. R. Soc.*, vol. 240, Ser. A.1.
L. Fox, 1962, *Numerical Solution of Ordinary and Partial differential Equations*. Addison-Wesley, Reading, Massachusetts.
S. P. Frankel, 1950. *Math. Tables Aids Comput.*, 4, pp. 65–75.
J. E. Fromm, 1968. *J. Comp. Phys.*, vol. 3, p. 176.
J. E. Fromm, 1971, *Proc. 2nd Int. Conf. Num. Math. Fluid Dynamics*, Springer-Verlag, Berlin, p. 120.
T. Fujii, O. Miyatake, M. Fujii, H. Tanaka, and K. Murakami, 1973. *Int. J. Heat and Mass Transfer*, vol. 16, pp. 2177–2187.
T. Fujii, O. Miyatake, M. Fujii, H. Tanaka, and K. Murakami, 1974. *Int. J. Heat and Mass Transfer*, vol. 17, pp. 149–154.
K. N. Ghia and R. T. Davis, 1974. *Computers and Fluids*, vol. 2, pp. 17–34.
A. D. Gosman, W. M. Pun, A. K. Runchal, D. B. Spalding, and M. Wolfshtein, 1969, *Heat and Mass Transfer in Recirculating Flows*, Academic Press, London.
A. W. Graves and I. Catton, 1973. *ASME J. Heat Transfer*, vol. 95, pp. 559–562.
P. F. Greenfield, 1974. *Chem. Engng. Sci.*, vol. 29, pp. 2115–2123.
D. Greenspan and D. Schultz, 1974. *Computer Methods in Appl. Mech. and Engng.*, vol. 3, pp. 1–10.
R. S. Gupta, 1974. *Computer Methods in Appl. Mech. and Engng.*, vol. 4, pp. 143–152.
C. Gutfinger, J. Isenberg, and M. A. Zeitlen, 1970. *Israel J. Tech.*, vol. 8, pp. 225–237.
N. E. Hardwick and E. K. Levy, 1973. *ASME J. Heat Transfer*, vol. 95, pp. 289–294.
J. D. Hellums, 1960, Finite difference calculation of natural convection heat transfer. *Ph.D. Thesis*, University of Michigan, U.S.A.
J. D. Hellums and S. W. Churchill, 1961. *Int. Dev. in Heat Transfer*, vol. 5, p. 985.
F. B. Hildebrand, 1968, *Finite Difference Equations and Simulations*. Prentice Hall, Englewood Cliffs, New Jersey.
F. B. Hildebrand, 1974. *Introduction to Numerical Analysis*, 2nd ed. McGraw-Hill, New York.

G. J. Hirasaki and J. D. Hellums, 1968. *Quart. Appl. Math.*, vol. 26, p. 331.
G. J. Hirasaki and J. D. Hellums, 1970. *Quart. Appl. Math.*, vol. 28, p. 293.
M. F. Hussein and W. Tabakoff, 1974. *Computers and Fluids,* vol. 2, pp. 1–15.
G. J. Hwang and J. P. Sheu, 1974. *Int. J. Heat and Mass Transfer,* vol. 17, pp. 372–375.
E. Isaacson and H. B. Keller, 1966, *Analysis of Numerical Methods.* John Wiley & Sons, Inc., New York.
J. Isenberg, 1975. An evaluation of finite difference schemes for parabolic–hyperbolic equations. Submitted for publication.
J. Isenberg and C. Gutfinger, 1973. *Int. J. Heat and Mass Transfer,* vol. 16, pp. 505–512.
J. Isenberg and C. Gutfinger, 1975. Solidification of a liquid film on a vertical wall. *Chem. Engng. Sci.,* vol. 30, pp. 327–334.
J. Isenberg and S. Malkin, 1975. Effects of variable thermal properties on moving source temperatures. Accepted for publication in *ASME J. Engng. for Industry.*
J. Isenberg and S. Sideman, 1970. *Int. J. Heat and Mass Transfer,* vol. 13, pp. 997–1011.
J. C. Jaeger, 1942. *Proc. Roy. Soc. N.S.W.*, vol. 76, p. 203.
D. R. Jones, 1974. *Int. J. Heat and Mass Transfer,* vol. 17, pp. 11–21.
P. K. Khosla and S. C. Rubin, 1974. *Computers and Fluids,* vol. 2, pp. 207–209.
D. B. Killeen, 1966, The numerical solution of equations describing and unsteady draining process. *Ph.D. Thesis,* Tulane University, U.S.A.
P. G. Kroeger, 1972. *Ph.D. Thesis,* Case Western Reserve University, Cleveland, Ohio.
P. G. Kroeger and S. Ostrach, 1974. *Int. J. Heat and Mass Transfer,* vol. 17, pp. 1191–1207.
H. L. Langhaar, 1942. *J. Appl. Mech.*, vol. 9, pp. A55–A58.
P. D. Lax and B. Wendroff, 1960. *Communications in Pure and Appl. Math.*, vol. 13, p. 217.
P. D. Lax and B. Wendroff, 1964. *Communications in Pure and Appl. Math.*, vol. 17, p. 381.
E. C. Lemmon and H. S. Heaton, 1969. *ASME Paper,* No. 69-WA/HT-35.
P. C. Mahata and O. McNary, 1974. *International Journal for Numerical Methods in Engineering,* vol. 8, pp. 547–567.
G. D. Mallinson, 1973. *Ph.D. Thesis,* University of New South Wales, Kensington, Australia.
G. D. Mallinson and G. de Vahl Davis, 1973. *J. Comp. Phys.*, vol. 12, p. 435.
A. R. Mitchell, 1969, *Computational Methods in Partial Differential Equations.* John Wiley, London.
O. Miyatake and T. Fujii, 1972. *Heat Transfer–Japanese Research,* vol. 1, No. 3, pp. 30–38.
O. Miyatake and T. Fujii, 1973. *Heat Transfer–Japanese Research,* vol. 2, No. 4, pp. 79–88.
O. Miyatake, T. Fujii, M. Fujii, and H. Tanaka, 1973. *Heat Transfer–Japanese Research,* vol. 2, No. 1, pp. 25–33.
H. Miyazaki, 1973. *ASME J. Heat Transfer,* vol. 95, pp. 64–71.
Y. Mori and W. Nakayama, 1965. *Int. J. Heat and Mass Transfer,* vol. 8, pp. 67–82.
Y. Mori and W. Nakayama, 1967. *Int. J. Heat and Mass Transfer,* vol. 10, pp. 37–59.
H. Motz, 1947. *Quart. Appl. Math.*, vol. 4, p. 371.
W. D. Murray and F. Landis, 1959. *ASME J. Heat Transfer,* vol. 81, pp. 106–112.
J. W. Ou, K. C. Cheng, and R. C. Lin, 1974. *Int. J. Heat and Mass Transfer,* vol. 17, pp. 835–843.
H. Ozoe and S. W. Churchill, 1972. *AIChE J.*, vol. 18, pp. 1196–1207.
H. Ozoe, H. Sayama, and S. W. Churchill, 1974. *Int. J. Heat and Mass Transfer,* vol. 17, pp. 401–406.
H. Ozoe, K. Yamamoto, H. Sayama, and S. W. Churchill, 1974. *Int. J. Heat and Mass Transfer,* vol. 17, pp. 1209–1217.
D. W. Peaceman and H. H. Rachford, 1955. *J. SIAM,* vol. 3, p. 28.
C. Quon, 1972. *Phys. Fluids,* vol. 15, p. 12.
G. D. Raithby and K. E. Torrance, 1974. *Computers and Fluids,* vol. 2, pp. 191–206.
L. F. Richardson, 1910. *Trans. Roy. Soc., London,* vol. A210, pp. 307–357.

L. F. Richardson, 1925. *Math. Gazette,* vol. 12, pp. 415–421.

L. F. Richardson, 1965, *Weather Prediction by Numerical Process.* Cambridge University Press, London, 1923. Republished Dover Publications Inc., New York.

L. F. Richardson and J. A. Gaunt, 1927. *Trans. Roy. Soc., London,* vol. A226, pp. 299–361.

R. D. Richtmyer and K. W. Morton, 1967, *Difference Methods of Initial Value Problems.* Wiley (Interscience), New York.

P. J. Roache, 1973, *Numerical Fluid Dynamics.* Hermosa Publishers, Albuquerque, New Mexico.

K. V. Roberts and N. O. Weiss, 1966. *Math. Comp.,* vol. 20, p. 272.

A. R. Rohani and C. L. Tien, 1973. *ASME J. Heat Transfer,* vol. 95, pp. 377–382.

A. Rubel and F. Landis, 1969. *Phys. Fluids,* vol. 12, (suppl. II), p. 208.

S. C. Rubin and B. Grossman, 1971. *Q. Appl. Math.,* vol. 29, pp. 169–186.

A. A. Samarskii, 1963. *USSR Comp. Math. Math. Phys.,* vol. 2, p. 894.

A. A. Samarskii and V. B. Andreev, 1963. *USSR Comp. Math. Math. Phys.,* vol. 3, p. 1373.

R. Siegel and E. M. Sparrow, 1959. *ASME J. Heat Transfer,* vol. 81, pp. 29–36.

G. D. Smith, 1965, *Numerical Solution of Partial Differential Equations.* Oxford Press, London.

J. E. Stack, 1967, The numerical solution of a two-dimensional moving boundary problem. *Ph.D. Thesis,* Tulane University, U.S.A.

P. P. Starling, 1963. *Brit. J. Appl. Phys.,* vol. 14, pp. 603–605.

A. Tamir and I. Rachmilev, 1974. *Int. J. Heat and Mass Transfer,* vol. 17, 1241–1251.

L. H. Thomas, 1949, *Elliptic problems in linear difference equations over a network.* Watson Scientific Computing Laboratory, Columbia University, New York.

R. W. Thomas and G. de Vahl Davis, 1971. *Heat Transfer 1970,* vol. 4, Paper NC 2.4, Elsevier, Amsterdam.

C. L. Tien and A. R. Rohani, 1974. *Int. J. Heat and Mass Transfer,* vol. 17, pp. 61–67.

V. N. Vedamurthy and V. M. K. Sastri, 1974. *Int. J. Heat and Mass Transfer,* vol. 17, pp. 1–9.

D. U. von Rosenberg, 1969, *Methods for the Numerical Solution of Partial Differential Equations.* American Elsevier, New York.

B. Wendroff, 1960. *J. Soc. Indust. Appl. Math.,* vol. 8, pp. 549–555.

J. O. Wilkes and S. W. Churchill, 1966. *AIChE J.,* vol. 12, pp. 161–166.

G. P. Williams, 1969. *J. Fluid Mech.,* vol. 37, p. 727.

D. D. Wittke, 1966, Collapse of vapor bubbles with translatory motion. *Ph.D. Thesis,* University of Illinois, U.S.A.

D. D. Wittke and B. T. Chao, 1967. *ASME J. Heat Transfer,* vol. 89, pp. 17–24.

M. Wolfshtein, 1968, Numerical smearing in one-sided difference approximations to the equations of non-viscous flow. Imperial College, Department of Mechanical Engineering, Report EF/TN/A/3.

L. C. Woods, 1954. *Aero. Quarterly,* vol. 5, p. 176.

R. V. S. Yalamanchili and S. C. Chu, 1973. *ASME J. Heat Transfer,* vol. 95, pp. 235–239.

H. K. Yang and C. P. Yu, 1974. *Int. J. Heat and Mass Transfer,* vol. 17, pp. 681–691.

D. M. Young, 1962, in J. Todd, Ed., *Survey of Numerical Analysis,* pp. 380–435, McGraw-Hill, New York.

6

Application of Asymptotic Methods to Laminar Flame Theory

G. Sivashinsky and C. Gutfinger

Faculty of Mechanical Engineering
Technion—Israel Institute of Technology
Haifa, Israel

CONTENTS

1 INTRODUCTION

The theory of a laminar flame in a gas, which combines the hydrodynamics of a compressible fluid with chemical kinetics, has now come to occupy a firm position in fluid mechanics. The mathematical problems arising in flame theory are so unique that the theory has every chance of becoming one of the most beautiful branches of mathematical physics. In flame theory, as in no other branch of fluid mechanics, one constantly encounters such problems as, for example, the existence of solutions representing the simple physical reality that the combustible mixture cannot always be ignited. In other words, will a flame be formed or not? The essential nonlinearity of many flame problems frequently faces one with the need to choose one of several possible solutions. It thus becomes important to investigate the stability of the possible steady solutions.

In his well-known review of the theory of combustion in laminar flows, Williams [1] mentioned a mathematical tool of considerable promise for the treatment of many flame problems: the matching principle for singular asymptotic expansions [2, 3] in which the large parameter is the activation energy of the chemical reaction:

$$\frac{E}{R^0 T_b} \gg 1 \tag{1.1}$$

Here E is the activation energy, R^0 the universal gas constant, and T_b the adiabatic temperature of the combustion products. High activation energy represents the basic physical reality that the rate of the chemical reaction is strongly temperature-dependent, $\sim \exp(-E/R^0 T)$.

Condition (1.1) is not only useful as indicating the applicability of the asymptotic approach, but is also physically justifiable for almost all cases of nonisothermal combustion. If this condition is not taken into consideration, one must allow for the rate of the chemical reaction upstream the flame, in the fresh mixture. But it is well known that the reaction rate in the cold mixture is extremely low (relative to the rate in the reaction zone), and its inclusion is devoid of any physical meaning.

On the other hand, inclusion of the reaction rate in the fresh mixture would not allow introduction of such a fundamental concept of flame theory as the normal propagation velocity of the flame, for in that case the propagation velocity could never (even asymptotically) approach a constant value.

This review constitutes an attempt to describe mathematical flame theory as a consistent asymptotic theory, with the quantity $R^0 T_b/E$ treated as a small parameter. Discussing various physical problems, we have also tried to devote some attention to the characteristic mathematical points that come up in an analysis of the equations.

The basic simplifying assumptions used in asymptotic flame theories are the following:

1. All the reactants are slight impurities in some substance M_0.
2. The effects of multicomponent diffusion and thermal diffusion are disregarded.
3. The specific heat as well as the ratios of transport coefficients (Lewis number L and Prandtl number σ) are constant.
4. The propagation velocity of the flame is small relative to the speed of sound. For a flame propagating in an open space, this condition is equivalent to dynamic incompressibility of the gaseous mixture. In other words, the density depends only on temperature. For a flame propagating in a closed volume, the pressure is no longer constant, but is nevertheless instantaneously equalized throughout the volume.

As far as the possibilities of matched asymptotic expansions are concerned, none of these assumptions is really essential, and if necessary, effects originating in the neglected processes may be included.

The basic equations describing the structure of a laminar flame in an open space may be written as follows:

The equations of the chemical reaction:

$$\sum_k \nu'_{nk} M_k \to \sum_k \nu''_{nk} M_k \qquad k = 1, \ldots, k^* \qquad n = 1, \ldots, n^* \tag{1.2}$$

The heat transfer (energy) equation:

$$c_p \frac{\partial \rho T}{\partial t} + c_p \frac{\partial \rho U_i T}{\partial x_i} = \frac{\partial}{\partial x_i} \kappa \frac{\partial T}{\partial x_i} + \sum_n Q_n \Phi_n \qquad \kappa = \kappa(T) \tag{1.3}$$

The diffusion equation for the reactants:

$$\frac{\partial \rho C_k}{\partial t} + \frac{\partial \rho U_i C_k}{\partial x_i} = \frac{\partial}{\partial x_i} D_k \rho \frac{\partial C_k}{\partial x_i} + \sum_n (\nu''_{nk} - \nu'_{nk}) \Phi_n \qquad \frac{\kappa}{c_p \rho D_k} = \mathrm{L}_k \quad \text{(Lewis number)} \tag{1.4}$$

The equations of chemical kinetics:

$$\Phi_n = z_n(T) \prod_k (\rho C_k)^{\nu'_{nk}} \exp\left(-\frac{E_n}{R^0 T}\right) \tag{1.5}$$

The continuity equation:

$$\frac{\partial \rho}{\partial t} + \frac{\partial \rho U_i}{\partial x_i} = 0 \tag{1.6}$$

The momentum equations (Navier-Stokes):

$$\frac{\partial \rho U_i}{\partial t} + \frac{\partial \rho U_i U_j}{\partial x_i} + \frac{\partial P}{\partial x_i} = \frac{\partial P_{ij}}{\partial x_j} \tag{1.7}$$

where $P_{ij} = \mu \left(\frac{\partial U_i}{\partial x_j} + \frac{\partial U_j}{\partial x_i} - \frac{2}{3} \delta_{ij} \frac{\partial U_m}{\partial x_m} \right)$

$\frac{\mu c_p}{\kappa} = \sigma$ (Prandtl number)

The equation of state (dynamic incompressibility):

$$\rho T = \text{const} \tag{1.8}$$

We conclude this section with a remark concerning a strong assumption frequently employed in flame theory. The diffusion and heat conduction equations in the system (1.2)–(1.8) are related to the equations of hydrodynamics through the temperature-dependence of density. If we adopt the "artificial" assumption that the density in the equation of state (1.8) is constant ($\rho = \rho_b$), the hydrodynamic problem may be solved separately from the problem of combustion proper. Following Zel'dovich et al. [4], we shall call a model ignoring thermal expansion a diffusional thermal model of the flame. If the combustible mixture is immobile at the initial time, the diffusional-thermal model does not allow any thermal perturbations in the flame to be transformed into hydrodynamic perturbations, and the mixture remains immobile. The equations of laminar flame theory then reduce to the "solid-state" system of diffusion and heat conduction equations.

$$\rho_b c_p \frac{\partial T}{\partial t} = \frac{\partial}{\partial x_i} \kappa \frac{\partial T}{\partial x_i} + \sum_n Q_n \Phi_n \tag{1.9}$$

$$\rho_b \frac{\partial C_k}{\partial t} = \rho_b \frac{\partial}{\partial x_i} D_k \frac{\partial C_k}{\partial x_i} + \sum_n (\nu''_{nk} - \nu'_{nk}) \Phi_n \tag{1.10}$$

In the framework of the diffusional thermal model, it is frequently possible to give a correct qualitative description of *strong* diffusional-thermal effects in the flame, when the perturbation of hydrodynamic flux is insignificant. In some cases, however, as we shall see in the next section, the hydrodynamic effects may compete with diffusional thermal effects and it is then necessary to consider the exact combustion equations.

2 NORMAL FLAME PROPAGATION

The propagation velocity of a steady plane flame of premixed fuel is called the normal velocity of the flame. It is assumed that the combustion takes place under adiabatic conditions.

Because of thermal expansion of the gas and the resulting acceleration of the gas flow, the normal propagation velocity of the combustion wave is measured relative to the combustion products.

We shall first confine ourselves to combustion described by a one-step monomolecular reaction:

$$\nu M \rightarrow \text{combustion products} \tag{2.1}$$

Formally, any one-step reaction in which one of the reactants (M) is present in small concentrations may be assumed to be monomolecular with respect to the reactant.

Our starting point is a diffusional-thermal model, ignoring compressibility. For reaction (2.1), system (1.9)-(1.10), (1.5) has the following form (one-dimensional flame propagation):

$$\rho_b \frac{\partial T}{\partial t} = \frac{\kappa_b}{c_p} \frac{\partial^2 T}{\partial x^2} + \frac{Q}{c_p} \Phi(C, T) \tag{2.2}$$

$$\rho_b \frac{\partial C}{\partial t} = \frac{\kappa_b}{\mathrm{L} c_p} \frac{\partial^2 C}{\partial x^2} - \nu \Phi(C, T) \tag{2.3}$$

$$\Phi = z_b {\rho_b}^\nu C^\nu \exp\left(-\frac{E}{R^0 T}\right) \tag{2.4}$$

ρ, κ, D, and z are constants and thus designated by the subscript b.

If the initial mass fraction (or concentration) of substance M in the mixture is C_0 and the initial temperature is T_0, it is easy to see that the temperature T_b attained in burning under adiabatic conditions is

$$T_b = T_0 + \frac{QC_0}{c_p \nu} \tag{2.5}$$

T_b is called the adiabatic temperature of the combustion products.

Suppose that at time $t = 0$ we have the following step distributions of mass concentration and temperature:

$$\begin{aligned} C(x, 0) &= C_0 \quad T(x, 0) = T_0 \quad \text{for} \quad x < 0 \\ C(x, 0) &= 0 \quad T(x, 0) = T_b \quad \text{for} \quad x > 0 \end{aligned} \tag{2.6}$$

At $t > 0$, both steps begin to erode owing to transport effects. Nonzero concentration appears in the high temperature region. The chemical reaction of combustion begins and must reach conditions in which the flame propagates as a wave travelling in the negative x-direction.

It is easy to see that the initial state $T = T_0, C = C_0$ is not an equilibrium state of system (2.2)-(2.4), since

$$\Phi(C_0, T_0) \neq 0$$

Thus the concentration of the reactant will decrease continuously at $-\infty$, while the temperature will increase. Consequently, the flame formed can never reach the state of steady-wave propagation. It can be shown, however, that if $E/R^0 T_b \gg 1$ one can ignore burn-out (i.e., the reaction) in the fresh mixture to within exponentially high accuracy [5]. Thus, for time intervals which are not too large (in a certain sense) one can expect the flame to tend asymptotically to the behavior of a wave propagating at a constant velocity U_b:

$$T\left(x, t, \frac{E}{R^0 T_b}\right) = T\left(x + U_b{}^t, \frac{E}{R^0 T}\right)$$
$$C\left(x, t, \frac{E}{R^0 T}\right) \cong C\left(x + U_b{}^t, \frac{E}{R^0 T}\right) \tag{2.7}$$

It may be shown, a posteriori, that the asymptotic wave behavior is valid over a fairly broad interval:

$$\frac{\kappa_b}{c_p U_b{}^2} \lesssim t \ll \frac{\kappa_b}{c_p U_b{}^2}\left(\frac{E}{R^0 T_b}\right)^{-1-\nu} \exp\left[\frac{E}{R^0 T_0 T_b}(T_b - T_0)\right]$$

We now turn to the as yet unknown definition of the velocity U_b of normal flame propagation. Transforming variables from $x + U_b t$ to x, we obtain from (2.2), (2.3), and (2.7) a system of ordinary differential equations:

$$\rho_b U_b \frac{dT}{dx} = \frac{\kappa_b}{c_p}\frac{d^2 T}{dx^2} + \frac{Q}{c_p}\Phi(C, T) \tag{2.8}$$

$$\rho_b U_b \frac{dC}{dx} = \frac{\kappa_b}{\mathrm{L}c_p}\frac{d^2 C}{dx^2} - \nu\Phi(C, T) \tag{2.9}$$

The solution of this system must satisfy four boundary conditions, to within exponentially high accuracy (in $R^0 T_b/E$):

$$T(-\infty) = T_0 \qquad C(-\infty) = C_0 \tag{2.10}$$

$$T(+\infty) = T_b \qquad C(+\infty) = 0 \tag{2.11}$$

In view of the form of Eqs. (2.8), (2.9), and the boundary conditions (2.10), (2.11), the solution is invariant with respect to translation along the x-axis. That is to say, the four boundary conditions must completely determine the remaining three undefined constants of the general solution. This is possible only for selected values of U_b.

Thus, solving the problem as a whole, one finds both the temperature and concentration distribution profiles in the combustion wave, and its propagation velocity.

We now briefly discuss the history of this problem since the theory of thermal flame propagation was first proposed by Zel'dovich and Frank-Kamenetskii [6, 7, 8].

To overcome the difficulty on the cold boundary, Zel'dovich suggested an artificial truncation of the heat generation function [6]. In other words, one selects, more or less arbitrarily, a certain ignition temperature $T_* > T_0$ such that $\Phi \equiv 0$ for $T_0 < T < T_*$.

This operation allowed the formulation of a problem which has a rigorous solution in the form of waves propagating at constant velocity. The problem has been thoroughly investigated on the assumption that the temperature and concentration fields are similar, i.e., the Lewis number, L, is equal to unity.

The very justification to neglect the reaction rate at low temperatures was due to a strong temperature-dependence of the reaction rate, or, in other words, a high activation energy. The most important corollaries of the above solution are the following:

1. The width of the reaction zone l_r is smaller (in order of magnitude) by a factor of $E/R^0 T_b$ than the width of the preheat zone l_T:

$$l_T \sim \frac{l_r E}{R^0 T_b} \qquad l_T = \frac{\kappa_b}{\rho_b U_b c_p} \tag{2.12}$$

2. The relationship between the flame propagation velocity and the activation energy is (assuming $E/RT_b \gg 1$):

$$U_b \sim U_* \left(\frac{E}{R^0 T_b}\right)^{(-1+\nu)/2} \exp\left(-\frac{E}{2R^0 T_b}\right) \tag{2.13}$$

   where $U_* = (z_b \kappa_b C_0{}^{\nu-1} \rho_b{}^{\nu-2} c_p{}^{-1})^{1/2}$ may be interpreted as the effective propagation velocity of the flame at zero activation energy (i.e., when every collision of reactant molecules is chemically effective). Equation (2.13) shows that because of the high activation energy the actual propagation velocity of the flame is an exponentially small quantity in comparison with U_*.

3. In the reaction zone:

$$\frac{T_b - T}{T_b} \lesssim \frac{R^0 T_b}{E} \tag{2.14}$$

In [7, 8], discussing the definition of the normal propagation velocity of the flame, Zel'dovich and Frank-Kamenetskii suggested dividing the flame into two regions: preheat zone and reaction zone. In the first region one neglects the reaction rate, in the second–convective heat transfer. The inner and outer solutions were matched on the basis of an intuitive matching principle. The authors of this approach also emphasized the legitimacy of all the operations by virtue of the strong temperature-dependence of the reaction rate. The division of the combustion zone into regions external and internal to the reaction zone made it possible to derive a formula for the flame propagation velocity even when the concentration and temperature fields were not similar ($L \neq 1$) [5]. It was shown that the propagation velocity is proportional to $L^{\nu/2}$.

The results of [7, 8] were immediately applied to the investigation of actual mechanisms of combustion reactions [9–11]. Zel'dovich then extended the idea of subdividing the reaction zone to flames in systems with chain reactions [12]. Accounts of the Zel'dovich-Frank-Kamenetskii method have been given in many review articles and monographs on combustion [5, 13, 14].

After appearance of the works of Lagerstrom, Cole, and Kaplun [15, 16], where the principle of asymptotic matching was first formulated with regard to singular perturbation problems in hydrodynamics, several authors evinced renewed interest in normal flame propagation as a singular perturbation problem, proposing to take the quantity $R^0 T_b/E$ as the small parameter. It was important to derive the propagation velocity of the flame from a rigorous asymptotic matching principle, and also to obtain higher-order approximations. A comparison of the well-known Frank-Kamenteskii–Zel'dovich formulas with some numerical results [17] nevertheless showed a marked discrepancy between the exact (numerical) solution and the asymptotic solution for relatively small activation energies ($E/R^0 T_b \sim 5$). The available approaches precluded any possibility of constructing higher approximations and sharpening the analytical results.

The first published work applying asymptotic expansions to normal flame propagation was the investigation of Bush and Fendell [18]. Then Berman and Ryazantsev [19] somewhat generalized the results of [18] to the case of variable density and transport coefficients. It was shown rigorously in [18] that, of all the existing intuitive matching principles [5], only the so-called "heat balance" method gave a correct asymptotic expression for the first approximation to the flame velocity. In addition, a first rigorous justification was provided for the expansion of the exponent in Arrhenius' law, introduced by Frank-Kamenetskii [5] in the steady-state theory of thermal explosion and later used in publications on flame theory [7, 8]. Bush and Fendell [18] have also calculated a second approximation to the flame velocity, which showed excellent agreement with the numerical solution (for $E/R^0 T_b > 2$). Berman and Ryazantsev then used the method previously developed [18, 19] for problems in normal flame propagation in a two-stage exothermic reaction, characterized by two high activation energies [20–22]. These papers opened up new vistas in flame propagation research in real systems, e.g., chemical reactions, which almost always involve complex successive and parallel stages.

We now discuss the basic arguments of [18] in more detail. As mentioned above, the density distribution of heat and mass sources in the combustion wave is such that the main heat generation and chemical transformation are concentrated in a narrow zone l_r, of the order of $l_T R^0 T_b/E$. This led to the development of two independent asymptotic expansions, one outside the reaction zone, in a region whose characteristic dimension depends on the thermal thickness l_T of the flame and the dimensionless space variable x/l_T,* and the other in the reaction zone, whose characteristic dimension is $l_r \sim l_T R^0 T_b/E$, the corresponding dimensionless space variable being $xE/R^0 T_b l_T$. In each of these regions, one looks for T and C as expansions in

$$\left(\frac{R^0 T_b}{E}\right)^0, \quad \left(\frac{R^0 T_b}{E}\right)^1, \quad \left(\frac{R^0 T_b}{E}\right)^2, \quad \cdots \tag{2.15}$$

The velocity U_b (see [2.13]) is sought as a series in

$$\begin{gathered}\left(\frac{E}{R^0 T_b}\right)^{-(1+\nu)/2} \exp\left(-\frac{E}{2R^0 T_b}\right) \\ \left(\frac{E}{R^0 T_b}\right)^{-(3+\nu)/2} \exp\left(-\frac{E}{2R^0 T_b}\right) \quad \cdots\end{gathered} \tag{2.16}$$

The asymptotic expansion outside the reaction zone must satisfy the boundary conditions at infinity (2.10), (2.11), while the expansion in the reaction zone must be asymptotically matched to the outer expansion, in accordance with the general technique of, e.g., [2, 3].

Let us eliminate Φ between Eqs. (2.8) and (2.9) and integrate the resulting equation with conditions (2.10), (2.11), and (2.5) to obtain:

$$\rho_b U_b \left[T - T_b + \frac{C}{C_0}(T_b - T_0)\right] = \frac{\kappa_b}{c_p}\frac{d}{dx}\left[T + \frac{1}{\mathrm{L}}\frac{C}{C_0}(T_b - T_0)\right] \tag{2.17}$$

We now introduce the dimensionless quantities and parameters

$$\begin{gathered}\widetilde{T} = \frac{T}{T_b} \quad \widetilde{C} = \frac{C}{C_0} \quad \epsilon = \frac{T_0}{T_b} \quad N = \frac{E}{R^0 T_b} \\ A = z_b C_0{}^{\nu-1} \rho_b{}^{\nu-2} \kappa_b U_b{}^{-2} c_p{}^{-1} N^{-1-\nu} \exp(-N)\end{gathered} \tag{2.18}$$

In view of (2.16), the asymptotic expansion of A is to be sought as a series in 1, $N^{-1}, N^{-2}, \ldots$. We now introduce a new dimensionless function:

$$P(\widetilde{T}) = l_T \frac{d\widetilde{T}}{dx} \qquad l_T = \frac{\kappa_b}{\rho_b U_b c_p} \tag{2.19}$$

*I.e., we assume that the reaction zone is concentrated in a neighborhood of $x = 0$.

Using the new notation (2.18), (2.19), we can reduce (2.8), (2.17), (2.10), (2.11) to the following:

$$P + \frac{1-\epsilon}{\mathrm{L}} P \frac{d\widetilde{C}}{d\widetilde{T}} = \widetilde{T} - 1 + (1-\epsilon)\widetilde{C} \tag{2.20}$$

$$P = P\frac{dP}{d\widetilde{T}} + AN^{1+\nu}\widetilde{C}^{\nu} \exp\left[N\left(1 - \frac{1}{\widetilde{T}}\right)\right] \tag{2.21}$$

$$P(1) = \widetilde{C}(1) = 0 \tag{2.22}$$

$$P(\epsilon) = 0 \qquad \widetilde{C}(\epsilon) = 1 \qquad (\epsilon \leqslant \widetilde{T} \leqslant 1) \tag{2.23}$$

One thus goes over from the independent space variable, x, to the independent variable T. The reaction zone corresponds to the region of $\widetilde{T} - 1 \sim N^{-1}$ (see Eq. [2.14]). We thus have the following (inner) characteristic variable for the reaction zone:

$$\theta = N(T-1) \tag{2.24}$$

We retain $\widetilde{T}$ as the characteristic variable for the preheat zone (outer variable).

The inner asymptotic expansion of the solution must satisfy conditions at $\widetilde{T} = 1$, the outer expansion at $\widetilde{T} = \epsilon$. In both zones, the expansions of the functions P, C, and the constant A are sought as series in integral powers of N^{-1}:

$$A = A_0 + \frac{1}{N}A_1 + \frac{1}{N^2}A_2 + \cdots \tag{2.25}$$

In the preheat zone:

$$\begin{aligned} P &= P_0(\widetilde{T}) + \frac{1}{N}P_1(\widetilde{T}) + \cdots \\ C &= C_0(\widetilde{T}) + \frac{1}{N}\widetilde{C}_1(\widetilde{T}) + \cdots \end{aligned} \tag{2.26}$$

In the reaction zone:

$$\begin{aligned} P &= P_0(\theta) + \frac{1}{N}P_1(\theta) + \cdots \\ \widetilde{C} &= \frac{1}{N}S = \frac{1}{N}S_0(\theta) + \frac{1}{N^2}S_1(\theta) + \cdots \end{aligned} \tag{2.27}$$

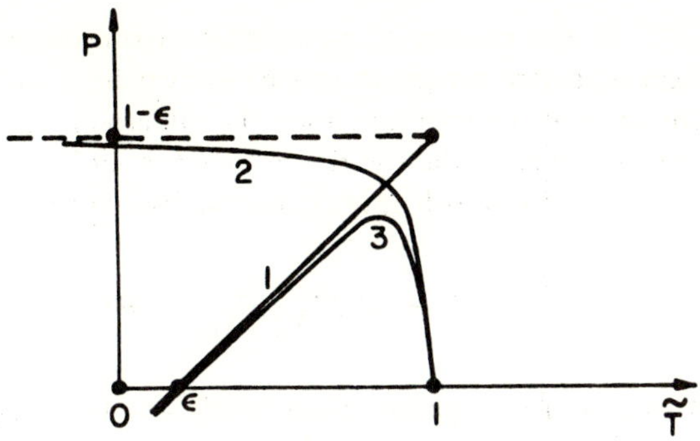

Figure 1 Asymptotic expansions in the phase plane $(\widetilde{T}, P)$: (1) asymptotic solution of outer problem (2) asymptotic solution of inner problem (zeroth approximation); (3) asymptotic solution of inner problem (first approximation).

In the outer problem one can neglect the chemical reaction rate with accuracy up to quantities which are exponentially small relative to N^{-1}. It follows at once that the zeroth approximation of the outer problem is the unique nontrivial asymptotic approximation (Figure 1, curve 1):

$$
\begin{aligned}
P(\widetilde{T}, N^{-1}) &\cong P_0(\widetilde{T}) = \widetilde{T} - \epsilon \\
\widetilde{C}(T, N^{-1}) &\cong \widetilde{C}_0(\widetilde{T}) = 1 - \left(\frac{\widetilde{T} - \epsilon}{1 - \epsilon}\right)^L
\end{aligned}
\tag{2.28}
$$

In the inner variables, system (2.20)-(2.22) has the form:

$$
P + \frac{1 - \epsilon}{\mathrm{L}} P \frac{dS}{d\theta} = \frac{1}{N} \theta + \frac{1 - \epsilon}{N} S \tag{2.29}
$$

$$
\frac{1}{N} P = P \frac{dP}{d\theta} + (1 - \epsilon) A S^{\nu} \exp \frac{\theta}{1 + N^{-1} \theta} \tag{2.30}
$$

$$
S(0) = P(0) = 0 \tag{2.31}
$$

Using the boundary conditions (2.23), we can replace Eq. (2.29) by:

$$
\theta + \frac{1}{\mathrm{L}} (1 - \epsilon) S = \frac{1}{N} \int_0^{\theta} \frac{\theta + (1 - \epsilon) S}{P} \, d\theta \tag{2.32}
$$

In the process of asymptotic expansion in powers of N^{-1}, the quantities figuring in the integrand will be known from previous approximations. Thus, at any step of the approximation the integro-differential system (2.30), (2.32), will in fact reduce to the solution of one first-order differential equation.

For the zeroth approximation, we have from (2.30), (2.32)

$$P_0 \frac{dP_0}{d\theta} + (1-\epsilon)A^0\left(\frac{\mathrm{L}}{1-\epsilon}\right)^{\nu}(-\theta)^{\nu}\exp\theta = 0 \tag{2.33}$$

This first-order equation must satisfy two boundary conditions:

$$P_0(0) = 0 \qquad P_0(-\infty) = 1-\epsilon \tag{2.34}$$

The second is the condition of asymptotic matching to the outer solution (2.28). Integrating (2.33) with conditions (2.34), we obtain*

$$A_0 = \frac{(1-\epsilon)^{1+\nu}}{2\nu!\mathrm{L}^{\nu}} \tag{2.35}$$

Once A_0 is known, the first approximation for U_b may be deduced from (2.18). Knowing the first approximation, one can now calculate A_1 in analogous fashion and derive the second approximation for the normal propagation velocity of the flame. In the second approximation, system (2.30), (2.32) assumes the form:

$$\begin{aligned} P_0 &= P_1 \frac{dP_0}{d\theta} + P_0 \frac{dP_1}{d\theta} + (1-\epsilon)A_1 {S_0}^{\nu}\exp\theta \\ &\quad + (1-\epsilon)A_0\nu S_1 {S_0}^{\nu-1}\exp\theta - (1-\epsilon)A_0 {S_0}^{\nu}\theta^2\exp\theta \end{aligned}$$

$$\frac{1-\epsilon}{\mathrm{L}} S_1 = \int_0^{\theta} \frac{\theta + (1-\epsilon)S_0}{P_0}\, d\theta \tag{2.36}$$

The boundary conditions for $P_1(\theta)$ are

$$P_1(0) = 0 \qquad P_1(\theta) \to \theta \quad \text{as} \quad \theta \to -\infty \tag{2.37}$$

Here again the second condition is the asymptotic matching to the solution (2.28).

Solution of problem (2.36), (2.37) yields the following values for A_1:

*Here and below the stoichiometric coefficients ν will be assumed to be integers. More general formulas may be found in [19].

$$A_1 = \frac{1}{1-\epsilon} A_0[(\mathrm{L}-1)J_2(\nu) + (1-\epsilon)(2+\nu)(1+\nu) - 2J_1(\nu)] \tag{2.38}$$

where, for integer ν, $J_1(\nu) = \displaystyle\int_0^\infty \left[1 - \sqrt{\frac{\gamma(\nu+1,z)}{\nu!}}\right] dz$

$$J_2(\nu) = \frac{\nu}{\sqrt{\nu!}} \int_0^\infty \frac{z\Gamma(\nu, z)}{\sqrt{\gamma(\nu+1,z)}} dz$$

$$\gamma(\nu, z) = (\nu-1)! - \Gamma(\nu, z)$$

$$\Gamma(\nu, z) = \int_z^\infty t^{\nu-1} e^{-t}\, dt$$

In the special case $\nu = 1$ and $\nu = 2$, the integrals J_1 and J_2 are

$$J_1(1) = 1.344 \qquad J_1(2) = 2.114 \qquad J_2(1) = 2 \qquad J_2(2) = 8.885$$

From (2.8) we obtain the second approximation for the flame velocity as

$$U_b \cong U_* N^{-(1+\nu)/2} \exp\left(-\frac{1}{2}N\right) A_0^{-1/2} \left(1 - \frac{1}{2}\frac{A_1}{A_0}\frac{1}{N}\right) \tag{2.39}$$

Transforming the results from the $(P, \tilde{T})$-plane to the plane of the variables T and x, one readily sees that the outer solution for the temperature profile (Figure 2) is

$$\left.\begin{aligned} T &= T_0 + (T_b - T_0)\exp\frac{x}{l_T} \\ C &= C_0 + \left(1 - \exp\frac{\mathrm{L}x}{l_T}\right) \end{aligned}\right\} \quad \text{for} \quad x < 0$$
$$T = T_b \qquad C = 0 \qquad \text{for} \quad x > 0 \tag{2.40}$$

That is to say, in the outer problem the reaction zone appears as a point source, playing the role of the reaction rate $Q\Phi$:

$$\Phi(C, T) \to \frac{\kappa_b C_0}{l_T}\, \delta(x) \tag{2.41}$$

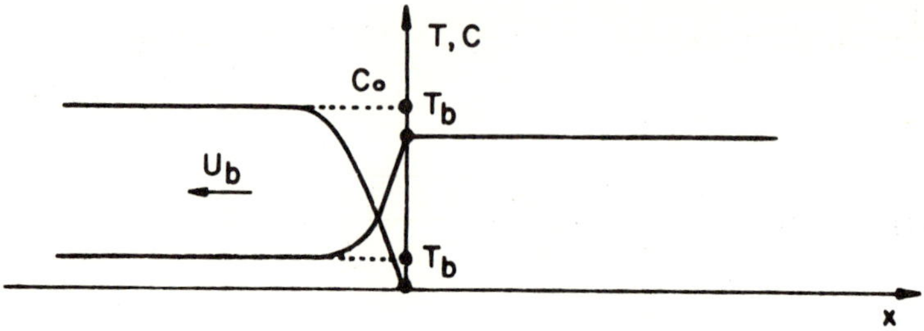

Figure 2 Temperature and concentration profiles in the combustion wave, outer expansion.

In [21, 22], Berman and Ryazantsev considered normal propagation of a flame whose chemical reaction is characterized by two successive exothermic stages:

$$M_1 \to M_2 \qquad M_2 \to \text{combustion products} \tag{2.42}$$

The flame structure in the steady combustion wave is described by the following equations (see Eqs. [1.3], [1.4]):

$$c_p \rho_b U_b \frac{dT}{dx} = \kappa_b \frac{d^2 T}{dx^2} + Q_1 \Phi_1 + Q_2 \Phi_2 \tag{2.43}$$

$$\rho_b U_b \frac{dC_1}{dx} = D_1 \rho_b \frac{d^2 C_1}{dx^2} - \Phi_1 \tag{2.44}$$

$$\rho_b U_b \frac{dC_2}{dx} = D_2 \rho_b \frac{d^2 C_2}{dx^2} + \Phi_1 - \Phi_2 \tag{2.45}$$

$$\Phi_n = z_{nb} \rho_b C_n \exp\left(-\frac{E_n}{R^0 T}\right) \qquad n = 1, 2 \tag{2.46}$$

It was assumed in [21, 22] that $D_1 = D_2 = D$. The case investigated in [21] was $D = 0$ (reaction in condensed phase), and that in [2] $D \neq 0$, i.e., propagation of flame in a gas (although compressibility was not included). Here we shall discuss [22], which includes all the main points of [21].

The boundary conditions are the following:

$$\begin{aligned} &T(-\infty) = T_0 \qquad C_1(-\infty) = C_{10} \\ &C_2(\pm\infty) = C_1(+\infty) = 0 \qquad T(+\infty) < \infty \end{aligned} \tag{2.47}$$

It was assumed that both reaction rates Φ_1 and Φ_2 are strongly temperature-dependent:

$$\frac{E_1}{R^0 T_b} \sim \frac{E_2}{R^0 T_b} \gg 1 \tag{2.48}$$

Three qualitatively different flame propagation regimes are distinguished:

Fusion regime $1 < (E_1/E_2) < \infty$ Here both reactions proceed simultaneously in one reaction zone. The combustion rate is fully determined by the kinetic parameters of the first reaction. The presence of the second reaction affects only the value of the burning temperature, T_b, which depends on the total heat released by the two reactions:

$$T_b = T_0 + \frac{C_0}{c_p}(Q_1 + Q_2) \tag{2.49}$$

$$U_b \rho_b \cong (2\mathrm{L} z_{1b} \kappa_b \rho_b c_p{}^{-1})^{1/2} \left(\frac{R^0 T_b}{E_1}\right)\left(\frac{T_b}{T_b + T_0}\right) \cdot \left\{1 + \left(\frac{R^0 T_b}{E}\right)\left[\frac{T_b}{T_b - T_0}(2.344 - \mathrm{L}) - 3\right]\right\} \exp\left(-\frac{E_1}{2R^0 T_b}\right) \tag{2.50}$$

Under conditions of fusion, the intermediate product M_2 remains within the reaction zone; in order of magnitude, the concentration of M_2 in the reaction zone is smaller by a factor of $\exp(1 - E_1/E_2)$ than that of the original substance M_1. Outside the reaction zone (outer problem), the temperature profile and mass fraction C_1 are entirely analogous to those shown in Figure 2.

Control regime $\{[T_0 + (Q_1 C_0/c_p)]/T_b\} < (E_1/E_2) < 1$ Here the zones of the two successive stages of chemical conversion are separated by a space interval and a temperature interval, connected by the flow of heat.

The temperature in the first reaction zone is $T_{b*} = (E_1/E_2)\, T_b$, that in the second reaction zone is T_b (2.49).

The propagation velocity of the flame depends on the chemical parameters of the second reaction:

$$\rho_b U_b = (2\mathrm{L} z_{2b} \kappa_b \rho_b c_p{}^{-1})^{1/2} \left(\frac{R^0 T_b}{E_2}\right)\left(\frac{c_p T_b}{Q_2}\right) \cdot \left\{1 + \left(\frac{R^0 T_b}{E_2}\right)\left[\frac{T_b}{Q_2}(2.344 - \mathrm{L}) - 3\right]\right\} \exp\left(-\frac{E_2}{2R^0 T_b}\right) \tag{2.51}$$

The temperature and mass fraction profiles of the components M_1 and M_2 are shown in Figure 3.

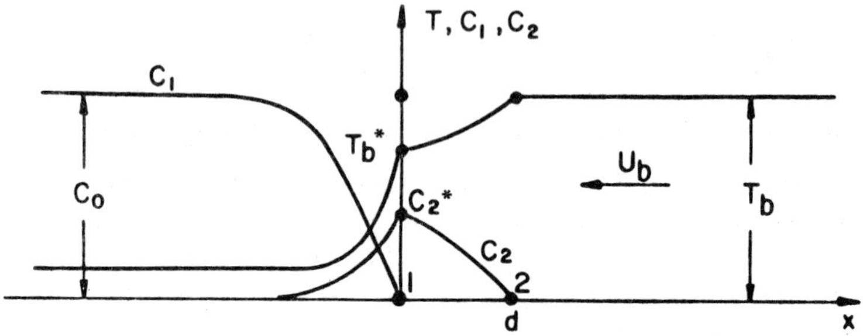

Figure 3 Temperature and concentration profiles in *control* regime. Points (1) and (2) correspond to first and second reaction zones.

Break-off regime $0 < (E_1/E_2) < \{[T_0 + (Q_1C_0/c_p)]/T_b\}$ In the fusion regime, the two reaction zones are separate but propagate as a single entity. In the break-off regime, the front of the first reaction propagates independently of that of the second, which proceeds under induction conditions.

The flame velocity is determined here by the chemical parameters of the first reaction and the temperature T_{b1}:

$$T_{b1} = T_0 + \frac{Q_1 C_{10}}{c_p}$$

$$\rho_b U_b \cong (2\mathrm{L} z_{1b} \kappa_b \rho_b c_p{}^{-1})^{1/2} \left(\frac{R^0 T_{b1}}{E_1}\right)\left(\frac{T_{b1}}{T_{b1} - T_0}\right) \cdot \left\{1 + \left(\frac{R^0 T_{b1}}{E_1}\right)\left[\frac{T_{b1}}{T_{b1} - T_0}(2.344 - \mathrm{L}) - 3\right]\right\} \exp\left(-\frac{E_1}{2R^0 T_{b1}}\right) \tag{2.52}$$

If one no longer defines a solution as a *single, steady* combustion wave satisfying (2.47), it is easy to see that the solution for the break-off regime may involve two independent flame fronts. The second front, corresponding to the reaction: $M_2 \to$ combustion products, moves more slowly than the first. The "cold mixture" region for the second front is an "inter-front" region in which the temperature is T_{b1} and the mass fraction of the intermediate product M_2 is C_{10}. For $T_{b1} < T_b$ and high activation energy E_2 ($E_2/R^0 T_b \gg 1$), burn-out of M_2 must be neglected in the region between the fronts.

The main mathematical arguments used in [22] are as follows. As in flame propagation in a one-step reaction, also here one constructs asymptotic expansions in the regions outside the reaction zones (the characteristic dimension of these regions depends on l_T Eq. [2.12]) and directly inside the reaction zones. The characteristic dimension of the latter regions is of the order of $(R^0 T_b/E) l_T$.

We first observe that, in full analogy with (2.17), the problem (2.43)–(2.47) yields after elimination of Φ_1 and Φ_2 and integration the resultant equation:

$$\frac{\kappa_b}{\rho_b U_b c_p}\frac{dT}{dx} = T - T_b + \frac{Q_1 + Q_2}{c_p}\left(C_1 - \frac{\kappa_b}{\mathrm{L}\rho_b U_b c_p}\frac{dC_1}{dx}\right) + \frac{Q_2}{c_p}\left(C_2 - \frac{\kappa_b}{\mathrm{L}\rho_b U_b c_p}\frac{dC_2}{dx}\right)$$

$$T_b = T_0 + \frac{Q_1 + Q_2}{c_p} C_{10} \tag{2.53}$$

The following dimensionless quantities and parameters are now introduced:

$$\widetilde{T} = \frac{T}{T_b} \qquad \widetilde{C}_i = \frac{C_i}{C_{10}} \qquad \epsilon = \frac{T_0}{T_b} \qquad 1 - \epsilon^* = \frac{Q_2 C_{10}}{c_p T_b}$$

$$N = \frac{E_1}{R^0 T_b} \qquad \frac{E_1}{E_2} = \psi \qquad P = \frac{l_T d\widetilde{T}}{d\widetilde{x}}$$

$$\frac{z_{2b}}{z_{1b}} = \mu \qquad (\epsilon^* > \epsilon)$$

In terms of these new variables, the integral (2.53) and the original Eqs. (2.43), (2.45) become

$$P = \widetilde{T} - 1 + (1-\epsilon)\left(\widetilde{C}_1 - \frac{1}{\mathrm{L}} P \frac{d\widetilde{C}_1}{d\widetilde{T}}\right) + (1-\epsilon^*)\left(\widetilde{C}_2 - \frac{1}{\mathrm{L}} P \frac{d\widetilde{C}_2}{d\widetilde{T}}\right) \tag{2.54}$$

$$P\frac{dP}{d\widetilde{T}} - P + (\epsilon^* - \epsilon) A \widetilde{C}_1 N^2 \exp N\left(1 - \frac{1}{\widetilde{T}}\right) + (1-\epsilon^*)\mu A \widetilde{C}_2 N^2 \exp N\left(1 - \frac{1}{\psi\widetilde{T}}\right) \tag{2.55}$$

$$\frac{1}{\mathrm{L}} P \frac{d}{d\widetilde{T}}\left(P\frac{d\widetilde{C}_1}{d\widetilde{T}}\right) - P\frac{d\widetilde{C}_1}{d\widetilde{T}} - A\widetilde{C}_1 N^2 \exp N\left(1 - \frac{1}{\widetilde{T}}\right) \tag{2.56}$$

$$\frac{1}{\mathrm{L}} P \frac{d}{d\widetilde{T}}\left(P\frac{d\widetilde{C}_2}{d\widetilde{T}}\right) - P\frac{d\widetilde{C}_2}{d\widetilde{T}} + A\widetilde{C}_1 N^2 \exp N\left(1 - \frac{1}{\widetilde{T}}\right) - \mu A \widetilde{C}_2 N^2 \exp N\left(1 - \frac{1}{\psi\widetilde{T}}\right) \tag{2.57}$$

$$\begin{aligned} &\epsilon \leqslant \widetilde{T} \leqslant 1 \quad P(\epsilon) = \widetilde{C}_2(\epsilon) = 0 \quad \widetilde{C}_1(\epsilon) = 1 \\ &P(1) = 0 \quad \widetilde{C}_2(1) = \widetilde{C}_1(1) = 0 \end{aligned} \tag{2.58}$$

$$A = (z_{b1}\kappa_b\rho_b{}^{-1}c_p{}^{-1}U_b{}^{-2})N^{-2}\exp(-N) \tag{2.59}$$

In the fusion solution ($\psi > 1$), the asymptotic expansion of the functions P, $\widetilde{C}_1$, $\widetilde{C}_2$ on the interval $\epsilon \leqslant \widetilde{T} \leqslant 1$ have two inhomogeneity zones: the reaction zone, contiguous to $\widetilde{T} = 1$, and the preheat zone, $\epsilon \leqslant \widetilde{T} \leqslant 1$.

In the reaction zone, we introduce the characteristic variable

$$\theta = N(\widetilde{T} - 1) \tag{2.60}$$

The asymptotic expansions in the reaction zone are sought as series

$$\begin{aligned} P &= P_0(\theta) + \frac{1}{N}P_1(\theta) + \cdots \\ \widetilde{C}_1 &= \frac{1}{N}\left[S_{10}(\theta) + \frac{1}{N}S_{11}(\theta) + \cdots\right] \\ \widetilde{C}_2 &= \frac{1}{N}\exp N\left(\frac{1}{\psi} - 1\right)\left[S_{20}(\theta) + \frac{1}{N}S_{21}(\theta) + \cdots\right] \end{aligned} \tag{2.61}$$

and the expansion of A as

$$A = A_0 + \frac{1}{N}A_1 + \cdots \tag{2.62}$$

From (2.54), using the boundary conditions at $\theta = 0$, we obtain

$$S_{10}(\theta) = -\frac{\mathrm{L}}{1-\epsilon}\theta \qquad (\theta < 0) \tag{2.63}$$

Using Eq. (2.57), we find the first approximation for the inner problem:

$$S_{20}(\theta) = \frac{1}{\mu}S_{10}(\theta)\exp\left(1 - \frac{1}{\psi}\right)\theta = -\frac{\mathrm{L}}{\mu(1-\epsilon)}\theta\exp\left(1 - \frac{1}{\psi}\right)\theta \tag{2.64}$$

Using (2.63) and (2.64) we obtain from (2.55), in the first approximation,

$$P_0\frac{dP_0}{d\theta} + A_0\mathrm{L}(-\theta)\exp\theta = 0 \tag{2.65}$$

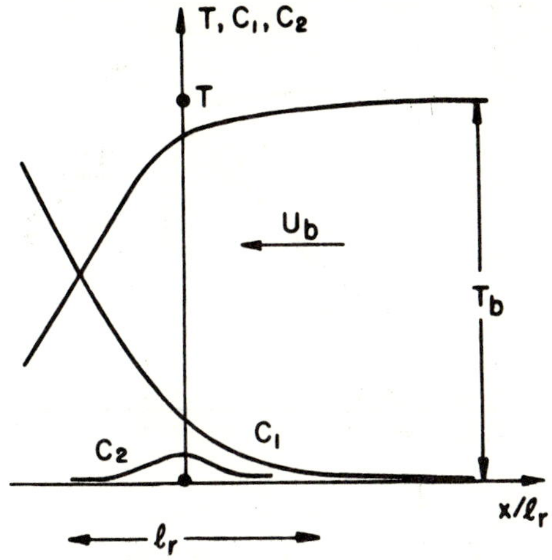

Figure 4 Temperature and mass fraction profiles of components M_1 and M_2 in reaction zone–*fusion* regime.

This is equivalent to Eq. (2.33) for $\nu = 1$. It is easy to see that the outer solution is

$$P^0(\widetilde{T}) = \widetilde{T} - \epsilon \qquad \widetilde{C}_1{}^0(\widetilde{T}) = 1 - \left(\frac{\widetilde{T} - \epsilon}{1 - \epsilon}\right)^{\mathrm{L}} \qquad \widetilde{C}_2{}^0(\widetilde{T}) = 0$$

From (2.65) and (2.66) (in analogy with Eq. [2.35]), we obtain

$$A_0 = \frac{(1 - \epsilon)^2}{2\mathrm{L}}$$

Knowing A_0, we can determine the principal term of the asymptotic expansion of U_b. The construction of higher-order approximations is similar. The structure of the reaction zone in the fusion regime is shown in Figure 4.

In the control regime ($\rightarrow^* < \sigma < 1$), the asymptotic expansion of P, $\widetilde{C}_1$, and $\widetilde{C}_2$ on the interval $\epsilon < \widetilde{T} < 1$ has three inhomogeneity regions: two reaction zones, concentrated in the vicinity of $\widetilde{T} = \psi$ and $\widetilde{T} = 1$, and a preheat zone ($\epsilon < \widetilde{T} < \psi$ and $\psi < \widetilde{T} < 1$) (see Figure 5).

Outside the reaction zones we have the following asymptotic relationships, accurate up to exponentially small quantities relative to N^{-1}:

$$P(\widetilde{T}) = \widetilde{T} - \epsilon \qquad (\epsilon \leqslant \widetilde{T} < \psi) \tag{2.66}$$

$$P(\widetilde{T}) = \widetilde{T} - \epsilon^* \qquad (\psi < \widetilde{T} < 1) \tag{2.67}$$

$$\widetilde{C}_1(\widetilde{T}) = 1 - \left(\frac{\widetilde{T} - \epsilon}{\psi - \epsilon}\right)^{\mathrm{L}} \qquad (\epsilon \leqslant \widetilde{T} < \psi) \tag{2.68}$$

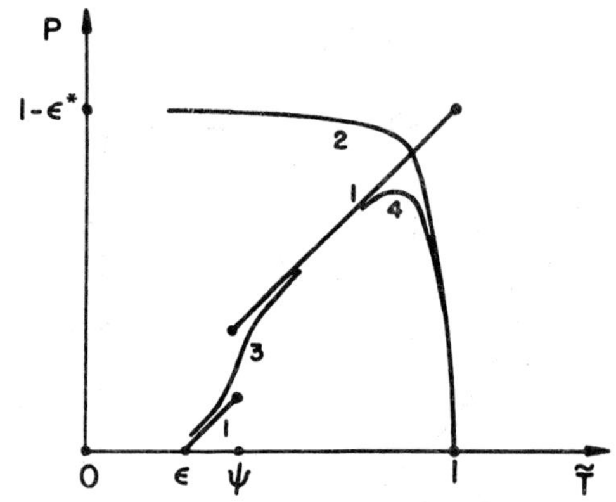

Figure 5 Asymptotic expansions in the phase plane for the control regime: (1) asymptotic curve for outer problem; (2) asymptotic curve for second reaction zone (first approximation); (3) asymptotic curve for first reaction zone; (4) asymptotic curve for second reaction zone (second approximation).

$$\widetilde{C}_1(\widetilde{T}) = 0 \qquad (\psi < T < 1) \tag{2.69}$$

$$\widetilde{C}_2(\widetilde{T}) = \left(\frac{1-\psi}{1-\epsilon^*}\right)\left[1 - \left(\frac{\widetilde{T}-\psi}{1-\psi}\right)^{\mathrm{L}}\right] \qquad (\psi < T < 1) \tag{2.70}$$

$$\widetilde{C}_2(\widetilde{T}) = \left(\frac{1-\psi}{1-\epsilon^*}\right)\left(\frac{\widetilde{T}-\epsilon}{\psi-\epsilon}\right)^{\mathrm{L}} \qquad (\epsilon \leqslant T < \psi) \tag{2.71}$$

Returning now from T to the space variable x, one readily determines the distance, d, between the two reaction zones:

$$d = l_T \ln\left(\frac{1-\epsilon^*}{\psi-\epsilon^*}\right) \tag{2.72}$$

This expression reveals the significance of the restriction $\epsilon^* < \psi$.

The asymptotic relationships (2.66)-(2.71) may be derived by direct approach, without considering the structure of the reaction zones. However, to determine the flame propagation velocity, one must investigate the structure of the reaction zones, as will be done below.

We introduce characteristic variables in the reaction zones:

$$\theta = N(\widetilde{T}-1) \qquad \text{and} \qquad \Theta = N(\widetilde{T}-\psi) \tag{2.73}$$

The asymptotic expansions in the reaction zones will be sought thus:

Zone $M_1 \to M_2$:

$$
\begin{aligned}
P &= P_0(\Theta) + \frac{1}{N} P_1(\Theta) + \cdots \\
\widetilde{C}_1 &= \frac{1}{N}\left[S_{10}(\Theta) + \frac{1}{N} S_{11}(\Theta) + \cdots\right] \\
\widetilde{C}_2 &= S_{20} + \frac{1}{N} S_{21}(\Theta)
\end{aligned}
\tag{2.74}
$$

Zone $M_2 \to$ combustion products:

$$
\begin{aligned}
P &= P_0(\theta) + \frac{1}{N} P_1(\theta) + \cdots \\
\widetilde{C}_2 &= \frac{1}{N}\left[S_{20}(\theta) + \frac{1}{N} S_{21}(\theta) + \cdots\right] \\
\widetilde{C}_1 &\equiv 0
\end{aligned}
\tag{2.75}
$$

The constant A is expanded as follows:

$$A = \exp N\left(\frac{1}{\psi} - 1\right)\left(A_0 + \frac{1}{N} A_1 + \cdots\right) \tag{2.76}$$

We first consider the first reaction zone. By (2.54)

$$S_{20}(\theta) = -\frac{\mathrm{L}}{1 - \epsilon^*}\,\theta \tag{2.77}$$

Now, by (2.55)

$$P_0 \frac{dP_0}{d\theta} + \mu \mathrm{L} A_0(-\theta) \exp\frac{\theta}{\psi} = 0 \tag{2.78}$$

Hence, since $P_0(0) = 0$:

$$P_0^2(-\infty) = 2\mathrm{L}\mu\psi^2 A_0 \tag{2.79}$$

Matching the inner solution $P_0(\theta)$ to the outer solution $P_0(\widetilde{T})$ (2.67), one obtains A_0:

$$A_0 = \frac{(1 - \epsilon^*)^2}{2\mathrm{L}\mu\psi^2} \tag{2.80}$$

Thus the principal term of the asymptotic expansion of the flame velocity is given by (2.51). The asymptotic expansions obtained for the control regime are depicted in Figure 5.

One may note that the flame velocity in the control regime is completely determined by the second reaction and is absolutely independent of the processes taking place in the first reaction zone. The structure of the first reaction zone is described in the first approximation by the following equations and boundary conditions, defined by the outer solution (2.66)-(2.69):

$$P_0 \frac{dP_0}{d\Theta} + (\epsilon^* - \epsilon)A_0 S_{10} \exp \frac{\Theta}{\psi} = 0 \tag{2.81}$$

$$P_0 + \frac{\epsilon^* - \epsilon}{\mathrm{L}} P_0 \frac{dS_{10}}{d\Theta} = \psi - \epsilon^* \tag{2.82}$$

$$P_0(+\infty) = \psi - \epsilon^* \qquad P_0(-\infty) = \psi - \epsilon \qquad S_{10}(+\infty) = 0 \tag{2.83}$$

Problem (2.81)-(2.83) is not overdetermined, since the third boundary condition of (2.83) is an elementary consequence of the first.

It is readily seen that the problem is invariant with respect to the transformation:

$$\Theta \to \Theta' + \text{const} \qquad A_0 \to A_0' \exp\left(-\frac{\text{const}}{\psi}\right) \tag{2.84}$$

Thus, the first reaction zone may always be adjusted for any flame propagation velocity.

In the break-off regime ($0 < \psi < \epsilon^*$), the flame velocity is determined by the first reaction, whose reaction zone breaks away ahead of the second one. The problem is almost completely analogous to the case of a one-step reaction with $\nu = 1$ (see beginning of this section).

3 ON THE PROPAGATION LIMIT OF A NONADIABATIC FLAME

The separation of a flame into characteristic zones at high activation energies may also be observed when there are heat losses.

Consider the propagation of a flame in a tube cooled by the ambient atmosphere. This celebrated problem of flame propagation in nonadiabatic conditions was solved by Zel'dovich [5, 23] (see also [1]) by dividing the flame into preheat and reaction zones. He showed that, other conditions remaining equal, the flame will be unable to propagate in sufficiently narrow tubes, owing to excessive heat loss from the combustion zone to the ambient space. If the tube diameter is fixed, the limit of existence of the flame appears at fairly low concentrations of the combustible component. Analysis of these results showed that the diameter of the tube at the flame propagation limit is of the order of $l_T\sqrt{E/R^0 T_b}$, i.e., it exceeds the thickness l_T of the flame structure. The main features of the solution of the problem may be obtained by carrying out an asymptotic analysis of the problem for tubes of near-critical diameter. We note that the flame propagation velocity U is no longer the velocity under adiabatic conditions, U_b. In the approximation of the one-dimensional model, the system of equations for flame propagation is

$$c_p \rho_b U \frac{dT}{dx} = \kappa_b \frac{d^2 T}{dx^2} + Q(\rho_b C)^\nu z_b \exp\left(-\frac{E}{R^0 T}\right) - h(T - T_0) \qquad (3.1)$$

$$\rho_b U \frac{dC}{dx} = \frac{\kappa_b}{Lc_p} \frac{d^2 C}{dx^2} - \nu(\rho_b C)^\nu \exp\left(-\frac{E}{R^0 T}\right) \qquad (3.2)$$

To simplify matters, these equations are given for a one-step reaction. Here h is the heat transfer coefficient. In the region of critical tube diameter, it is convenient to express h as

$$h = \frac{\alpha \kappa_b}{N l_T^2} \qquad N = \frac{E}{R^0 T_b} \qquad (3.3)$$

(α depends on the cross-section of the tube). The conditions that must hold at infinity are: thermal equilibrium with the external medium, and also concentration conditions:

$$T(+\infty) = T_0 \qquad C(-\infty) = C_0 \qquad C(+\infty) = 0 \qquad (3.4)$$

The temperature T_b and dimension l_T are determined by the problem of normal adiabatic flame propagation (i.e., when $k = 0$). Problem (3.2)-(3.4) defines the temperature and mass fraction profiles and the velocity U.

Whereas normal propagation was characterized by two regions, for the present problem we have three regions, each with a characteristic asymptotic expansion.

The first region is the reaction zone, with characteristic dimension $l_r \sim l_T/N$ and characteristic variable x/N. The second region is the preheat zone (Characteristic dimension, l_T, characteristic variable x). The third is caused by heat exchange with the environment; its characteristic width is of the order of Nl_T (characteristic variable xN).

The asymptotic expansions of the solutions in each of these regions are assumed to be series in N^{-1}.

Once the solutions in adjacent regions have been matched, we have the following relationship to zero order:

$$\left(\frac{U}{U_b}\right)^2 \ln\left(\frac{U}{U_b}\right)^2 = \alpha\left(1 - \frac{T_0}{T_b}\right) \tag{3.5}$$

Treating (3.5) as an equation for U/U_b, one easily sees that a flame can propagate only provided α is sufficiently small (i.e., the tube diameter is sufficiently large). If $\alpha < \alpha^* = T_b/e(T_b - T_0)$, Eq. (3.5) has two solutions (see Figure 6).

Physical meaning may be attached to the solution whose limiting behavior as $\alpha \to 0$ describes the normal propagation of a flame with velocity U_b. In order to answer this question rigorously, one must investigate the stability of both solutions as has been done by Spalding [24].

As the tube diameter decreases (α increases), the temperature in the reaction zone falls continuously to the flame-existence limit T_{f*}:

$$T_{f*} = T_b\left(1 - \frac{R^0 T_b}{E}\right) \tag{3.6}$$

In the region external to the reaction zone, the function describing the reaction rate should be treated (as in normal propagation) as a point source. Therefore, to first order in $R^0 T_b/E$ we have the following correspondence (see Eq. [2.41]):

$$\Phi(C, T) \to \frac{\kappa_b C_0}{l_T} \exp\left[\frac{E}{2R^0 T_b}\left(\frac{T - T_b}{T_b}\right)\right]\delta(x) \tag{3.7}$$

where $T - T_b \sim R^0 T_b^2/E$.

The asymptotic analysis of problem (3.1)-(3.4) raises an essential point, not encountered in our discussion of normal propagation: the deviation of

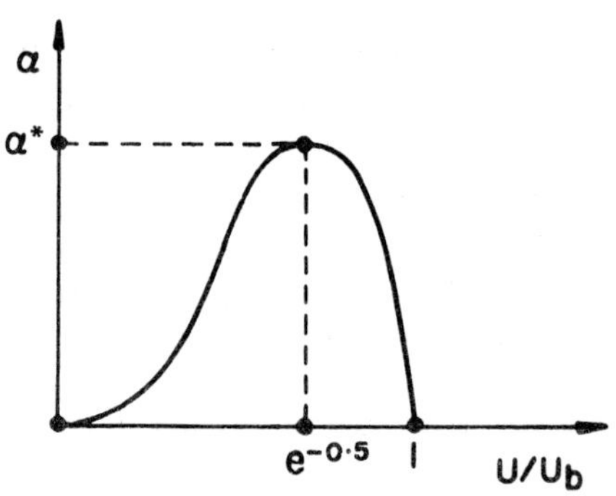

Figure 6 Flame propagation velocity vs. heat exchange with environment.

temperature from adiabatic at the exit from the reaction zone. Since we shall frequently meet with this phenomenon in the context of other, more complicated problems, it is worth giving a brief account of the main solution steps.

Since the temperature is no longer a monotonic function of the coordinates, it makes sense to maintain the natural space variables.

In addition to (2.18), we introduce the following characteristic dimensionless space variables and parameters:

$$\frac{x}{l_T} = \tilde{x} \qquad \frac{Nx}{l_T} = X \qquad \frac{x}{Nl_T} = \zeta \qquad \lambda = \frac{U}{U_b} \tag{3.8}$$

In terms of the space variables $\tilde{x}$, problem (3.1)-(3.4) becomes:

$$\lambda \frac{d\tilde{T}}{d\tilde{x}} = \frac{d^2\tilde{T}}{d\tilde{x}^2} + (1-\epsilon)AN^{1+\nu}\tilde{C}^\nu \exp N\left(1 - \frac{1}{\tilde{T}}\right) - \frac{\alpha}{N}(\tilde{T} - \epsilon) \tag{3.9}$$

$$\lambda \frac{d\tilde{C}}{d\tilde{x}} = \frac{1}{\mathrm{L}} \frac{d\tilde{C}}{d\tilde{x}^2} - AN^{1+\nu}\tilde{C}^\nu \exp N\left(1 - \frac{1}{\tilde{T}}\right) \tag{3.10}$$

$$T(\pm\infty) = \epsilon \qquad C(-\infty) = 1 \qquad C(+\infty) = 0 \tag{3.11}$$

In each of the homogeneity regions, the asymptotic expansions are sought as follows:

$$\tilde{T}(\tilde{x}) = \tilde{T}^0(\tilde{x}) + \frac{1}{N}\tilde{T}^1(\tilde{x}) + \cdots \qquad \tilde{C}(\tilde{x}) = \tilde{C}^0(\tilde{x}) + \frac{1}{N}\tilde{C}^1(\tilde{x}) + \cdots \tag{3.12}$$

$$\tilde{T}(\zeta) = \tilde{T}^0(\zeta) + \frac{1}{N}\tilde{T}^1(\zeta) + \cdots \qquad \tilde{C}(\zeta) = \tilde{C}^0(\zeta) + \frac{1}{N}\tilde{C}^1(\zeta) + \cdots \tag{3.13}$$

$$\tilde{T}(X) = 1 + \frac{1}{N}\theta_1(X) + \cdots \qquad \tilde{C}(x) = \frac{1}{N}S_1(X) + \cdots \tag{3.14}$$

$$\lambda = \lambda_0 + \frac{1}{N}\lambda_1 + \cdots \tag{3.15}$$

In the zeroth approximation, the outer solution (with respect to the reaction zone) is

$$\begin{aligned} \widetilde{T}^0(\widetilde{x}) &= \epsilon + (1-\epsilon)\exp \lambda_0 \widetilde{x} \qquad (\widetilde{x} < 0) \\ &= 1 \qquad (x > 0) \\ \widetilde{C}^0(\widetilde{x}) &= 1 - \exp \mathrm{L}\lambda_0 \widetilde{x} \qquad (\widetilde{x} < 0) \\ &= 0 \qquad (x > 0) \end{aligned} \tag{3.16}$$

$$\begin{aligned} &\widetilde{T}^0(\zeta) = \epsilon \quad (\widetilde{\zeta} < 0) \quad \widetilde{T}^0(\zeta) = \epsilon + (1-\epsilon)\exp\left(-\frac{\alpha\zeta}{\lambda_0}\right) \quad (\zeta > 0) \\ &\widetilde{C}^0(\zeta) = 1 \quad (\widetilde{\zeta} < 0) \quad \widetilde{C}^0(\zeta) = 0 \quad (\zeta > 0) \end{aligned} \tag{3.17}$$

In view of what follows, it is useful to have some information about the first approximation for $\widetilde{x} > 0$ or

$$\widetilde{T}^1(\widetilde{x}) = \widetilde{T}^1(0) \qquad \text{for} \qquad \widetilde{x} > 0 \tag{3.18}$$

By (3.9), (3.10),

$$\lambda \frac{d}{d\widetilde{x}}[\widetilde{T} - 1 + (1-\epsilon)\widetilde{C}] = \frac{d^2}{d\widetilde{x}^2}\left(\widetilde{T} + \frac{1-\epsilon}{\mathrm{L}}\widetilde{C}\right) - \frac{\alpha}{N}(\widetilde{T} - \epsilon) \tag{3.19}$$

Hence, integrating with respect to $\widetilde{x}$ from $-\infty$ to 0, we have

$$\lambda[\widetilde{T}(0) - 1] = \frac{d}{d\widetilde{x}}\left[\widetilde{T}(-0) + \frac{1-\epsilon}{\mathrm{L}}\widetilde{C}(-0)\right] - \frac{\alpha}{N}\int_{-\infty}^{0}(\widetilde{T} - \epsilon)\, d\widetilde{x} \tag{3.20}$$

The continuity condition holds in the neighborhood of $\widetilde{x} = 0$:

$$\frac{d}{d\widetilde{x}}\left[\widetilde{T}(-0) + \frac{1-\epsilon}{\mathrm{L}}\widetilde{C}(-0)\right] = \frac{d}{d\widetilde{x}}\left[\widetilde{T}(+0) + \frac{1-\epsilon}{\mathrm{L}}\widetilde{C}(+0)\right] \tag{3.21}$$

$$\lambda_0 \widetilde{T}_1(0) = -\alpha \int_{-\infty}^{0} [\widetilde{T}^0(\widetilde{x}) - \epsilon]\; d\widetilde{x} \tag{3.22}$$

Using the integral relationship (3.20), together with (3.21), (3.18), we obtain, to first order in $1/N$:

$$\lambda_0^2 \widetilde{T}_1(0) = -\alpha(1-\epsilon) \tag{3.23}$$

or, in view of (3.16) and (3.7), $\lambda_0 = \exp \frac{1}{2}\, T^1(0)$, which together with (3.23) yields

$$\lambda_0^2 \ln \lambda_0^2 = \alpha(1-\epsilon) \tag{3.5}$$

Equation (3.23) shows quite clearly that if $\alpha \neq 0$ the temperature $T_b(1 + 1/N \, \widetilde{T}_1 \, [0])$ at the exit from the reaction zone is less than the adiabatic temperature. Figure 7 presents the outer solution temperature and concentration profiles for this case.

To a first approximation, the structure of the reaction zone is described by the system of equations:

$$\begin{aligned} &\frac{d^2\theta_1}{dX^2} + (1-\epsilon)A_0 S_1{}^{\nu} \exp\theta_1 = 0 \\ &\frac{1}{\mathrm{L}}\frac{d^2 S_1}{dX^2} - A_0 S_1{}^{\nu} \exp\theta_1 = 0 \end{aligned} \tag{3.24}$$

The boundary conditions for Eqs. (3.24) are conditions of matching to the outer solution (3.12), (3.13):

$$\begin{aligned} &\frac{d\theta_1(-\infty)}{dX} = \frac{d\widetilde{T}^0(-0)}{d\widetilde{x}} = \lambda_0(1-\epsilon) \qquad \theta_1(+\infty) = \widetilde{T}_1(0) \\ &\frac{dS_1(-\infty)}{dX} = \frac{dC^0(-0)}{d\widetilde{x}} = -\lambda_0 \mathrm{L} \qquad S_1(+\infty) = 0 \end{aligned} \tag{3.25}$$

Using (3.24), (3.25), (A_0 is defined in Eq. [2.35]), we have

$$\frac{d\theta_1(-\infty)}{dX} = (1-\epsilon)\exp\frac{1}{2}\,\theta_1(+\infty) \tag{3.26}$$

or

$$\lambda_0 = \exp\frac{1}{2}\,\widetilde{T}^1(0) \tag{3.27}$$

Figure 7 Distribution of temperature and mass fraction for flame propagating in a cooled tube.

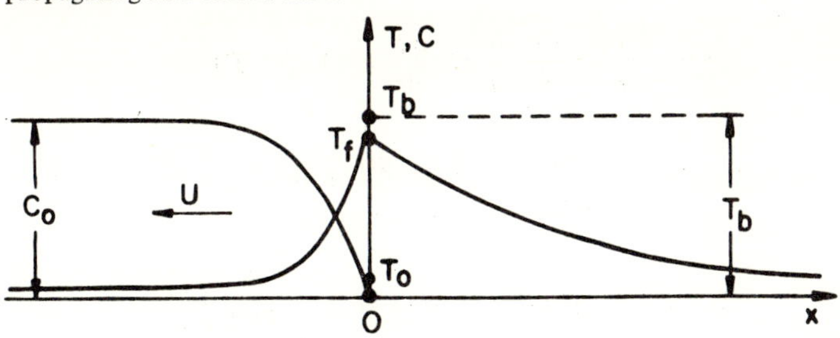

The relation (3.5) follows from (3.20), (3.24). The validity of (3.7) also follows from (3.23), (3.22).

Finally, we note that in the first approximation

$$\frac{d\widetilde{T}}{dX} \to 0 \quad \text{as} \quad X \to \infty \tag{3.28}$$

That is to say, the heat flux from the reaction zone into the burnt-gas region is zero (although the temperature in the reaction zone is less than adiabatic). Thus, (in the first approximation), the state in the reaction zone may be referred to as *quasi-adiabatic.* This state is characterized by a finite relationship between $\widetilde{T}$ and $\widetilde{C}$, invariant with respect to displacement along the X-axis:

$$\theta_1(X) - \theta_1(+\infty) = -\frac{1-\epsilon}{\mathrm{L}} S_1(X) \tag{3.29}$$

The existence of the integral (3.29) yields an explicit relationship between the heat flux in the reaction zone and the maximum temperature in the reaction zone (3.26).

4 COMBUSTION OF A MONOPROPELLANT FUEL DROP

Combustion of a single-component fuel drop has much in common with combustion of a premixed gaseous mixture. The main distinctive features of this type of combustion are its spherical symmetry and the appearance of a condensed phase [1]. An asymptotic analysis of such spherically symmetric combustion was carried out by Fendell [25]. Considering the case of adiabatic combustion, Fendell determined the first two terms in the asymptotic expansion of the flame velocity (relative to the gas), and also the time dependence of the burning velocity of the drop.

Because of the high density of the drop in comparison with that of the gas evaporating from its surface, the combustion process (as characterized by the variation of temperature and mass fraction profiles) may be assumed to be quasi-adiabatic. The flame equations are

$$c_p \rho_b v_r \frac{\partial T}{\partial r} = \frac{\kappa_b}{r^2} \frac{\partial}{\partial r} r^2 \frac{\partial T}{\partial r} + Q\Phi(C, T) \tag{4.1}$$

$$\rho_b v_r \frac{\partial C}{\partial r} = \frac{\kappa_b}{\mathrm{L} c_p} \frac{1}{r^2} \frac{\partial}{\partial r} r^2 \frac{\partial C}{\partial r} + \Phi(C, T) \tag{4.2}$$

$$\Phi(C, T) = z_b \rho_b (C_0 - C) \exp\left(-\frac{E}{R^0 T}\right) \tag{4.3}$$

Here C is the concentration of the combustion product and C_0 its concentration far outside the flame. We consider a one-step monomolecular reaction.

If $a(t)$ is the radius of the evaporating fuel drop (which we assume to be spherical), its mass is $m(t)$

$$m(t) = \frac{4}{3}\pi\rho_* a^3(t) \qquad \frac{dm}{dt} = 4\pi\rho_b v_r a^2 \tag{4.4}$$

where ρ_* is the density of the drop and ρ_b the density of the fuel outside the drop, which we assume to be constant, $\rho_* \gg \rho_b$.

On the drop surface $r = a(t)$ we have the following evaporation conditions:

$$T[a(t), t] = T_{0*} \qquad 4\pi\kappa_b a^2 \frac{\partial T}{\partial r} = q_\lambda \frac{dm}{dt} \tag{4.5}$$

where T_{0*} is the evaporation temperature and q_λ the heat of phase transition. The drop surface is an impenetrable barrier for the reaction products. Hence, we have the following conditions for C on the surface:

$$\rho_b v_r C = \frac{1}{\mathrm{L}} \frac{\kappa_b}{c_p} \frac{\partial C}{\partial r} \qquad \text{for} \qquad r = a(t) \tag{4.6}$$

The mass fraction of the combustion products at infinity is C_0:

$$C(+\infty) = C_0 \tag{4.7}$$

In adiabatic combustion, the conductive heat flux must vanish at infinity:

$$4\pi r^2 \frac{\kappa_b}{\mathrm{L}c_p} \frac{\partial T}{\partial r} \to 0 \qquad \text{as} \qquad r \to \infty \tag{4.8}$$

It can be shown that condition (4.8) dictates the following temperature T_b of combustion products:

$$T_b = T_0 + \frac{QC_0 - q_\lambda}{c_p} \qquad T(+\infty, t) = T_b \tag{4.9}$$

If U_b is the velocity of the gas relative to the spherical front, then, in analogy to normal propagation, one can introduce the width of the preheat zone $l_T = \kappa_b/\rho_b c_p U_b$. Then if $N \gg 1$ the width of the reaction zone is of the order of l_T/N and it is concentrated in the vicinity of the surface $r = R(t)$ $(R \sim l_T)$. The characteristic dimensionless space variable for the region external to the reaction zone should be r/l_T, that for the region of the reaction zone $N(r - R)/l_T$.

In the outer problem, the heat and mass fluxes are discontinuous across the surface $r = R$.

To a first approximation, the propagation velocity of the flame relative to the gas is not sensitive to curvature and is determined by the well-known formula for normal propagation (Sec. 2). Moreover, following Williams [14], for large drops ($a/l_T \gg 1$) one may write

$$\frac{da}{dt} \cong \frac{\rho_b}{\rho_*} U_b \tag{4.10}$$

$$\frac{da}{dt} \sim \frac{\rho_b U_b}{\rho_*}\left(\frac{l_T}{a}\right) \tag{4.11}$$

A case of considerable practical interest is that of drop combustion under nonadiabatic conditions, i.e., when heat exchange takes place with the ambient medium. One must then replace condition (4.8) (or [4.9]) by

$$T(\infty, t) = T_0$$

where T_0 is the temperature of the cold atmosphere surrounding the drop.

However, the attempt to carry out a rigorous analytical and consistent asymptotic analysis of this problem comes up against serious mathematical difficulties. The point is that in nonadiabatic drop combustion, in contradiction to flame propagation in a cooled tube, the width of the zone characterized by heat exchange with the environment is of the same order of magnitude as that of the preheat region (i.e., $\sim l_T$). Thus, even to first order in $1/N$, the conductive heat flux from the reaction zone to the combustion products is not zero. As a result, there is no longer a "quasi-adiabatic" integral of type (3.29), and this makes it impossible to establish a finite relationship between the heat flux from the reaction zone and the temperature drop there.

It would be highly interesting to carry out a thorough qualitative investigation of the equations arising here and to derive some kind of quantitative estimates. It can be shown that because of the sharp temperature drop beyond the flame front $r > R(t)$ the reaction comes to an end before all the combustible component burns up. The flux of unreacted fuel is of order $\sim 1/N$. In other words, in nonadiabatic combustion the reaction is limited only by temperature, and so part of the fuel may "leak through" the reaction zone into the ambient space. It may even be true that nonadiabatic combustion of sufficiently small drops is not possible at all.

5 ON DIFFUSION FLAMES

The framework of ideas developed above for thermal flames in a premixed fuel may be successfully applied to describe exothermic diffusion flames, where burning is caused by mixing the fuel with oxidizer directly in the reaction zone, the reactants entering the reaction zone from opposite directions. In diffusion flames there is no concept of normal propagation velocity. Nevertheless, here too if the activation energy is high the reaction zone turns out to be narrow ($\sim 1/N$) compared with the characteristic diffusional-thermal layer formed outside the reaction zone. The simplest structure of a diffusion flame (as far as ease of investigation is concerned) corresponds to a point diffusion source of fuel immersed in an oxidizing atmosphere [26]. In a certain respect, this model provides a description of diffusional combustion of a drop from whose surface the combustible component is evaporating. It turns out that if the diffusion source (droplet) is sufficiently small, combustion becomes impossible and the flame is quenched. However, even under conditions of combustion the combustible component is not fully used up on the reaction zone. Some of the fuel "leaks through" into the surrounding space, without having reacted with the oxidizer. The effect of the fuel loss becomes weaker, the stronger the diffusion source (i.e., the larger the drop).

The equations describing the structure of a diffusion flame for a bi-molecular reaction may be written in this case as

$$\frac{\kappa_b}{r^2}\frac{d}{dr}r^2\frac{dT}{dr} + Q\Phi(C_1, C_2, T) = 0 \tag{5.1}$$

$$\frac{\kappa_b}{L_1 c_p}\frac{1}{r^2}\frac{d}{dr}r^2\frac{dC_1}{dr} - \Phi(C_1, C_2, T) = 0 \tag{5.2}$$

$$\frac{\kappa_b}{L_2 c_p}\frac{1}{r^2}\frac{d}{dr}r^2\frac{dC_2}{dr} - \Phi(C_1, C_2, T) = 0 \tag{5.3}$$

$$\Phi = z_b \rho_b{}^2 C_1 C_2 \exp\left(-\frac{E}{R^0 T}\right) \tag{5.4}$$

where L_1 and L_2 are the Lewis numbers of fuel and oxidizer, respectively.

The temperature of the ambient atmosphere and the mass fraction of the oxidizer in the atmosphere are assumed to be given:

$$T(+\infty) = T_0 \qquad C_2(+\infty) = C_0 \tag{5.5}$$

The concentration of the combustible component falls to zero with increasing distance from the flame surface (this does not imply, though, that the flux of unburned combustible component must also vanish):

$$C_1(+\infty) = 0 \tag{5.6}$$

As $r \to 0$,

$$-\frac{4\pi\kappa_b}{Lc_p} r^2 \frac{dC}{dr} \to q \tag{5.7}$$

where q is the strength of the diffusion source of combustible component, assumed constant. On the other hand, we are assuming that the diffusion source is neither a source nor a sink of heat. In other words,

$$4\pi\kappa_b r^2 \frac{dT}{dr} \to 0 \quad \text{as} \quad r \to 0 \tag{5.8}$$

Actually, in order to achieve a more adequate description of how the drop itself burns, one must allow for a certain heat loss due to evaporation (see Sec. 4).

As shown in the solution of the problem, condition (5.8) may be written equivalently as

$$T \to T_b = T_0 + \frac{QC_0}{c_p L_2} \quad \text{as} \quad r \to 0 \tag{5.9}$$

Thus, the temperature reached near the diffusion source depends on the mobility (L_2) of the oxidizer.

An asymptotic analysis of the problem (5.1)-(5.7), (5.9) shows that the radius of the spherical flame formed when $N = E/R^0 T_b \gg 1$ has the order of magnitude

$$R(t) \sim l_T = N \exp \frac{1}{2} N \sqrt{\kappa_b z_b^{-1} C_0^{-1} c_p^{-1} \rho_b^{-2}} \tag{5.10}$$

where $\sqrt{\kappa_b z_b^{-1} c_p^{-1} C_0^{-1} \rho_b^{-2}}$ has the physical meaning of a characteristic radius of the flame front at "zero" activation energy.

The zone of major heat evolution and mass conversion is located in the vicinity of $r = R$. It may be verified by using asymptotic matching that the width of this zone is of the order of $1/N$ relative to the diffusional-thermal thickness l_T of the flame. At considerable distances from the reaction, where $|r - R/R| \gg 1/N$, the reaction zone may be viewed as a source concentrated on the surface of a sphere of radius $r = R$. Thus, the reaction rate in Eqs. (5.1)-(5.3) may be replaced by a suitable surface δ-function.

Temperature and concentration profiles are shown in Figure 8. Note that in general the concentration of the combustible component does not vanish when it leaves the reaction zone—the component combusts only partially ($C_1(R) \neq 0$). By contrast, the oxidizer C_2 must combust fully in the reaction zone. Otherwise one would have all necessary conditions for the reaction to continue at $r < R$; high temperature and nonzero concentrations of both reactants. At $r > R$, however, nonzero concentration of fuel and oxidizer is not a sufficient

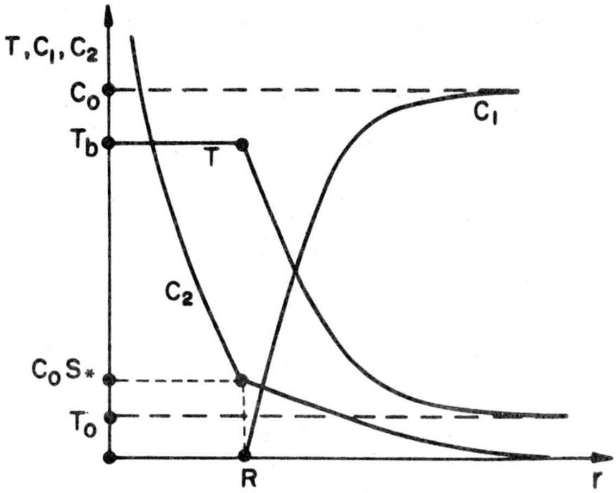

Figure 8 Temperature and concentration profiles in a diffusion flame. Outer expansion.

condition for the reaction to take place, because of the sharp temperature drop. Thus, the presence of combustible component outside the reaction zone does not contradict the equilibrium conditions.

To determine R and $C_1(R)$, we must investigate the inner problem, i.e., the reaction zone. In the reaction zone, we introduce the dimensionless space variable

$$X = \frac{r - R}{l_T} N \tag{5.11}$$

The temperature and mass fraction are sought as asymptotic expansions

$$\begin{aligned} T &= T_b \left[1 + \frac{1}{N}\theta(X) + \cdots\right] \\ C_2 &= \frac{1}{N} C_0 S_2(X) + \cdots \\ C_1 &= C_0 \left[S_* + \frac{1}{N} S_1(X) + \cdots\right] \end{aligned} \tag{5.12}$$

where $S_* = C_1(R)/C_0 = \text{const}$.

With this notation, system (5.1)-(5.3) assumes the following form for the first approximation of the inner problems:

$$\frac{d^2\theta}{dX^2} + \mathrm{L}_2(1 - \epsilon)S_* S_2 \exp\theta = 0 \tag{5.13}$$

$$\frac{d^2 S_2}{dX^2} - L_2 S_* S_2 \exp\theta = 0 \tag{5.14}$$

$$\frac{d^2 S_1}{dX^2} - L_1 S_* S_2 \exp\theta = 0 \tag{5.15}$$

where $\epsilon = T_0/T_b$.

The boundary conditions obtained from matching to the outer solution are ($r < R$)

$$S_2(-\infty) = \theta(-\infty) = 0 \tag{5.16}$$

Then Eqs. (5.13), (5.14) are integrated to yield

$$\theta(X) + (1-\epsilon)S_2(X) = 0 \tag{5.17}$$

Due to the existence of the relation (5.17), our problem is largely similar to normal flame propagation (see Sec. 2).

Matching the solutions of the inner and outer problems gives an equation for the radius of the flame front:

$$\frac{l_T}{R} = \frac{\tilde{q} L_1 L_2}{(1-\epsilon)^2} - \sqrt{\frac{\tilde{q}^2 {L_1}^2 {L_2}^2}{(1-\epsilon)^4} - \frac{2L_1}{(1-\epsilon)^2}} \tag{5.18}$$

It is readily seen from (5.18) that the flame front radius decreases with decreasing the source strength, q. At a certain critical strength, q_*, and the corresponding radius, R_*, the problem no longer has a steady solution and the flame is extinguished.

The flux of the unreacted part of the fuel through the reaction zone does not vanish; it is

$$q - \frac{\kappa_b}{L_2 c_p} 4\pi R C_0 \tag{5.19}$$

With increasing q, the loss of combustible component disappears and the relationship between q and R becomes simpler:

$$R \cong \frac{q c_p L_2}{4\pi\kappa_b C_0} \tag{5.20}$$

An incomparably more complicated problem is that of the structure of a plane diffusion flame, to which a considerable part of Williams' review [1] is devoted. A flame of this type may be formed, for example, in the mixture zone of parallel fuel and oxidizer streams.

When $N \gg 1$, the width of the diffusional-thermal layer (assuming a bi-molecular reaction) is of the order of $N^{3/2} \exp \frac{1}{2} N$ relative to the same layer at "zero" activation energy.

If there is complete symmetry (and the dimensionless variables are chosen accordingly) the flame structure equations may be written

$$\frac{d^2\theta}{dX^2} + S_1 S_2 \exp \theta = 0$$

$$S_1 = \theta(X) - \beta X \tag{5.21}$$

$$S_2 = \theta(X) + \beta X$$

with the boundary conditions:

$$\frac{d\theta}{dX}(\pm\infty) = \beta \qquad \frac{dS_2}{dX}(+\infty) = \frac{dS_2}{dX}(-\infty) = 0 \tag{5.22}$$

It follows directly from (5.21), (5.22) that

$$\frac{dS_2}{dX}(-\infty) = -\frac{dS_1}{dX}(+\infty) = 2\beta \tag{5.23}$$

where β is the dimensionless heat flux from the reaction zone into the surrounding diffusional-thermal layer. As in the case of nonadiabatic drop combustion (Sec. 4), problem (5.21)-(5.23) cannot be solved in closed form. A qualitative analysis of the corresponding integral equation, however, may yield the following estimate:

$$e^{-\alpha}(\alpha^2 - 2) + e^{-2\alpha}(\alpha^2 + \alpha + 2) < \beta^2 < 2e^{-\alpha}(\alpha^2 + 2\alpha) \tag{5.24}$$

where $\alpha = -\theta(0) > 0$ corresponds to the maximum temperature in the reaction zone. It follows that, qualitatively speaking, the function $\beta^2(\alpha)$ has the shape of the curve shown in Figure 9.

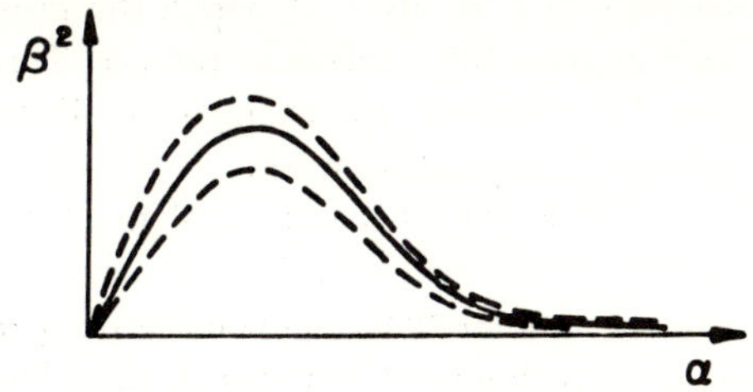

Figure 9 Relationship between maximum temperature in reaction zone and heat flux into surrounding diffusional-thermal layer. Dashed curves—estimating functions (5.24).

The estimate (5.24) implies the following conclusions:

1. If sufficient heat is removed from the reaction zone, the flame cannot exist and is quenched.
2. If heat removal is sufficiently small, at least two combustion regimes are possible. Most probably, the stable regime is that corresponding to less than maximum temperature in the reaction zone.

It is also to be expected that $S_1(-\infty) = S_2(+\infty) > 0$. That is to say, because of the sharp temperature drop outside the reactants are not fully consumed by combustion.

The question of extinction of a diffusion flame was first examined by Zel'dovich [27], who used arguments of dimensional analysis. Zel'dovich showed that the critical flux exhibits the same dependence on the thermal and kinetic parameters of the reaction as the propagation velocity of a laminar flame. Spalding [28] considered the extinction of diffusion flames in connection with Potter's theory of burners [29], in which a diffusion flame is formed when oppositely directed fuel and oxidizer streams meet. Spalding found a critical condition defining the velocity of the stream at which the diffusion flame is extinguished owing to excess supply of reactants to the reaction zone. The computations in [28] are done by numerical methods, but the results are in good agreement with [27]. A detailed account of the theory of a diffusion flame in laminar flow may be found in Linan [30].

6 CURVATURE EFFECT. CONVERGING SPHERICAL COMBUSTION WAVE

The preceding examples provide clear-cut evidence that if the activation energy is high the reaction zone will be narrow compared with the region of the flame associated with transport processes. One is thus able to give separate consideration to the reaction zone and to the diffusional-thermal layer outside the reaction zone. It turns out, that apart from this narrowness of the reaction zone, a high activation energy also entails the appearance of a new characteristic dimension Nl_T in distorted premixed flames. The dimension Nl_T, which considerably exceeds the thermal thickness of the flame, is related to the curvature of the front at which the propagation velocity of the flame begins to vary appreciably relative to the gas. This second effect of high activation energy may be observed, for example, in Bunsen and cellular flames, in a converging spherical combustion wave.

We begin with the problem of a converging spherical flame front [31]. Combustion waves of this type may be observed in large-scale turbulent flames when a laminar flame splits up into separate combustion centers each containing a small volume of unburned gas. The burning of "islets" of fuel may be described in terms of the laminar theory as suggested by Shelkin and Troshin [32]. As shown in [31], the nature of the combustion in a converging spherical flame depends essentially on the relationship between the processes of diffusion in the combustible component and heat transfer in the mixture characterized by the Lewis number. If the Lewis number is greater than 1, the temperature in the reaction zone increases monotonically as the flame front approaches the center of symmetry and the flame is accelerated. If the Lewis number is less than 1, the flame propagation velocity drops near the center together with the temperature in the reaction zone, until the combustion becomes impossible at a certain distance from the center and the flame is extinguished without full combustion of the mixture having occurred.

We thus have the possibility of the flame being extinguished in *globally* adiabatic conditions, the only reason being excessive local heat removal from the combustible sections of the gaseous mixture to colder sections.

A substantial change in the propagation velocity of the flame sets in when the radius of the flame is of the order of Nl_T. It is clear that for such curvatures the flame is quasi-planar and quasi-steady relative to the changes taking place in its diffusional-thermal processes.

The main idea of the asymptotic analysis will now be described briefly. In this review we are restricting attention to an "incompressible" diffusional-thermal model of the flame. The hydrodynamic formulation and solution of the problem may be found in [31].

From the standpoint of asymptotic analysis, the appropriate singular perturbation problem exhibits four homogeneity regions of asymptotic solutions: reaction zone ($\sim l_T/N$), preheat zone ($\sim l_T$), a zone determined by the radius of the front ($\sim Nl_T$), and a zone in which the temperature perturbations are dissipated ($\sim N^2 l_T$).

The flame structure equations for the spherically symmetric case are

$$c_p \rho_b \frac{\partial T}{\partial t} = \frac{\kappa_b}{r^2} \frac{\partial}{\partial r}\left(r^2 \frac{\partial T}{\partial r}\right) + Q\Phi \tag{6.1}$$

$$\rho_b \frac{\partial C}{\partial t} = \frac{\kappa_b}{\mathrm{L}c_p} \frac{1}{r^2} \frac{\partial}{\partial r}\left(r^2 \frac{\partial C}{\partial r}\right) - \nu\Phi \tag{6.2}$$

where Φ is defined by (2.4).

Boundary conditions for system (6.1), (6.2) are taken as follows:

$$T(0, t) = T_0 \qquad C(0, t) = C_0 \qquad T(\infty, t) = T_b = T_0 + \frac{QC_0}{\nu c_p} \qquad C(\infty, t) = 0 \tag{6.3}$$

As time origin ($t = 0$) we take the focusing time. Time prior to the focusing time is assumed to be negative.

The main energy release is concentrated in the vicinity of the surface $r = R(t)$, where $R \sim Nl_T$. At distances $|r - R(t)| \gg l_T/N$, i.e., outside the reaction zone, the latter may be treated as a point source. As shown in [31] (see also Sec. 3), the strength of this point source is of order of magnitude

$$\exp \frac{1}{2} N\left(\frac{T - T_b}{T_b}\right) \qquad \frac{T - T_b}{T_b} \sim \frac{1}{N}$$

That is to say, regardless of the deviation of the temperature in the reaction zone from adiabatic, the heat flux into the combustion-product region vanishes in the first approximation.

The combustion wave is conveniently handled in a coordinate system $(\tilde{x},\tau)$ attached to the flame front:

$$\tilde{x} = \frac{r}{l_T} - N\phi \qquad \tau = \frac{tU_b}{Nl_T} \qquad \phi = \frac{R}{Nl_T} \tag{6.4}$$

Outside, the reaction zone, the equation system (6.1), (6.2) becomes

$$\frac{1}{N}\frac{\partial \tilde{T}}{\partial \tau} - \frac{d\phi}{d\tau}\frac{\partial \tilde{T}}{\partial \tilde{x}} = \left(1 + \frac{\tilde{x}}{N\phi}\right)^{-2} \frac{\partial}{\partial \tilde{x}} \left(1 + \frac{\tilde{x}}{N\phi}\right)^{2} \frac{\partial \tilde{T}}{\partial \tilde{x}}$$
$$+ (1 - \epsilon) \exp \frac{1}{2} N(\tilde{T} - 1)\delta(\tilde{x}) \tag{6.5}$$

$$\frac{1}{N}\frac{\partial \tilde{C}}{\partial \tau} - \frac{d\phi}{d\tau}\frac{\partial \tilde{C}}{\partial x} = \frac{1}{\mathrm{L}}\left(1 + \frac{\tilde{x}}{N\phi}\right)^{-2} \frac{\partial}{\partial \tilde{x}} \left(1 + \frac{\tilde{x}}{N\phi}\right)^{2} \frac{\partial \tilde{C}}{\partial \tilde{x}}$$
$$- \exp \frac{1}{2} N(\tilde{T} - 1)\delta(\tilde{x}) \tag{6.6}$$

where $\tilde{T} = T/T_b$, $\tilde{C} = C/C_0$.

System (6.5), (6.6) is solved in terms of series in powers of $1/N$. In the zeroth approximation, Eqs. (6.5), (6.6) do not involve the derivatives $\partial\tilde{T}/\partial\tau$ and $\partial\tilde{C}/\partial\tau$. Thus, the process does indeed prove to be quasi-steady in the zeroth approximation. This remarkable fact reduces the problem at each order of approximation to the solution of a system of ordinary differential equations in which only the coefficients depend on time. Thus, the outer problem may be solved by elementary techniques. The method whereby one relates the solution of the outer problem to that of the inner is largely analogous to the treatment of the same problem for flame propagation in a cooled tube (Sec. 3).

Appropriate arguments yield the following nonlinear equation (to zero order in $1/N$) and boundary condition for the radius of the flame front ϕ_0:

$$I\frac{d^2\phi_0}{d\tau^2} + 2\left(\frac{d\phi_0}{d\tau}\right)^3 \ln\left(\frac{d\phi_0}{d\tau}\right) = \frac{2}{\phi_0}\left(\frac{d\phi_0}{d\tau}\right)^2 \tag{6.7}$$

where $$I = (1-\epsilon)\left[\frac{1}{\mathrm{L}}(1-\epsilon)^{\mathrm{L}} - 1\right]$$

$$\frac{d\phi_0}{d\tau} \to 1 \quad \text{as} \quad \tau \to -\infty$$

Equation (6.7) incorporates all the physical data on converging spherical flames specified at the beginning of this section. A complete analysis of the equation was given in [31].

7 ON THE STRUCTURE OF A BUNSEN FLAME

An effect resembling that just described may also be observed in the cone of a bunsen flame [33]. It turns out that near the tip of the cone, at a distance ~ Nl_T from the symmetry axis, the temperature in the reaction zone begins to change, thus entailing a change in the propagation velocity of the flame with respect to the gas. At the same time, the flame surface begins to bend.

If $\mathrm{L} > 1$, the temperature in the reaction zone rises continuously as one approaches the tip of the cone, and the velocity increases. The flame takes on the shape of a smooth, continuous surface with a rounded tip, the radius of curvature of the tip is also of the order of Nl_T.

If $\mathrm{L} < 1$, the temperature of the flame drops near the tip, together with the flame propagation velocity with respect to the gas. Sufficiently near the axis of symmetry, combustion becomes impossible and the flame is quenched. Such a flame with an open tip is well known from experiments. Lewis and von Elbe [34] give a description of this kind of flame for a hydrogen-air mixture, when the reaction is limited by hydrogen, the lighter component of the combustible mixture, for which the condition $\mathrm{L} < 1$ is clearly valid.

Irrespective of the different equations, the idea of the analysis and the physical effects relevant to the structure of a Bunsen cone are analogous in many respects to those in the problem of converging spherical flames (Sec. 6). Once again one finds four characteristic zones: reaction, preheat, flame-curvature, and dissipation of temperature perturbations.

The discussion of [33] was conducted in terms of a purely diffusional-thermal flame model, i.e., the density was assumed to be independent of temperature. Thus, the model did not describe the refraction of flow lines at the flame front due to thermal expansion of the gas. The velocity U_∞ of the gas in the flame-front coordinate system was assumed to be constant everywhere and directed along the symmetry axis of the flame.

In cases of cylindrical symmetry, the system of equations for the structure of the bunsen flame is as follows:

$$U_\infty \rho_b c_p \frac{\partial T}{\partial z} = \kappa_b \left(\frac{\partial^2 T}{\partial r^2} + \frac{1}{r}\frac{\partial T}{\partial r} + \frac{\partial^2 T}{\partial z^2} \right) + Q\Phi(C, T) \tag{7.1}$$

$$U_\infty \rho_b \frac{\partial C}{\partial z} = \frac{1}{\mathrm{L}} \frac{\kappa_b}{c_p} \left(\frac{\partial^2 C}{\partial r^2} + \frac{1}{r}\frac{\partial C}{\partial r} + \frac{\partial^2 C}{\partial z^2} \right) - \nu\Phi(C, T) \tag{7.2}$$

with the boundary conditions:

$$\begin{aligned} T(-\infty, r) &= T_0 \qquad C(-\infty, r) = C_0 \\ T(z, \infty) &= T_0 \qquad C(z, \infty) = C_0 \end{aligned} \tag{7.3}$$

The reaction zone is concentrated in the vicinity of a surface $z = F(r)$, which is not known a priori but must be determined along with C and T.

The problem is conveniently handled in a coordinate system (y,s) attached to the flame front:

$$y = \frac{z}{l_T} - N\phi(s) \qquad s = \frac{r}{Nl_T} \qquad \phi = \frac{F}{Nl_T} \tag{7.4}$$

Now the variables (s, Ny) will serve to describe the structure of the reaction zone, the variables (s,y) characterize the preheat zone, and $(s, y/N^2)$ the region of dissipation of the temperature perturbations caused by the temperature changes that take place in the reaction zone as it is distorted.

In terms of the variables (7.4) outside the reaction zone, Eq. (7.1), for example, becomes

$$\begin{aligned} \frac{1}{\sin\alpha}\frac{\partial \widetilde{T}}{\partial y} &= \frac{1}{N^2}\frac{\partial^2 \widetilde{T}}{\partial s^2} - \frac{1}{N}\left(\frac{d^2\phi}{ds^2} - \frac{1}{s}\frac{d\phi}{ds}\right)\frac{\partial \widetilde{T}}{\partial y} \\ &+ \left[1 + \left(\frac{d\phi}{ds}\right)^2\right]\frac{\partial^2 \widetilde{T}}{\partial y^2} - \frac{2}{N}\frac{d\phi}{ds}\frac{\partial^2 \widetilde{T}}{\partial s \partial y} + \frac{1}{N^2}\frac{1}{s}\frac{\partial \widetilde{T}}{\partial s} \\ &+ (1-\epsilon)\sqrt{1 + \left(\frac{d\phi}{ds}\right)^2} \exp\frac{1}{2}N(\widetilde{T} - 1)\delta(y) \end{aligned} \tag{7.5}$$

Here $1/\sin\alpha = U_\infty/U_b$, where α is the inclination of the flame-front surface to be symmetry axis as $z \to \infty$, $T = T/T_b$, $\epsilon = T_0/T_b$.

To zero order in $1/N$, the terms with $\partial^2 \tilde{T}/\partial s^2$ and $\partial \tilde{T}/\partial s$ drop out of Eq. (7.5). Thus, in the zeroth approximation the structure of the flame in a bunsen cone proves to be quasi-onedimensional. It follows that at each order of approximation one is actually dealing with ordinary differential equations (see Sec. 6).

An analysis similar to that of Sec. 6 yields the following nonlinear differential equation for the flame-front surface:

$$\frac{\sin \alpha(1-\epsilon)(1-\mathrm{L})}{2\mathrm{L}}\left(\frac{d^2\phi_0}{ds^2}+\frac{1}{s}\frac{d\phi_0}{ds}\right)=\ln\left[\sin\alpha\sqrt{1+\left(\frac{d\phi_0}{ds}\right)^2}\right]$$
$$\frac{d\phi_0}{ds}(\pm\infty)=\frac{1}{\tan\alpha} \qquad (7.6)$$

The deviation of temperature from adiabatic in the reaction zone is given by:

$$\frac{T-T_b}{T_b}N=-2\ln\left[\sin\alpha\sqrt{1+\left(\frac{d\phi_0}{ds}\right)^2}\right] \qquad (7.7)$$

Flame-front surfaces $z = F(r)$ corresponding to solutions of Eq. (7.6) are illustrated in Figure 10.

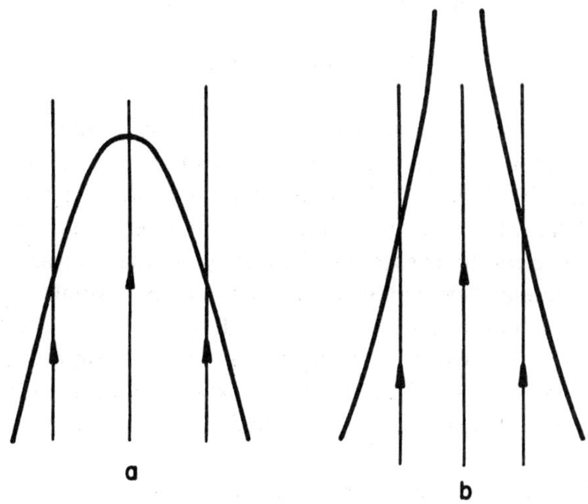

Figure 10 Conical flame front surfaces: (a) corresponds to the case $\mathrm{L} > 1$; (b) corresponds to the case $\mathrm{L} < 1$. (The arrows show the direction of the gas stream.)

The situation is completely different in a wedge-shaped bunsen flame, which may be stabilized on a burner of elongated rectangular cross section. If $L > 1$, the flame will again have a smooth, rounded vertex. However, whereas the bunsen cone is characterized at large distances from the tip by the asymptotic relationship:

$$\frac{d\phi}{ds} + \frac{1}{\tan \alpha} \sim \frac{1}{s} \qquad s \to \infty \tag{7.8}$$

in a wedge-shaped flame the effect of the tip vanishes incomparably more rapidly:

$$\frac{d\phi}{ds} + \frac{1}{\tan \alpha} \sim \exp\left[-\frac{2Ls}{(1-\epsilon)(L-1)}\right] \tag{7.9}$$

If $L < 1$, however, the characteristic curvature is of the order of $l_T/\sqrt{N}$ and the flame structure at the vertex becomes essentially two-dimensional. In the vicinity of the vertex, the flame is fully described by two Laplace equations (for temperature and concentration) with boundary conditions on an unknown surface (the flame front). Because of this, the mathematical problem for $L < 1$ is quite reminiscent of the treatment of gravitational waves on water.

We have been discussing premixed flames, in which the reaction is limited by some reactant present in small concentrations. In other words, the original combustible mixture is nonstoichiometric with respect to its components. In [35], which also studies the structure of a bunsen flame, there is a discussion of the flame of a bi-molecular reaction in which the stoichiometry condition holds upstream, in the fresh mixture. The equation of a bimolecular reaction is

$$M_1 + M_2 \rightarrow \text{combustion products}$$

For stoichiometry it is necessary that the mass fractions of both components upstream be equal. Thus, in normal flame propagation, the reaction will be limited simultaneously by both reactants. In the distorted flames obtained on Bunsen burners stoichiometry upstream of the flame is no longer sufficient to ensure stoichiometry of the mixture near the reaction zone. The different diffusivities of the reactants may violate the stoichiometry condition as the reaction zone is approached. It turns out that in that case the mixture is always deficient in the lighter of the two components.* An asymptotic analysis presented in [35] provided an explanation for this phenomenon, which has long been familiar to experimentalists [36]. It was also shown that if both

*This effect is known as diffusion stratification.

components have Lewis numbers greater than unity the flame front is continuous. The temperature at the exit from the reaction zone then exceeds the adiabatic temperature of the combustion products. But if the lighter component has a Lewis number of less than one, there may be either a flame with continuous front, in which the temperature of the reaction zone is *less* than adiabatic, or a flame with an open tip.

Asymptotic analysis of this problem is in general analogous to the nonstoichiometric case considered above.

The flame structure is now described not by two but by three equations: two for each of the reactants, plus the heat conduction equation. The chemical reaction rate is taken to be

$$\Phi = z\rho_b^2 C_1 C_2 \exp\left(-\frac{E}{R^0 T}\right)$$

Outside the reaction zone Φ may be replaced by a suitable surface δ-function, whose intensity for $N \gg 1$ is proportional to the quantity

$$\sqrt{1 + \frac{1}{2} N \frac{1-\epsilon}{L_2} \widetilde{C}_2 \exp \frac{1}{2} N(\widetilde{T} - 1)} \qquad (7.10)$$

where $\widetilde{C}_2 = \dfrac{C_2}{C_0}$

$$\widetilde{T} = \frac{T}{T_b}$$

with $\widetilde{T} - 1 \sim \widetilde{C}_2 \sim 1/N$ in the reaction zone (C_2 is the mass fraction of the heavy component: $L_2 > L_1$).

Equation (7.10) clearly shows that under conditions of stoichimetry and large activation energy the reaction rate is as sensitive to changes of composition in the reaction zone as it is to temperature changes. The diffusion stratification effect always increases the reaction rate and under certain conditions may even suppress the downward tendency of the reaction rate due to falling temperature. The diffusion stratification effect is apparent only in reactants which limit the reaction in normal flame propagation. In distorted flames if the stoichiometry conditions are not satisfied in the fresh mixture, upstream of the flame, there may also be a change of composition in the vicinity of the reaction zone. In this case, however, diffusion stratification can have no real effect on the reaction rate. The change of flame velocity here will depend only on the diffusivity of the limiting component and on the temperature.

8 STABILITY OF LAMINAR FLAMES

It is known from experiment that a plane steady flame on top of a burner of sufficiently large diameter becomes unstable [36], [37].

The hydrodynamic instability of the flame front cannot be described by means of a model that makes no allowance for thermal expansion of the gas, like the model we have been using hitherto. For a mathematical description of these phenomena, therefore, we must appeal to the equations of hydrodynamics (1.6)-(1.8).

The problem of flame-front stability for a plane laminar flame was first studied by Landau [38] and Darrieus [39]. They treated the problem in terms of an unstructured flame model, regarding the front as a discontinuity surface of density and temperature in an ideal fluid. The velocity of the front surface relative to the gas was assumed to be a given, constant quantity. Employing the method of small harmonic perturbations, Landau and Darrieus arrived at the paradoxical conclusion that laminar flames are always unstable. This was unexpected, since numerous experiments had indicated the existence of stable laminar flames. In order to make the theory accord with the known experimental data, many authors began to look for possible stabilizing effects not incorporated in the unstructured Landau-Darrieus model. Important work in this direction was done by Markstein [40], who, without going beyond the unstructured model, suggested that the flame propagation velocity depends on the curvature of the front. The new parameter appearing here, of the dimension of length, and the precise form of the velocity as a function of front curvature, remained undetermined and were considered to be given a priori. In the linear approximation (small perturbations), it was shown that if the above parameter (the Markstein constant) has a suitable sign, the flame will be stable with respect to sufficiently short-wave perturbations and unstable for long-wave perturbations. In fact, there exists a wavelength Λ^* corresponding to neutral stability, i.e., to a steady corrugated flame front of small amplitude. Thus, Marksteins' model implied stability of a plane flame front in sufficiently narrow tubes, which do not allow long-wave perturbations to affect the flame.

Markstein's approach was purely phenomenological. Moreover, in contradistinction to the Landau-Darrieus model, it displays a certain lack of consistency [37]. Namely, it was clear from dimensional arguments that Markstein's parameter should be proportional to the flame-structure width l_T. But the magnitude of l_T is determined by such transport effects as diffusion and conduction, whereas the Landau-Darrieus-Markstein models are unstructured, treat the gas as ideal, and completely neglect these effects. Viscosity effects also have the same order of magnitude as diffusional-thermal effects. Thus it was not clear how consistent it is to include transport effects in the conditions of the flame-front surface but to ignore them elsewhere.

The first model proposed in order to pinpoint the role of diffusion and heat condition vis-à-vis flame-front stability neglected the thermal expansion of the gas [4]. It is shown that for small harmonic perturbations, of wavelength much longer than the thermal width of the flame ($\Lambda \gg l_T$) the characteristic "frequency" ω of the perturbations has an order of magnitude

$$\omega \sim N(1 - \mathrm{L})k^2 l_T U_b \qquad N \gg 1 \tag{8.1}$$

where $k = 2\pi/\Lambda$ is the wave number. Thus, when $\mathrm{L} > 1$ the processes of diffusion and conduction have a tendency to stabilize the flame. Since ω is proportional to the large factor N, Librovich and Istratov [37] conjectured that Markstein's parameter also contains the product Nl_T. This justified neglecting transport effects outside the flame and incorporated them (through Markstein's constant) only in the conditions at the jump.

Istratov and Librovich [41] then studied flame stability in the framework of a hydrodynamic model, with consistent inclusion of compressibility and all transport effects (viscosity, diffusion, conduction). Their method was to match outer and inner asymptotic expansions, as proposed by Germain and Guiraud [42]. The small parameter in the expansions was the ratio of the width of the preheat zone to the wavelength of the perturbation ($l_T/\Lambda \ll 1$). As first approximation, the authors of [41] chose the well-known Landau-Darrieus solution, and then sought a second approximation for the characteristic frequency ω as an expansion in powers of l_T/Λ.

Istratov and Librovich [41] isolated the general stabilizing tendency of transport effects. They derived the explicit form of Markstein's parameter as a function of activation energy and transport coefficients, and were thereby able to confirm their earlier conjecture as to the role of N. They did not achieve a complete solution to the stability problem, however, since the discussion was confined to longwave perturbations. The asymptotic expansion

$$\omega = \alpha_1\left(\frac{l_T}{\Lambda}\right) + \alpha_2\left(\frac{l_T}{\Lambda}\right)^2 + \cdots \tag{8.2}$$

obtained in [41] was strictly speaking not applicable to determine the boundary of the stability region, where $\omega = 0$ and, consequently (see Eq. [8.2]), $l_T/\Lambda \sim 1$. To obtain some approximate value for the critical wavelength, the authors of [41] were forced to extrapolate the relationship (8.2) to the short-wave region, where ω is formally equal to zero.

Sivashinsky [43] suggested that the asymptotic analysis of stability be conducted from a somewhat different standpoint. His choice of a small parameter was not the ratio l_T/Λ, but the dimensionless parameter $1/N$, characteristic of high activation energy. This makes it possible to study the stability problem uniformly with respect to all wavelengths and to avoid the indeterminate factors in finding critical wavelengths.

The research done in [43] established that in terms of wavelength the flame structure has four inhomogeneity regions with respect to asymptotic expansion in powers of $1/N$:

$\Lambda \sim Nl_T$	long-wave perturbations
$\Lambda \sim \frac{l_T}{\sqrt{N}}$	short-wave perturbations
$\Lambda \sim l_T$	medium-wave perturbations
$\Lambda \sim \frac{l_T}{N}$	ultrashort-wave perturbations, of wavelength comparable with the width of the reaction zone

The analysis of [43] was carried through for the first two types of perturbation, i.e., on the assumption that it is legitimate to treat the *reaction zone* as a geometric surface.

The spatial structure of the flame is essentially different in each of the above regions.

If $x = F(y)$ is the equation of the steady state slightly corrugated flame front (the surface in whose vicinity the reaction zone is concentrated), then in the case of long-wave perturbations one can distinguish four characteristic regions near the flame front:

$x - F \sim \frac{l_T}{N}$	reaction zone
$x - F \sim l_T$	preheat zone
$x - F \sim l_T N$	ideal fluid zone (in this zone transport effects may be neglected to a first approximation)
$x - F \sim l_T N^2$	zone of dissipation of temperature and hydrodynamic perturbations (here effects such as viscosity, diffusion, and heat conduction again become significant)

For medium-wave perturbations, only the first two of these regions exist.

For short-wave perturbations, a new region forms between the reaction and heat zones, of width depending on the wavelength of the perturbation:

$$x - F \sim \frac{l_T}{\sqrt{N}}$$

In this case, then, there are three inhomogeneity regions.

Each of the above-mentioned zones is characterized by specific asymptotic expansions.

In the long-wave case the critical wavelength is:

$$\Lambda = \frac{2\pi}{k^*} = \frac{\pi N I \epsilon (3 - \epsilon) l_T}{1 - \epsilon} \tag{8.3}$$

where

$$I = \int_{\epsilon}^{1} \frac{\tilde{\kappa}(\tilde{T})}{\tilde{T}} \left[1 - \left(\frac{\tilde{T} - \epsilon}{1 - \epsilon} \right)^{L-1} d\tilde{T} \right] d\tilde{T} \tag{8.4}$$

and $\tilde{\kappa}$ is the dimensionless thermal conductivity, defined as κ/κ_b.

In the short-wave case the critical wavelength is:

$$\Lambda = \frac{2\pi}{k^{**}} = \frac{8\pi l_T}{\sqrt{N(1-\epsilon)(1-L)}} \qquad L < 1 \tag{8.5}$$

It is worthy of note that, in the first approximation, the Prandtl number does not figure in the above asymptotic relations (8.3), (8.5). The role of viscosity does not become apparent until one proceeds to the second approximation with respect to $1/N$. Another reason for the importance of the second approximation is that (as known from the one-dimensional problem [18]) it provides a good approximation to the exact solution for relatively small activation energies.

9 SUMMARY

In the preceding sections an attempt has been made to present the mathematical theory of a laminar flame as an asymptotic theory, with the quantity $1/N = R^0 T_b/E$ playing the role of small parameter. When $N \gg 1$ the reaction zone of an exothermic reaction is much smaller, in order of magnitude, than the surrounding diffusional-thermal layer. Thanks to this property of the flame, one can study the problem separately in the reaction zone, where the equations are significantly simpler than the initial equations (regardless of nonlinearity), and outside the reaction zone, where the reaction zone itself may be treated as a point source.

A consistent asymptotic analysis yields rigorous asymptotic relationships for the normal propagation velocity of the flame, and also makes it possible to solve several important problems of the theory of premixed and diffusion flames. We have also brought out the additional role of the strong temperature dependence of the reaction rate, which not only implies a narrow reaction zone but also yields a new characteristic dimension of the flame, related to its curvature. Having observed this fact, one can describe such phenomena as open tips of bunsen cones, diffusion stratification in distorted flames, and hydrodynamic stability of a plane flame front.

A final remark: the use of $E/R^0 T_b$ as a large parameter proves helpful in the study of detonation flames in gases [44] as well as in describing weakly turbulent flames in which the magnitude of the pulsations is much larger than the width of the flame structure [45].

NOTATION

A	defined in Eq. (2.18)
a	radius of droplet
C	mass fraction (concentration)
C_0	initial mass fraction
$\widetilde{C}$	nondimensional mass fraction, C/C_0
c_p	specific heat at constant pressure
D	diffusion coefficient
d	defined in Eq. (2.72)
E	activation energy
F	flame front surface
I	defined in Eq. (8.4)
h	heat transfer coefficient
k	wave number, $2\pi/\Lambda$
L	Lewis number
l_T	preheating flame width
l_r	reaction zone width
l_K	dissipation zone width
m	mass of droplet
M	chemical symbol of reactant
N	nondimensional activation energy
P	pressure (Sec. 1)
P	defined in Eq. (2.9)
Q	heat release
q_λ	defined in Eq. (4.5), latent heat of vaporization
$\widetilde{q}$	intensity of point source (Sec. 5)
q_0	defined in Eq. (5.18)
R	gas constant
R	radius of flame front
r	radial coordinate
S	defined in Eq. (2.27)
s	defined in Eq. (7.4)
T	temperature

T_0	initial temperature
T_b	adiabatic temperature
T_f	flame front temperature
$\widetilde{T}$	nondimensional temperature, T/T_b
t	time
U_i	gas velocity components
U_b	normal flame propagation velocity
v_r	radial component of gas velocity
V	velocity component
x,y,z	cartesian coordinates
$\widetilde{x}$	defined in Eq. (3.8) Sec. 3
$\widetilde{x}$	defined in Eq. (6.4) Sec. 6
X	characteristic variable of reaction zone
y	defined in Eq. (7.4)
z	pre-exponential factor
z_b	pre-exponential factor at $T = T_b$
$\widetilde{z}$	nondimensional pre-exponential factor, Z/Z_b

Greek symbols

α	angle of inclination (Sec. 7)
α_i	defined in Eq. (8.2)
β	defined in Eq. (5.24)
ϵ	T_0/T_b
ζ	defined in Eq. (3.8)
Θ	defined in Eq. (2.24)
θ	defined in Eq. (2.74)
κ	conductivity
κ_b	conductivity at $T = T_b$
Λ	wave length
λ	defined in Eq. (3.8)
λ	defined in Eq. (5.21)
μ	viscosity
ν	stoichiometric coefficient
ρ	density
ρ_*	liquid drop density
ρ_b	density at $T = T_b$
σ	Prandtl number
τ	dimensionless time (defined in Eq. [6.4])
Φ	reaction rate
ϕ	flame front surface
ψ	E_1/E_2 (Sec. 2)
ω	characteristic frequency

REFERENCES

1. F. A. Williams, Theory of combustion in laminar flows. *Ann. Rev. Fl. Mech.*, vol. 3, pp. 171–188, 1973.
2. M. Van Dyke, *Perturbation Methods in Fluid Mechanics.* Academic, New York, 1964.
3. J. C. Cole, *Perturbation Methods in Applied Mathematics.* Blaisdell, Waltham, Mass., 1968.
4. Ya. B. Zel'dovich, G. I. Barenblatt, A. G. Istratov, On diffusion-heat stability of laminar flames (in Russian). *Zurn. Prikl. Mekh. Tekh. Fiz.*, No. 4, 1962.
5. D. A. Frank-Kamenetskii, *Diffusion and Heat Transfer in Chemical Kinetics,* 2d ed., Plenum, New York, 1969.
6. Ya. B. Zel'dovich, On the theory of flame propagation (in Russian). *Zurn. Fiz. Khim.*, vol. 22, no. 1, 1948.
7. Ya. B. Zel'dovich and D. A. Frank-Kamenetskii, On the theory of steady flame propagation (in Russian). *Dokl. AN SSSR,* vol. 19, p. 693, 1938.
8. Ya. B. Zel'dovich and D. A. Frank-Kamenetskii, Theory of thermal flame propagation (in Russian). *Zurn. Fiz. Khim.*, vol. 12, p. 100, 1938.
9. A. F. Beliaev, On nitroglycol combustion (in Russian). *Zurn. Fiz. Khim.*, vol. 14, no. 8, p. 1009, 1940.
10. Ya. B. Zel'dovich and N. N. Semionov, Kinetics of chemical reactions in flames (in Russian). *Zurn. Eksp. Teor. Fiz.*, vol. 10, pp. 1116–1427, 1940.
11. Ya. B. Zel'dovich and N. N. Semionov, Kinetics of chemical reactions in flames. Mechanism of CO combustion (in Russian). *Zurn. Eksp. Teor. Fiz.*, vol. 10, p. 1427, 1940.
12. Ya. B. Zel'dovich, Chain reactions in hot flames. Approximate theory of flame velocity (in Russian). *Kinetika i Kataliz,* vol. 2, no. 3, 1961.
13. J. F. Wehner, Flame processes–theoretical and experimental, in T. B. Drew and J. W. Hoopes, Jr., Eds., *Advances in Chemical Engineering,* vol. 5, Academic, New York, 1964.
14. F. A. Williams, *Combustion Theory.* Addison-Wesley, Palo Alto, Calif., 1965.
15. P. A. Lagerstrom and J. D. Cole, Examples illustrating expansion procedures for the Navier-Stokes equations. *J. Rat. Mech. Ann.*, vol. 4, pp. 817–882, 1955.
16. S. Kaplun and P. A. Lagerstrom, Asymptotic expansions of Navier-Stokes solutions for small Reynolds numbers. *J. Math. Mech.*, vol. 6, pp. 585–593, 1957.
17. R. Friedman and E. Burke, A theoretical model of a gaseous combustion wave governed by a first-order reaction. *J. Chem. Phys.*, vol. 21, pp. 710–714, 1953.
18. W. B. Bush and F. E. Fendell, Asymptotic analysis of laminar flame propagation for general Lewis numbers, *Combust. Sci. Tech.*, vol. 1, pp. 421–428, 1970.
19. V. S. Berman and U. S. Ryazantsev, On analysis of the problem of thermal propagation of a flame by the method of matched asymptotic expansions (in Russian). *Prikl. Math. Mekh.*, vol. 36, no. 4, pp. 659–666, 1972.
20. V. S. Berman and U. S. Ryazantsev, Use of the method of matched asymptotic expansions to calculate the steady thermal propagation of the front of an exothermic reaction in a condensed medium (in Russian). *Prikl. Mekh. Tekh. Fiz.*, no. 5, pp. 106–111, 1972.
21. V. S. Berman and U. S. Ryazantsev, Asymptotic analysis of steady propagation of the front of a two-step exothermic reaction in a gas (in Russian). *Prikl. Mekh. Tekh. Fiz.*, no. 1, pp. 75–87, 1973.
22. V. S. Berman and U. S. Ryazantsev, Asymptotic analysis of steady propagation of the front of a two-step exothermic reaction in a gas (in Russian). *Prikl. Math. Mekh.*, vol. 37, pp. 1049–1058, 1973.

23. Ya. B. Zel'dovich, On the quiet flame propagation (in Russian). *Zurn. Eksp. Teor. Fiz.*, vol. 11, p. 159, 1941.
24. D. B. Spalding, *Proc. Roy. Soc. London,* A240, vol. 83, 1957.
25. F. E. Fendell, Asymptotic analysis of premixed burning with large activation energy. *J. Fl. Mech.*, vol. 56, pt. 1, pp. 81–95, 1972.
26. G. I. Sivashinsky and C. Gutfinger, The extinction of spherical diffusion flames, in N. H. Afgan and J. M. Beer, Eds., *Heat Transfer in Flames,* Hemisphere, Washington, D.C., 1974, pp. 495–501.
27. Ya. B. Zel'dovich, On combustion of unpremixed gases (in Russian). *Zurn. Tekh. Fiz.*, vol. 19, no. 9, 1949.
28. D. B. Spalding, A theory of the extinction of diffusion flames. *Fuel,* vol. 33, no. 3, pp. 255–273, 1953.
29. A. E. Potter and J. N. Butler, A novel combustion measurement based on the extinguishment of diffusion flames. *ARS J.*, vol. 29, no. 1, pp. 54–56, 1959.
30. A. Linan, On the structure of laminar diffusion flames. *Inst. Nac. Techn. Aeronaut. Techn. Rept.*, NFM 63-2 (Clearinghouse Feder. Scient. and Techn. Inform NAD-432822), Madrid, 1963.
31. G. I. Sivashinsky, On a converging spherical flame front. *Intern. J. Heat Mass Trans.*, vol. 17, pp. 1499–1511, 1974.
32. K. I. Shelkin and Ya. K. Troshin, Gas dynamics of combustion, *AN SSSR,* Moscow, 1963.
33. G. I. Sivashinsky, The structure of bunsen flames. *J. Chem. Phys.*, vol. 62, p. 638, 1975.
34. B. Lewis and G. Von Elbe, *Combustion, Flames, and Explosions of Gases.* Academic, New York, 1961.
35. G. I. Sivashinsky, The diffusion stratification effect in bunsen flames. *J. Heat Trans., ASME,* vol. 96, Series C, pp. 530–535, 1974.
36. G. H. Markstein, Ed., *Non-steady Flame Propagation.* Pergamon, New York, 1964.
37. A. G. Istratov and V. B. Librovich, On stability of spherical flame propagation (in Russian). *Zh. Prikl. Mech. and Tekh. Fiz.*, no. 1, pp. 67–78, 1966.
38. L. D. Landau, On the theory of slow combustion. *Zh. Exsp. Teor. Fiz.*, vol. 14, no. 6, 1944. (Also see *Acta Physiochim,* USSR, vol. 19, p. 77, 1944.)
39. G. Darrieus, Propagation d'un front de flamme. Essai de theorie de vittesses anomales de deflagration par developpement spontane de la turbulence. *6th Intern. Congr. Appl. Mech.*, Paris, 1946.
40. G. H. Markstein, Experimental and theoretical studies of flame front stability. *J. Aero. Sci.*, vol. 18, p. 199, 1951.
41. A. G. Istratov and V. B. Librovich, On influence of transport processes on stability of laminar plane flame front (in Russian). *Prikl. Math. Mekh.*, vol. 30, no. 3, pp. 451–466, 1966.
42. P. Germain and J. P. Guiraud, Conditions de choc et structure des ondes de choc dans un ecoulement stationaire de fluide dissipatif. *Publ. ONERA*, no. 105, 45 pp., 1962.
43. G. I. Sivashinsky, On a steady corrugated flame front. *Astronautica Acta,* vol. 18, pp. 253–260, 1973.
44. W. B. Bush and F. E. Fendell, Asymptotic analysis of the structure of a steady planar detonation. *Combust. Sci. Tech.*, pp. 271–285, 1971.
45. F. A. Williams, A weakly turbulent flame with high activation energy. *Tech. Rept.*, UCSD-4-PU. Thermal Science and Propulstion Center, Purdue University, West Lafayette, Ind., 1973.

Author Index

Subject Index

NOTES

NOTES

NOTES

NOTES

NOTES

NOTES

NOTES

NOTES